Springer Handbook
of Nanotechnology

Bharat Bhushan (Ed.)

3rd revised and extended edition

Springer Handbook of Nanotechnology

Since 2004 and with the 2nd edition in 2006, the Springer Handbook of Nanotechnology has established itself as the definitive reference in the nanoscience and nanotechnology area. It integrates the knowledge from nanofabrication, nanodevices, nanomechanics, nanotribology, materials science, and reliability engineering in just one volume. Beside the presentation of nanostructures, micro/nanofabrication, and micro/nanodevices, special emphasis is on scanning probe microscopy, nanotribology and nanomechanics, molecularly thick films, industrial applications and microdevice reliability, and on social aspects. In its 3rd edition, the book grew from 8 to 9 parts now including a part with chapters on biomimetics. More information is added to such fields as bionanotechnology, nanorobotics, and (bio) MEMS/NEMS, bio/nanotribology and bio/nanomechanics. The book is organized by an experienced editor with a universal knowledge and written by an international team of over 145 distinguished experts. It addresses mechanical and electrical engineers, materials scientists, physicists and chemists who work either in the nano area or in a field that is or will be influenced by this new key technology.

"The strong point is its focus on many of the practical aspects of nanotechnology... Anyone working in or learning about the field of nanotechnology would find this an excellent working handbook."

IEEE Electrical Insulation Magazine

"Outstandingly succeeds in its aim... It really is a magnificent volume and every scientific library and nanotechnology group should have a copy."

Materials World

"The integrity and authoritativeness... is guaranteed by an experienced editor and an international team of authors which have well summarized in their chapters information on fundamentals and applications."

Polymer News

List of Abbreviations

1 Introduction to Nanotechnology

Part A Nanostructures, Micro-/Nanofabrication and Materials

2 Nanomaterials Synthesis and Applications: Molecule-Based Devices
3 Introduction to Carbon Nanotubes
4 Nanowires
5 Template-Based Synthesis of Nanorod or Nanowire Arrays
6 Templated Self-Assembly of Particles
7 Three-Dimensional Nanostructure Fabrication by Focused Ion Beam Chemical Vapor Deposition
8 Introduction to Micro-/Nanofabrication
9 Nanoimprint Lithography-Patterning of Resists Using Molding
10 Stamping Techniques for Micro- and Nanofabrication
11 Material Aspects of Micro- and Nanoelectromechanical Systems

Part B MEMS/NEMS and BioMEMS/NEMS

12 MEMS/NEMS Devices and Applications
13 Next-Generation DNA Hybridization and Self-Assembly Nanofabrication Devices
14 Single-Walled Carbon Nanotube Sensor Concepts
15 Nanomechanical Cantilever Array Sensors
16 Biological Molecules in Therapeutic Nanodevices
17 G-Protein Coupled Receptors: Progress in Surface Display and Biosensor Technology
18 Microfluidic Devices and Their Applications to Lab-on-a-Chip
19 Centrifuge-Based Fluidic Platforms
20 Micro-/Nanodroplets in Microfluidic Devices

Part C Scanning-Probe Microscopy

21 Scanning Probe Microscopy-Principle of Operation, Instrumentation, and Probes
22 General and Special Probes in Scanning Microscopies
23 Noncontact Atomic Force Microscopy and Related Topics
24 Low-Temperature Scanning Probe Microscopy
25 Higher Harmonics and Time-Varying Forces in Dynamic Force Microscopy
26 Dynamic Modes of Atomic Force Microscopy
27 Molecular Recognition Force Microscopy: From Molecular Bonds to Complex Energy Landscapes

Part D Bio-/Nanotribology and Bio-/Nanomechanics

28 Nanotribology, Nanomechanics, and Materials Characterization
29 Surface Forces and Nanorheology of Molecularly Thin Films
30 Friction and Wear on the Atomic Scale
31 Computer Simulations of Nanometer-Scale Indentation and Friction
32 Force Measurements with Optical Tweezers
33 Scale Effect in Mechanical Properties and Tribology
34 Structural, Nanomechanical, and Nanotribological Characterization of Human Hair Using Atomic Force Microscopy and Nanoindentation
35 Cellular Nanomechanics
36 Optical Cell Manipulation
37 Mechanical Properties of Nanostructures

Part E Molecularly Thick Films for Lubrication

38 Nanotribology of Ultrathin and Hard Amorphous Carbon Films
39 Self-Assembled Monolayers for Nanotribology and Surface Protection
40 Nanoscale Boundary Lubrication Studies

Part F Biomimetics

41 Multifunctional Plant Surfaces and Smart Materials
42 Lotus Effect: Surfaces with Roughness-Induced Superhydrophobicity, Self-Cleaning, and Low Adhesion
43 Biological and Biologically Inspired Attachment Systems
44 Gecko Feet: Natural Hairy Attachment Systems for Smart Adhesion

Part G Industrial Applications

45 The *Millipede*-A Nanotechnology-Based AFM Data-Storage System
46 Nanorobotics

Part H Micro-/Nanodevice Reliability

47 MEMS/NEMS and BioMEMS/BioNEMS: Materials, Devices, and Biomimetics
48 Friction and Wear in Micro- and Nanomachines
49 Failure Mechanisms in MEMS/NEMS Devices
50 Mechanical Properties of Micromachined Structures
51 High-Volume Manufacturing and Field Stability of MEMS Products
52 Packaging and Reliability Issues in Micro-/Nanosystems

Part I Technological Convergence and Governing Nanotechnology

53 Governing Nanotechnology: Social, Ethical and Human Issues

Subject Index

使用说明

1.《纳米技术手册》原版为一册,分为A~I部分。考虑到使用方便以及内容一致,影印版分为7册:第1册—Part A,第2册—Part B,第3册—Part C,第4册—Part D,第5册—Part E,第6册—Part F,第7册—Part G、H、I。

2.各册在页脚重新编排页码,该页码对应中文目录。保留了原书页眉及页码,其页码对应原书目录及主题索引。

3.各册均给出完整7册书的章目录。

4.作者及其联系方式、缩略语表各册均完整呈现。

5.主题索引安排在第7册。

6.目录等采用中英文对照形式给出,方便读者快速浏览。

材料科学与工程图书工作室

联系电话　0451-86412421
　　　　　0451-86414559
邮　　箱　yh_bj@yahoo.com.cn
　　　　　xuyaying81823@gmail.com
　　　　　zhxh6414559@yahoo.com.cn

Springer 手册精选系列

纳米技术手册

纳米结构、微纳米材料及其制备

【第1册】

Springer
Handbook of
Nanotechnology

〔美〕Bharat Bhushan　主编

（第三版影印版）

哈尔滨工业大学出版社
HARBIN INSTITUTE OF TECHNOLOGY PRESS

黑版贸审字 08-2013-001 号

Reprint from English language edition:
Springer Handbook of Nanotechnology
by Bharat Bhushan
Copyright © 2010 Springer Berlin Heidelberg
Springer Berlin Heidelberg is a part of Springer Science+Business Media
All Rights Reserved

This reprint has been authorized by Springer Science & Business Media for distribution in China Mainland only and not for export there from.

图书在版编目（CIP）数据

纳米技术手册：第3版. 1, 纳米结构、微纳米材料及其制备 = Handbook of Nanotechnology.1,Nanostructures,Micro-/Nanofabrication and Materials：英文 /（美）布尚（Bhushan.B.）主编. —哈尔滨：哈尔滨工业大学出版社，2013.1
（Springer手册精选系列）
ISBN 978-7-5603-3947-4

Ⅰ.①纳… Ⅱ.①布… Ⅲ.①纳米技术 – 手册 – 英文 Ⅳ.①TB303-62

中国版本图书馆CIP数据核字(2013)第004421号

材料科学与工程
图书工作室

责任编辑 杨 桦 许雅莹 张秀华
出版发行 哈尔滨工业大学出版社
社　　址 哈尔滨市南岗区复华四道街10号 邮编 150006
传　　真 0451-86414749
网　　址 http://hitpress.hit.edu.cn
印　　刷 哈尔滨市石桥印务有限公司
开　　本 787mm×960mm 1/16 印张 25
版　　次 2013年1月第1版 2013年1月第1次印刷
书　　号 ISBN 978-7-5603-3947-4
定　　价 78.00元

（如因印刷质量问题影响阅读，我社负责调换）

Foreword by Neal Lane

In a January 2000 speech at the California Institute of Technology, former President W.J. Clinton talked about the exciting promise of *nanotechnology* and the importance of expanding research in nanoscale science and engineering and, more broadly, in the physical sciences. Later that month, he announced in his State of the Union Address an ambitious US$ 497 million federal, multiagency national nanotechnology initiative (NNI) in the fiscal year 2001 budget; and he made the NNI a top science and technology priority within a budget that emphasized increased investment in US scientific research. With strong bipartisan support in Congress, most of this request was appropriated, and the NNI was born. Often, federal budget initiatives only last a year or so. It is most encouraging that the NNI has remained a high priority of the G.W. Bush Administration and Congress, reflecting enormous progress in the field and continued strong interest and support by industry.

Nanotechnology is the ability to manipulate individual atoms and molecules to produce nanostructured materials and submicron objects that have applications in the real world. Nanotechnology involves the production and application of physical, chemical and biological systems at scales ranging from individual atoms or molecules to about 100 nm, as well as the integration of the resulting nanostructures into larger systems. Nanotechnology is likely to have a profound impact on our economy and society in the early 21st century, perhaps comparable to that of information technology or cellular and molecular biology. Science and engineering research in nanotechnology promises breakthroughs in areas such as materials and manufacturing, electronics, medicine and healthcare, energy and the environment, biotechnology, information technology and national security. Clinical trials are already underway for nanomaterials that offer the promise of cures for certain cancers. It is widely felt that nanotechnology will be the next industrial revolution.

Nanometer-scale features are built up from their elemental constituents. Micro- and nanosystems components are fabricated using batch-processing techniques that are compatible with integrated circuits and range in size from micro- to nanometers. Micro- and nanosystems include micro/nanoelectro-mechanical systems (MEMS/NEMS), micromechatronics, optoelectronics, microfluidics and systems integration. These systems can sense, control, and activate on the micro/nanoscale and can function individually or in arrays to generate effects on the macroscale. Due to the enabling nature of these systems and the significant impact they can have on both the commercial and defense applications, industry as well as the federal government have taken special interest in seeing growth nurtured in this field. Micro- and nanosystems are the next logical step in the *silicon revolution*.

The discovery of novel materials, processes, and phenomena at the nanoscale and the development of new experimental and theoretical techniques for research provide fresh opportunities for the development of innovative nanosystems and nanostructured materials. There is an increasing need for a multidisciplinary, systems-oriented approach to manufacturing micro/nanodevices which function reliably. This can only be achieved through the cross-fertilization of ideas from different disciplines and the systematic flow of information and people among research groups.

Nanotechnology is a broad, highly interdisciplinary, and still evolving field. Covering even the most important aspects of nanotechnology in a single book that reaches readers ranging from students to active researchers in academia and industry is an enormous challenge. To prepare such a wide-ranging book on nanotechnology, Prof. Bhushan has harnessed his own knowledge and experience, gained in several industries and universities, and has assembled internationally recognized authorities from four continents to write chapters covering a wide array of nanotechnology topics, including the latest advances. The authors come from both academia and industry. The topics include major advances in many fields where nanoscale science and engineering is being pursued and illustrate how the field of nanotechnology has continued to emerge and blossom. Given the accelerating pace of discovery and applications in nanotechnology, it is a challenge to cap-

Prof. Neal Lane
Malcolm Gillis University Professor,
Department of Physics and Astronomy,
Senior Fellow,
James A. Baker III Institute for Public Policy
Rice University
Houston, Texas

Served in the Clinton Administration as Assistant to the President for Science and Technology and Director of the White House Office of Science and Technology Policy (1998–2001) and, prior to that, as Director of the National Science Foundation (1993–1998). While at the White House, he was a key figure in the creation of the NNI.

ture it all in one volume. As in earlier editions, professor Bhushan does an admirable job.

Professor Bharat Bhushan's comprehensive book is intended to serve both as a textbook for university courses as well as a reference for researchers. The first and second editions were timely additions to the literature on nanotechnology and stimulated further interest in this important new field, while serving as invaluable resources to members of the international scientific and industrial community. The increasing demand for up-to-date information on this fast moving field led to this third edition. It is increasingly important that scientists and engineers, whatever their specialty, have a solid grounding in the fundamentals and potential applications of nanotechnology. This third edition addresses that need by giving particular attention to the widening audience of readers. It also includes a discussion of the social, ethical and political issues that tend to surround any emerging technology.

The editor and his team are to be warmly congratulated for bringing together this exclusive, timely, and useful nanotechnology handbook.

Foreword by James R. Heath

Nanotechnology has become an increasingly popular buzzword over the past five years or so, a trend that has been fueled by a global set of publicly funded nanotechnology initiatives. Even as researchers have been struggling to demonstrate some of the most fundamental and simple aspects of this field, the term nanotechnology has entered into the public consciousness through articles in the popular press and popular fiction. As a consequence, the expectations of the public are high for nanotechnology, even while the actual public definition of nanotechnology remains a bit fuzzy.

Why shouldn't those expectations be high? The late 1990s witnessed a major information technology (IT) revolution and a minor biotechnology revolution. The IT revolution impacted virtually every aspect of life in the western world. I am sitting on an airplane at 30 000 feet at the moment, working on my laptop, as are about half of the other passengers on this plane. The plane itself is riddled with computational and communications equipment. As soon as we land, many of us will pull out cell phones, others will check e-mail via wireless modem, some will do both. This picture would be the same if I was landing in Los Angeles, Beijing, or Capetown. I will probably never actually print this text, but will instead submit it electronically. All of this was unthinkable a dozen years ago. It is therefore no wonder that the public expects marvelous things to happen quickly. However, the science that laid the groundwork for the IT revolution dates back 60 years or more, with its origins in fundamental solid-state physics.

By contrast, the biotech revolution was relatively minor and, at least to date, not particularly effective. The major diseases that plagued mankind a quarter century ago are still here. In some third-world countries, the average lifespan of individuals has actually decreased from where it was a full century ago. While the costs of electronics technologies have plummeted, health care costs have continued to rise. The biotech revolution may have a profound impact, but the task at hand is substantially more difficult than what was required for the IT revolution. In effect, the IT revolution was based on the advanced engineering of two-dimensional digital circuits constructed from relatively simple components – extended solids. The biotech revolution is really dependent upon the ability to reverse engineer three-dimensional analog systems constructed from quite complex components – proteins. Given that the basic science behind biotech is substantially younger than the science that has supported IT, it is perhaps not surprising that the biotech revolution has not really been a proper revolution yet, and it likely needs at least another decade or so to come into fruition.

Where does nanotechnology fit into this picture? In many ways, nanotechnology depends upon the ability to engineer two- and three-dimensional systems constructed from complex components such as macromolecules, biomolecules, nanostructured solids, etc. Furthermore, in terms of patents, publications, and other metrics that can be used to gauge the birth and evolution of a field, nanotech lags some 15–20 years behind biotech. Thus, now is the time that the fundamental science behind nanotechnology is being explored and developed. Nevertheless, progress with that science is moving forward at a dramatic pace. If the scientific community can keep up this pace and if the public sector will continue to support this science, then it is possible, and even perhaps likely, that in 20 years we may be speaking of the nanotech revolution.

The first edition of Springer Handbook of Nanotechnology was timely to assemble chapters in the broad field of nanotechnology. Given the fact that the second edition was in press one year after the publication of the first edition in April 2004, it is clear that the handbook has shown to be a valuable reference for experienced researchers as well as for a novice in the field. The third edition has one Part added and an expanded scope should have a wider appeal.

Prof. James R. Heath

Department of Chemistry
California Institute of Technology
Pasadena, California

Worked in the group of Nobel Laureate Richard E. Smalley at Rice University (1984–88) and co-invented Fullerene molecules which led to a revolution in Chemistry including the realization of nanotubes. The work on Fullerene molecules was cited for the 1996 Nobel Prize in Chemistry. Later he joined the University of California at Los Angeles (1994–2002), and co-founded and served as a Scientific Director of The California Nanosystems Institute.

Preface to the 3rd Edition

On December 29, 1959 at the California Institute of Technology, Nobel Laureate Richard P. Feynman gave at talk at the Annual meeting of the American Physical Society that has become one of the 20th century classic science lectures, titled *There's Plenty of Room at the Bottom*. He presented a technological vision of extreme miniaturization in 1959, several years before the word *chip* became part of the lexicon. He talked about the problem of manipulating and controlling things on a small scale. Extrapolating from known physical laws, Feynman envisioned a technology using the ultimate toolbox of nature, building nanoobjects atom by atom or molecule by molecule. Since the 1980s, many inventions and discoveries in fabrication of nanoobjects have been testament to his vision. In recognition of this reality, National Science and Technology Council (NSTC) of the White House created the Interagency Working Group on Nanoscience, Engineering and Technology (IWGN) in 1998. In a January 2000 speech at the same institute, former President W.J. Clinton talked about the exciting promise of *nanotechnology* and the importance of expanding research in nanoscale science and technology, more broadly. Later that month, he announced in his State of the Union Address an ambitious US$ 497 million federal, multi-agency national nanotechnology initiative (NNI) in the fiscal year 2001 budget, and made the NNI a top science and technology priority. The objective of this initiative was to form a broad-based coalition in which the academe, the private sector, and local, state, and federal governments work together to push the envelop of nanoscience and nanoengineering to reap nanotechnology's potential social and economic benefits.

The funding in the US has continued to increase. In January 2003, the US senate introduced a bill to establish a National Nanotechnology Program. On December 3, 2003, President George W. Bush signed into law the 21st Century Nanotechnology Research and Development Act. The legislation put into law programs and activities supported by the National Nanotechnology Initiative. The bill gave nanotechnology a permanent home in the federal government and authorized US$ 3.7 billion to be spent in the four year period beginning in October 2005, for nanotechnology initiatives at five federal agencies. The funds would provide grants to researchers, coordinate R&D across five federal agencies (National Science Foundation (NSF), Department of Energy (DOE), NASA, National Institute of Standards and Technology (NIST), and Environmental Protection Agency (EPA)), establish interdisciplinary research centers, and accelerate technology transfer into the private sector. In addition, Department of Defense (DOD), Homeland Security, Agriculture and Justice as well as the National Institutes of Health (NIH) also fund large R&D activities. They currently account for more than one-third of the federal budget for nanotechnology.

European Union (EU) made nanosciences and nanotechnologies a priority in Sixth Framework Program (FP6) in 2002 for a period of 2003–2006. They had dedicated small funds in FP4 and FP5 before. FP6 was tailored to help better structure European research and to cope with the strategic objectives set out in Lisbon in 2000. Japan identified nanotechnology as one of its main research priorities in 2001. The funding levels increases sharply from US$ 400 million in 2001 to around US$ 950 million in 2004. In 2003, South Korea embarked upon a ten-year program with around US$ 2 billion of public funding, and Taiwan has committed around US$ 600 million of public funding over six years. Singapore and China are also investing on a large scale. Russia is well funded as well.

Nanotechnology literally means any technology done on a nanoscale that has applications in the real world. Nanotechnology encompasses production and application of physical, chemical and biological systems at scales, ranging from individual atoms or molecules to submicron dimensions, as well as the integration of the resulting nanostructures into larger systems. Nanotechnology is likely to have a profound impact on our economy and society in the early 21st century, comparable to that of semiconductor technology, information technology, or cellular and molecular biology. Science and technology research in nanotechnology promises breakthroughs in areas such as materials and manufacturing, nanoelectronics, medicine and healthcare, energy, biotechnology, information technology and national security. It is widely felt that nanotechnology will be the next industrial revolution.

There is an increasing need for a multidisciplinary, system-oriented approach to design and manufacturing of micro/nanodevices which function reliably. This can only be achieved through the cross-fertilization of ideas from different disciplines and the systematic flow of information and people among research groups. Reliability is a critical technology for many micro- and nanosystems and nanostructured materials. A broad based handbook was needed, and the first edition of Springer Handbook of Nanotechnology was published in April 2004. It presented an overview of nanomaterial synthesis, micro/nanofabrication, micro- and nanocomponents and systems, scanning probe microscopy, reliability issues (including nanotribology and nanomechanics) for nanotechnology, and industrial applications. When the handbook went for sale in Europe, it was sold out in ten days. Reviews on the handbook were very flattering.

Given the explosive growth in nanoscience and nanotechnology, the publisher and the editor decided to develop a second edition after merely six months of publication of the first edition. The second edition (2007) came out in December 2006. The publisher and the editor again decided to develop a third edition after six month of publication of the second edition. This edition of the handbook integrates the knowledge from nanostructures, fabrication, materials science, devices, and reliability point of view. It covers various industrial applications. It also addresses social, ethical, and political issues. Given the significant interest in biomedical applications, and biomimetics a number of additional chapters in this arena have been added. The third edition consists of 53 chapters (new 10, revised 28, and as is 15). The chapters have been written by 139 internationally recognized experts in the field, from academia, national research labs, and industry, and from all over the world.

This handbook is intended for three types of readers: graduate students of nanotechnology, researchers in academia and industry who are active or intend to become active in this field, and practicing engineers and scientists who have encountered a problem and hope to solve it as expeditiously as possible. The handbook should serve as an excellent text for one or two semester graduate courses in nanotechnology in mechanical engineering, materials science, applied physics, or applied chemistry.

We embarked on the development of third edition in June 2007, and we worked very hard to get all the chapters to the publisher in a record time of about 12 months. I wish to sincerely thank the authors for offering to write comprehensive chapters on a tight schedule. This is generally an added responsibility in the hectic work schedules of researchers today. I depended on a large number of reviewers who provided critical reviews. I would like to thank Dr. Phillip J. Bond, Chief of Staff and Under Secretary for Technology, US Department of Commerce, Washington, D.C. for suggestions for chapters as well as authors in the handbook. Last but not the least, I would like to thank my secretary Caterina Runyon-Spears for various administrative duties and her tireless efforts are highly appreciated.

I hope that this handbook will stimulate further interest in this important new field, and the readers of this handbook will find it useful.

February 2010   Bharat Bhushan
Editor

Preface to the 2nd Edition

On 29 December 1959 at the California Institute of Technology, Nobel Laureate Richard P. Feynman gave at talk at the Annual meeting of the American Physical Society that has become one of the 20th century classic science lectures, titled "There's Plenty of Room at the Bottom." He presented a technological vision of extreme miniaturization in 1959, several years before the word "chip" became part of the lexicon. He talked about the problem of manipulating and controlling things on a small scale. Extrapolating from known physical laws, Feynman envisioned a technology using the ultimate toolbox of nature, building nanoobjects atom by atom or molecule by molecule. Since the 1980s, many inventions and discoveries in the fabrication of nanoobjects have been a testament to his vision. In recognition of this reality, the National Science and Technology Council (NSTC) of the White House created the Interagency Working Group on Nanoscience, Engineering and Technology (IWGN) in 1998. In a January 2000 speech at the same institute, former President W. J. Clinton talked about the exciting promise of "nanotechnology" and the importance of expanding research in nanoscale science and, more broadly, technology. Later that month, he announced in his State of the Union Address an ambitious $497 million federal, multiagency national nanotechnology initiative (NNI) in the fiscal year 2001 budget, and made the NNI a top science and technology priority. The objective of this initiative was to form a broad-based coalition in which the academe, the private sector, and local, state, and federal governments work together to push the envelope of nanoscience and nanoengineering to reap nanotechnology's potential social and economic benefits.

The funding in the U.S. has continued to increase. In January 2003, the U. S. senate introduced a bill to establish a National Nanotechnology Program. On 3 December 2003, President George W. Bush signed into law the 21st Century Nanotechnology Research and Development Act. The legislation put into law programs and activities supported by the National Nanotechnology Initiative. The bill gave nanotechnology a permanent home in the federal government and authorized $3.7 billion to be spent in the four year period beginning in October 2005, for nanotechnology initiatives at five federal agencies. The funds would provide grants to researchers, coordinate R&D across five federal agencies (National Science Foundation (NSF), Department of Energy (DOE), NASA, National Institute of Standards and Technology (NIST), and Environmental Protection Agency (EPA)), establish interdisciplinary research centers, and accelerate technology transfer into the private sector. In addition, Department of Defense (DOD), Homeland Security, Agriculture and Justice as well as the National Institutes of Health (NIH) would also fund large R&D activities. They currently account for more than one-third of the federal budget for nanotechnology.

The European Union made nanosciences and nanotechnologies a priority in the Sixth Framework Program (FP6) in 2002 for the period of 2003-2006. They had dedicated small funds in FP4 and FP5 before. FP6 was tailored to help better structure European research and to cope with the strategic objectives set out in Lisbon in 2000. Japan identified nanotechnology as one of its main research priorities in 2001. The funding levels increased sharply from $400 million in 2001 to around $950 million in 2004. In 2003, South Korea embarked upon a ten-year program with around $2 billion of public funding, and Taiwan has committed around $600 million of public funding over six years. Singapore and China are also investing on a large scale. Russia is well funded as well.

Nanotechnology literally means any technology done on a nanoscale that has applications in the real world. Nanotechnology encompasses production and application of physical, chemical and biological systems at scales, ranging from individual atoms or molecules to submicron dimensions, as well as the integration of the resulting nanostructures into larger systems. Nanotechnology is likely to have a profound impact on our economy and society in the early 21st century, comparable to that of semiconductor technology, information technology, or cellular and molecular biology. Science and technology research in nanotechnology promises breakthroughs in areas such as materials and manufacturing, nanoelectronics, medicine and healthcare, energy, biotechnology, information technology and national security. It is widely felt that nanotechnology will be the next industrial revolution.

There is an increasing need for a multidisciplinary, system-oriented approach to design and manufacturing of micro/nanodevices that function reliably. This can only be achieved through the cross-fertilization of ideas from different disciplines and the systematic flow of information and people among research groups. Reliability is a critical technology for many micro- and nanosystems and nanostructured materials. A broad-based handbook was needed, and thus the first edition of Springer Handbook of Nanotechnology was published in April 2004. It presented an overview of nanomaterial synthesis, micro/nanofabrication, micro- and nanocomponents and systems, scanning probe microscopy, reliability issues (including nanotribology and nanomechanics) for nanotechnology, and industrial applications. When the handbook went for sale in Europe, it sold out in ten days. Reviews on the handbook were very flattering.

Given the explosive growth in nanoscience and nanotechnology, the publisher and the editor decided to develop a second edition merely six months after publication of the first edition. This edition of the handbook integrates the knowledge from the nanostructure, fabrication, materials science, devices, and reliability point of view. It covers various industrial applications. It also addresses social, ethical, and political issues. Given the significant interest in biomedical applications, a number of chapters in this arena have been added. The second edition consists of 59 chapters (new: 23; revised: 27; unchanged: 9). The chapters have been written by 154 internationally recognized experts in the field, from academia, national research labs, and industry.

This book is intended for three types of readers: graduate students of nanotechnology, researchers in academia and industry who are active or intend to become active in this field, and practicing engineers and scientists who have encountered a problem and hope to solve it as expeditiously as possible. The handbook should serve as an excellent text for one or two semester graduate courses in nanotechnology in mechanical engineering, materials science, applied physics, or applied chemistry.

We embarked on the development of the second edition in October 2004, and we worked very hard to get all the chapters to the publisher in a record time of about 7 months. I wish to sincerely thank the authors for offering to write comprehensive chapters on a tight schedule. This is generally an added responsibility to the hectic work schedules of researchers today. I depended on a large number of reviewers who provided critical reviews. I would like to thank Dr. Phillip J. Bond, Chief of Staff and Under Secretary for Technology, US Department of Commerce, Washington, D.C. for chapter suggestions as well as authors in the handbook. I would also like to thank my colleague, Dr. Zhenhua Tao, whose efforts during the preparation of this handbook were very useful. Last but not the least, I would like to thank my secretary Caterina Runyon-Spears for various administrative duties; her tireless efforts are highly appreciated.

I hope that this handbook will stimulate further interest in this important new field, and the readers of this handbook will find it useful.

May 2005   Bharat Bhushan
Editor

Preface to the 1st Edition

On December 29, 1959 at the California Institute of Technology, Nobel Laureate Richard P. Feynman gave a talk at the Annual meeting of the American Physical Society that has become one classic science lecture of the 20th century, titled "There's Plenty of Room at the Bottom." He presented a technological vision of extreme miniaturization in 1959, several years before the word "chip" became part of the lexicon. He talked about the problem of manipulating and controlling things on a small scale. Extrapolating from known physical laws, Feynman envisioned a technology using the ultimate toolbox of nature, building nanoobjects atom by atom or molecule by molecule. Since the 1980s, many inventions and discoveries in fabrication of nanoobjects have been a testament to his vision. In recognition of this reality, in a January 2000 speech at the same institute, former President W. J. Clinton talked about the exciting promise of "nanotechnology" and the importance of expanding research in nanoscale science and engineering. Later that month, he announced in his State of the Union Address an ambitious $497 million federal, multi-agency national nanotechnology initiative (NNI) in the fiscal year 2001 budget, and made the NNI a top science and technology priority. Nanotechnology literally means any technology done on a nanoscale that has applications in the real world. Nanotechnology encompasses production and application of physical, chemical and biological systems at size scales, ranging from individual atoms or molecules to submicron dimensions as well as the integration of the resulting nanostructures into larger systems. Nanofabrication methods include the manipulation or self-assembly of individual atoms, molecules, or molecular structures to produce nanostructured materials and sub-micron devices. Micro- and nanosystems components are fabricated using top-down lithographic and nonlithographic fabrication techniques. Nanotechnology will have a profound impact on our economy and society in the early 21st century, comparable to that of semiconductor technology, information technology, or advances in cellular and molecular biology. The research and development in nanotechnology will lead to potential breakthroughs in areas such as materials and manufacturing, nanoelectronics, medicine and healthcare, energy, biotechnology, information technology and national security. It is widely felt that nanotechnology will lead to the next industrial revolution.

Reliability is a critical technology for many micro- and nanosystems and nanostructured materials. No book exists on this emerging field. A broad based handbook is needed. The purpose of this handbook is to present an overview of nanomaterial synthesis, micro/nanofabrication, micro- and nanocomponents and systems, reliability issues (including nanotribology and nanomechanics) for nanotechnology, and industrial applications. The chapters have been written by internationally recognized experts in the field, from academia, national research labs and industry from all over the world.

The handbook integrates knowledge from the fabrication, mechanics, materials science and reliability points of view. This book is intended for three types of readers: graduate students of nanotechnology, researchers in academia and industry who are active or intend to become active in this field, and practicing engineers and scientists who have encountered a problem and hope to solve it as expeditiously as possible. The handbook should serve as an excellent text for one or two semester graduate courses in nanotechnology in mechanical engineering, materials science, applied physics, or applied chemistry.

We embarked on this project in February 2002, and we worked very hard to get all the chapters to the publisher in a record time of about 1 year. I wish to sincerely thank the authors for offering to write comprehensive chapters on a tight schedule. This is generally an added responsibility in the hectic work schedules of researchers today. I depended on a large number of reviewers who provided critical reviews. I would like to thank Dr. Phillip J. Bond, Chief of Staff and Under Secretary for Technology, US Department of Commerce, Washington, D.C. for suggestions for chapters as well as authors in the handbook. I would also like to thank my colleague, Dr. Huiwen Liu, whose efforts during the preparation of this handbook were very useful.

I hope that this handbook will stimulate further interest in this important new field, and the readers of this handbook will find it useful.

September 2003   Bharat Bhushan
                                                                                 Editor

Editors Vita

Dr. Bharat Bhushan received an M.S. in mechanical engineering from the Massachusetts Institute of Technology in 1971, an M.S. in mechanics and a Ph.D. in mechanical engineering from the University of Colorado at Boulder in 1973 and 1976, respectively, an MBA from Rensselaer Polytechnic Institute at Troy, NY in 1980, Doctor Technicae from the University of Trondheim at Trondheim, Norway in 1990, a Doctor of Technical Sciences from the Warsaw University of Technology at Warsaw, Poland in 1996, and Doctor Honouris Causa from the National Academy of Sciences at Gomel, Belarus in 2000. He is a registered professional engineer. He is presently an Ohio Eminent Scholar and The Howard D. Winbigler Professor in the College of Engineering, and the Director of the Nanoprobe Laboratory for Bio- and Nanotechnology and Biomimetics (NLB²) at the Ohio State University, Columbus, Ohio. His research interests include fundamental studies with a focus on scanning probe techniques in the interdisciplinary areas of bio/nanotribology, bio/nanomechanics and bio/nanomaterials characterization, and applications to bio/nanotechnology and biomimetics. He is an internationally recognized expert of bio/nanotribology and bio/nanomechanics using scanning probe microscopy, and is one of the most prolific authors. He is considered by some a pioneer of the tribology and mechanics of magnetic storage devices. He has authored 6 scientific books, more than 90 handbook chapters, more than 700 scientific papers (h factor $- 45+$; ISI Highly Cited in Materials Science, since 2007), and more than 60 technical reports, edited more than 45 books, and holds 17 US and foreign patents. He is co-editor of Springer NanoScience and Technology Series and co-editor of Microsystem Technologies. He has given more than 400 invited presentations on six continents and more than 140 keynote/plenary addresses at major international conferences.

Dr. Bhushan is an accomplished organizer. He organized the first symposium on Tribology and Mechanics of Magnetic Storage Systems in 1984 and the first international symposium on Advances in Information Storage Systems in 1990, both of which are now held annually. He is the founder of an ASME Information Storage and Processing Systems Division founded in 1993 and served as the founding chair during 1993–1998. His biography has been listed in over two dozen Who's Who books including Who's Who in the World and has received more than two dozen awards for his contributions to science and technology from professional societies, industry, and US government agencies. He is also the recipient of various international fellowships including the Alexander von Humboldt Research Prize for Senior Scientists, Max Planck Foundation Research Award for Outstanding Foreign Scientists, and the Fulbright Senior Scholar Award. He is a foreign member of the International Academy of Engineering (Russia), Byelorussian Academy of Engineering and Technology and the Academy of Triboengineering of Ukraine, an honorary member of the Society of Tribologists of Belarus, a fellow of ASME, IEEE, STLE, and the New York Academy of Sciences, and a member of ASEE, Sigma Xi and Tau Beta Pi.

Dr. Bhushan has previously worked for the R&D Division of Mechanical Technology Inc., Latham, NY; the Technology Services Division of SKF Industries Inc., King of Prussia, PA; the General Products Division Laboratory of IBM Corporation, Tucson, AZ; and the Almaden Research Center of IBM Corporation, San Jose, CA. He has held visiting professor appointments at University of California at Berkeley, University of Cambridge, UK, Technical University Vienna, Austria, University of Paris, Orsay, ETH Zurich and EPFL Lausanne.

List of Authors

Chong H. Ahn
University of Cincinnati
Department of Electrical
and Computer Engineering
Cincinnati, OH 45221, USA
e-mail: *chong.ahn@uc.edu*

Boris Anczykowski
nanoAnalytics GmbH
Münster, Germany
e-mail: *anczykowski@nanoanalytics.com*

W. Robert Ashurst
Auburn University
Department of Chemical Engineering
Auburn, AL 36849, USA
e-mail: *ashurst@auburn.edu*

Massood Z. Atashbar
Western Michigan University
Department of Electrical
and Computer Engineering
Kalamazoo, MI 49008-5329, USA
e-mail: *massood.atashbar@wmich.edu*

Wolfgang Bacsa
University of Toulouse III (Paul Sabatier)
Laboratoire de Physique des Solides (LPST),
UMR 5477 CNRS
Toulouse, France
e-mail: *bacsa@ramansco.ups-tlse.fr;
bacsa@lpst.ups-tlse.fr*

Kelly Bailey
University of Adelaide
CSIRO Human Nutrition
Adelaide SA 5005, Australia
e-mail: *kelly.bailey@csiro.au*

William Sims Bainbridge
National Science Foundation
Division of Information, Science and Engineering
Arlington, VA, USA
e-mail: *wsbainbridge@yahoo.com*

Antonio Baldi
Institut de Microelectronica de Barcelona (IMB)
Centro National Microelectrónica (CNM-CSIC)
Barcelona, Spain
e-mail: *antoni.baldi@cnm.es*

Wilhelm Barthlott
University of Bonn
Nees Institute for Biodiversity of Plants
Meckenheimer Allee 170
53115 Bonn, Germany
e-mail: *barthlott@uni-bonn.de*

Roland Bennewitz
INM – Leibniz Institute for New Materials
66123 Saarbrücken, Germany
e-mail: *roland.bennewitz@inm-gmbh.de*

Bharat Bhushan
Ohio State University
Nanoprobe Laboratory for Bio- and
Nanotechnology and Biomimetics (NLB²)
201 W. 19th Avenue
Columbus, OH 43210-1142, USA
e-mail: *bhushan.2@osu.edu*

Gerd K. Binnig
Definiens AG
Trappentreustr. 1
80339 Munich, Germany
e-mail: *gbinnig@definiens.com*

Marcie R. Black
Bandgap Engineering Inc.
1344 Main St.
Waltham, MA 02451, USA
e-mail: *marcie@alum.mit.edu;
marcie@bandgap.com*

Donald W. Brenner
Department of Materials Science and Engineering
Raleigh, NC, USA
e-mail: *brenner@ncsu.edu*

Jean-Marc Broto
Institut National des Sciences Appliquées
of Toulouse
Laboratoire National
des Champs Magnétiques Pulsés (LNCMP)
Toulouse, France
e-mail: *broto@lncmp.fr*

Guozhong Cao
University of Washington
Dept. of Materials Science and Engineering
302M Roberts Hall
Seattle, WA 98195-2120, USA
e-mail: *gzcao@u.washington.edu*

Edin (I-Chen) Chen
National Central University
Institute of Materials Science and Engineering
Department of Mechanical Engineering
Chung-Li, 320, Taiwan
e-mail: *ichen@ncu.edu.tw*

Yu-Ting Cheng
National Chiao Tung University
Department of Electronics Engineering
& Institute of Electronics
1001, Ta-Hsueh Rd.
Hsinchu, 300, Taiwan, R.O.C.
e-mail: *ytcheng@mail.nctu.edu.tw*

Giovanni Cherubini
IBM Zurich Research Laboratory
Tape Technologies
8803 Rüschlikon, Switzerland
e-mail: *cbi@zurich.ibm.com*

Mu Chiao
Department of Mechanical Engineering
6250 Applied Science Lane
Vancouver, BC V6T 1Z4, Canada
e-mail: *muchiao@mech.ubc.ca*

Jin-Woo Choi
Louisiana State University
Department of Electrical
and Computer Engineering
Baton Rouge, LA 70803, USA
e-mail: *choi@ece.lsu.edu*

Tamara H. Cooper
University of Adelaide
CSIRO Human Nutrition
Adelaide SA 5005, Australia
e-mail: *tamara.cooper@csiro.au*

Alex D. Corwin
GE Global Research
1 Research Circle
Niskayuna, NY 12309, USA
e-mail: *corwin@ge.com*

Maarten P. de Boer
Carnegie Mellon University
Department of Mechanical Engineering
5000 Forbes Avenue
Pittsburgh, PA 15213, USA
e-mail: *mpdebo@andrew.cmu.edu*

Dietrich Dehlinger
Lawrence Livermore National Laboratory
Engineering
Livermore, CA 94551, USA
e-mail: *dehlinger1@llnl.gov*

Frank W. DelRio
National Institute of Standards and Technology
100 Bureau Drive, Stop 8520
Gaithersburg, MD 20899-8520, USA
e-mail: *frank.delrio@nist.gov*

Michel Despont
IBM Zurich Research Laboratory
Micro- and Nanofabrication
8803 Rüschlikon, Switzerland
e-mail: *dpt@zurich.ibm.com*

Lixin Dong
Michigan State University
Electrical and Computer Engineering
2120 Engineering Building
East Lansing, MI 48824-1226, USA
e-mail: *ldong@egr.msu.edu*

Gene Dresselhaus
Massachusetts Institute of Technology
Francis Bitter Magnet Laboratory
Cambridge, MA 02139, USA
e-mail: *gene@mgm.mit.edu*

Mildred S. Dresselhaus
Massachusetts Institute of Technology
Department of Electrical Engineering
and Computer Science
Department of Physics
Cambridge, MA, USA
e-mail: *millie@mgm.mit.edu*

Urs T. Dürig
IBM Zurich Research Laboratory
Micro-/Nanofabrication
8803 Rüschlikon, Switzerland
e-mail: *drg@zurich.ibm.com*

Andreas Ebner
Johannes Kepler University Linz
Institute for Biophysics
Altenberger Str. 69
4040 Linz, Austria
e-mail: *andreas.ebner@jku.at*

Evangelos Eleftheriou
IBM Zurich Research Laboratory
8803 Rüschlikon, Switzerland
e-mail: *ele@zurich.ibm.com*

Emmanuel Flahaut
Université Paul Sabatier
CIRIMAT, Centre Interuniversitaire de Recherche
et d'Ingénierie des Matériaux, UMR 5085 CNRS
118 Route de Narbonne
31062 Toulouse, France
e-mail: *flahaut@chimie.ups-tlse.fr*

Anatol Fritsch
University of Leipzig
Institute of Experimental Physics I
Division of Soft Matter Physics
Linnéstr. 5
04103 Leipzig, Germany
e-mail: *anatol.fritsch@uni-leipzig.de*

Harald Fuchs
Universität Münster
Physikalisches Institut
Münster, Germany
e-mail: *fuchsh@uni-muenster.de*

Christoph Gerber
University of Basel
Institute of Physics
National Competence Center for Research
in Nanoscale Science (NCCR) Basel
Klingelbergstr. 82
4056 Basel, Switzerland
e-mail: *christoph.gerber@unibas.ch*

Franz J. Giessibl
Universität Regensburg
Institute of Experimental and Applied Physics
Universitätsstr. 31
93053 Regensburg, Germany
e-mail: *franz.giessibl@physik.uni-regensburg.de*

Enrico Gnecco
University of Basel
National Center of Competence in Research
Department of Physics
Klingelbergstr. 82
4056 Basel, Switzerland
e-mail: *enrico.gnecco@unibas.ch*

Stanislav N. Gorb
Max Planck Institut für Metallforschung
Evolutionary Biomaterials Group
Heisenbergstr. 3
70569 Stuttgart, Germany
e-mail: *s.gorb@mf.mpg.de*

Hermann Gruber
University of Linz
Institute of Biophysics
Altenberger Str. 69
4040 Linz, Austria
e-mail: *hermann.gruber@jku.at*

Jason Hafner
Rice University
Department of Physics and Astronomy
Houston, TX 77251, USA
e-mail: *hafner@rice.edu*

Judith A. Harrison
U.S. Naval Academy
Chemistry Department
572 Holloway Road
Annapolis, MD 21402-5026, USA
e-mail: *jah@usna.edu*

Martin Hegner
CRANN – The Naughton Institute
Trinity College, University of Dublin
School of Physics
Dublin, 2, Ireland
e-mail: *martin.hegner@tcd.ie*

Thomas Helbling
ETH Zurich
Micro and Nanosystems
Department of Mechanical
and Process Engineering
8092 Zurich, Switzerland
e-mail: *thomas.helbling@micro.mavt.ethz.ch*

Michael J. Heller
University of California San Diego
Department of Bioengineering
Dept. of Electrical and Computer Engineering
La Jolla, CA, USA
e-mail: *mjheller@ucsd.edu*

Seong-Jun Heo
Lam Research Corp.
4650 Cushing Parkway
Fremont, CA 94538, USA
e-mail: *seongjun.heo@lamrc.com*

Christofer Hierold
ETH Zurich
Micro and Nanosystems
Department of Mechanical
and Process Engineering
8092 Zurich, Switzerland
e-mail: *christofer.hierold@micro.mavt.ethz.ch*

Peter Hinterdorfer
University of Linz
Institute for Biophysics
Altenberger Str. 69
4040 Linz, Austria
e-mail: *peter.hinterdorfer@jku.at*

Dalibor Hodko
Nanogen, Inc.
10498 Pacific Center Court
San Diego, CA 92121, USA
e-mail: *dhodko@nanogen.com*

Hendrik Hölscher
Forschungszentrum Karlsruhe
Institute of Microstructure Technology
Linnéstr. 5
76021 Karlsruhe, Germany
e-mail: *hendrik.hoelscher@imt.fzk.de*

Hirotaka Hosoi
Hokkaido University
Creative Research Initiative Sousei
Kita 21, Nishi 10, Kita-ku
Sapporo, Japan
e-mail: *hosoi@cris.hokudai.ac.jp*

Katrin Hübner
Staatliche Fachoberschule Neu-Ulm
89231 Neu-Ulm, Germany
e-mail: *katrin.huebner1@web.de*

Douglas L. Irving
North Carolina State University
Materials Science and Engineering
Raleigh, NC 27695-7907, USA
e-mail: *doug_irving@ncsu.edu*

Jacob N. Israelachvili
University of California
Department of Chemical Engineering
and Materials Department
Santa Barbara, CA 93106-5080, USA
e-mail: *jacob@engineering.ucsb.edu*

Guangyao Jia
University of California, Irvine
Department of Mechanical
and Aerospace Engineering
Irvine, CA, USA
e-mail: *gjia@uci.edu*

Sungho Jin
University of California, San Diego
Department of Mechanical
and Aerospace Engineering
9500 Gilman Drive
La Jolla, CA 92093-0411, USA
e-mail: *jin@ucsd.edu*

Anne Jourdain
Interuniversity Microelectronics Center (IMEC)
Leuven, Belgium
e-mail: *jourdain@imec.be*

Yong Chae Jung
Samsung Electronics C., Ltd.
Senior Engineer Process Development Team
San #16 Banwol-Dong, Hwasung-City
Gyeonggi-Do 445-701, Korea
e-mail: *yc423.jung@samsung.com*

Harold Kahn
Case Western Reserve University
Department of Materials Science and Engineering
Cleveland, OH, USA
e-mail: *kahn@cwru.edu*

Roger Kamm
Massachusetts Institute of Technology
Department of Biological Engineering
77 Massachusetts Avenue
Cambridge, MA 02139, USA
e-mail: *rdkamm@mit.edu*

Ruti Kapon
Weizmann Institute of Science
Department of Biological Chemistry
Rehovot 76100, Israel
e-mail: *ruti.kapon@weizmann.ac.il*

Josef Käs
University of Leipzig
Institute of Experimental Physics I
Division of Soft Matter Physics
Linnéstr. 5
04103 Leipzig, Germany
e-mail: *jkaes@physik.uni-leipzig.de*

Horacio Kido
University of California at Irvine
Mechanical and Aerospace Engineering
Irvine, CA, USA
e-mail: *hkido@uci.edu*

Tobias Kießling
University of Leipzig
Institute of Experimental Physics I
Division of Soft Matter Physics
Linnéstr. 5
04103 Leipzig, Germany
e-mail: *Tobias.Kiessling@uni-leipzig.de*

Jitae Kim
University of California at Irvine
Department of Mechanical
and Aerospace Engineering
Irvine, CA, USA
e-mail: *jitaekim@uci.edu*

Jongbaeg Kim
Yonsei University
School of Mechanical Engineering
1st Engineering Bldg.
Seoul, 120-749, South Korea
e-mail: *kimjb@yonsei.ac.kr*

Nahui Kim
Samsung Advanced Institute of Technology
Research and Development
Seoul, South Korea
e-mail: *nahui.kim@samsung.com*

Kerstin Koch
Rhine-Waal University of Applied Science
Department of Life Science, Biology
and Nanobiotechnology
Landwehr 4
47533 Kleve, Germany
e-mail: *kerstin.koch@hochschule.rhein-waal.de*

Jing Kong
Massachusetts Institute of Technology
Department of Electrical Engineering
and Computer Science
Cambridge, MA, USA
e-mail: *jingkong@mit.edu*

Tobias Kraus
Leibniz-Institut für Neue Materialien gGmbH
Campus D2 2
66123 Saarbrücken, Germany
e-mail: *tobias.kraus@inm-gmbh.de*

Anders Kristensen
Technical University of Denmark
DTU Nanotech
2800 Kongens Lyngby, Denmark
e-mail: *anders.kristensen@nanotech.dtu.dk*

Ratnesh Lal
University of Chicago
Center for Nanomedicine
5841 S Maryland Av
Chicago, IL 60637, USA
e-mail: *rlal@uchicago.edu*

Jan Lammerding
Harvard Medical School
Brigham and Women's Hospital
65 Landsdowne St
Cambridge, MA 02139, USA
e-mail: *jlammerding@rics.bwh.harvard.edu*

Hans Peter Lang
University of Basel
Institute of Physics, National Competence Center
for Research in Nanoscale Science (NCCR) Basel
Klingelbergstr. 82
4056 Basel, Switzerland
e-mail: *hans-peter.lang@unibas.ch*

Carmen LaTorre
Owens Corning Science and Technology
Roofing and Asphalt
2790 Columbus Road
Granville, OH 43023, USA
e-mail: *carmen.latorre@owenscorning.com*

Christophe Laurent
Université Paul Sabatier
CIRIMAT UMR 5085 CNRS
118 Route de Narbonne
31062 Toulouse, France
e-mail: *laurent@chimie.ups-tlse.fr*

Abraham P. Lee
University of California Irvine
Department of Biomedical Engineering
Department of Mechanical
and Aerospace Engineering
Irvine, CA 92697, USA
e-mail: *aplee@uci.edu*

Stephen C. Lee
Ohio State University
Biomedical Engineering Center
Columbus, OH 43210, USA
e-mail: *lee@bme.ohio-state.edu*

Wayne R. Leifert
Adelaide Business Centre
CSIRO Human Nutrition
Adelaide SA 5000, Australia
e-mail: *wayne.leifert@csiro.au*

Liwei Lin
UC Berkeley
Mechanical Engineering Department
5126 Etcheverry
Berkeley, CA 94720-1740, USA
e-mail: *lwlin@me.berkeley.edu*

Yu-Ming Lin
IBM T.J. Watson Research Center
Nanometer Scale Science & Technology
1101 Kitchawan Road
Yorktown Heigths, NY 10598, USA
e-mail: *yming@us.ibm.com*

Marc J. Madou
University of California Irvine
Department of Mechanical and Aerospace
and Biomedical Engineering
Irvine, CA, USA
e-mail: *mmadou@uci.edu*

Othmar Marti
Ulm University
Institute of Experimental Physics
Albert-Einstein-Allee 11
89069 Ulm, Germany
e-mail: *othmar.marti@uni-ulm.de*

Jack Martin
66 Summer Street
Foxborough, MA 02035, USA
e-mail: *jack.martin@alumni.tufts.edu*

Shinji Matsui
University of Hyogo
Laboratory of Advanced Science
and Technology for Industry
Hyogo, Japan
e-mail: *matsui@lasti.u-hyogo.ac.jp*

Mehran Mehregany
Case Western Reserve University
Department of Electrical Engineering
and Computer Science
Cleveland, OH 44106, USA
e-mail: *mxm31@cwru.edu*

Etienne Menard
Semprius, Inc.
4915 Prospectus Dr.
Durham, NC 27713, USA
e-mail: *etienne.menard@semprius.com*

Ernst Meyer
University of Basel
Institute of Physics
Basel, Switzerland
e-mail: *ernst.meyer@unibas.ch*

Robert Modliñski
Baolab Microsystems
Terrassa 08220, Spain
e-mail: *rmodlinski@gmx.com*

Mohammad Mofrad
University of California, Berkeley
Department of Bioengineering
Berkeley, CA 94720, USA
e-mail: *mofrad@berkeley.edu*

Marc Monthioux
CEMES – UPR A-8011 CNRS
Carbones et Matériaux Carbonés,
Carbons and Carbon-Containing Materials
29 Rue Jeanne Marvig
31055 Toulouse 4, France
e-mail: *monthiou@cemes.fr*

Markus Morgenstern
RWTH Aachen University
II. Institute of Physics B and JARA-FIT
52056 Aachen, Germany
e-mail: *mmorgens@physik.rwth-aachen.de*

Seizo Morita
Osaka University
Department of Electronic Engineering
Suita-City
Osaka, Japan
e-mail: *smorita@ele.eng.osaka-u.ac.jp*

Koichi Mukasa
Hokkaido University
Nanoelectronics Laboratory
Sapporo, Japan
e-mail: *mukasa@nano.eng.hokudai.ac.jp*

Bradley J. Nelson
Swiss Federal Institute of Technology (ETH)
Institute of Robotics and Intelligent Systems
8092 Zurich, Switzerland
e-mail: *bnelson@ethz.ch*

Michael Nosonovsky
University of Wisconsin-Milwaukee
Department of Mechanical Engineering
3200 N. Cramer St.
Milwaukee, WI 53211, USA
e-mail: *nosonovs@uwm.edu*

Hiroshi Onishi
Kanagawa Academy of Science and Technology
Surface Chemistry Laboratory
Kanagawa, Japan
e-mail: *oni@net.ksp.or.jp*

Alain Peigney
Centre Inter-universitaire de Recherche
sur l'Industrialisation des Matériaux (CIRIMAT)
Toulouse 4, France
e-mail: *peigney@chimie.ups-tlse.fr*

Oliver Pfeiffer
Individual Computing GmbH
Ingelsteinweg 2d
4143 Dornach, Switzerland
e-mail: *oliver.pfeiffer@gmail.com*

Haralampos Pozidis
IBM Zurich Research Laboratory
Storage Technologies
Rüschlikon, Switzerland
e-mail: *hap@zurich.ibm.com*

Robert Puers
Katholieke Universiteit Leuven
ESAT/MICAS
Leuven, Belgium
e-mail: *bob.puers@esat.kuleuven.ac.be*

Calvin F. Quate
Stanford University
Edward L. Ginzton Laboratory
450 Via Palou
Stanford, CA 94305-4088, USA
e-mail: *quate@stanford.edu*

Oded Rabin
University of Maryland
Department of Materials Science and Engineering
College Park, MD, USA
e-mail: *oded@umd.edu*

Françisco M. Raymo
University of Miami
Department of Chemistry
1301 Memorial Drive
Coral Gables, FL 33146-0431, USA
e-mail: *fraymo@miami.edu*

Manitra Razafinimanana
University of Toulouse III (Paul Sabatier)
Centre de Physique des Plasmas
et leurs Applications (CPPAT)
Toulouse, France
e-mail: *razafinimanana@cpat.ups-tlse.fr*

Ziv Reich
Weizmann Institute of Science Ha'Nesi Ha'Rishon
Department of Biological Chemistry
Rehovot 76100, Israel
e-mail: *ziv.reich@weizmann.ac.il*

John A. Rogers
University of Illinois
Department of Materials Science and Engineering
Urbana, IL, USA
e-mail: *jrogers@uiuc.edu*

Cosmin Roman
ETH Zurich
Micro and Nanosystems Department of Mechanical and Process Engineering
8092 Zurich, Switzerland
e-mail: *cosmin.roman@micro.mavt.ethz.ch*

Marina Ruths
University of Massachusetts Lowell
Department of Chemistry
1 University Avenue
Lowell, MA 01854, USA
e-mail: *marina_ruths@uml.edu*

Ozgur Sahin
The Rowland Institute at Harvard
100 Edwin H. Land Blvd
Cambridge, MA 02142, USA
e-mail: *sahin@rowland.harvard.edu*

Akira Sasahara
Japan Advanced Institute
of Science and Technology
School of Materials Science
1-1 Asahidai
923-1292 Nomi, Japan
e-mail: *sasahara@jaist.ac.jp*

Helmut Schift
Paul Scherrer Institute
Laboratory for Micro- and Nanotechnology
5232 Villigen PSI, Switzerland
e-mail: *helmut.schift@psi.ch*

André Schirmeisen
University of Münster
Institute of Physics
Wilhelm-Klemm-Str. 10
48149 Münster, Germany
e-mail: *schirmeisen@uni-muenster.de*

Christian Schulze
Beiersdorf AG
Research & Development
Unnastr. 48
20245 Hamburg, Germany
e-mail: *christian.schulze@beiersdorf.com; christian.schulze@uni-leipzig.de*

Alexander Schwarz
University of Hamburg
Institute of Applied Physics
Jungiusstr. 11
20355 Hamburg, Germany
e-mail: *aschwarz@physnet.uni-hamburg.de*

Udo D. Schwarz
Yale University
Department of Mechanical Engineering
15 Prospect Street
New Haven, CT 06520-8284, USA
e-mail: *udo.schwarz@yale.edu*

Philippe Serp
Ecole Nationale Supérieure d'Ingénieurs
en Arts Chimiques et Technologiques
Laboratoire de Chimie de Coordination (LCC)
118 Route de Narbonne
31077 Toulouse, France
e-mail: *philippe.serp@ensiacet.fr*

Huamei (Mary) Shang
GE Healthcare
4855 W. Electric Ave.
Milwaukee, WI 53219, USA
e-mail: *huamei.shang@ge.com*

Susan B. Sinnott
University of Florida
Department of Materials Science and Engineering
154 Rhines Hall
Gainesville, FL 32611-6400, USA
e-mail: *ssinn@mse.ufl.edu*

Anisoara Socoliuc
SPECS Zurich GmbH
Technoparkstr. 1
8005 Zurich, Switzerland
e-mail: *socoliuc@nanonis.com*

Olav Solgaard
Stanford University
E.L. Ginzton Laboratory
450 Via Palou
Stanford, CA 94305-4088, USA
e-mail: *solgaard@stanford.edu*

Dan Strehle
University of Leipzig
Institute of Experimental Physics I
Division of Soft Matter Physics
Linnéstr. 5
04103 Leipzig, Germany
e-mail: *dan.strehle@uni-leipzig.de*

Carsten Stüber
University of Leipzig
Institute of Experimental Physics I
Division of Soft Matter Physics
Linnéstr. 5
04103 Leipzig, Germany
e-mail: *stueber@rz.uni-leipzig.de*

Yu-Chuan Su
ESS 210
Department of Engineering and System Science 101
Kuang-Fu Road
Hsinchu, 30013, Taiwan
e-mail: *ycsu@ess.nthu.edu.tw*

Kazuhisa Sueoka
Graduate School of Information Science
and Technology
Hokkaido University
Nanoelectronics Laboratory
Kita-14, Nishi-9, Kita-ku
060-0814 Sapporo, Japan
e-mail: *sueoka@nano.isthokudai.ac.jp*

Yasuhiro Sugawara
Osaka University
Department of Applied Physics
Yamada-Oka 2-1, Suita
565-0871 Osaka, Japan
e-mail: *sugawara@ap.eng.osaka-u.ac.jp*

Benjamin Sullivan
TearLab Corp.
11025 Roselle Street
San Diego, CA 92121, USA
e-mail: *bdsulliv@TearLab.com*

Paul Swanson
Nexogen, Inc.
Engineering
8360 C Camino Santa Fe
San Diego, CA 92121, USA
e-mail: *pswanson@nexogentech.com*

Yung-Chieh Tan
Washington University School of Medicine
Department of Medicine
Division of Dermatology
660 S. Euclid Ave.
St. Louis, MO 63110, USA
e-mail: *ytanster@gmail.com*

Shia-Yen Teh
University of California at Irvine
Biomedical Engineering Department
3120 Natural Sciences II
Irvine, CA 92697-2715, USA
e-mail: *steh@uci.edu*

W. Merlijn van Spengen
Leiden University
Kamerlingh Onnes Laboratory
Niels Bohrweg 2
Leiden, CA 2333, The Netherlands
e-mail: *spengen@physics.leidenuniv.nl*

Peter Vettiger
University of Neuchâtel
SAMLAB
Jaquet-Droz 1
2002 Neuchâtel, Switzerland
e-mail: *peter.vettiger@unine.ch*

Franziska Wetzel
University of Leipzig
Institute of Experimental Physics I
Division of Soft Matter Physics
Linnéstr. 5
04103 Leipzig, Germany
e-mail: *franziska.wetzel@uni-leipzig.de*

Heiko Wolf
IBM Research GmbH
Zurich Research Laboratory
Säumerstr. 4
8803 Rüschlikon, Switzerland
e-mail: *hwo@zurich.ibm.com*

Darrin J. Young
Case Western Reserve University
Department of EECS, Glennan 510
10900 Euclid Avenue
Cleveland, OH 44106, USA
e-mail: *djy@po.cwru.edu*

Babak Ziaie
Purdue University
Birck Nanotechnology Center
1205 W. State St.
West Lafayette, IN 47907-2035, USA
e-mail: *bziaie@purdue.edu*

Christian A. Zorman
Case Western Reserve University
Department of Electrical Engineering
and Computer Science
10900 Euclid Avenue
Cleveland, OH 44106, USA
e-mail: *caz@case.edu*

Jim V. Zoval
Saddleback College
Department of Math and Science
28000 Marguerite Parkway
Mission Viejo, CA 92692, USA
e-mail: *jzoval@saddleback.edu*

Acknowledgements

A.3 Introduction to Carbon Nanotubes
by Marc Monthioux, Philippe Serp, Emmanuel Flahaut, Manitra Razafinimanana, Christophe Laurent, Alain Peigney, Wolfgang Bacsa, Jean-Marc Broto

The authors wish to acknowledge their membership to the European Research Group *Sciences and Applications of Nanotubes* (contact: annick.loiseau@onera.fr) and the French Carbon Group (contact: marc.monthioux @cemes.fr).

A.4 Nanowires
by Mildred S. Dresselhaus, Yu-Ming Lin, Oded Rabin, Marcie R. Black, Jing Kong, Gene Dresselhaus

The authors gratefully acknowledge the stimulating discussions with Professors Charles Lieber, Gang Chen, S.T. Lee, Arun Majumdar, Peidong Yang, and Jean-Paul Issi, Dr. Joseph Heremans and Ted Harman. The authors are grateful for support for this work by the ONR Grant No. 000140-21-0865, the MURI program subcontract PO No. 0205-G-7A114-01 through UCLA, and DARPA contract No. N66001-00-1-8603.

A.10 Stamping Techniques for Micro- and Nanofabrication
by Etienne Menard, John A. Rogers

The authors extend their deepest thanks to all of the collaborators who contributed the work described here.

目 录

缩略语

1. 纳米技术介绍 1
    1.1 纳米技术——定义和实例 1
    1.2 背景和研究阶段的费用 4
    1.3 源于自然的课程（仿生学） 6
    1.4 不同领域中的应用 9
    1.5 各种问题 10
    1.6 研究培训 11
    1.7 手册的编排 11
    参考文献 12

Part A 纳米结构、微/纳米制备和材料

2. 纳米材料的合成与应用：分子元器件 17
    2.1 制备纳米结构材料的化学方法 18
    2.2 分子开关与逻辑门 22
    2.3 固态器件 30
    2.4 结论与展望 42
    参考文献 43

3. 碳纳米管介绍 47
    3.1 碳纳米管的结构 48
    3.2 碳纳米管的合成 53
    3.3 碳纳米管的生长机制 70
    3.4 碳纳米管的特性 74
    3.5 基于纳米对象的碳纳米管 80
    3.6 碳纳米管的应用 85
    3.7 碳纳米管的毒理与环境影响 99
    3.8 结束语 100
    参考文献 101

4. 纳米线 119
    4.1 合成 121

4.2 纳米线的特征和物理特性　130
　　4.3 应　用　152
　　4.4 结束语　159
　　参考文献　159
5. 基于模板法的纳米棒或纳米线阵列的合成　169
　　5.1 模板法　170
　　5.2 电化学淀积　171
　　5.3 电泳淀积　175
　　5.4 模板填充　180
　　5.5 从可动模板转换　182
　　5.6 总结与结束语　182
　　参考文献　183
6. 粒子的自组装模板　187
　　6.1 组装工艺　189
　　6.2 经典定向粒子组装　194
　　6.3 模　板　202
　　6.4 工艺和设备　205
　　6.5 结　论　206
　　参考文献　207
7. 利用聚焦离子束化学气相淀积制造三维纳米结构　211
　　7.1 三维纳米结构的制备　212
　　7.2 纳米电子力学　215
　　7.3 纳米光学　223
　　7.4 纳米生物学　224
　　7.5 总　结　228
　　参考文献　228
8. 微/纳米制备介绍　231
　　8.1 基本的微制备技术　232
　　8.2 微机电系统制备技术　244
　　8.3 纳米制备技术　256
　　8.4 总结和结论　265
　　参考文献　265
9. 纳米压印光刻技术——使用印模刻印　271
　　9.1 新型纳米版图方法　273

9.2 纳米压印工艺 …… 277
9.3 纳米压印的工具和材料 …… 288
9.4 纳米压印的应用 …… 294
9.5 结论与展望 …… 302
参考文献 …… 304

10. 微/纳米制造中的冲压技术 …… 313
10.1 高分辨率冲压 …… 314
10.2 微米接触印刷 …… 316
10.3 纳米传输印刷 …… 318
10.4 应 用 …… 322
10.5 结 论 …… 329
参考文献 …… 330

11. 微/纳米机电系统的材料特性 …… 333
11.1 硅 …… 333
11.2 锗基材料 …… 340
11.3 金 属 …… 341
11.4 恶劣环境下的半导体 …… 343
11.5 砷化镓，磷化铟，Ⅲ-Ⅴ族化合材料 …… 349
11.6 铁电材料 …… 350
11.7 聚合物材料 …… 351
11.8 未来趋势 …… 352
参考文献 …… 353

Contents

List of Abbreviations

1 Introduction to Nanotechnology
Bharat Bhushan 1
1.1 Nanotechnology – Definition and Examples 1
1.2 Background and Research Expenditures 4
1.3 Lessons from Nature (Biomimetics) 6
1.4 Applications in Different Fields 9
1.5 Various Issues 10
1.6 Research Training 11
1.7 Organization of the Handbook 11
References 12

Part A Nanostructures, Micro-/Nanofabrication and Materials

2 Nanomaterials Synthesis and Applications: Molecule-Based Devices
Francisco M. Raymo 17
2.1 Chemical Approaches to Nanostructured Materials 18
2.2 Molecular Switches and Logic Gates 22
2.3 Solid State Devices 30
2.4 Conclusions and Outlook 42
References 43

3 Introduction to Carbon Nanotubes
Marc Monthioux, Philippe Serp, Emmanuel Flahaut,
Manitra Razafinimanana, Christophe Laurent, Alain Peigney,
Wolfgang Bacsa, Jean-Marc Broto 47
3.1 Structure of Carbon Nanotubes 48
3.2 Synthesis of Carbon Nanotubes 53
3.3 Growth Mechanisms of Carbon Nanotubes 70
3.4 Properties of Carbon Nanotubes 74
3.5 Carbon Nanotube-Based Nano-Objects 80
3.6 Applications of Carbon Nanotubes 85
3.7 Toxicity and Environmental Impact of Carbon Nanotubes 99
3.8 Concluding Remarks 100
References 101

4 **Nanowires**
Mildred S. Dresselhaus, Yu-Ming Lin, Oded Rabin, Marcie R. Black,
Jing Kong, Gene Dresselhaus 119
  4.1 Synthesis 121
  4.2 Characterization and Physical Properties of Nanowires 130
  4.3 Applications 152
  4.4 Concluding Remarks 159
  References 159

5 **Template-Based Synthesis of Nanorod or Nanowire Arrays**
Huamei (Mary) Shang, Guozhong Cao 169
  5.1 Template-Based Approach 170
  5.2 Electrochemical Deposition 171
  5.3 Electrophoretic Deposition 175
  5.4 Template Filling 180
  5.5 Converting from Reactive Templates 182
  5.6 Summary and Concluding Remarks 182
  References 183

6 **Templated Self-Assembly of Particles**
Tobias Kraus, Heiko Wolf 187
  6.1 The Assembly Process 189
  6.2 Classes of Directed Particle Assembly 194
  6.3 Templates 202
  6.4 Processes and Setups 205
  6.5 Conclusions 206
  References 207

7 **Three-Dimensional Nanostructure Fabrication
  by Focused Ion Beam Chemical Vapor Deposition**
Shinji Matsui 211
  7.1 Three-Dimensional Nanostructure Fabrication 212
  7.2 Nanoelectromechanics 215
  7.3 Nanooptics: Brilliant Blue Observation
      from a *Morpho* Butterfly Scale Quasistructure 223
  7.4 Nanobiology 224
  7.5 Summary 228
  References 228

8 **Introduction to Micro-/Nanofabrication**
Babak Ziaie, Antonio Baldi, Massood Z. Atashbar 231
  8.1 Basic Microfabrication Techniques 232
  8.2 MEMS Fabrication Techniques 244
  8.3 Nanofabrication Techniques 256
  8.4 Summary and Conclusions 265
  References 265

9 Nanoimprint Lithography – Patterning of Resists Using Molding
Helmut Schift, Anders Kristensen 271
- 9.1 Emerging Nanopatterning Methods 273
- 9.2 Nanoimprint Process 277
- 9.3 Tools and Materials for Nanoimprinting 288
- 9.4 Nanoimprinting Applications 294
- 9.5 Conclusions and Outlook 302
- **References** 304

10 Stamping Techniques for Micro- and Nanofabrication
Etienne Menard, John A. Rogers 313
- 10.1 High-Resolution Stamps 314
- 10.2 Microcontact Printing 316
- 10.3 Nanotransfer Printing 318
- 10.4 Applications 322
- 10.5 Conclusions 329
- **References** 330

11 Material Aspects of Micro- and Nanoelectromechanical Systems
Christian A. Zorman, Mehran Mehregany 333
- 11.1 Silicon 333
- 11.2 Germanium-Based Materials 340
- 11.3 Metals 341
- 11.4 Harsh-Environment Semiconductors 343
- 11.5 GaAs, InP, and Related III–V Materials 349
- 11.6 Ferroelectric Materials 350
- 11.7 Polymer Materials 351
- 11.8 Future Trends 352
- **References** 353

List of Abbreviations

μCP	microcontact printing
1-D	one-dimensional
18-MEA	18-methyl eicosanoic acid
2-D	two-dimensional
2-DEG	two-dimensional electron gas
3-APTES	3-aminopropyltriethoxysilane
3-D	three-dimensional

A

a-BSA	anti-bovine serum albumin
a-C	amorphous carbon
A/D	analog-to-digital
AA	amino acid
AAM	anodized alumina membrane
ABP	actin binding protein
AC	alternating-current
AC	amorphous carbon
ACF	autocorrelation function
ADC	analog-to-digital converter
ADXL	analog devices accelerometer
AFAM	atomic force acoustic microscopy
AFM	atomic force microscope
AFM	atomic force microscopy
AKD	alkylketene dimer
ALD	atomic layer deposition
AM	amplitude modulation
AMU	atomic mass unit
AOD	acoustooptical deflector
AOM	acoustooptical modulator
AP	alkaline phosphatase
APB	actin binding protein
APCVD	atmospheric-pressure chemical vapor deposition
APDMES	aminopropyldimethylethoxysilane
APTES	aminopropyltriethoxysilane
ASIC	application-specific integrated circuit
ASR	analyte-specific reagent
ATP	adenosine triphosphate

B

BAP	barometric absolute pressure
BAPDMA	behenyl amidopropyl dimethylamine glutamate
bcc	body-centered cubic
BCH	brucite-type cobalt hydroxide
BCS	Bardeen–Cooper–Schrieffer
BD	blu-ray disc
BDCS	biphenyldimethylchlorosilane
BE	boundary element
BFP	biomembrane force probe
BGA	ball grid array
BHF	buffered HF
BHPET	1,1'-(3,6,9,12,15-pentaoxapentadecane-1,15-diyl)bis(3-hydroxyethyl-1H-imidazolium-1-yl) di[bis(trifluoromethanesulfonyl)imide]
BHPT	1,1'-(pentane-1,5-diyl)bis(3-hydroxyethyl-1H-imidazolium-1-yl) di[bis(trifluoromethanesulfonyl)imide]
BiCMOS	bipolar CMOS
bioMEMS	biomedical microelectromechanical system
bioNEMS	biomedical nanoelectromechanical system
BMIM	1-butyl-3-methylimidazolium
BP	bit pitch
BPAG1	bullous pemphigoid antigen 1
BPT	biphenyl-4-thiol
BPTC	cross-linked BPT
BSA	bovine serum albumin
BST	barium strontium titanate
BTMAC	behentrimonium chloride

C

CA	constant amplitude
CA	contact angle
CAD	computer-aided design
CAH	contact angle hysteresis
cAMP	cyclic adenosine monophosphate
CAS	Crk-associated substrate
CBA	cantilever beam array
CBD	chemical bath deposition
CCD	charge-coupled device
CCVD	catalytic chemical vapor deposition
CD	compact disc
CD	critical dimension
CDR	complementarity determining region
CDW	charge density wave
CE	capillary electrophoresis
CE	constant excitation
CEW	continuous electrowetting
CG	controlled geometry
CHO	Chinese hamster ovary
CIC	cantilever in cantilever
CMC	cell membrane complex
CMC	critical micelle concentration
CMOS	complementary metal–oxide–semiconductor
CMP	chemical mechanical polishing

CNF	carbon nanofiber
CNFET	carbon nanotube field-effect transistor
CNT	carbon nanotube
COC	cyclic olefin copolymer
COF	chip-on-flex
COF	coefficient of friction
COG	cost of goods
CoO	cost of ownership
COS	CV-1 in origin with SV40
CP	circularly permuted
CPU	central processing unit
CRP	C-reactive protein
CSK	cytoskeleton
CSM	continuous stiffness measurement
CTE	coefficient of thermal expansion
Cu-TBBP	Cu-tetra-3,5 di-tertiary-butyl-phenyl porphyrin
CVD	chemical vapor deposition

D

DBR	distributed Bragg reflector
DC-PECVD	direct-current plasma-enhanced CVD
DC	direct-current
DDT	dichlorodiphenyltrichloroethane
DEP	dielectrophoresis
DFB	distributed feedback
DFM	dynamic force microscopy
DFS	dynamic force spectroscopy
DGU	density gradient ultracentrifugation
DI	FESPdigital instrument force modulation etched Si probe
DI	TESPdigital instrument tapping mode etched Si probe
DI	digital instrument
DI	deionized
DIMP	diisopropylmethylphosphonate
DIP	dual inline packaging
DIPS	industrial postpackaging
DLC	diamondlike carbon
DLP	digital light processing
DLVO	Derjaguin–Landau–Verwey–Overbeek
DMD	deformable mirror display
DMD	digital mirror device
DMDM	1,3-dimethylol-5,5-dimethyl
DMMP	dimethylmethylphosphonate
DMSO	dimethyl sulfoxide
DMT	Derjaguin–Muller–Toporov
DNA	deoxyribonucleic acid
DNT	2,4-dinitrotoluene
DOD	Department of Defense
DOE	Department of Energy
DOE	diffractive optical element
DOF	degree of freedom
DOPC	1,2-dioleoyl-sn-glycero-3-phosphocholine
DOS	density of states
DP	decylphosphonate
DPN	dip-pen nanolithography
DRAM	dynamic random-access memory
DRIE	deep reactive ion etching
ds	double-stranded
DSC	differential scanning calorimetry
DSP	digital signal processor
DTR	discrete track recording
DTSSP	3,3'-dithio-bis(sulfosuccinimidylproprionate)
DUV	deep-ultraviolet
DVD	digital versatile disc
DWNT	double-walled CNT

E

EAM	embedded atom method
EB	electron beam
EBD	electron beam deposition
EBID	electron-beam-induced deposition
EBL	electron-beam lithography
ECM	extracellular matrix
ECR-CVD	electron cyclotron resonance chemical vapor deposition
ED	electron diffraction
EDC	1-ethyl-3-(3-diamethylaminopropyl) carbodiimide
EDL	electrostatic double layer
EDP	ethylene diamine pyrochatechol
EDTA	ethylenediamine tetraacetic acid
EDX	energy-dispersive x-ray
EELS	electron energy loss spectra
EFM	electric field gradient microscopy
EFM	electrostatic force microscopy
EHD	elastohydrodynamic
EO	electroosmosis
EOF	electroosmotic flow
EOS	electrical overstress
EPA	Environmental Protection Agency
EPB	electrical parking brake
ESD	electrostatic discharge
ESEM	environmental scanning electron microscope
EU	European Union
EUV	extreme ultraviolet
EW	electrowetting
EWOD	electrowetting on dielectric

F

F-actin	filamentous actin
FA	focal adhesion
FAA	formaldehyde–acetic acid–ethanol
FACS	fluorescence-activated cell sorting

FAK	focal adhesion kinase		HDT	hexadecanethiol
FBS	fetal bovine serum		HDTV	high-definition television
FC	flip-chip		HEK	human embryonic kidney 293
FCA	filtered cathodic arc		HEL	hot embossing lithography
fcc	face-centered cubic		HEXSIL	hexagonal honeycomb polysilicon
FCP	force calibration plot		HF	hydrofluoric
FCS	fluorescence correlation spectroscopy		HMDS	hexamethyldisilazane
FD	finite difference		HNA	hydrofluoric-nitric-acetic
FDA	Food and Drug Administration		HOMO	highest occupied molecular orbital
FE	finite element		HOP	highly oriented pyrolytic
FEM	finite element method		HOPG	highly oriented pyrolytic graphite
FEM	finite element modeling		HOT	holographic optical tweezer
FESEM	field emission SEM		HP	hot-pressing
FESP	force modulation etched Si probe		HPI	hexagonally packed intermediate
FET	field-effect transistor		HRTEM	high-resolution transmission electron microscope
FFM	friction force microscope		HSA	human serum albumin
FFM	friction force microscopy		HtBDC	hexa-*tert*-butyl-decacyclene
FIB-CVD	focused ion beam chemical vapor deposition		HTCS	high-temperature superconductivity
FIB	focused ion beam		HTS	high throughput screening
FIM	field ion microscope		HUVEC	human umbilical venous endothelial cell
FIP	feline coronavirus			
FKT	Frenkel–Kontorova–Tomlinson		**I**	
FM	frequency modulation			
FMEA	failure-mode effect analysis		IBD	ion beam deposition
FP6	Sixth Framework Program		IC	integrated circuit
FP	fluorescence polarization		ICA	independent component analysis
FPR	*N*-formyl peptide receptor		ICAM-1	intercellular adhesion molecules 1
FS	force spectroscopy		ICAM-2	intercellular adhesion molecules 2
FTIR	Fourier-transform infrared		ICT	information and communication technology
FV	force–volume		IDA	interdigitated array
			IF	intermediate filament
G			IF	intermediate-frequency
			IFN	interferon
GABA	γ-aminobutyric acid		IgG	immunoglobulin G
GDP	guanosine diphosphate		IKVAV	isoleucine–lysine–valine–alanine–valine
GF	gauge factor		IL	ionic liquid
GFP	green fluorescent protein		IMAC	immobilized metal ion affinity chromatography
GMR	giant magnetoresistive		IMEC	Interuniversity MicroElectronics Center
GOD	glucose oxidase		IR	infrared
GPCR	G-protein coupled receptor		ISE	indentation size effect
GPS	global positioning system		ITO	indium tin oxide
GSED	gaseous secondary-electron detector		ITRS	International Technology Roadmap for Semiconductors
GTP	guanosine triphosphate		IWGN	Interagency Working Group on Nanoscience, Engineering, and Technology
GW	Greenwood and Williamson			
			J	
H				
			JC	jump-to-contact
HAR	high aspect ratio		JFIL	jet-and-flash imprint lithography
HARMEMS	high-aspect-ratio MEMS		JKR	Johnson–Kendall–Roberts
HARPSS	high-aspect-ratio combined poly- and single-crystal silicon			
HBM	human body model			
hcp	hexagonal close-packed			
HDD	hard-disk drive			

K

KASH	Klarsicht, ANC-1, Syne Homology
KPFM	Kelvin probe force microscopy

L

LA	lauric acid
LAR	low aspect ratio
LB	Langmuir–Blodgett
LBL	layer-by-layer
LCC	leadless chip carrier
LCD	liquid-crystal display
LCoS	liquid crystal on silicon
LCP	liquid-crystal polymer
LDL	low-density lipoprotein
LDOS	local density of states
LED	light-emitting diode
LFA-1	leukocyte function-associated antigen-1
LFM	lateral force microscope
LFM	lateral force microscopy
LIGA	Lithographie Galvanoformung Abformung
LJ	Lennard-Jones
LMD	laser microdissection
LMPC	laser microdissection and pressure catapulting
LN	liquid-nitrogen
LoD	limit-of-detection
LOR	lift-off resist
LPC	laser pressure catapulting
LPCVD	low-pressure chemical vapor deposition
LSC	laser scanning cytometry
LSN	low-stress silicon nitride
LT-SFM	low-temperature scanning force microscope
LT-SPM	low-temperature scanning probe microscopy
LT-STM	low-temperature scanning tunneling microscope
LT	low-temperature
LTM	laser tracking microrheology
LTO	low-temperature oxide
LTRS	laser tweezers Raman spectroscopy
LUMO	lowest unoccupied molecular orbital
LVDT	linear variable differential transformer

M

MALDI	matrix assisted laser desorption ionization
MAP	manifold absolute pressure
MAPK	mitogen-activated protein kinase
MAPL	molecular assembly patterning by lift-off
MBE	molecular-beam epitaxy
MC	microcantilever
MC	microcapillary
MCM	multi-chip module
MD	molecular dynamics
ME	metal-evaporated
MEMS	microelectromechanical system
MExFM	magnetic exchange force microscopy
MFM	magnetic field microscopy
MFM	magnetic force microscope
MFM	magnetic force microscopy
MHD	magnetohydrodynamic
MIM	metal–insulator–metal
MIMIC	micromolding in capillaries
MLE	maximum likelihood estimator
MOCVD	metalorganic chemical vapor deposition
MOEMS	microoptoelectromechanical system
MOS	metal–oxide–semiconductor
MOSFET	metal–oxide–semiconductor field-effect transistor
MP	metal particle
MPTMS	mercaptopropyltrimethoxysilane
MRFM	magnetic resonance force microscopy
MRFM	molecular recognition force microscopy
MRI	magnetic resonance imaging
MRP	molecular recognition phase
MscL	mechanosensitive channel of large conductance
MST	microsystem technology
MT	microtubule
mTAS	micro total analysis system
MTTF	mean time to failure
MUMP	multiuser MEMS process
MVD	molecular vapor deposition
MWCNT	multiwall carbon nanotube
MWNT	multiwall nanotube
MYD/BHW	Muller–Yushchenko–Derjaguin/Burgess–Hughes–White

N

NA	numerical aperture
NADIS	nanoscale dispensing
NASA	National Aeronautics and Space Administration
NC-AFM	noncontact atomic force microscopy
NEMS	nanoelectromechanical system
NGL	next-generation lithography
NHS	N-hydroxysuccinimidyl
NIH	National Institute of Health
NIL	nanoimprint lithography
NIST	National Institute of Standards and Technology
NMP	no-moving-part
NMR	nuclear magnetic resonance
NMR	nuclear mass resonance
NNI	National Nanotechnology Initiative

NOEMS	nanooptoelectromechanical system		PET	poly(ethyleneterephthalate)
NP	nanoparticle		PETN	pentaerythritol tetranitrate
NP	nanoprobe		PFDA	perfluorodecanoic acid
NSF	National Science Foundation		PFDP	perfluorodecylphosphonate
NSOM	near-field scanning optical microscopy		PFDTES	perfluorodecyltriethoxysilane
NSTC	National Science and Technology Council		PFM	photonic force microscope
NTA	nitrilotriacetate		PFOS	perfluorooctanesulfonate
nTP	nanotransfer printing		PFPE	perfluoropolyether
			PFTS	perfluorodecyltricholorosilane
			PhC	photonic crystal
			PI3K	phosphatidylinositol-3-kinase

O

ODA	octadecylamine
ODDMS	n-octadecyldimethyl(dimethylamino)silane
ODMS	n-octyldimethyl(dimethylamino)silane
ODP	octadecylphosphonate
ODTS	octadecyltrichlorosilane
OLED	organic light-emitting device
OM	optical microscope
OMVPE	organometallic vapor-phase epitaxy
OS	optical stretcher
OT	optical tweezers
OTRS	optical tweezers Raman spectroscopy
OTS	octadecyltrichlorosilane
oxLDL	oxidized low-density lipoprotein

P

P–V	peak-to-valley
PAA	poly(acrylic acid)
PAA	porous anodic alumina
PAH	poly(allylamine hydrochloride)
PAPP	p-aminophenyl phosphate
Pax	paxillin
PBC	periodic boundary condition
PBS	phosphate-buffered saline
PC	polycarbonate
PCB	printed circuit board
PCL	polycaprolactone
PCR	polymerase chain reaction
PDA	personal digital assistant
PDMS	polydimethylsiloxane
PDP	2-pyridyldithiopropionyl
PDP	pyridyldithiopropionate
PE	polyethylene
PECVD	plasma-enhanced chemical vapor deposition
PEEK	polyetheretherketone
PEG	polyethylene glycol
PEI	polyethyleneimine
PEN	polyethylene naphthalate
PES	photoemission spectroscopy
PES	position error signal

PI	polyisoprene
PID	proportional–integral–differential
PKA	protein kinase
PKC	protein kinase C
PKI	protein kinase inhibitor
PL	photolithography
PLC	phospholipase C
PLD	pulsed laser deposition
PMAA	poly(methacrylic acid)
PML	promyelocytic leukemia
PMMA	poly(methyl methacrylate)
POCT	point-of-care testing
POM	polyoxy-methylene
PP	polypropylene
PPD	p-phenylenediamine
PPMA	poly(propyl methacrylate)
PPy	polypyrrole
PS-PDMS	poly(styrene-b-dimethylsiloxane)
PS/clay	polystyrene/nanoclay composite
PS	polystyrene
PSA	prostate-specific antigen
PSD	position-sensitive detector
PSD	position-sensitive diode
PSD	power-spectral density
PSG	phosphosilicate glass
PSGL-1	P-selectin glycoprotein ligand-1
PTFE	polytetrafluoroethylene
PUA	polyurethane acrylate
PUR	polyurethane
PVA	polyvinyl alcohol
PVD	physical vapor deposition
PVDC	polyvinylidene chloride
PVDF	polyvinyledene fluoride
PVS	polyvinylsiloxane
PWR	plasmon-waveguide resonance
PZT	lead zirconate titanate

Q

QB	quantum box
QCM	quartz crystal microbalance
QFN	quad flat no-lead
QPD	quadrant photodiode
QWR	quantum wire

R

RBC	red blood cell
RCA	Radio Corporation of America
RF	radiofrequency
RFID	radiofrequency identification
RGD	arginine–glycine–aspartic
RH	relative humidity
RHEED	reflection high-energy electron diffraction
RICM	reflection interference contrast microscopy
RIE	reactive-ion etching
RKKY	Ruderman–Kittel–Kasuya–Yoshida
RMS	root mean square
RNA	ribonucleic acid
ROS	reactive oxygen species
RPC	reverse phase column
RPM	revolutions per minute
RSA	random sequential adsorption
RT	room temperature
RTP	rapid thermal processing

S

SAE	specific adhesion energy
SAM	scanning acoustic microscopy
SAM	self-assembled monolayer
SARS-CoV	syndrome associated coronavirus
SATI	self-assembly, transfer, and integration
SATP	(S-acetylthio)propionate
SAW	surface acoustic wave
SB	Schottky barrier
SCFv	single-chain fragment variable
SCM	scanning capacitance microscopy
SCPM	scanning chemical potential microscopy
SCREAM	single-crystal reactive etching and metallization
SDA	scratch drive actuator
SEcM	scanning electrochemical microscopy
SEFM	scanning electrostatic force microscopy
SEM	scanning electron microscope
SEM	scanning electron microscopy
SFA	surface forces apparatus
SFAM	scanning force acoustic microscopy
SFD	shear flow detachment
SFIL	step and flash imprint lithography
SFM	scanning force microscope
SFM	scanning force microscopy
SGS	small-gap semiconducting
SICM	scanning ion conductance microscopy
SIM	scanning ion microscope
SIP	single inline package
SKPM	scanning Kelvin probe microscopy
SL	soft lithography
SLIGA	sacrificial LIGA
SLL	sacrificial layer lithography
SLM	spatial light modulator
SMA	shape memory alloy
SMM	scanning magnetic microscopy
SNOM	scanning near field optical microscopy
SNP	single nucleotide polymorphisms
SNR	signal-to-noise ratio
SOG	spin-on-glass
SOI	silicon-on-insulator
SOIC	small outline integrated circuit
SoS	silicon-on-sapphire
SP-STM	spin-polarized STM
SPM	scanning probe microscope
SPM	scanning probe microscopy
SPR	surface plasmon resonance
sPROM	structurally programmable microfluidic system
SPS	spark plasma sintering
SRAM	static random access memory
SRC	sampling rate converter
SSIL	step-and-stamp imprint lithography
SSRM	scanning spreading resistance microscopy
STED	stimulated emission depletion
SThM	scanning thermal microscope
STM	scanning tunneling microscope
STM	scanning tunneling microscopy
STORM	statistical optical reconstruction microscopy
STP	standard temperature and pressure
STS	scanning tunneling spectroscopy
SUN	Sad1p/UNC-84
SWCNT	single-wall carbon nanotube
SWCNT	single-walled carbon nanotube
SWNT	single wall nanotube
SWNT	single-wall nanotube

T

TA	tilt angle
TASA	template-assisted self-assembly
TCM	tetracysteine motif
TCNQ	tetracyanoquinodimethane
TCP	tricresyl phosphate
TEM	transmission electron microscope
TEM	transmission electron microscopy
TESP	tapping mode etched silicon probe
TGA	thermogravimetric analysis
TI	Texas Instruments
TIRF	total internal reflection fluorescence
TIRM	total internal reflection microscopy
TLP	transmission-line pulse
TM	tapping mode
TMAH	tetramethyl ammonium hydroxide
TMR	tetramethylrhodamine
TMS	tetramethylsilane

TMS	trimethylsilyl	**V**		
TNT	trinitrotoluene			
TP	track pitch	VBS	vinculin binding site
TPE-FCCS	two-photon excitation fluorescence cross-correlation spectroscopy	VCO	voltage-controlled oscillator
		VCSEL	vertical-cavity surface-emitting laser
TPI	threads per inch	vdW	van der Waals
TPMS	tire pressure monitoring system	VHH	variable heavy–heavy
TR	torsional resonance	VLSI	very large-scale integration
TREC	topography and recognition	VOC	volatile organic compound
TRIM	transport of ions in matter	VPE	vapor-phase epitaxy
TSDC	thermally stimulated depolarization current	VSC	vehicle stability control
TTF	tetrathiafulvalene	**X**	
TV	television		
		XPS	x-ray photon spectroscopy
U		XRD	x-ray powder diffraction
UAA	unnatural AA	**Y**	
UHV	ultrahigh vacuum		
ULSI	ultralarge-scale integration	YFP	yellow fluorescent protein
UML	unified modeling language	**Z**	
UNCD	ultrananocrystalline diamond		
UV	ultraviolet	Z-DOL	perfluoropolyether
UVA	ultraviolet A		

1. Introduction to Nanotechnology

Bharat Bhushan

A biological system can be exceedingly small. Many of the cells are very tiny, but they are very active; they manufacture various substances; they walk around; they wiggle; and they do all kinds of marvelous things – all on a very small scale. Also, they store information. Consider the possibility that we too can make a thing very small which does what we want – that we can manufacture an object that maneuvers at that level.

(From the talk *There's Plenty of Room at the Bottom*, delivered by Richard P. Feynman at the annual meeting of the American Physical Society at the California Institute of Technology; Pasadena, December 29, 1959).

1.1	Nanotechnology – Definition and Examples	1
1.2	Background and Research Expenditures	4
1.3	Lessons from Nature (Biomimetics)	6
1.4	Applications in Different Fields	9
1.5	Various Issues	10
1.6	Research Training	11
1.7	Organization of the Handbook	11
References		12

1.1 Nanotechnology – Definition and Examples

Nanotechnology literally means any technology on a nanoscale that has applications in the real world. Nanotechnology encompasses the production and application of physical, chemical, and biological systems at scales ranging from individual atoms or molecules to submicron dimensions, as well as the integration of the resulting nanostructures into larger systems. Nanotechnology is likely to have a profound impact on our economy and society in the early 21st century, comparable to that of semiconductor technology, information technology, or cellular and molecular biology. Science and technology research in nanotechnology promises breakthroughs in areas such as materials and manufacturing, nanoelectronics, medicine and healthcare, energy, biotechnology, information technology, and national security. It is widely felt that nanotechnology will be the next Industrial Revolution.

Nanometer-scale features are mainly built up from their elemental constituents. Examples include chemical synthesis, spontaneous self-assembly of molecular clusters (molecular self-assembly) from simple reagents in solution, biological molecules (e.g., DNA) used as building blocks for production of three-dimensional nanostructures, and quantum dots (nanocrystals) of arbitrary diameter (about $10-10^5$ atoms). The definition of a nanoparticle is an aggregate of atoms bonded together with a radius between 1 and 100 nm. It typically consists of $10-10^5$ atoms. A variety of vacuum deposition and nonequilibrium-plasma chemistry techniques are used to produce layered nanocomposites and nanotubes. Atomically controlled structures are produced using molecular-beam epitaxy and organometallic vapor-phase epitaxy. Micro- and nanosystem components are fabricated using top-down lithographic and nonlithographic fabrication techniques and range in size from micro- to nanometers. Continued improvements in lithography for use in the production of nanocomponents have resulted in line widths as small as 10 nm in experimental prototypes. The nanotechnology field, in addition to the fabrication of nanosystems, provides impetus for the development of experimental and computational tools.

The discovery of novel materials, processes, and phenomena at the nanoscale and the development of new experimental and theoretical techniques for research provide fresh opportunities for the development of innovative nanosystems and nanostructured materials. The properties of materials at the nanoscale can be very different from those at a larger scale. When the dimension of a material is reduced from a large size, the properties remain the same at first, then small changes occur, until finally when the size drops below 100 nm, dramatic changes in properties can occur. If only one length of a three-dimensional nanostructure is of nanodimension, the structure is referred to as a quantum well; if two sides are of nanometer length, the structure is referred to as a quantum wire. A quantum dot has all three dimensions in the nano range. The term *quantum* is associated with these three types of nanostructures because the changes in properties arise from the quantum-mechanical nature of physics in the domain of the ultrasmall. Materials can

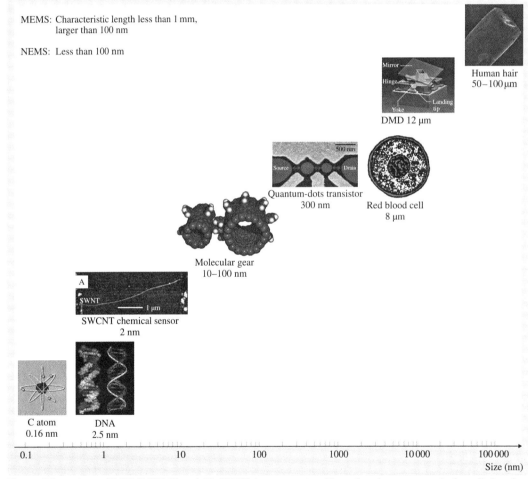

Fig. 1.1 Dimensions of MEMS/NEMS and BioNEMS in perspective. Examples shown are a single-walled carbon nanotube (SWCNT) chemical sensor [1.1], molecular dynamic simulations of carbon-nanotube-based gears [1.2], quantum-dot transistor obtained from [1.3], and digital microdevice (DMD) obtained from www.dlp.com. For comparison, dimensions and weights of various biological objects found in nature are also presented

Table 1.1 Characteristic dimensions and weights in perspective

Characteristic dimensions in perspective	
NEMS characteristic length	< 100 nm
MEMS characteristic length	< 1 mm and > 100 nm
SWCNT chemical sensor	≈ 2 nm
Molecular gear	≈ 10 nm
Quantum-dot transistor	300 nm
Digital micromirror	12 000 nm
Individual atoms	Typically a fraction of a nm in diameter
DNA molecules	≈ 2.5 nm wide
Biological cells	In the range of thousands of nm in diameter
Human hair	≈ 75 000 nm in diameter
Weight in perspective	
NEMS built with cross-sections of about 10 nm	As low as 10^{-20} N
Micromachine silicon structure	As low as 1 nN
Eyelash	≈ 100 nN
Water droplet	≈ 10 µN

be nanostructured for new properties and novel performance. This field is opening new avenues in science and technology.

Micro- and nanosystems include micro/nanoelectromechanical systems (MEMS/NEMS). MEMS refers to microscopic devices that have a characteristic length of less than 1 mm but more than 100 nm and that combine electrical and mechanical components. NEMS refers to nanoscopic devices that have a characteristic length of less than 100 nm and that combine electrical and mechanical components. In mesoscale devices, if the functional components are on the micro- or nanoscale, they may be referred to as MEMS or NEMS, respectively. These are referred to as intelligent miniaturized systems, comprising sensing, processing, and/or actuating functions and combining electrical and mechanical components. The acronym MEMS originated in the USA. The term commonly used in Europe is *microsystem technology* (MST), and in Japan the term *micromachines* is used. Another term generally used is micro/nanodevices. The terms MEMS/NEMS are also now used in a broad sense and include electrical, mechanical, fluidic, optical, and/or biological function. MEMS/NEMS for optical applications are referred to as micro/nanooptoelectromechanical systems (MOEMS/NOEMS). MEMS/NEMS for electronic applications are referred to as radiofrequency MEMS/NEMS (RF-MEMS/RF-NEMS). MEMS/NEMS for biological applications are referred to as bioMEMS/bioNEMS.

To put the dimensions of MEMS/NEMS and BioNEMS in perspective, see Fig. 1.1 and Table 1.1. Individual atoms are typically a fraction of a nanometer in diameter, DNA molecules are about 2.5 nm wide, biological cells are in the range of thousands of nm in diameter, and human hair is about 75 µm in diameter. The smallest length of BioNEMS shown in the figure is about 2 nm, NEMS ranges in size from 10 to 300 nm, and the size of MEMS is 12 000 nm. The mass of a micromachined silicon structure can be a low as 1 nN, and NEMS can be built with mass as low as 10^{-20} N with cross-sections of about 10 nm. In comparison, the mass

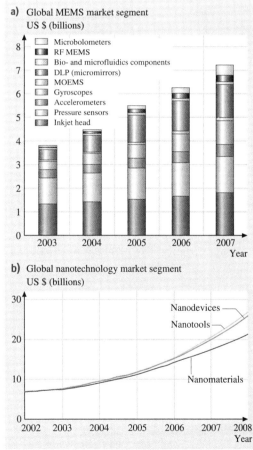

Fig. 1.2 Global MEMS and nanotechnology market segments (DLP – digital light processing)

of a drop of water is about 10 μN, and the mass of an eyelash is about 100 nN.

MEMS/NEMS and BioMEMS/BioNEMS are expected to have a major impact on our lives, comparable to that of semiconductor technology, information technology, or cellular and molecular biology [1.4, 5]. MEMS/NEMS and BioMEMS/BioNEMS are used in electromechanical, electronics, information/communication, chemical, and biological applications. The MEMS industry in 2004 was worth about US$ 4.5 billion, with a projected annual growth rate of 17% (Fig. 1.2) [1.6]. The NEMS industry was worth about US$ 10 billion in 2004, mostly in nanomaterials (Fig. 1.2) [1.7]. Growth of Si-based MEMS/NEMS may slow down and that of nonsilicon MEMS may pick up during the next decade. It is expected to expand in this decade, for nanomaterials and biomedical applications as well as nanoelectronics or molecular electronics. For example, miniaturized diagnostics could be implanted for early diagnosis of illness. Targeted drug-delivery devices are under development. Due to the enabling nature of these systems and because of the significant impact they can have on both commercial and defense applications, industry as well as federal governments have taken special interest in seeing growth in this field nurtured. MEMS/NEMS and BioMEMS/BioNEMS are the next logical step in the *silicon revolution*.

1.2 Background and Research Expenditures

On December 29, 1959 at the California Institute of Technology, Nobel Laureate *Richard P. Feynman* gave a talk at the Annual Meeting of the American Physical Society that has become one of the 20th century's classic science lectures, entitled *There's Plenty of Room at the Bottom* [1.8]. He presented a technological vision of extreme miniaturization in 1959, several years before the word *chip* became part of the lexicon. He talked about the problem of manipulating and controlling things on a small scale. Extrapolating from known physical laws, Feynman envisioned a technology using the ultimate toolbox of nature, building nanoobjects atom by atom or molecule by molecule. Since the 1980s, many inventions and discoveries in the fabrication of nanoobjects have been testaments to his vision. In recognition of this reality, the National Science and Technology Council (NSTC) of the White House created the Interagency Working Group on Nanoscience, Engineering, and Technology (IWGN) in 1998. In a January 2000 speech at the same institute, President W. J. Clinton talked about the exciting promise of *nanotechnology* and the importance of expanding research in nanoscale science and technology more broadly. Later that month, he announced in his State of the Union Address an ambitious US$ 497 million federal, multiagency National Nanotechnology Initiative (NNI) in the fiscal year 2001 budget, and made the NNI a top science and technology priority [1.9, 10]. The objective of this initiative was to form a broad-based coalition in which academia, the private sector, and local, state, and federal governments work together to push the envelop of nanoscience and nanoengineering to reap nanotechnology's potential social and economic benefits.

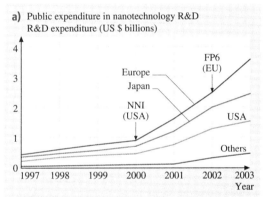

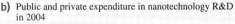

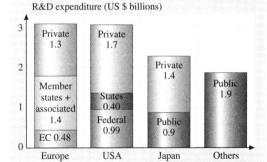

Fig. 1.3a,b Breakdown of expenditure in nanotechnology R&D (**a**) around the world (source: European Commission, 2003), and (**b**) by public and private resources in 2004 (source: European Commission, 2005; figures for private sources based upon data from Lux Research)

Funding in the USA has continued to increase. In January 2003, the US Senate introduced a bill to establish a National Nanotechnology Program. On December 3, 2003, President George W. Bush signed into law the 21st Century Nanotechnology Research and Development Act. This legislation put into law programs and activities supported by the National Nanotechnology Initiative. The bill gave nanotechnology a permanent home in the federal government and authorized US$ 3.7 billion to be spent in the 4 year period beginning in October 2005 for nanotechnology initiatives at five federal agencies. The funds would provide grants to researchers, coordinate research and development (R&D) across five federal agencies [the National Science Foundation (NSF), the Department of Energy (DOE), the National Aeronautics and Space Administration (NASA), the National Institute of Standards and Technology (NIST), and the Environmental Protection Agency (EPA)], establish interdisciplinary research centers, and accelerate technology transfer into the private sector. In addition, the Departments of Defense (DOD), Homeland Security, Agriculture, and Justice as well as the National Institutes of Health (NIH) also fund large R&D activities. They currently account

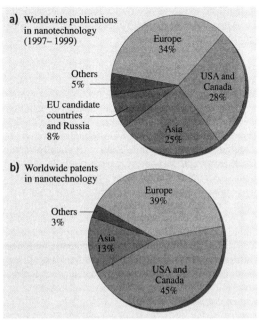

Fig. 1.5a,b Breakdown of (**a**) worldwide publications and (**b**) worldwide patents (source: European Commission, 2003)

for more than one-third of the federal budget for nanotechnology.

The European Union (EU) made nanosciences and nanotechnologies a priority in the Sixth Framework Program (FP6) in 2002 for the period 2003–2006. There were also small dedicated funds in FP4 and FP5 before. FP6 was tailored to help better structure European research and to cope with the strategic objectives set out in Lisbon in 2000. Japan identified nanotechnology as one of its main research priorities in 2001. The funding levels increased sharply from US$ 400 million in 2001 to around US$ 950 million in 2004. In 2003, South Korea embarked upon a 10 year program with around US$ 2 billion of public funding, and Taiwan has committed around US$ 600 million of public funding over 6 years. Singapore and China are also investing on a large scale. Russia is well funded as well.

Figure 1.3a shows the public expenditure breakdown of nanotechnology R&D around the world, with about US$ 5 billion in 2004, coming approximately equally from the USA, Japan, and Europe. Next we compare public expenditure on a per-capita basis. The average expenditures per capita for the USA, the EU-

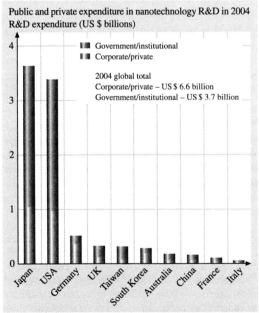

Fig. 1.4 Breakdown of public and private expenditures in nanotechnology R&D in 2004 in various countries (after [1.7])

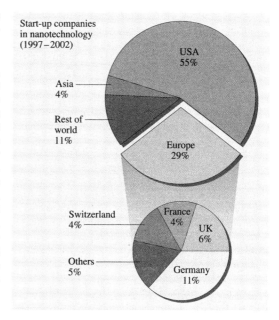

Fig. 1.6 Breakdown of start-up companies around the world (1997–2002) (source: CEA, Bureau d'Etude Marketing) ▶

25, and Japan are about US$ 3.7 billion, US$ 2.4 billion, and US$ 6.2 billion, respectively [1.11]. Figure 1.3b shows the breakdown of expenditure in 2004 by public and private sources, with more than US$ 10 billion spent in nanotechnology research. Two-thirds of this came from corporate and private funding. Private expenditure in the USA and Japan was slightly larger than that from public sources, whereas in Europe it was about one-third. Figure 1.4 shows the public and private expenditure breakdown in 2004 in various countries. Japan and USA had the largest expenditure, followed by Germany, Taiwan, South Korea, the UK, Australia, China, France, and Italy. Figure 1.5 shows a breakdown of worldwide publications and patents. USA and Canada led, followed by Europe and Asia. Figure 1.6 shows the breakdown in start-up companies around the world (1997–2002). Entrepreneurship in USA is clearly evident, followed by Europe.

1.3 Lessons from Nature (Biomimetics)

The word nanotechnology is a relatively new word, but it is not an entirely new field. Nature has gone through evolution over the 3.8 billion years since life is estimated to have appeared on Earth. Nature has many materials, objects, and processes which function from the macroscale to nanoscale [1.9]. Understanding the functions provided by these objects and processes can guide us to imitate and produce nanomaterials, nanodevices, and processes. Biologically inspired design, adaptation or derivation from nature is referred to as *biomimetics*, a term coined by the polymath Otto Schmitt in 1957. Biomimetics is derived from the Greek word *biomimesis*. Other terms used include bionics, biomimicry, and biognosis. The term biomimetics is relatively new; however, our ancestors looked to nature for inspiration and the development of various materials and devices many centuries ago [1.12, 13]. There are a large number of objects, including bacteria, plants, land and aquatic animals, seashells, and spider web, with properties of commercial interest. Figure 1.7 provides an overview of various objects from nature and their selected functions. Figure 1.8 shows a montage of some examples from nature, which serve as the inspiration for various technological developments.

The flagella of bacteria rotate at over 10 000 rpm [1.14]. This is an example of a biological molecular machine. The flagella motor is driven by the proton flow caused by the electrochemical potential differences across the membrane. The diameter of the bearing is about 20–30 nm, with an estimated clearance of ≈ 1 nm.

Several billions years ago, molecules began organizing into complex structures that could support life. Photosynthesis harnesses solar energy to support plant life. Molecular ensembles present in plant leaves, which include light-harvesting molecules such as chlorophyll, arranged within the cells (on the nanometer to micrometer scales), capture light energy and convert it into the chemical energy that drives the biochemical machinery of plant cells. Live organs use chemical energy in the body. This technology is being exploited for solar energy applications.

Some natural surfaces, including the leaves of water-repellent plants such as lotus, are known to be superhydrophobic and self-cleaning due to hierarchical roughness (microbumps superimposed with nanostructure) and the presence of a wax coating [1.15–19]. Roughness-induced superhydrophobic

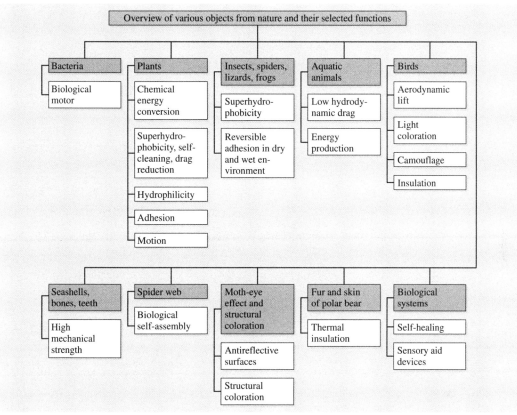

Fig. 1.7 Overview of various objects from nature and their selected function (after [1.13])

and self-cleaning surfaces are of interest in various applications, including self-cleaning windows, windshields, exterior paints for buildings and navigation ships, utensils, roof tiles, textiles, and applications requiring a reduction of drag in fluid flow, e.g., in micro/nanofluidics. Superhydrophobic surfaces can also be used for energy conversion and conservation [1.20]. Nonwetting surfaces also reduce stiction at contacting interfaces in machinery [1.21, 22].

The leg attachment pads of several creatures, including many insects (e.g., beetles and flies), spiders, and lizards (e.g., geckoes), are capable of attaching to a variety of surfaces and are used for locomotion [1.23]. Biological evolution over a long period of time has led to the optimization of their leg attachment systems. The attachment pads have the ability to cling to different smooth and rough surfaces and detach at will [1.24, 25]. This dynamic attachment ability is referred to as reversible adhesion or smart adhesion. Replication of the characteristics of gecko feet would enable the development of a superadhesive polymer tape capable of clean, dry adhesion which is reversible [1.25–27]. (It should be noted that common manmade adhesives such as tape or glue involve the use of wet adhesives that permanently attach two surfaces.) The reusable gecko-inspired adhesives have the potential for use in everyday objects such as tapes, fasteners, and toys, and in high technology such as microelectronic and space applications. Replication of the dynamic climbing and peeling ability of geckoes could find use in the treads of wall-climbing robots. Incidentally, *Velcro* was invented based on the action of the hooked seeds of the burdock plant [1.28].

Many aquatic animals can move in water at high speeds with low energy input. Drag is a major hindrance to movement. Most shark species move through water with high efficiency and maintain buoyancy. Through its ingenious design, their skin turns out to be an essen-

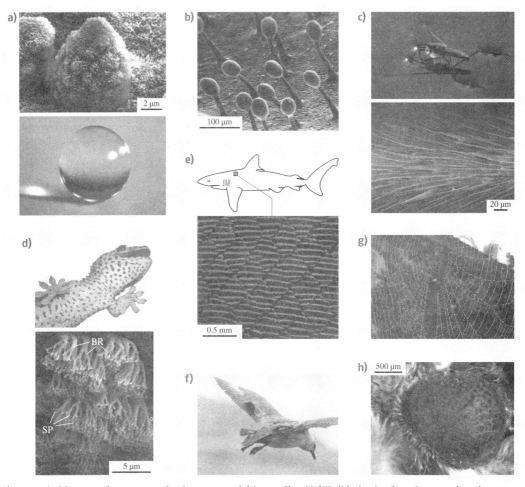

Fig. 1.8a–h Montage of some examples from nature: (**a**) lotus effect [1.30], (**b**) glands of carnivorous plant that secrete adhesive to trap insects [1.17], (**c**) water strider walking on water [1.31], (**d**) gecko foot exhibits reversible adhesion [1.32] (BR – branch, SP – spatula), (**e**) scale structure of shark reduces drag [1.33], (**f**) wings of a bird in landing approach, (**g**) spider web made of silk material [1.12], (**h**) moth's eyes are antireflective [1.34]

tial aid to this behavior by reducing friction drag and autocleaning ectoparasites from their surface [1.29]. The very small individual tooth-like scales of shark skin, called dermal denticles, are ribbed with longitudinal grooves, which result in water moving very efficiently over their surface. The scales also minimize the collection of barnacles and algae. Speedo created the whole-body swimsuit called Fastskin, modeled on shark skin, for elite swimming. Boat, ship, and aircraft manufacturers are trying to mimic shark skin to reduce friction drag and minimize the attachment of organisms to their bodies. In addition, mucus on the skin of aquatic animals, including sharks, acts as an osmotic barrier against the salinity of seawater and protects the creature from parasites and infections. It also acts as a drag-reducing agent. Artificial derivatives of fish mucus (polymer additives) are used to propel crude oil in the Alaska pipeline. The compliant skin of dolphins allows them to swim at high speed. By interacting with the water flowing over the body's surface it stabilizes the flow and delays the transition to turbulence. Dolphins possess an optimum shape for drag reduction of submerged bodies. Submarines use the shape of dolphins. The streamlined form of boxfish (*Ostracion meleagris*) has inspired Mercedes Benz's bionic concept car with low aerodynamic drag. The beak of the kingfisher was used to model the nose cone of the Japanese Shinkansen bullet train. Power is generated by the scalloped edges of a humpback whale, and this design is exploited for wind turbine blades.

Bird feathers make the body water repellant, and movable flaps create wing and tail for aerodynamic lift during flying [1.29]. Birds and butterflies create brilliant hues by refracting light through millions of repeated structures that bend light to make certain colors. Seashells are natural nanocomposites with a laminated structure and exhibit superior mechanical properties. Spider web consists of silk fiber which is very strong. The materials and structures used in these objects have led to the development of various materials and fibers with high mechanical strength. Moth eyes have a multifaceted surface on the nanoscale and are structured to reduce reflection. This antireflective design led to the discovery of antireflective surfaces [1.35].

A remarkable property of biological tissues is their ability for self-healing. In biological systems, chemical signals released at the site of a fracture initiate a systemic response that transports repair agents to the site of an injury and promotes healing. Various artificial self-healing materials are being developed [1.36]. Human skin is sensitive to impact, leading to purple-colored marks in areas that are hit. This idea has led to the development of coatings indicating impact damage [1.12]. Another interesting and promising idea involves the application of an array of sensors to develop an *artificial nose* or an *artificial tongue*.

Other lessons from nature include the wings of flying insects, abalone shell with high-impact ceramic properties, strong spider silk, ultrasonic detection by bats, infrared detection by beetles, and silent flying of owls because of frayed feathers on the edges of their wings.

1.4 Applications in Different Fields

Science and technology continue to move forward in making the fabrication of micro/nanodevices and systems possible for a variety of industrial, consumer, and biomedical applications [1.37, 38]. A variety of MEMS devices have been produced, and some are in commercial use [1.39–48]. A variety of sensors are used in industrial, consumer, defense, and biomedical applications. Various micro/nanostructures and micro/nanocomponents are used in microinstruments and other industrial applications such as micromirror arrays. The largest *killer* MEMS applications include accelerometers (some 90 million units installed in vehicles in 2004), silicon-based piezoresistive pressure sensors for manifold absolute pressure sensing for engines and for disposable blood pressure sensors (about 30 million and 25 million units, respectively), capacitive pressure sensors for tire pressure measurements (about 37 million units in 2005), thermal inkjet printheads (about 500 million units in 2004), micromirror arrays for digital projection displays (about US$ 700 million revenue in 2004), and optical cross-connections in telecommunications. Other applications of MEMS devices include chemical/biosensors and gas sensors, microresonators, infrared detectors and focal-plane arrays for Earth observation, space science, and missile defense applications, picosatellites for space applications, fuel cells, and many hydraulic, pneumatic, and other consumer products. MEMS devices are also being pursued for use in magnetic storage systems [1.49], where they are being developed for supercompact and ultrahigh-recording-density magnetic disk drives.

NEMS are produced by nanomachining in a typical top–down approach and bottom–up approach, largely relying on nanochemistry [1.50–56]. Examples of NEMS include microcantilevers with integrated sharp nanotips for scanning tunneling microscopy (STM) and atomic force microscopy (AFM), quantum corrals formed using STM by placing atoms one by one, AFM cantilever arrays for data storage, AFM tips for nanolithography, dip-pen lithography for printing molecules, nanowires, carbon nanotubes, quantum wires (QWRs), quantum boxes (QBs), quantum-dot transistors, nanotube-based sensors, biological (DNA) motors, molecular gears formed by attaching benzene molecules to the outer walls of carbon nanotubes, devices incorporating nm-thick films [e.g., in giant magnetoresistive (GMR) read/write magnetic heads and magnetic media] for magnetic rigid disk drives and magnetic tape drives, nanopatterned magnetic rigid disks, and nanoparticles (e.g., nanoparticles in magnetic tape substrates and magnetic particles in magnetic tape coatings).

Nanoelectronics can be used to build computer memory using individual molecules or nanotubes to store bits of information, molecular switches, molecular or nanotube transistors, nanotube flat-panel displays, nanotube integrated circuits, fast logic gates, switches, nanoscopic lasers, and nanotubes as electrodes in fuel cells.

BioMEMS/BioNEMS are increasingly used in commercial and defense applications [1.57–63]. They are used for chemical and biochemical analyses (biosensors) in medical diagnostics (e.g., DNA, RNA, proteins, cells, blood pressure and assays, and toxin identification) [1.63, 64], tissue engineering [1.65], and implantable pharmaceutical drug delivery [1.66, 67]. Biosensors, also referred to as biochips, deal with liquids and gases. There are two types of biosensors. A large variety of biosensors are based on micro/nanofluidics. Micro/nanofluidic devices offer the ability to work with smaller reagent volumes and shorter reaction times, and perform analyses multiple times at once. The second type of biosensors includes micro/nanoarrays which perform one type of analysis thousands of times. Micro/nanoarrays are a tool used in biotechnology research to analyze DNA or proteins to diagnose diseases or discover new drugs. Also called DNA arrays, they can identify thousands of genes simultaneously [1.60]. They include a microarray of silicon nanowires, roughly a few nm in size, to selectively bind and detect even a single biological molecule, such as DNA or protein, by using nanoelectronics to detect the slight electrical charge caused by such binding, or a microarray of carbon nanotubes to electrically detect glucose.

After the tragedy of September 11, 2001, concern about biological and chemical warfare has led to the development of handheld units with bio- and chemical sensors for detection of biological germs, chemical or nerve agents, and mustard agents, and chemical precursors to protect subways, airports, water supplies, and the population at large [1.68].

BioMEMS/BioNEMS are also being developed for minimal invasive surgery, including endoscopic surgery, laser angioplasty, and microscopic surgery. Other applications include implantable drug-delivery devices (micro/nanoparticles with drug molecules encapsulated in functionalized shells for site-specific targeting applications) and a silicon capsule with a nanoporous membrane filled with drugs for long-term delivery.

1.5 Various Issues

There is an increasing need for a multidisciplinary, system-oriented approach to the manufacture of micro/nanodevices which function reliably. This can only be achieved through the cross-fertilization of ideas from different disciplines and the systematic flow of information and people among research groups. Common potential failure mechanisms for MEMS/NEMS requiring relative motion that need to be addressed in order to increase their reliability are: adhesion, friction, wear, fracture, fatigue, and contamination [1.21, 22, 69, 70]. Surface micro/nanomachined structures often include smooth and chemically active surfaces. Due to the large surface area to volume ratio in MEMS/NEMS, they are particularly prone to stiction (high static friction) as part of normal operation. Fracture occurs when the load on a microdevice is greater than the strength of the material. Fracture is a serious reliability concern, particularly for brittle materials used in the construction of these components, since it can immediately or would eventually lead to catastrophic failures. Additionally, debris can be formed from the fracturing of microstructures, leading to other failure processes. For less brittle materials, repeated loading over a long period of time causes fatigue that can also lead to the breaking and fracturing of the device. In principle, this failure mode is relatively easy to observe and simple to predict. However, the materials properties of thin films are often not known, making fatigue predictions error prone.

Many MEMS/NEMS devices operate near their thermal dissipation limit. They may encounter hot spots that may cause failures, particularly in weak structures such as diaphragms or cantilevers. Thermal stressing and relaxation caused by thermal variations can create material delamination and fatigue in cantilevers. When exposed to large temperature changes, as experienced in the space environment, bimetallic beams will also experience warping due to mismatched coefficients of thermal expansion. Packaging has been a big problem. The contamination that probably happens in packaging and during storage also can strongly influence the reliability of MEMS/NEMS. For example, a dust particle that lands on one of the electrodes of a comb drive can cause catastrophic failure. There are no MEMS/NEMS fabrications standards, which make it difficult to transfer fabrication steps in MEMS/NEMS between foundries.

Obviously, studies of the determination and suppression of active failure mechanisms affecting this new and promising technology are critical to high reliability of MEMS/NEMS and are determining factors for successful practical application.

Adhesion between a biological molecular layer and the substrate, referred to as *bioadhesion*, and reduction of friction and wear of biological layers, biocompatibility, and biofouling for BioMEMS/BioNEMS are important.

Mechanical properties are known to exhibit a dependence on specimen size. Mechanical property evaluation of nanoscale structures is carried out to help design reliable systems since good mechanical properties are of critical importance in such applications. Some of the properties of interest are: Young's modulus of elasticity, hardness, bending strength, fracture toughness, and fatigue life. Finite-element modeling is carried out to study the effects of surface roughness and scratches on stresses in nanostructures. When nanostructures are smaller than a fundamental physical length scale, conventional theory may no longer apply, and new phenomena emerge. Molecular mechanics is used to simulate the behavior of a nanoobject.

The societal, ethical, political, and health/safety implications of nanotechnology are also attracting major attention [1.11]. One of the prime reasons is to avoid some of the public skepticism that surrounded the debate over biotechnology advances such as genetically modified foods, while at the same time dispelling some of the misconceptions the public may already have about nanotechnology. Health/safety issues need to be addressed as well. For example, one key question is what happens to nanoparticles (such as buckyballs or nanotubes) in the environment and whether they are toxic in the human body, if digested.

1.6 Research Training

With the decreasing number of people in Western countries going into science and engineering and the rapid progress being made in nanoscience and nanotechnology, the problem of ensuring a trained workforce is expected to become acute. Education and training are essential to produce a new generation of scientists, engineers, and skilled workers with the flexible and interdisciplinary R&D approach necessary for rapid progress in the nanosciences and nanotechnology [1.71]. The question is being asked: is the traditional separation of academic disciplines into physics, chemistry, biology, and various engineering disciplines meaningful at the nanolevel? Generic skills and entrepreneurship are needed to transfer scientific knowledge into products. Scientists and engineers in cooperation with relevant experts should address the societal, ethical, political, and health/safety implications of their work for society at large.

To increase the pool of students interested in science and technology, science needs to be projected to be exciting at the high-school level. Interdisciplinary curricula relevant for nanoscience and nanotechnology need to be developed. This requires the revamping of education, the development of new courses and course material including textbooks [1.47, 56, 70, 72–74] and instruction manuals, and the training of new instructors.

1.7 Organization of the Handbook

This Handbook integrates knowledge from the fabrication, mechanics, materials science, and reliability points of view. Organization of the Handbook is straightforward. The Handbook is divided into nine parts. The first part of the book includes an introduction to nanostructures, micro/nanofabrication, methods, and materials. The second part introduces various MEMS/NEMS and BioMEMS/BioNEMS devices. The third part introduces scanning probe microscopy. The fourth part provides an overview of bio/nanotribology and bio/nanomechanics, which will prepare the reader to understand the interface reliability in industrial applications. The fifth part provides an overview of molecularly thick films for lubrication. The sixth part focuses on the emerging field of biomimetics, in which one mimics nature to develop products and processes of interest. The seventh part focuses on industrial applications, and the eighth part focuses on micro/nanodevice reliability. The final part focuses on technological convergence from the nanoscale as well as social, ethical, and political implications of nanotechnology.

References

1.1 R.J. Chen, H.C. Choi, S. Bangsaruntip, E. Yenilmex, X. Tang, Q. Wang, Y.L. Chang, H. Dai: An investigation of the mechanisms of electrode sensing of protein adsorption on carbon nanotube devices, J. Am. Chem. Soc. **126**, 1563–1568 (2004)

1.2 D. Srivastava: Computational nanotechnology of carbon nanotubes. In: *Carbon Nanotubes: Science and Applications*, ed. by M. Meyyappan (CRC, Boca Raton 2004) pp. 25–36

1.3 W.G. van der Wiel, S. De Franceschi, J.M. Elzerman, T. Fujisawa, S. Tarucha, L.P. Kauwenhoven: Electron transport through double quantum dots, Rev. Mod. Phys. **75**, 1–22 (2003)

1.4 Anonymous: *Microelectromechanical Systems: Advanced Materials and Fabrication Methods* (National Academy Press, Washington 1997), NMAB-483

1.5 M. Roukes: Nanoelectromechanical systems face the future, Phys. World **14**, 25–31 (2001)

1.6 J.C. Eloy: *Status of the MEMS Industry* (Yole Developpement, Lyon 2005), presented at SPIE Photonics West, San Jose (2005)

1.7 S. Lawrence: Nanotech grows up, Technol. Rev. **108**(6), 31 (2005)

1.8 R.P. Feynman: There's plenty of room at the bottom, Eng. Sci. **23**, 22–36 (1960), http://www.zyvex.com/nanotech/feynman.html

1.9 I. Amato: *Nanotechnology* (2000), http://www.ostp.gov/nstc/html/iwgn/iwgn.public.brochure/welcome.htm or http://www.nsf.gov/home/crssprgm/nano/nsfnnireports.htm

1.10 Anonymous: *National Nanotechnology Initiative* (2000), http://www.ostp.gov/nstc/html/iwgn.fy01budsuppl/nni.pdf or http://www.nsf.gov/home/crssprgm/nano/nsfnnireports.htm

1.11 Anonymous: *Towards a European Strategy for Nanotechnology* (European Commission Research Directorate General, Brussels 2004)

1.12 Y. Bar-Cohen (Ed.): *Biomimetics – Biologically Inspired Technologies* (CRC, Boca Raton 2005)

1.13 B. Bhushan: Biomimetics: Lessons from nature – An overview, Philos. Trans. R. Soc. Lond. Ser. A **367**, 1445–1486 (2009)

1.14 C.J. Jones, S. Aizawa: The bacterial flagellum and flagellar motor: Structure, assembly, and functions, Adv. Microb. Physiol. **32**, 109–172 (1991)

1.15 B. Bhushan, Y.C. Jung: Wetting, adhesion and friction of superhydrophobic and hydrophilic leaves and fabricated micro-/nanopatterned surfaces, J. Phys. D **20**, 225010 (2008)

1.16 K. Koch, B. Bhushan, W. Barthlott: Diversity of structure, morphology, and wetting of plant surfaces, Soft Matter **4**, 1943–1963 (2008)

1.17 K. Koch, B. Bhushan, W. Barthlott: Multifunctional surface structures of plants: An inspiration for biomimetics (invited), Prog. Mater. Sci. **54**, 137–178 (2009)

1.18 M. Nosonovsky, B. Bhushan: *Multiscale Dissipative Mechanisms and Hierarchical Surfaces: Friction, Superhydrophobicity, and Biomimetics* (Springer, Berlin, Heidelberg 2008)

1.19 M. Nosonovsky, B. Bhushan: Roughness-induced superhydrophobicity: A way to design non-adhesive surfaces, J. Phys. D **20**, 225009 (2008)

1.20 M. Nosonovsky, B. Bhushan: Multiscale effects and capillary interactions in functional biomimetic surfaces for energy conversion and green engineering, Philos. Trans. R. Soc. Lond. Ser. A **367**, 1511–1539 (2009)

1.21 B. Bhushan: *Principles and Applications of Tribology* (Wiley, New York 1999)

1.22 B. Bhushan (Ed.): *Introduction to Tribology* (Wiley, New York 2002)

1.23 S. Gorb (Ed.): *Attachment Devices of Insect Cuticle* (Kluwer, Dordrecht 2001)

1.24 K. Autumn, Y.A. Liang, S.T. Hsieh, W. Zesch, W.P. Chan, T.W. Kenny, R. Fearing, R.J. Full: Adhesive force of a single gecko foot-hair, Nature **405**, 681–685 (2000)

1.25 B. Bhushan: Adhesion of multi-level hierarchical attachment systems in gecko feet, J. Adhes. Sci. Technol. **21**, 1213–1258 (2007)

1.26 A.K. Geim, S.V. Dubonos, I.V. Grigorieva, K.S. Novoselov, A.A. Zhukov, S.Y. Shapoval: Microfabricated adhesive mimicking gecko foot-hair, Nat. Mater. **2**, 461–463 (2003)

1.27 B. Bhushan, R.A. Sayer: Surface characterization and friction of a bio-inspired reversible adhesive tape, Microsyst. Technol. **13**, 71–78 (2007)

1.28 S.A. Velcro: Improvements in or relating to a method and a device for producing velvet type fabric, Switzerland Patent 721338 (1995)

1.29 D.W. Bechert, M. Bruse, W. Hage, R. Meyer: Fluid mechanics of biological surfaces and their technological application, Naturwissenschaften **87**, 157–171 (2000)

1.30 B. Bhushan, Y.C. Jung, K. Koch: Micro-, nano-, and hierarchical structures for superhydrophobicity, self-cleaning, and low adhesion, Philos. Trans. R. Soc. Lond. Ser. A **367**, 1631–1672 (2009)

1.31 X.F. Gao, L. Jiang: Biophysics: Water-repellent legs of water striders, Nature **432**, 36 (2004)

1.32 H. Gao, X. Wang, H. Yao, S. Gorb, E. Arzt: Mechanics of hierarchical adhesion structures of geckos, Mech. Mater. **37**, 275–285 (2005)

1.33 W.E. Reif: *Squamation and Ecology of Sharks*, Courier Forschungsinst. Senckenberg, Vol. 78 (Schweizerbart, Stuttgart 1985)

1.34 J. Genzer, K. Efimenko: Recent developments in superhydrophobic surfaces and their relevance to

1.35 C.G. Bernhard, W.H. Miller, A.R. Möller: The insect corneal nipple array: A biological, broad-band impedance transformer that acts as a antireflection coating, Acta Physiol. Scand. **63**, 1–79 (1965)

marine fouling: A review, Biofouling **22**, 339–360 (2006)

1.36 J.P. Youngblood, N.R. Sottos: Bioinspired materials for self-cleaning and self-healing, MRS Bulletin **33**, 732–738 (2008)

1.37 Anonymous: *Small Tech 101 – An Introduction to Micro and Nanotechnology* (Small Times, 2003)

1.38 M. Schulenburg: *Nanotechnology – Innovation for Tomorrow's World* (European Commission Research Directorate General, Brussels 2004)

1.39 R.S. Muller, R.T. Howe, S.D. Senturia, R.L. Smith, R.M. White: *Microsensors* (IEEE, New York 1991)

1.40 I. Fujimasa: *Micromachines: A New Era in Mechanical Engineering* (Oxford Univ. Press, Oxford 1996)

1.41 W.S. Trimmer (Ed.): *Micromachines and MEMS, Classical and Seminal Papers to 1990* (IEEE, New York 1997)

1.42 B. Bhushan: *Tribology Issues and Opportunities in MEMS* (Kluwer, Dordrecht 1998)

1.43 G.T.A. Kovacs: *Micromachined Transducers Sourcebook* (WCB McGraw-Hill, Boston 1998)

1.44 M. Elwenspoek, R. Wiegerink: *Mechanical Microsensors* (Springer, Berlin Heidelberg 2001)

1.45 S.D. Senturia: *Microsystem Design* (Kluwer, Boston 2000)

1.46 T.R. Hsu: *MEMS and Microsystems: Design and Manufacture* (McGraw-Hill, Boston 2002)

1.47 M. Madou: *Fundamentals of Microfabrication: The Science of Miniaturization*, 2nd edn. (CRC, Boca Raton 2002)

1.48 A. Hierlemann: *Integrated Chemical Microsensor Systems in CMOS Technology* (Springer, Berlin Heidelberg 2005)

1.49 B. Bhushan: *Tribology and Mechanics of Magnetic Storage Devices*, 2nd edn. (Springer, Berlin Heidelberg 1996)

1.50 K.E. Drexler: *Nanosystems: Molecular Machinery, Manufacturing and Computation* (Wiley, New York 1992)

1.51 G. Timp (Ed.): *Nanotechnology* (Springer, New York 1999)

1.52 M.S. Dresselhaus, G. Dresselhaus, P. Avouris: *Carbon Nanotubes – Synthesis, Structure, Properties, and Applications* (Springer, Berlin Heidelberg 2001)

1.53 E.A. Rietman: *Molecular Engineering of Nanosystems* (Springer, Berlin, Heidelberg 2001)

1.54 W.A. Goddard, D.W. Brenner, S.E. Lyshevski, G.J. Iafrate (Eds.): *Handbook of Nanoscience, Engineering, and Technology* (CRC, Boca Raton 2002)

1.55 H.S. Nalwa (Ed.): *Nanostructured Materials and Nanotechnology* (Academic, San Diego 2002)

1.56 C.P. Poole, F.J. Owens: *Introduction to Nanotechnology* (Wiley, New York 2003)

1.57 A. Manz, H. Becker (Eds.): *Microsystem Technology in Chemistry and Life Sciences*, Top. Curr. Chem., Vol. 194 (Springer, Berlin, Heidelberg 1998)

1.58 J. Cheng, L.J. Kricka (Eds.): *Biochip Technology* (Harwood, Philadephia 2001)

1.59 M.J. Heller, A. Guttman (Eds.): *Integrated Microfabricated Biodevices* (Marcel Dekker, New York 2001)

1.60 C. Lai Poh San, E.P.H. Yap (Eds.): *Frontiers in Human Genetics* (World Scientific, Singapore 2001)

1.61 C.H. Mastrangelo, H. Becker (Eds.): *Microfluidics and BioMEMS*, Proc. SPIE, Vol. 4560 (SPIE, Bellingham 2001)

1.62 H. Becker, L.E. Locascio: Polymer microfluidic devices, Talanta **56**, 267–287 (2002)

1.63 A. van der Berg (Ed.): *Lab-on-a-Chip: Chemistry in Miniaturized Synthesis and Analysis Systems* (Elsevier, Amsterdam 2003)

1.64 P. Gravesen, J. Branebjerg, O.S. Jensen: Microfluidics – A review, J. Micromech. Microeng. **3**, 168–182 (1993)

1.65 R.P. Lanza, R. Langer, J. Vacanti (Eds.): *Principles of Tissue Engineering*, 2nd edn. (Academic, San Diego 2000)

1.66 K. Park (Ed.): *Controlled Drug Delivery: Challenges and Strategies* (American Chemical Society, Washington 1997)

1.67 P.Å. Öberg, T. Togawa, F.A. Spelman: *Sensors in Medicine and Health Care* (Wiley, New York 2004)

1.68 M. Scott: MEMS and MOEMS for national security applications, Proc. SPIE **4980**, xxxvii–xliv (2003)

1.69 B. Bhushan: *Handbook of Micro/Nanotribology* (CRC, Boca Raton 1999)

1.70 B. Bhushan (Ed.): *Nanotribology and Nanomechanics – An Introduction*, 2nd edn. (Springer, Berlin, Heidelberg 2008)

1.71 Anonymous: Current status and future needs, Proc. Workshop Res. Train. Nanosci. Nanotechnol. (European Commission Research Directorate General, Brussels 2005)

1.72 M. Di Ventra, S. Evoy, J.R. Heflin: *Introduction to Nanoscale Science and Technology* (Springer, Berlin Heidelberg 2004)

1.73 A. Hett: *Nanotechnology – Small Matter, Many Unknowns* (Swiss Reinsurance Company, Zurich 2004)

1.74 M. Köhler, W. Fritzsche: *Nanotechnology* (Wiley, New York 2004)

Part A Nanostructures, Micro-/Nanofabrication and Materials

2 Nanomaterials Synthesis and Applications: Molecule-Based Devices
Francisco M. Raymo, Coral Gables, USA

3 Introduction to Carbon Nanotubes
Marc Monthioux, Toulouse, France
Philippe Serp, Toulouse, France
Emmanuel Flahaut, Toulouse, France
Manitra Razafinimanana, Toulouse, France
Christophe Laurent, Toulouse, France
Alain Peigney, Toulouse, France
Wolfgang Bacsa, Toulouse, France
Jean-Marc Broto, Toulouse, France

4 Nanowires
Mildred S. Dresselhaus, Cambridge, USA
Yu-Ming Lin, Yorktown Heigths, USA
Oded Rabin, College Park, USA
Marcie R. Black, Waltham, USA
Jing Kong, Cambridge, USA
Gene Dresselhaus, Cambridge, USA

5 Template-Based Synthesis of Nanorod or Nanowire Arrays
Huamei (Mary) Shang, Milwaukee, USA
Guozhong Cao, Seattle, USA

6 Templated Self-Assembly of Particles
Tobias Kraus, Saarbrücken, Germany
Heiko Wolf, Rüschlikon, Switzerland

7 Three-Dimensional Nanostructure Fabrication by Focused Ion Beam Chemical Vapor Deposition
Shinji Matsui, Hyogo, Japan

8 Introduction to Micro-/Nanofabrication
Babak Ziaie, West Lafayette, USA
Antonio Baldi, Barcelona, Spain
Massood Z. Atashbar, Kalamazoo, USA

9 Nanoimprint Lithography – Patterning of Resists Using Molding
Helmut Schift, Villigen PSI, Switzerland
Anders Kristensen, Kongens Lyngby, Denmark

10 Stamping Techniques for Micro- and Nanofabrication
Etienne Menard, Durham, USA
John A. Rogers, Urbana, USA

11 Material Aspects of Micro- and Nanoelectromechanical Systems
Christian A. Zorman, Cleveland, USA
Mehran Mehregany, Cleveland, USA

2. Nanomaterials Synthesis and Applications: Molecule-Based Devices

Françisco M. Raymo

The constituent components of conventional devices are carved out of larger materials relying on physical methods. This top-down approach to engineered building blocks becomes increasingly challenging as the dimensions of the target structures approach the nanoscale. Nature, on the other hand, relies on chemical strategies to assemble nanoscaled biomolecules. Small molecular building blocks are joined to produce nanostructures with defined geometries and specific functions. It is becoming apparent that nature's bottom-up approach to functional nanostructures can be mimicked to produce artificial molecules with nanoscaled dimensions and engineered properties. Indeed, examples of artificial nanohelices, nanotubes, and molecular motors are starting to be developed. Some of these fascinating chemical systems have intriguing electrochemical and photochemical properties that can be exploited to manipulate chemical, electrical, and optical signals at the molecular level. This tremendous opportunity has led to the development of the molecular equivalent of conventional logic gates. Simple logic operations, for example, can be reproduced with collections of molecules operating in solution. Most of these chemical systems, however, rely on bulk addressing to execute combinational and sequential logic operations. It is essential to devise methods to reproduce these useful functions in solid-state configurations and, eventually, with single molecules. These challenging objectives are stimulating the design of clever devices that interface small assemblies of organic molecules with macroscaled and nanoscaled electrodes. These strategies have already produced rudimentary examples of diodes, switches, and transistors based on functional molecular components. The rapid

2.1	Chemical Approaches to Nanostructured Materials..................	18
	2.1.1 From Molecular Building Blocks to Nanostructures.........................	18
	2.1.2 Nanoscaled Biomolecules: Nucleic Acids and Proteins............	18
	2.1.3 Chemical Synthesis of Artificial Nanostructures............	20
	2.1.4 From Structural Control to Designed Properties and Functions............................	20
2.2	Molecular Switches and Logic Gates	22
	2.2.1 From Macroscopic to Molecular Switches...................	22
	2.2.2 Digital Processing and Molecular Logic Gates............	23
	2.2.3 Molecular AND, NOT, and OR Gates..	24
	2.2.4 Combinational Logic at the Molecular Level..................	25
	2.2.5 Intermolecular Communication......	26
2.3	Solid State Devices...............................	30
	2.3.1 From Functional Solutions to Electroactive and Photoactive Solids.................	30
	2.3.2 Langmuir–Blodgett Films.............	31
	2.3.3 Self-Assembled Monolayers...........	35
	2.3.4 Nanogaps and Nanowires.............	38
2.4	Conclusions and Outlook	42
References		43

and continuous progress of this exploratory research will, we hope, lead to an entire generation of molecule-based devices that might ultimately find useful applications in a variety of fields, ranging from biomedical research to information technology.

2.1 Chemical Approaches to Nanostructured Materials

The fabrication of conventional devices relies on the assembly of macroscopic building blocks with specific configurations. The shapes of these components are carved out of larger materials by exploiting physical methods. This top-down approach to engineered building blocks is extremely powerful and can deliver effectively and reproducibly microscaled objects. This strategy becomes increasingly challenging, however, as the dimensions of the target structures approach the nanoscale. Indeed, the physical fabrication of nanosized features with subnanometer precision is a formidable technological challenge.

2.1.1 From Molecular Building Blocks to Nanostructures

Nature efficiently builds nanostructures by relying on chemical approaches. Tiny molecular building blocks are assembled with a remarkable degree of structural control in a variety of nanoscaled materials with defined shapes, properties, and functions. In contrast to the top-down physical methods, small components are connected to produce larger objects in these bottom-up chemical strategies. It is becoming apparent that the limitations of the top-down approach to artificial nanostructures can be overcome by mimicking nature's bottom-up processes. Indeed, we are starting to see emerge beautiful and ingenious examples of molecule-based strategies to fabricate chemically nanoscaled building blocks for functional materials and innovative devices.

2.1.2 Nanoscaled Biomolecules: Nucleic Acids and Proteins

Nanoscaled macromolecules play a fundamental role in biological processes [2.1]. Nucleic acids, for example, ensure the transmission and expression of genetic information. These particular biomolecules are linear polymers incorporating nucleotide repeating units (Fig. 2.1a). Each nucleotide has a phosphate bridge and a sugar residue. Chemical bonds between the phosphate of one nucleotide and the sugar of the next ensures the propagation of a polynucleotide strand from the 5′ to the 3′ end. Along the sequence of alternating sugar and phosphate fragments, an extended chain of robust covalent bonds involving carbon, oxygen, and phosphorous atoms forms the main backbone of the polymeric strand.

Every single nucleotide of a polynucleotide strand carries one of the four heterocyclic bases shown in Fig. 2.1b. For a strand incorporating 100 nucleotide repeating units, a total of 4^{100} unique polynucleotide sequences are possible. It follows that nature can fabricate a huge number of closely related nanostructures relying only on four building blocks. The heterocyclic bases appended to the main backbone of alternating phosphate and sugar units can sustain hydrogen bonding and $[\pi \cdots \pi]$ stacking interactions. Hydrogen bonds, formed between [N—H] donors and either N or O acceptors, encourage the pairing of adenine (A) with thymine (T) and of guanine (G) with cytosine (C). The stacking interactions involve attractive contacts between the extended π-surfaces of heterocyclic bases.

In the B conformation of deoxyribonucleic acid (DNA), the synergism of hydrogen bonds and $[\pi \cdots \pi]$ stacking glues pairs of complementary polynucleotide strands in fascinating double helical supermolecules (Fig. 2.1c) with precise structural control at the subnanometer level. The two polynucleotide strands wrap around a common axis to form a right-handed double helix with a diameter of ≈ 2 nm. The hydrogen bonded and $[\pi \cdots \pi]$ stacked base pairs lie at the core of the helix with their π-planes perpendicular to the main axis of the helix. The alternating phosphate and sugar units define the outer surface of the double helix. In B-DNA, ≈ 10 base pairs define each helical turn corresponding to a rise per turn or helical pitch of ≈ 3 nm. Considering that these molecules can incorporate up to $\approx 10^{11}$ base pairs, extended end-to-end lengths spanning from only few nanometers to hundreds of meters are possible.

Nature's operating principles to fabricate nanostructures are not limited to nucleic acids. Proteins are also built joining simple molecular building blocks, the amino acids, by strong covalent bonds [2.1]. More precisely, nature relies on 20 amino acids differing in their side chains to assemble linear polymers, called polypeptides, incorporating an extended backbone of robust [C—N] and [C—C] bonds (Fig. 2.2a). For a single polymer strand of 100 repeating amino acid units, a total of 20^{100} unique combinations of polypeptide sequences are possible. Considering that proteins can incorporate more than one polypetide chain with over 4000 amino acid residues each, it is obvious that nature can assemble an enormous number of different biomolecules relying on the same fabrication strategy and a relatively small pool of building blocks.

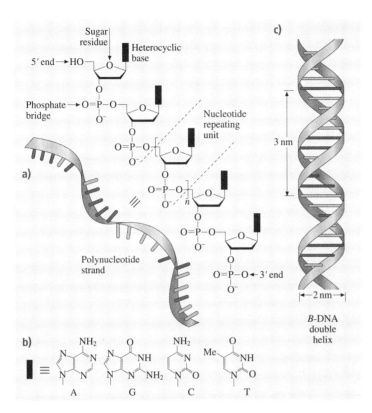

Fig. 2.1a–c A polynucleotide strand (**a**) incorporates alternating phosphate and sugar residues joined by covalent bonds. Each sugar carries one of four heterocyclic bases (**b**). Noncovalent interactions between complementary bases in two independent polynucleotide strands encourage the formation of nanoscaled double helixes (**c**)

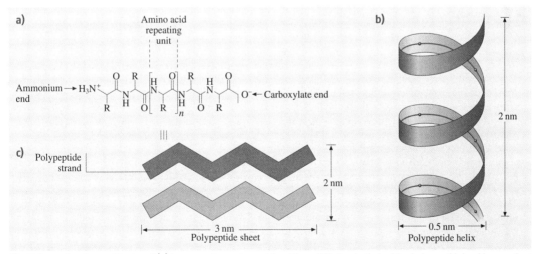

Fig. 2.2a–c A polypeptide strand (**a**) incorporates amino acid residues differing in their side chains and joined by covalent bonds. Hydrogen bonding interactions curl a single polypeptide strand into a helical arrangement (**b**) or lock pairs of strands into nanoscaled sheets (**c**)

The covalent backbones of the polypeptide strands form the main skeleton of a protein molecule. In addition, a myriad of secondary interactions, involving noncovalent contacts between portions of the amino acid residues, control the arrangement of the individual polypeptide chains. Intrastrand hydrogen bonds curl single polypeptide chains around a longitudinal axis in a helical fashion to form tubular nanostructures ≈ 0.5 nm wide and ≈ 2 nm long (Fig. 2.2b). Similarly, interstrand hydrogen bonds can align from 2 up to 15 parallel or antiparallel polypeptide chains to form nanoscaled sheets with average dimensions of 2×3 nm^2 (Fig. 2.2c). Multiple nanohelices and/or nanosheets combine into a unique three-dimensional arrangement dictating the overall shape and dimensions of a protein.

2.1.3 Chemical Synthesis of Artificial Nanostructures

Nature fabricates complex nanostructures relying on simple criteria and a relatively small pool of molecular building blocks. Robust chemical bonds join the basic components into covalent scaffolds. Noncovalent interactions determine the three-dimensional arrangement and overall shape of the resulting assemblies. The multitude of unique combinations possible for long sequences of chemically connected building blocks provides access to huge libraries of nanoscaled biomolecules.

Modern chemical synthesis has evolved considerably over the past few decades [2.2]. Experimental procedures to join molecular components with structural control at the picometer level are available. A multitude of synthetic schemes to encourage the formation of chemical bonds between selected atoms in reacting molecules have been developed. Furthermore, the tremendous progress of crystallographic and spectroscopic techniques has provided efficient and reliable tools to probe directly the structural features of artificial inorganic and organic compounds. It follows that designed molecules with engineered shapes and dimensions can be now prepared in a laboratory relying on the many tricks of chemical synthesis and the power of crystallographic and spectroscopic analyses.

The high degree of sophistication reached in this research area translates into the possibility of mimicking the strategies successfully employed by nature to fabricate chemically nanostructures [2.3]. Small molecular building blocks can be synthesized and joined covalently following routine laboratory procedures. It is even possible to design the stereoelectronic properties of the assembling components in order to shape the geometry of the final product with the assistance of noncovalent interactions. For example, five bipyridine building blocks (Fig. 2.3) can be connected in five synthetic steps to produce an oligobipyridine strand [2.4]. The five repeating units are bridged by C−O bonds and can chelate metal cations in the bay regions defined by their two nitrogen atoms. The spontaneous assembly of two organic strands in a double helical arrangement occurs in the presence of inorganic cations. In the resulting helicate, the two oligobipyridine strands wrap around an axis defined by five Cu(I) centers. Each inorganic cation coordinates two bipyridine units with a tetrahedral geometry imposing a diameter of ≈ 0.6 nm on the nanoscaled helicate [2.5]. The overall length from one end of the helicate to the other is ≈ 3 nm [2.6]. The analogy between this artificial double helix and the B-DNA double helix shown in Fig. 2.1c is obvious. In both instances, a supramolecular glue combines two independent molecular strands into nanostructures with defined shapes and dimensions.

The chemical synthesis of nanostructures can borrow nature's design criteria as well as its molecular building blocks. Amino acids, the basic components of proteins, can be assembled into artificial macrocycles. In the example of Fig. 2.4, eight amino acid residues are joined through the formation of C−N bonds in multiple synthetic steps [2.7]. The resulting covalent backbone defines a circular cavity with a diameter of ≈ 0.8 nm [2.8]. In analogy to the polypeptide chains of proteins, the amino acid residues of this artificial oligopeptide can sustain hydrogen bonding interactions. It follows that multiple macrocycles can pile on top of each other to form tubular nanostructures. The walls of the resulting nanotubes are maintained in position by the cooperative action of at least eight primary hydrogen bonding contacts per macrocycle. These noncovalent interactions maintain the mean planes of independent macrocycles in an approximately parallel arrangement with a plane-to-plane separation of ≈ 0.5 nm.

2.1.4 From Structural Control to Designed Properties and Functions

The examples in Figs. 2.3 and 2.4 demonstrate that modular building blocks can be assembled into target compounds with precise structural control at the picometer level through programmed sequences of synthetic steps. Indeed, modern chemical synthesis offers access to complex molecules with nanoscaled dimensions and, thus, provides cost-effective strategies for the production and characterization of billions of engineered nanostructures in parallel. Furthermore, the high degree of structural control is accompanied by the possibility of designing specific properties into the target nanostructures. Electroactive and photoactive components can be integrated chemically into functional molecular machines [2.9]. Extensive electrochemical investigations have demonstrated that inorganic and organic compounds can exchange electrons with macroscopic electrodes [2.10]. These studies have unraveled the processes responsible for the oxidation and reduction of numerous functional groups and indicated viable design criteria to adjust the ability of molecules to accept or donate electrons [2.11]. Similarly, detailed photochemical and photophysical investigations have elucidated the mechanisms responsible for the absorption and emission of photons at the molecular level [2.12]. The vast knowledge established on the interactions between light and molecules offers the opportunity to engineer chromophoric and fluorophoric functional groups with defined absorption and emission properties [2.11, 13].

The power of chemical synthesis to deliver functional molecules is, perhaps, better illustrated by the molecular motor shown in Fig. 2.5. The preparation of this [2]rotaxane requires 12 synthetic steps starting from known precursors [2.14]. This complex molecule incorporates a Ru(II)-trisbipyridine stopper bridged to a linear tetracationic fragment by a rigid triaryl spacer. The other end of the tetracationic portion is terminated by a bulky tetraarylmethane stopper. The bipyridinium unit of this dumbbell-shaped compound is encircled by a macrocyclic polyether. No covalent bonds join the macrocyclic and linear components. Rather, hydrogen bonding and [π ··· π] stacking interactions maintain the

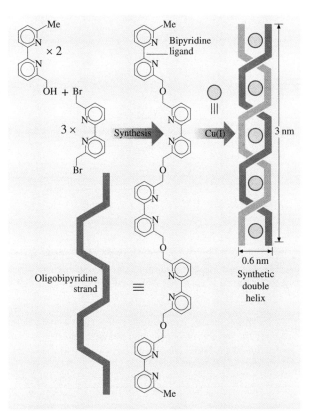

Fig. 2.3 An oligobipyridine strand can be synthesized joining five bipyridine subunits by covalent bonds. The tetrahedral coordination of pairs of bipyridine ligands by Cu(I) ions encourages the assembly two oligobipyridine strands into a double helical arrangement

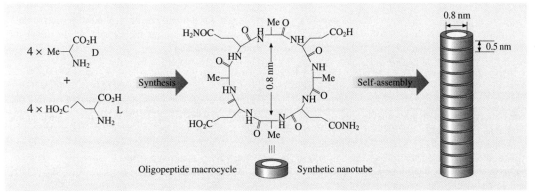

Fig. 2.4 Cyclic oligopeptides can be synthesized joining eight amino acid residues by covalent bonds. The resulting macrocycles self-assemble into nanoscaled tubelike arrays

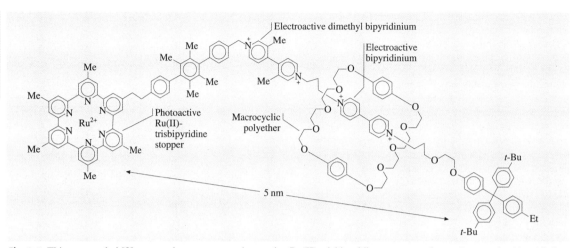

Fig. 2.5 This nanoscaled [2]rotaxane incorporates a photoactive Ru(II)-trisbipyridine stopper and two electroactive bipyridinium units. Photoinduced electron transfer from the photoactive stopper to the encircled electroactive unit forces the macrocyclic polyether to shuttle to the adjacent bipyridinium dication

macrocyclic polyether around the bipyridinium unit. In addition, mechanical constrains associated with the bulk of the two terminal stoppers prevent the macrocycle to slip off the thread. The approximate end-to-end distance for this [2]rotaxane is ≈ 5 nm.

The bipyridinium and the 3,3′-dimethyl bipyridinium units within the dumbbell-shaped component undergo two consecutive and reversible monoelectronic reductions [2.14]. The two methyl substituents on the 3,3′-dimethyl bipyridinium dication make this electroactive unit more difficult to reduce. In acetonitrile, its redox potential is ≈ 0.29 V more negative than that of the unsubstituted bipyridinium dication. Under irradiation at 436 nm in degassed acetonitrile, the excitation of the Ru(II)-trisbipyridine stopper is followed by electron transfer to the unsubstituted bipyridinium unit. In the presence of a sacrificial electron donor (triethanolamine) in solution, the photogenerated hole in the photoactive stopper is filled, and undesired back electron transfer is suppressed. The permanent and light-induced reduction of the dicationic bipyridinium unit to a radical cation depresses significantly the magnitude of the noncovalent interactions holding the macrocyclic polyether in position. As a result, the macrocycle shuttles from the reduced unit to the adjacent dicationic 3,3′-dimethyl bipyridinum. After the diffusion of molecular oxygen into the acetonitrile solution, oxidation occurs restoring the dicationic form of the bipyridinium unit and its ability to sustain strong noncovalent bonds. As a result, the macrocyclic polyether shuttles back to its original position. This amazing example of a molecular shuttle reveals that dynamic processes can be controlled reversibly at the molecular level relying on the clever integration of electroactive and photoactive fragments into functional and nanoscaled molecules.

2.2 Molecular Switches and Logic Gates

Everyday, we routinely perform dozens of switching operations. We turn on and off our personal computers, cellular phones, CD players, radios, or simple light bulbs at a click of a button. Every single time, our finger exerts a mechanical stimulation on a control device, namely a switch. The external stimulus changes the physical state of the switch closing or opening an electric circuit and enabling or preventing the passage of electrons. Overall, the switch transduces a mechanical input into an electrical output.

2.2.1 From Macroscopic to Molecular Switches

The use of switching devices is certainly not limited to electric circuits. For example, a switch at the junction of a railroad can divert trains from one track to another. Similarly, a faucet in a lavatory pipe can block or release the flow of water. Of course, the nature of the control stimulations and the character of the final outcome vary significantly from case to case, but the operating principle behind each switching device is the same. In all cases, input stimulations reach the switch changing its physical state and producing a specific output.

The development of nanoscaled counterparts to conventional switches is expected to have fundamental scientific and technological implications. For instance, one can envisage practical applications for ultraminiaturized switches in areas ranging from biomedical research to information technology. The major challenge in the quest for nanoswitches, however, is the identification of reliable design criteria and operating principles for these innovative and fascinating devices. Chemical approaches to implement molecule-sized switches appear to be extremely promising. The intrinsically small dimensions of organic molecules coupled with the power of chemical synthesis are the main driving forces behind these exploratory investigations.

Certain organic molecules adjust their structural and electronic properties when stimulated with chemical, electrical, or optical inputs. Generally, the change is accompanied by an electrochemical or spectroscopic response. Overall, these nanostructures transduce input stimulations into detectable outputs and, appropriately, are called molecular switches [2.15, 16]. The chemical transformations associated with these switching processes are often reversible. The chemical system returns to the original state when the input signal is turned off. The interconverting states of a molecular switch can be isomers, an acid and its conjugated base, the oxidized and reduced forms of a redox active molecule, or even the complexed and uncomplexed forms of a receptor [2.9, 13, 15, 16]. The output of a molecular switch can be a chemical, electrical, and/or optical signal that varies in intensity with the interconversion process. For example, changes in absorbance, fluorescence, pH, or redox potential can accompany the reversible transformation of a molecular switch.

2.2.2 Digital Processing and Molecular Logic Gates

In present computer networks, data are elaborated electronically by microprocessor systems [2.17] and are exchanged optically between remote locations [2.18]. Data processing and communication require the encoding of information in electrical and optical signals in the form of binary digits. Using arbitrary assumptions, logic thresholds can be established for each signal and, then, 0 and 1 digits can be encoded following simple conventions. Sequences of electronic devices manipulate the encoded bits executing logic functions as a result of basic switching operations.

The three basic AND, NOT, and OR operators combine binary inputs into binary outputs following precise logic protocols [2.17]. The NOT operator converts an input signal into an output signal. When the input is 0, the output is 1. When the input is 1, the output is 0. Because of the inverse relationship between the input and output values, the NOT gate is often called *inverter* [2.19]. The OR operator combines two input signals into a single output signal. When one or both inputs are 1, the output is 1. When both inputs are 0, the output is 0. The AND gate also combines two input signals into one output signal. In this instance, however, the output is 1 only when both inputs are 1. When at least one input is 0, the output is 0.

The output of one gate can be connected to one of the inputs of another operator. A NAND gate, for example, is assembled connecting the output of an AND operator to the input of a NOT gate. Now the two input signals are converted into the final output after two consecutive logic operations. In a similar fashion, a NOR gate can be assembled connecting the output of an OR operator to the input of a NOT gate. Once again, two consecutive logic operations determine the relation between two input signals and a single output. The NAND and NOR operations are termed universal functions because any conceivable logic operation can be implemented relying only on one of these two gates [2.17]. In fact, digital circuits are fabricated routinely interconnecting exclusively NAND or exclusively NOR operators [2.19].

The logic gates of conventional microprocessors are assembled interconnecting transistors, and their input and output signals are electrical [2.19]. But the concepts of binary logic can be extended to chemical, mechanical, optical, pneumatic, or any other type of signal. First it is necessary to design devices that can respond to these stimulations in the same way transistors respond to electrical signals. Molecular switches respond to a variety of input stimulations producing specific outputs and can, therefore, be exploited to implement logic functions [2.13, 20, 21].

2.2.3 Molecular AND, NOT, and OR Gates

More than a decade ago, researchers proposed a potential strategy to execute logic operations at the molecular level [2.22]. Later, the analogy between molecular switches and logic gates was recognized in a seminal article [2.23], in which it was demonstrated that AND, NOT, and OR operations can be reproduced with fluorescent molecules. The pyrazole derivative **1** (Fig. 2.6) is a molecular NOT gate. It imposes an inverse relation between a chemical input (concentration of H^+) and an optical output (emission intensity). In a mixture of methanol and water, the fluorescence quantum yield of **1** is 0.13 in the presence of only 0.1 equivalents of H^+ [2.23]. The quantum yield drops to 0.003 when the equivalents of H^+ are 1000. Photoinduced electron transfer from the central pyrazoline unit to the pendant benzoic acid quenches the fluorescence of the protonated form. Thus, a change in H^+ concentration (I) from a low to a high value switches the emission intensity (O) from a high to a low value. The inverse relationship between the chemical input I and the optical output O translates into the truth table of a NOT operation if a positive logic convention (low = 0, high = 1) is applied to both signals. The emission intensity is high (O = 1) when the concentration of H^+ is low (I = 0). The emission intensity is low (O = 0) when the concentration of H^+ is high (I = 1).

The anthracene derivative **2** (Fig. 2.6) is a molecular OR gate. It transduces two chemical inputs (concentrations of Na^+ and K^+) into an optical output (emission intensity). In methanol, the fluorescence quantum yield is only 0.003 in the absence of metal cations [2.23]. Photoinduced electron transfer from the nitrogen atom of the azacrown fragment to the anthracene fluorophore quenches the emission. After the

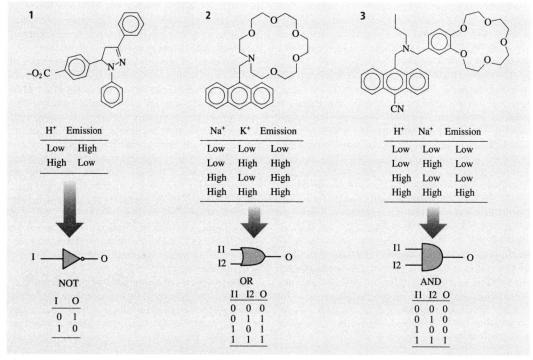

Fig. 2.6 The fluorescence intensity of the pyrazoline derivative **1** is high when the concentration of H^+ is low, and vice versa. The fluorescence intensity of the anthracene derivative **2** is high when the concentration of Na^+ and/or K^+ is high. The emission is low when both concentrations are low. The fluorescence intensity of the anthracene **3** is high only when the concentrations of H^+ and Na^+ are high. The emission is low in the other three cases. The signal transductions of the molecular switches **1**, **2**, and **3** translate into the truth tables of NOT, OR, and AND gates, respectively, if a positive logic convention is applied to all inputs and outputs (low = 0, high = 1)

addition of 1000 equivalents of either Na^+ or K^+, the quantum yield raises to 0.053 and 0.14, respectively. Similarly, the quantum yield is 0.14 when both metal cations are present in solution. The complexation of one of the two metal cations inside the azacrown receptor depresses the efficiency of the photoinduced electron transfer enhancing the fluorescence. Thus, changes in the concentrations of Na^+ (I1) and/or K^+ (I2) from low to high values switch the emission intensity (O) from a low to a high value. The relationship between the chemical inputs I1 and I2 and the optical output O translates into the truth table of an OR operation if a positive logic convention (low = 0, high = 1) is applied to all signals. The emission intensity is low (O = 0) only when the concentration of Na^+ and K^+ are low (I1 = 0, I2 = 0). The emission intensity is high (O = 1) for the other three input combinations.

The anthracene derivative **3** (Fig. 2.6) is a molecular AND gate. It transduces two chemical inputs (concentrations of H^+ and Na^+) into an optical output (emission intensity). In a mixture of methanol and *iso*propanol, the fluorescence quantum yield is only 0.011 in the absence of H^+ or Na^+ [2.23]. Photoinduced electron transfer from either the tertiary amino group or the catechol fragment to the anthracene fluorophore quenches the emission. After the addition of either 100 equivalents of H^+ or 1000 equivalents of Na^+, a modest change of the quantum yield to 0.020 and 0.011, respectively, is observed. Instead, the quantum yield increases to 0.068 when both species are present in solution. The protonation of the amino group and the insertion of the metal cation in the benzocrown ether receptor depress the efficiency of the photoinduced electron transfer processes enhancing the fluorescence. Thus, changes in the concentrations of H^+ (I1) and Na^+ (I2) from low to high values switch the emission intensity (O) from a low to a high value. The relationship between the chemical inputs I1 and I2 and the optical output O translates into the truth table of an AND operation if a positive logic convention (low = 0, high = 1) is applied to all signals. The emission intensity is high (O = 1) only when the concentration of H^+ and Na^+ are high (I1 = 1, I2 = 1). The emission intensity is low (O = 0) for the other three input combinations.

2.2.4 Combinational Logic at the Molecular Level

The fascinating molecular AND, NOT, and OR gates illustrated in Fig. 2.6 have stimulated the design of related chemical systems able to execute the three basic logic operations and simple combinations of them [2.13, 20, 21]. Most of these molecular switches convert chemical inputs into optical outputs. But the implementation of logic operations at the molecular level is not limited to the use of chemical inputs. For example, electrical signals and reversible redox processes can be exploited to modulate the output of a molecular switch [2.24]. The supramolecular assembly **4** (Fig. 2.7) executes a XNOR function relying on these operating principles. The π-electron rich tetrathiafulvalene (TTF) guest threads the cavity of a π-electron deficient bipyridinium (BIPY) host. In acetonitrile, an absorption band associated with the charge-transfer interactions between the complementary π-surfaces is observed at 830 nm. Electrical stimulations alter the redox state of either the TTF or the BIPY units encouraging the separation of the two components of the complex and the disappearance of the charge-transfer band. Electrolysis at a potential of $+0.5$ V oxidizes the neutral TTF unit to a monocationic state. The now cationic guest is expelled from the cavity of the tetracationic host as a result of electrostatic repulsion. Consistently, the absorption band at 830 nm disappears. The charge-transfer band, however, is restored after the exhaustive back reduction of the TTF unit at a potential of 0 V. Similar changes in the absorption properties can be induced addressing the BIPY units. Electrolysis at -0.3 V reduces the dicationic BIPY units to their monocationic forms encouraging the separation of the two components of the complex and the disappearance of the absorption band. The original absorption spectrum is restored after the exhaustive back oxidation of the BIPY units at a potential of 0 V. Thus, this supramolecular system responds to electrical stimulations producing an optical output. One of the electrical inputs (I1) controls the redox state of the TTF unit switching between 0 and $+0.5$ V. The other (I2) determines the redox state of the bipyridinium units switching between -0.3 and 0 V. The optical output (O) is the absorbance of the charge-transfer band. A positive logic convention (low = 0, high = 1) can be applied to the input I1 and output O. A negative logic convention (low = 1, high = 0) can be applied to the input I2. The resulting truth table corresponds to that of a XNOR circuit (Fig. 2.7). The charge-transfer absorbance is high (O = 1) only when one voltage input is low and the other is high (I1 = 0, I2 = 0) or vice versa (I1 = 1, I2 = 1). It is important to note that the input string with both I1 and I2 equal to 1 implies that input potentials of $+0.5$ and -0.3 V are applied simultaneously to a solution containing the supramolecular assembly **4** and not to an individual complex. Of course,

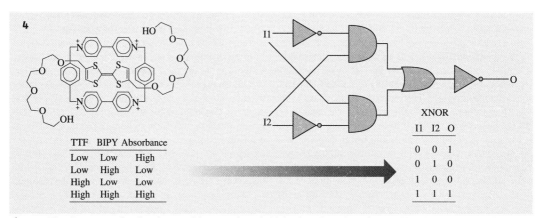

Fig. 2.7 The charge-transfer absorbance of the complex **4** is high when the voltage input addressing the tetrathiafulvalene (TTF) unit is low and that stimulating the bipyridinium (BIPY) units is high and vice versa. If a positive logic convention is applied to the TTF input and to the absorbance output (low = 0, high = 1) while a negative logic convention is applied to the BIPY input (low = 0, high = 1), the signal transduction of **4** translates into the truth table of a XNOR circuit

the concomitant oxidation of the TTF guest and reduction of the BIPY units in the very same complex would be unrealistic. In bulk solution, instead, some complexes are oxidized while others are reduced, leaving the average solution composition unaffected. Thus, the XNOR operation executed by this supramolecular system is a consequence of bulk properties and not a result of unimolecular signal transduction.

Optical inputs can be employed to operate the three-state molecular switch of Fig. 2.8 in acetonitrile solution [2.25]. This chemical system responds to three inputs producing two outputs. The three input stimulations are ultraviolet light (I1), visible light (I2), and the concentration of H^+ (I3). One of the two optical outputs is the absorbance at 401 nm (O1), which is high when the molecular switch is in the yellow-green state **6** and low in the other two cases. The other optical output is the absorbance at 563 nm (O2), which is high when the molecular switch is in the purple state **7** and low in the other two cases. The colorless spiropyran state **5** switches to the merocyanine form **7** upon irradiation with ultraviolet light. It switches to the protonated merocyanine from **6** when treated with H^+. The colored state **7** isomerizes back to **5** in the dark or upon irradiation with visible light. Alternatively, **7** switches to **6** when treated with H^+. The colored state **6** switches to **5**, when irradiated with visible light, and to **7**, after the removal of H^+. In summary, this three-state molecular switch responds to two optical inputs (I1 and I2) and one chemical input (I3) producing two optical outputs (O1 and O2). Binary digits can be encoded on each signal applying positive logic conventions (low = 0, high = 1). It follows that the three-state molecular switch converts input strings of three binary digits into output strings of two binary digits. The corresponding truth table (Fig. 2.8) reveals that the optical output O1 is high (O1 = 1) when only the input I3 is applied (I1 = 0, I2 = 0, I3 = 1), when only the input I2 is not applied (I1 = 1, I2 = 0, I3 = 0), or when all three inputs are applied (I1 = 1, I2 = 0, I3 = 0). The optical output O2 is high (O2 = 1) when only the input I1 is applied (I1 = 1, I2 = 0, I3 = 0) or when only the input I3 is not applied (I1 = 1, I2 = 0, I3 = 0). The combinational logic circuit (Fig. 2.8) equivalent to this truth table shows that all three inputs determine the output O1, while only I1 and I3 control the value of O2.

2.2.5 Intermolecular Communication

The combinational logic circuits in Figs. 2.7 and 2.8 are arrays of interconnected AND, NOT, and OR operators. The digital communication between these basic logic elements ensures the execution of a sequence of simple logic operations that results in the complex logic function processed by the entire circuit. It follows that the logic function of a given circuit can be adjusted altering the number and type of basic gates and their interconnection protocol [2.17]. This modular approach to combinational logic circuits is extremely powerful. Any logic function can be implemented connecting the appropriate combination of simple AND, NOT, and OR gates.

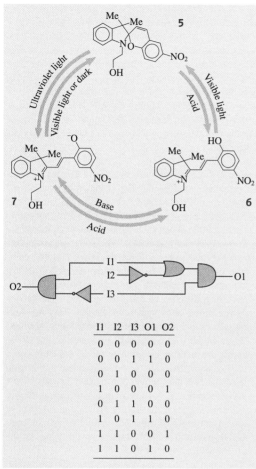

Fig. 2.8 Ultraviolet light (I1), visible light (I2), and H⁺ (I3) inputs induce the interconversion between the three states **5**, **6**, and **7**. The colorless state **5** does not absorb in the visible region. The yellow-green state **6** absorbs at 401 nm (O1). The purple state **7** absorbs at 563 nm (O2). The truth table illustrates the conversion of input strings of three binary digits (I1, I2, and I3) into output strings of two binary digits (O1 and O2) operated by this three-state molecular switch. A combinational logic circuit incorporating nine AND, NOT, and OR operators correspond to this particular truth table

The strategies followed so far to implement complex logic functions with molecular switches are based on the careful design of the chemical system and on the judicious choice of the inputs and outputs [2.13, 20, 21]. A specific sequence of AND, NOT, and OR operations is programmed in a single molecular switch. No digital communication between distinct gates is needed since they are built in the same molecular entity. Though extremely elegant, this strategy does not have the same versatility of a modular approach. A different molecule has to be designed, synthesized, and analyzed every single time a different logic function has to be realized. In addition, the degree of complexity that can be achieved with only one molecular switch is fairly limited. The connection of the input and output terminals of independent molecular AND, NOT, and OR operators, instead, would offer the possibility of assembling any combinational logic circuit from three basic building blocks.

In digital electronics, the communication between two logic gates can be realized connecting their terminals with a wire [2.19]. Methods to transmit binary data between distinct molecular switches are not so obvious and must be identified. Recently we developed two strategies to communicate signals between compatible molecular components. In one instance, a chemical signal is communicated between two distinct molecular switches [2.26]. They are the three-state switch illustrated in Fig. 2.8 and the two-state switch of Fig. 2.9. The merocyanine form **7** is a photogenerated base. Its p-nitrophenolate fragment, produced upon irradiation of the colorless state **5** with ultraviolet light, can abstract a proton from an acid present in the same solution. The resulting protonated form **6** is a photoacid. It releases a proton upon irradiation with visible light and can protonate a base co-dissolved in the same medium. The orange azopyridine **8** switches to the red-purple azopyridinium **9** upon protonation. This process is reversible, and the addition of a base restores the orange state **8**. It follows that photoinduced proton transfer can be exploited to communicate a chemical signal from **6** to **8** and from **9** to **7**. The two colored states **8** and **9** have different absorption properties in the visible region. In acetonitrile, the orange state **8** absorbs at 422 nm, and the red-purple state **9** absorbs at 556 nm. The changes in absorbance of these two bands can be exploited to monitor the photoinduced exchange of protons between the two communicating molecular switches.

The three-state molecular switch and the two-state molecular switch can be operated sequentially when dissolved in the same acetonitrile solution. In the presence of one equivalent of H⁺, the two-state molecular switch is in state **9** and the absorbance at 556 nm is high (O = 1). Upon irradiation with ultraviolet light (I1 = 0), **5** switches to **7**. The photogenerated base deprotonates **9** producing **8** and **6**. As a result, the absorbance at 556 nm decreases (O = 0). Upon irradiation

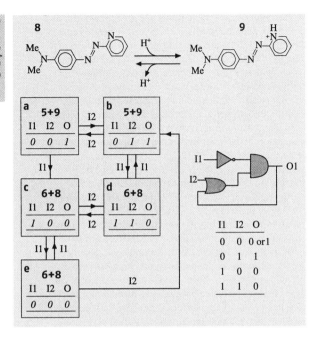

Fig. 2.9 The concentration of H^+ controls the reversible interconversion between the two states **8** and **9**. In response to ultraviolet (I1) and visible (I2) inputs, the three-state molecular switch in Fig. 2.7 modulates the ratio between these two forms and the absorbance (O) of **9** through photoinduced proton transfer. The truth table and sequential logic circuit illustrate the signal transduction behavior of the two communicating molecular switches. The interconversion between the five three-digit strings of input (I1 and I2) and output (O) data is achieved varying the input values in steps ◄

with visible light (I2 = 1), **6** switches to **5** releasing H^+. The result is the protonation of **8** to form **9** and restore the high absorbance at 556 nm (O = 1). In summary, the three-state molecular switch transduces two optical inputs (I1 = ultraviolet light, I2 = visible light) into a chemical signal (proton transfer) that is communicated to the two-state molecular switch and converted into a final optical output (O = absorbance at 556 nm).

The logic behavior of the two communicating molecular switches is significantly different from those of the chemical systems illustrated in Figs. 2.6–2.8 [2.26]. The truth table in Fig. 2.9 lists the four possible combinations of two-digit input strings and the corresponding one-digit output. The output digit O for the input strings 01, 10, and 11 can take only one value. In fact, the input string 01 is transduced into a 1, and the input strings 10 and 11 are converted into 0. Instead, the output digit O for the input string 00 can be either 0 or 1. The sequence of events leading to the input string 00 determines the value of the output. The boxes **a**–**e** in Fig. 2.9 illustrates this effect. They correspond to the five three-digit input/output strings. The transformation of one box into any of the other four is achieved in one or two steps by changing the values of I1 and/or I2. In two instances (**a** and **b**), the two-state molecular switch is in state **9**, and the output signal is high (O = 1). In the other three cases (**c**, **d**, and **e**), the two-state molecular switch is in state **8**, and the output signal is low (O = 0). The strings 000 (**e**) and 001 (**a**) correspond to the first entry of the truth table. They share the same input digits but differ in the output value. The string 000 (**e**) can be obtained only from the string 100 (**c**) varying the value of I1. Similarly, the string 001 (**a**) can be accessed only from the string 011 (**b**) varying the value of I2. In both transformations, the output digit remains unchanged. Thus, the value of O1 in the parent string is memorized and maintained in the daughter string when both inputs become 0. This memory effect is the fundamental operating principle of sequential logic circuits [2.17], which are used extensively to assemble the memory elements of modern microprocessors. The sequential logic circuit equivalent to the truth table of the two communicating molecular switches is also shown in Fig. 2.9. In this circuit, the input data I1 and I2 are combined through NOT, OR, and AND operators. The output of the AND gate O is also an input of the OR gate and controls, together with I1 and I2, the signal transduction behavior.

The other strategy for digital transmission between molecules is based on the communication of optical signals between the three-state molecular switch (Fig. 2.8) and fluorescent compounds [2.27]. In the optical network of Fig. 2.10, three optical signals travel from an excitation source to a detector after passing through two quartz cells. The first cell contains an equimolar acetonitrile solution of naphthalene, anthracene, and tetracene. The second cell contains an acetonitrile solution of the three-state molecular switch. The excitation source sends three consecutive monochromatic light beams to the first cell stimulating the emission of the three fluorophores. The light emitted in the direction perpendicular to the exciting beam reaches the second cell. When the molecular switch is in state **5**, the naphthalene emission at 335 nm is absorbed and a low intensity output (O1) reaches the detector. Instead, the anthracene and tetracene emissions at 401 and 544 nm,

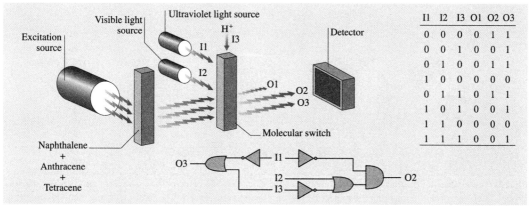

Fig. 2.10 The excitation source sends three monochromatic light beams (275, 357, and 441 nm) to a quartz cell containing an equimolar acetonitrile solution of naphthalene, anthracene and tetracene. The three fluorophores absorb the exciting beams and reemit at 305, 401, and 544 nm, respectively. The light emitted in the direction perpendicular to the exciting beams passes through another quartz cell containing an acetonitrile solution of the three-state molecular switch shown in Fig. 2.7. Ultraviolet (I1), visible (I2), and H$^+$ (I3) inputs control the interconversion between the three states of the molecular switch. They determine the intensity of the optical outputs reaching the detector and correspond to the naphthalene (O1), anthracene (O2), and tetracene (O3) emissions. The truth table and equivalent combinational logic illustrate the relation between the three inputs and the three outputs. The output O1 is always 0, and it is not influenced by the three inputs. Only two inputs determine the value of O3, while all of them control the output O2

respectively, pass unaffected and high intensity outputs (O2 and O3) reach the detector. When the molecular switch is in state **6**, the naphthalene and anthracene emissions are absorbed and only the tetracene emission reaches the detector (O1 = 0, O2 = 0, O3 = 1). When the molecular switch is state **7**, the emission of all three fluorophores is absorbed (O1 = 0, O2 = 0, O3 = 0). The interconversion of the molecular switch between the three states is induced addressing the second cell with ultraviolet (I1), visible (I2) and H$^+$ (I3) inputs. Thus, three independent optical outputs (O1, O2 and O3) can be modulated stimulating the molecular switch with two optical and one chemical input. The truth table in Fig. 2.10 illustrates the relation between the three inputs (I1, I2 and I3) and the three outputs (O1, O2 and O3), when positive logic conventions are applied to all signals. The equivalent logic circuit shows that all three inputs control the anthracene channel O2, but only I1 and I3 influence the tetracene channel O3. Instead, the intensity of the naphthalene channel O1 is always low, and it is not affected by the three inputs.

The operating principles of the optical network in Fig. 2.10 can be simplified to implement all-optical logic gates. The chemical input inducing the formation of the protonated form **6** of the molecular switch can be eliminated. The interconversion between the remaining two states **5** and **7** can be controlled relying exclusively on ultraviolet inputs. Indeed, ultraviolet irradiation induces the isomerization of the colorless form **5** to the colored species **7**, which reisomerizes to the original state in the dark. Thus, a single ultraviolet source is sufficient to control the switching from **5** to **7** and vice versa. On the basis of these considerations, all-optical NAND, NOR, and NOT gates can be implemented operating sequentially or in parallel from one to three independent switching elements [2.28]. For example, the all-optical network illustrated in Fig. 2.11 is a three-input NOR gate. A monochromatic optical signal travels from a visible source to a detector. Three switching elements are aligned along the path of the traveling light. They are quartz cells containing an acetonitrile solution of the molecular switch shown in Fig. 2.8. The interconversion of the colorless form **5** into the purple isomer **7** is induced stimulating the cell with an ultraviolet input. The reisomerization from **7** to **5** occurs spontaneously, as the ultraviolet sources is turned off. Using three distinct ultraviolet sources, the three switching elements can be controlled independently.

The colorless form **5** does not absorb in the visible region, while the purple isomer **7** has a strong absorption band at 563 nm. Thus, a 563 nm optical signal leaving the visible source can reach the detector

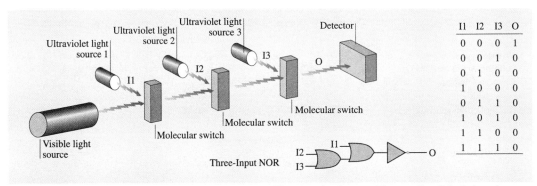

Fig. 2.11 The visible source sends a monochromatic beam (563 nm) to the detector. The traveling light is forced to pass through three quartz cells containing the molecular switch illustrated in Fig. 2.7. The three switching elements are operated by independent ultraviolet inputs. When at least one of them is on, the associated molecular switch is in the purple form **7**, which can absorb and block the traveling light. The truth table and equivalent logic circuit illustrate the relation between the three inputs I1, I2, and I3 and the optical output O

unaffected only if all three switching elements are in the nonabsorbing state **5**. If one of the three ultraviolet inputs I1, I2, or I3 is turned on, the intensity of the optical output O drops to 3–4% of its original value. If two or three ultraviolet inputs are turned on simultaneously, the optical output drops to 0%. Indeed, the photogenerated state **7** absorbs and blocks the traveling light. Applying positive logic conventions to all signals, binary digits can be encoded in the three optical inputs and in the optical output. The resulting truth table is illustrated in Fig. 2.11. The output O is 1 only if all three inputs I1, I2, or I3 are 0. The output O is 0 if at least one of the three inputs I1, I2, or I3 is 1. This signal transduction corresponds to that executed by a three-input NOR gate, which is a combination of one NOT and two OR operators.

2.3 Solid State Devices

The fascinating chemical systems illustrated in Figs. 2.6–2.11 demonstrate that logic functions can be implemented relying on the interplay between designed molecules and chemical, electrical and/or optical signals [2.13, 20, 21].

2.3.1 From Functional Solutions to Electroactive and Photoactive Solids

These molecular switches, however, are operated exclusively in solution and remain far from potential applications in information technology at this stage. The integration of liquid components and volatile organic solvents in practical digital devices is hard to envisage. Furthermore, the logic operations executed by these chemical systems rely on bulk addressing. Although the individual molecular components have nanoscaled dimensions, macroscopic collections of them are employed for digital processing. In some instances, the operating principles cannot even be scaled down to the unimolecular level. Often bulk properties are responsible for signal transduction. For example, a single fluorescent compound **2** cannot execute an OR operation. Its azacrown appendage can accommodate only one metal cation. As a result, an individual molecular switch can respond to only one of the two chemical inputs. It is a collection of numerous molecular switches dissolved in an organic solvent that responds to both inputs enabling an OR operation.

The development of miniaturized molecule-based devices requires the identification of methods to transfer the switching mechanisms developed in solution to the solid state [2.29]. Borrowing designs and fabrication strategies from conventional electronics, researchers are starting to explore the integration of molecular components into functional circuits and devices [2.30–33]. Generally, these strategies combine lithography and surface chemistry to assemble nanometer-thick organic films on the surfaces of microscaled or nanoscaled electrodes. Two main approaches for the deposition of organized molecular arrays on inorganic supports have emerged so far. In one instance, amphiphilic molecular building blocks are compressed into organized monolayers at air/water interfaces. The resulting films can be transferred on supporting solids employing the Langmuir–Blodgett technique [2.34]. Alternatively, certain molecules can be designed to adsorb spontaneously on the surfaces of compatible solids from liquid or vapor phases. The result is the self-assembly of organic layers on inorganic supports [2.35].

2.3.2 Langmuir–Blodgett Films

Films of amphiphilic molecules can be deposited on a variety of solid supports employing the Langmuir–Blodgett technique [2.34]. This method can be extended to electroactive compounds incorporating hydrophilic and hydrophobic groups. For example, the amphiphile **10** (Fig. 2.12) has a hydrophobic hexadecyl tail attached to a hydrophilic bipyridinium dication [2.36, 37]. This compound dissolves in mixtures of chloroform and methanol, but it is not soluble in moderately concentrated aqueous solutions of sodium perchlorate. Thus the spreading of an organic solution of **10** on an aqueous sodium perchlorate subphase affords a collection of disorganized amphiphiles floating on the water surface (Fig. 2.12), after the organic solvent has evaporated. The molecular building blocks can be compressed into a monolayer with the aid of a moving barrier. The hydrophobic tails align away from the aqueous phase. The hydrophilic dicationic heads and the accompanying perchlorate counterions pack to form an organized monolayer at the air/water interface. The compression process can be monitored recording the surface pressure (π)-area per molecule (A) isotherm, which indicates a limiting molecular area of $\approx 50\,\text{Å}^2$. This value is larger than the projected area of an oligomethylene chain. It correlates reasonably, however, with the overall

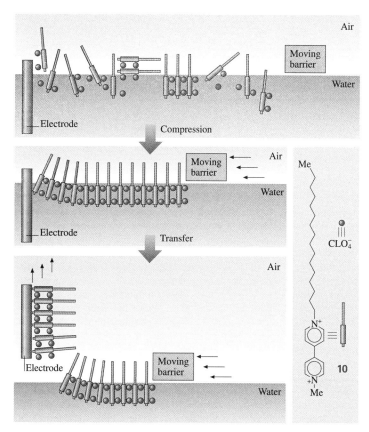

Fig. 2.12 The compression of the amphiphilic dication **10** with a moving barrier results in the formation of a packed monolayer at the air/water interface. The lifting of an electrode pre-immersed in the aqueous subphase encourages the transfer of part of the monolayer on the solid support

area of a bipyridinium dication plus two perchlorate anions.

The monolayer prepared at the air/water interface (Fig. 2.12) can be transferred on the surface of a indium-tin oxide electrode pre-immersed in the aqueous phase. The slow lifting of the solid support drags the monolayer away from the aqueous subphase. The final result is the coating of the electrode with an organic film containing electroactive bipyridinium building blocks. The modified electrode can be integrated in a conventional electrochemical cell to probe the redox response of the electroactive layer. The resulting cyclic voltammograms reveal the characteristic waves for the first reduction process of the bipyridinium dications, confirming the successful transfer of the electroactive amphiphiles from the air/water interface to the electrode surface. The integration of the redox waves indicates a surface coverage of $\approx 4 \times 10^{10}\,\text{mol cm}^{-2}$. This value corresponds to a molecular area of $\approx 40\,\text{Å}^2$ and is in excellent agreement with the limiting molecular area of the π–A isotherm.

These seminal experiments demonstrate that electroactive amphiphiles can be organized into uniform monolayers at the air/water interface and then transferred efficiently on the surface of appropriate substrates to produce electrode/monolayer junctions. The resulting electroactive materials can become the functional components of molecule-based devices. For example, bipyridinium-based photodiodes can be fabricated following this approach [2.38,39]. Their operating principles rely on photoinduced electron transfer from chromophoric units to bipyridinium acceptors. The electroactive and photoactive amphiphile **11** (Fig. 2.13) incorporates hydrophobic ferrocene and pyrene tails and a hydrophilic bipyridinium head. Chloroform solutions of **11** containing ten equivalents of arachidic acid can be spread on an aqueous calcium chloride subphase in a Langmuir trough. The amphiphiles can be compressed into a mixed monolayer, after the evaporation of the organic solvent. Pronounced steps in the corresponding π–A isotherm suggest that the bulky ferrocene and pyrene groups are squeezed away from the water surface. In the final arrangement, both photoactive groups align above the hydrophobic dication.

A mixed monolayer of **11** and arachidic acid can be transferred from the air/water interface to the surface of a transparent gold electrode following the methodology illustrated for the system in Fig. 2.12. The coated electrode can be integrated in a conventional electrochemical cell. Upon irradiation at 330 nm under an inert atmosphere, an anodic photocurrent of ≈ 2 nA devel-

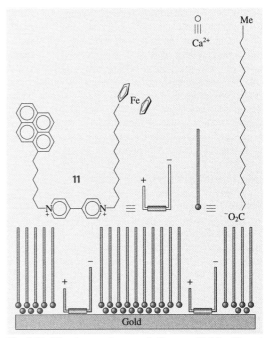

Fig. 2.13 Mixed monolayers of the amphiphile **11** and arachidic acid can be transferred from the air/water interface to the surface of an electrode to generate a molecule-based photodiode

ops at a potential of 0 V relative to a saturated calomel electrode. Indeed, the illumination of the electroactive monolayer induces the electron transfer from the pyrene appendage to the bipyridinium acceptor and then from the reduced acceptor to the electrode. A second intramolecular electron transfer from the ferrocene donor to the oxidized pyrene fills its photogenerated hole. Overall, a unidirectional flow of electrons across the monolayer/electrode junction is established under the influence of light.

The ability to transfer electroactive monolayers from air/water interfaces to electrode surfaces can be exploited to fabricate molecule-based electronic devices. In particular, arrays of interconnected electrode/monolayer/electrode tunneling junctions can be assembled combining the Langmuir–Blodgett technique with electron beam evaporation [2.33]. Figure 2.14 illustrates a schematic representation of the resulting devices. Initially, parallel fingers are patterned on a silicon wafer with a silicon dioxide overlayer by electron beam evaporation. The bottom electrodes deposited on the support can be either aluminum wires covered by an aluminum oxide or n-doped silicon lines with silicon dioxide overlayers. Their widths are ≈ 6 or $7\,\mu$m, respectively. The patterned silicon chip is immersed in the aqueous subphase of a Langmuir trough prior to monolayer formation. After the compression of electroactive amphiphiles at the air/water interface, the substrate is pulled out of the aqueous phase to encourage the transfer of the molecular layer on the parallel bottom electrodes as well as on the gaps between them. Then, a second set of electrodes orthogonal to the first is deposited through a mask by electron beam evaporation. They consist of a titanium underlayer plus an aluminum overlayer. Their thicknesses are ≈ 0.05 and $1\,\mu$m, respectively, and their width is $\approx 10\,\mu$m. In the final assembly, portions of the molecular layer become sandwiched between the bottom and top electrodes. The active areas of these electrode/monolayer/electrode junctions are $\approx 60-70\,\mu\text{m}^2$ and correspond to $\approx 10^6$ molecules.

The [2]rotaxane **12** (Fig. 2.14) incorporates a macrocyclic polyether threaded onto a bipyridinium-based backbone [2.40, 41]. The two bipyridinium dications are bridged by a m-phenylene spacer and terminated by tetraarylmethane appendages. These two bulky groups trap mechanically the macrocycle preventing its dissociation from the tetracationic backbone. In addition, their hydrophobicity complements the hydrophilicity of the two bipyridinium dications imposing amphiphilic character on the overall molecular assembly. This compound does not dissolve in aqueous solutions and can be compressed into organized monolayers at air/water interfaces. The corresponding π–A isotherm reveals a limiting molecular area of $\approx 130\,\text{Å}^2$. This large value is a consequence of the bulk associated with the hydrophobic tetraarylmethane tails and the macrocycle encircling the tetracationic backbone.

Monolayers of the [2]rotaxane **12** can be transferred from the air/water interface to the surfaces of the bottom aluminum/aluminum oxide electrodes of a patterned silicon chip with the hydrophobic tetraarylmethane groups pointing away from the supporting substrate. The subsequent assembly of a top titanium/aluminum electrode

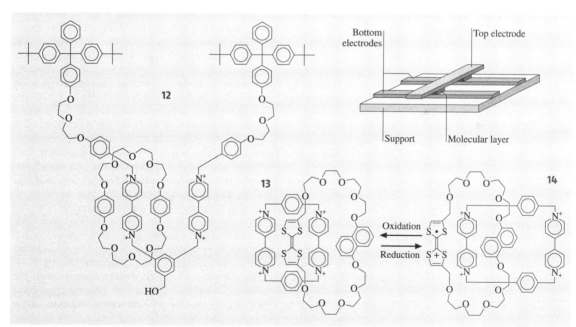

Fig. 2.14 The [2]rotaxane **12** and the [2]catenane **13** can be compressed into organized monolayers at air/water interfaces. The resulting monolayers can be transferred on the bottom electrodes of a patterned silicon support. After the deposition of a top electrode, electrode/monolayer/electrode junctions can be assembled. Note that only the portion of the monolayer sandwiched between the top and bottom electrodes is shown in the diagram. The oxidation of the tetrathiafulvalene unit of the [2]catenane **13** is followed by the circumrotation of the macrocyclic polyether to afford the [2]catenane **14**. The process is reversible, and the reduction of the cationic tetrathiafulvalene unit restores the original state

affords electrode/monolayer/electrode junctions. Their current/voltage signature can be recorded grounding the top electrode and scanning the potential of the bottom electrode. A pronounced increase in current is observed when the potential is lowered below -0.7 V. Under these conditions, the bipyridinium-centered LUMOs mediate the tunneling of electrons from the bottom to the top electrode leading to a current enhancement. A similar current profile is observed if the potential is returned to 0 and then back to -2 V. Instead, a modest increase in current in the opposite direction is observed when the potential is raised above $+0.7$ V. Presumably, this trend is a result of the participation of the phenoxy-centered HOMOs in the tunneling process. After a single positive voltage pulse, however, no current can be detected if the potential is returned to negative values. In summary, the positive potential scan suppresses irreversibly the conducting ability of the electrode/molecule/electrode junction. The behavior of this device correlates with the redox response of the [2]rotaxane **12** in solution. Cyclic voltammograms reveal reversible monoelectronic reductions of the bipyridinium dications. But they also show two irreversible oxidations associated, presumably, with the phenoxy rings of the macrocycle and tetraarylmethane groups. These observations suggest that a positive voltage pulse applied to the electrode/monolayer/electrode junction oxidizes irreversibly the sandwiched molecules suppressing their ability to mediate the transfer of electrons from the bottom to the top electrode under a negative bias.

The device incorporating the [2]rotaxane **13** can be exploited to implement simple logic operations [2.40]. The two bottom electrodes can be stimulated with voltage inputs (I1 and I2) while measuring a current output (O) at the common top electrode. When at least one of the two inputs is high (0 V), the output is low (< 0.7 nA). When both inputs are low (-2 V), the output is high (≈ 4 nA). If a negative logic convention is applied to the voltage inputs (low $= 1$, high $= 0$) and a positive logic convention is applied to the current output (low $= 0$, high $= 1$), the signal transduction behavior translates into the truth table of an AND gate. The output O is 1 only when both inputs are 1. Instead, an OR operation can be executed if the logarithm of the current is considered as the output. The logarithm of the current is -12 when both voltage inputs are 0 V. It raises to ≈ -9 when one or both voltage inputs are lowered to -2 V. This signal transduction behavior translates into the truth table of an OR gate if a negative logic convention is applied to the voltage inputs (low $= 1$, high $= 0$) and a positive logic convention is applied to the current output (low $= 0$, high $= 1$). The output O is 1 when at least one of the two inputs is 1.

The [2]catenane **13** (Fig. 2.14) incorporates a macrocyclic polyether interlocked with a tetracationic cyclophane [2.42, 43]. Organic solutions of the hexafluorophosphate salt of this [2]catenane and six equivalents of the sodium salt of dimyristoylphosphatidic acid can be co-spread on the water surface of a Langmuir trough [2.44]. The sodium hexafluorophosphate formed dissolves in the supporting aqueous phase, while the hydrophilic bipyridinium cations and the amphiphilic anions remain at the interface. Upon compression, the anions align their hydrophobic tails away from the water surface forming a compact monolayer above the cationic bipyridinium derivatives. The corresponding π–A isotherm indicates limiting molecular areas of ≈ 125 Å^2. This large value is a consequence of the bulk associated with the two interlocking macrocycles.

Monolayers of the [2]catenane **13** can be transferred from the air/water interface to the surfaces of the bottom n-doped silicon/silicon dioxide electrodes of a patterned silicon chip with the hydrophobic tails of the amphiphilic anions pointing away from the supporting substrate [2.45, 46]. The subsequent assembly of a top titanium/aluminum electrode affords electrode/monolayer/electrode arrays. Their junction resistance can be probed grounding the top electrode and maintaining the potential of the bottom electrode at $+0.1$ V. If a voltage pulse of $+2$ V is applied to the bottom electrode before the measurement, the junction resistance probed is ≈ 0.7 GΩ. After a pulse of -2 V applied to the bottom electrode, the junction resistance probed at $+0.1$ V drops ≈ 0.3 GΩ. Thus, alternating positive and negative voltage pulses can switch reversibly the junction resistance between high and low values. This intriguing behavior is a result of the redox and dynamic properties of the [2]catenane **13**.

Extensive spectroscopic and crystallographic studies [2.42, 43] demonstrated that the tetrathiafulvalene unit resides preferentially inside the cavity of the tetracationic cyclophane of the [2]catenane **13** (Fig. 2.14). Attractive [$\pi \cdots \pi$] stacking interactions between the neutral tetrathiafulvalene and the bipyridinium dications are responsible for this co-conformation. Oxidation of the tetrathiafulvalene generates a cationic form that is expelled from the cavity of the tetracationic cyclophane. After the circumrotation of the macrocyclic polyether, the oxidized tetrathiafulvalene is exchanged with the neutral 1,5-dioxynaphthalene producing the [2]catenane **14** (Fig. 2.14). The reduction of the tetrathiafulvalene back to its neutral state is followed by the circumrotation of the macrocyclic polyether, which restores the original state **14**. The voltage pulses applied to the bottom electrode of the electrode/monolayer/junction oxidize and reduce the tetrathiafulvalene unit inducing the interconversion between the forms **13** and **14**. The difference in the stereoelectronic properties of these two states translates into distinct current/voltage signatures. Indeed, their ability to mediate the tunneling of electrons across the junction differs significantly. As a result, the junction resistance probed at a low voltage after an oxidizing pulse is significantly different from that determined under the same conditions after a reducing pulse.

2.3.3 Self-Assembled Monolayers

In the examples illustrated in Figs. 2.12–2.14, monolayers of amphiphilic and electroactive derivatives are assembled at air/water interfaces and then transferred on the surfaces of appropriate substrates. An alternative strategy to coat electrodes with molecular layers relies on the ability of certain compounds to adsorb spontaneously on solid supports from liquid or vapor phases [2.35]. In particular, the affinity of certain sulfurated functional groups for gold can be exploited to encourage the self-assembly of organic molecules on microscaled and nanoscaled electrodes.

The electrode/monolayer/electrode junction in Fig. 2.15 incorporates a molecular layer between two gold electrodes mounted on a silicon nitride support. This device can be fabricated combining chemical vapor deposition, lithography, anisotropic etching, and self-assembly [2.47]. Initially, a silicon wafer is coated with a 50 nm thick layer of silicon nitride by low pressure chemical vapor deposition. Then, a square of $400 \times 400\,\mu m^2$ is patterned on one side of the coated wafer by optical lithography and reactive ion etching. Anisotropic etching of the exposed silicon up to the other side of the wafer leaves a suspended silicon nitride membrane of $40 \times 40\,\mu m^2$. Electron beam lithography and reactive ion etching can be used to carve a bowl-shaped hole (diameter = 30–50 nm) in the membrane. Evaporation of gold on the membrane fills the pore producing a bowl-shaped electrode. Immersion of the substrate in a solution of the thiol **15** results in the self-assembly of a molecular layer on the narrow part of the bowl-shaped electrode. The subsequent evaporation of a gold film on the organic monolayer produces an electrode/monolayer/electrode junction (Fig. 2.15) with a contact area of less than $2000\,nm^2$ and ≈ 1000 molecules.

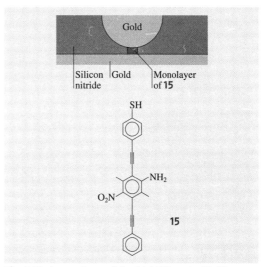

Fig. 2.15 A monolayer of the thiol **15** is embedded between two gold electrodes maintained in position by a silicon nitride support

Under the influence of voltage pulses applied to one of the two gold electrodes in Fig. 2.15, the conductivity of the sandwiched monolayer switches reversibly between low and high values [2.48]. In the initial state, the monolayer is in a low conducting mode. A current output of only 30 pA is detected, when a probing voltage of $+0.25\,V$ is applied to the bowl-shaped electrode. If the same electrode is stimulated with a short voltage pulse of $+5\,V$, the monolayer switches to a high conducting mode. Now a current output of 150 pA is measured at the same probing voltage of $+0.25\,V$. Repeated probing of the current output at various intervals of time indicates that the high conducting state is memorized by the molecule-based device, and it is retained for more than 15 min. The low conducting mode is restored after either a relatively long period of time or the stimulation of the bowl-shaped electrode with a reverse voltage pulse of $-5\,V$. Thus the current output switches from a low to a high value, if a high voltage input is applied. It switches from a high to a low value, under the influence of a low voltage pulse. This behavior offers the opportunity to store and erase binary data in analogy to a conventional random access memory [2.17]. Binary digits can be encoded on the current output of the molecule-based device applying a positive logic convention (low = 0, high = 1). It follows that a binary 1 can be stored in the molecule-based device applying a high voltage

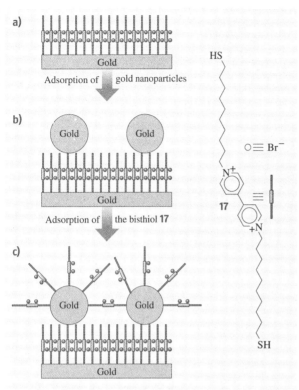

Fig. 2.16 (a) The bisthiol **16** self-assembles on gold electrodes as a result of thiolate–gold bond formation. (b) Gold nanoparticles adsorb spontaneously on the molecular layer. (c) Exposure of the composite assembly to a solution of **16** results in the formation of an additional molecular layer on the surface of the gold nanoparticles

material. Following these procedures, up to ten alternating organic and inorganic layers can be deposited on the electrode surface. The resulting assembly can mediate the unidirectional electron transfer from the supporting electrode to redox active species in solution. For example, the cyclic voltammogram of the $[Ru(NH_3)_6]^{3+/2+}$ couple recorded with a bare gold electrode reveals a reversible reduction process. In the presence of ten alternating molecular and nanoparticle layers on the electrode surface, the reduction potential shifts by ≈ -0.2 V and the back oxidation wave disappears. The pronounced potential shift indicates that $[Ru(NH_3)_6]^{3+}$ accepts electrons only after the surface-confined bipyridinium dications have been reduced. The lack of reversibility indicates that the back oxidation to the bipyridinium dications inhibits the transfer of electrons from the $[Ru(NH_3)_6]^{2+}$ to the electrode. Thus the electroactive multilayer allows the flow of electrons in one direction only in analogy to conventional diodes.

The current/voltage behavior of individual nanoparticles in Fig. 2.16b can be probed by scanning tunneling spectroscopy in an aqueous electrolyte under an inert atmosphere [2.51]. The platinum-iridium tip of a scanning tunneling microscope is positioned above one of the gold particles. The voltage of the gold substrate relative to the tip is maintained at -0.2 V while that relative to a reference electrode immersed in the same electrolyte is varied to control the redox state of the electroactive units. Indeed, the bipyridinium dications in the molecular layer can be reduced reversibly to a monocationic state. The resulting monocations can be reduced further and, once again, reversibly to a neutral form. Finally, the current flowing from the gold support to the tip of the scanning tunneling microscope is monitored as the tip–particle distance increases. From the distance dependence of the current, inverse length decays of ≈ 16 and $7\,\mathrm{nm}^{-1}$ for the dicationic and monocationic states, respectively, of the molecular spacer can be determined. The dramatic decrease indicates that the reduction of the electroactive unit facilitates the tunneling of electrons through the gold/molecule/nanoparticle/tip junction. In summary, a change in the redox state of the bipyridinium components can be exploited to gate reversibly the current flowing through this nanoscaled device.

Similar nanostructured materials, combining molecular and nanoparticles layers, can be prepared on layers on indium-tin oxide electrodes following multistep procedures [2.52]. The hydroxylated surfaces of indium-tin oxide supports can be functionalized with 3-ammoniumpropylysilyl groups and then exposed to

input, and it can be erased applying a low voltage input [2.48].

The ability of thiols to self-assemble on the surface of gold can be exploited to fabricate nanocomposite materials integrating organic and inorganic components. For example, the bisthiol **16** forms monolayers (Fig. 2.16a) on gold electrodes with surface coverages of $\approx 4.1 \times 10^{10}\,\mathrm{mol\,cm^2}$ [2.49, 50]. The formation of a thiolate–gold bond at one of the two thiol ends of **16** is responsible for adsorption. The remaining thiol group points away from the supporting surface and can be exploited for further functionalization. Gold nanoparticles adsorb on the molecular layer (Fig. 2.16b), once again, as a result of thiolate–gold bond formation. The immersion of the resulting material in a methanol solution of **16** encourages the adsorption of an additional organic layer (Fig. 2.16c) on the composite

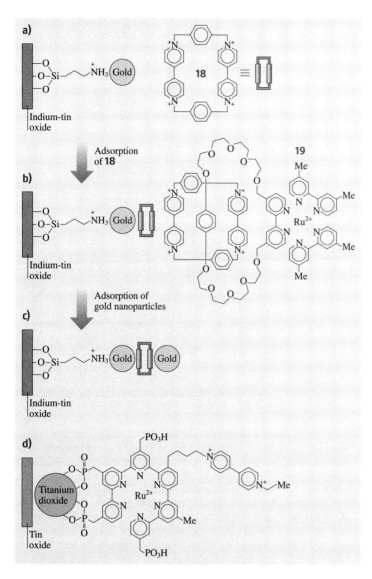

Fig. 2.17 (a) Gold nanoparticles assemble spontaneously on prefunctionalized indium-tin oxide electrodes. (b) Electrostatic interactions encourage the adsorption of the tetracationic cyclophane **17** on the surface-confined nanoparticles. (c) An additional layer of nanoparticles assembles on the cationic organic coating. Similar composite films can be prepared using the tetracationic [2]catenane **18** instead of the cyclophane **17**. (d) Phosphonate groups can be used to anchor molecular building blocks to titanium dioxide nanoparticles

gold nanoparticles having a diameter of ≈ 13 nm [2.53, 54]. Electrostatic interactions promote the adsorption of the nanoparticles on the organic layer (Fig. 2.17a). The treatment of the composite film with the bipyridinium cyclophane **17** produces an organic layer on the gold nanoparticles (Fig. 2.18b). Following this approach, alternating layers of inorganic nanoparticles and organic building blocks can be assembled on the indium-tin oxide support. Cyclic voltammograms of the resulting materials show the oxidation of the gold nanoparticles and the reduction of the bipyridinium units. The peak current for both processes increases with the number of alternating layers. Comparison of these values indicates that the ratio between the number of tetracationic cyclophanes and that of the nanoparticles is $\approx 100:1$.

The tetracationic cyclophane **17** binds dioxyarenes in solution [2.55, 56]. Attractive supramolecular forces between the electron deficient bipyridinium units and the electron rich guests are responsible for complexation. This recognition motif can be exploited to probe the ability of the composite films in Fig. 2.17b,c to sense electron rich analytes. In particular, hydroquinone is expected to enter the electron deficient cavities of the surface-confined cyclophanes. Cyclic voltammograms consistently reveal the redox waves associated with the reversible oxidation of hydroquinone even when very small amounts of the guest ($\approx 1 \times 10^{-5}$ M) are added to the electrolyte solution [2.53, 54]. No redox response can be detected with a bare indium-tin oxide electrode under otherwise identical conditions. The supramolecular association of the guest and the surface confined cyclophanes increases the local concentration of hydroquinone at the electrode/solution interface enabling its electrochemical detection.

Following a related strategy, the [2]catenane **18** (Fig. 2.17) can be incorporated into similar composite arrays [2.57, 58]. This interlocked molecule incorporates a Ru(II)/trisbipyridine sensitizer and two bipyridinium acceptors. Upon irradiation of the composite material at 440 nm, photoinduced electron transfer from the sensitizer to the appended acceptors occurs. The photogenerated hole in the sensitizer is filled after the transfer of an electron from a sacrificial electron donor present in the electrolyte solution. Under a positive voltage bias applied to the supporting electrode, an electron flow from the bipyridinium acceptors to the indium-tin oxide support is established. The resulting current switches between high and low values as the light source is turned on and off.

Another photoresponsive device, assembled combining inorganic nanoparticles with molecular building blocks, is illustrated in Fig. 2.17d. Phosphonate groups can be used to anchor a Ru(II)/trisbipyridine complex with an appended bipyridinium dication to titanium dioxide nanoparticles deposited on a doped tin oxide electrode [2.59, 60]. The resulting composite array can be integrated in a conventional electrochemical cell filled with an aqueous electrolyte containing triethanolamine. Under a bias voltage of -0.45 V and irradiation at 532 nm, 95% of the excited ruthenium centers transfer electrons to the titanium dioxide nanoparticles. The other 5% donate electrons to the bipyridinium dications. All the electrons transferred to the bipyridinium acceptors return to the ruthenium centers, while only 80% of those accepted by the nanoparticles return to the transition metal complexes. The remaining 15% reach the bipyridinium acceptors, while electron transfer from sacrificial triethanolamine donors fills the photogenerated holes left in the ruthenium sensitizers. The photoinduced reduction of the bipyridinium dication is accompanied by the appearance of the characteristic band of the radical cation in the absorption spectrum. This band persists for hours under open circuit conditions. But it fades in ≈ 15 s under a voltage bias of $+1$ V, as the radical cation is oxidized back to the dicationic form. In summary, an optical stimulation accompanied by a negative voltage bias reduces the bipyridinium building block. The state of the photogenerated form can be read optically, recording the absorption spectrum in the visible region, and erased electrically, applying a positive voltage pulse.

2.3.4 Nanogaps and Nanowires

The operating principles of the electroactive and photoactive devices illustrated in Figs. 2.12–2.17 exploit the ability of small collections of molecular components to manipulate electrons and photons. Designed molecules are deposited on relatively large electrodes and can be addressed electrically and/or optically by controlling the voltage of the support and/or illuminating its surface. The transition from devices relying on collections of molecules to unimolecular devices requires the identification of practical methods to contact single molecules. This fascinating objective demands the rather challenging miniaturization of contacting electrodes to the nanoscale.

A promising approach to unimolecular devices relies on the fabrication of nanometer-sized gaps in metallic features followed by the insertion of individual molecules between the terminals of the gap. This strategy permits the assembly of nanoscaled three-terminal devices equivalent to conventional transistors [2.61–63]. A remarkable example is illustrated in Fig. 2.18a [2.61]. It incorporates a single molecule in the nanogap generated between two gold electrodes. Initially electron beam lithography is used to pattern a gold wire on a doped silicon wafer covered by an insulating silicon dioxide layer. Then the gold feature is broken by electromigration to generate the nanogap. The lateral size of the separated electrodes is ≈ 100 nm and their thickness is ≈ 15 nm. Scanning electron microcopy indicates that the facing surfaces of the separated electrodes are not uniform and that tiny gaps between their protrusions are formed. Current/voltage measurements suggest that the size of the smallest nanogap is ≈ 1 nm. When the breakage of the gold feature is preceded by the deposition of a dilute toluene solution of C_{60} (**19**), junctions with enhanced conduction are obtained. This particular molecule has a diameter of ≈ 0.7 nm and can insert in the nanogap facilitating the flow of electrons across the junction.

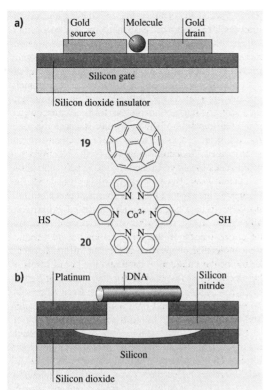

Fig. 2.18 (a) Nanoscaled transistors can be fabricated inserting a single molecule (**19** or **20**) between source and drain electrodes mounted on a silicon/silicon dioxide support. (b) A DNA nanowire can bridge nanoelectrodes suspended above a silicon dioxide support

The unique configuration of the molecule-based device in Fig. 2.18a can reproduce the functions of a conventional transistor [2.19] at the nanoscale. The two gold terminals of the junction are the drain and source of this nanotransistor, and the underlying silicon wafer is the gate. At a temperature of 1.5 K, the junction conductance is very small, when the gate bias is low, and increases in steps at higher voltages [2.61]. The conductance gap is a consequence of the finite energy required to oxidize/reduce the single C_{60} positioned in the junction. It is interesting that the zero-conductance window also changes with the gate voltage and can be opened and closed reversibly adjusting the gate bias.

A similar strategy can be employed to fabricate a nanoscaled transistor incorporating the Co(II) complex **20** shown in Fig. 2.18 [2.63]. In this instance, a silicon dioxide layer with a thickness of $\approx 30\,\text{nm}$ is grown thermally on a doped silicon substrate. Then a gold wire with a width of $\approx 200\,\text{nm}$ and a thicknesses of $\approx 10\text{--}15\,\text{nm}$ is patterned on the silicon dioxide overlayer by electron beam lithography. After extensive washing of the substrate with acetone and methylene chloride and cleaning with oxygen plasma, the gold wire is exposed to a solution of the bisthiol **20**. The formation of thiolate–gold bonds promotes the self-assembly of the molecular building block on the gold surface. At this point, electromigration-induced breakage produces a gap of 1–2 nm in the gold wire. The surface-confined bisthiol **20** is only 0.24 nm long and, therefore, it can insert in the nanogap producing an electrode/molecule/electrode junction.

The cobalt center in **20** can be oxidized/reduced reversibly between Co(II) and Co(III) [2.63]. When this electroactive molecule is inserted in a nanogap (Fig. 2.18a), its ability to accept and donate electrons dictates the current/voltage profile of the resulting electrode/molecule/electrode junction. More precisely, no current flows across the junction below a certain voltage threshold. As the source voltage is raised above this particular value, the drain current increases in steps. The threshold associated with the source voltage varies in magnitude with the gate voltage. This intriguing behavior is a consequence of the finite energy necessary to oxidize/reduce the cobalt center and of a change in the relative stabilities of the oxidized and reduced forms Co(II) and Co(III) with the gate voltage. In summary, the conduction of the electrode/molecule/electrode junction can be tuned adjusting the voltage of the silicon support. The behavior of this molecule-based nanoelectronic device is equivalent to that of a conventional transistor [2.19]. In both instances, the gate voltage regulates the current flowing from the source to the drain.

The electromigration-induced breakage of preformed metallic features successfully produces nanogaps by moving apart two fragments of the same wire. Alternatively, nanogaps can be fabricated reducing the separation of the two terminals of much larger gaps. For example, gold electrodes separated by a distance of 20–80 nm can be patterned on a silicon/silicon dioxide substrate by electron beam lithography [2.64]. The relatively large gap between them can be reduced significantly by the electrochemical deposition of gold on the surfaces of both electrodes. The final result is the fabrication of two nanoelectrodes separated by $\approx 1\,\text{nm}$ and with a radius of curvature of 5–15 nm. The two terminals of this nanogap can be *contacted* by organic nanowires grown between them [2.65]. In particular, the electropolymerization of aniline produces polyaniline bridges between the gold nanoelectrodes. The conductance of the resulting junction can be probed immersing the overall assembly in an electrolyte solution. Employing a bipotentiostat, the bias voltage of the two terminals of the junction can be maintained at 20 mV, while their potentials are scanned relative to that of a silver/silver chloride reference electrode. Below ≈ 0.15 V, the polymer wire is in an insulating state and the current flowing across the junction is less than 0.05 nA. At this voltage threshold, however, the current raises abruptly to ≈ 30 nA. This value corresponds to a conductivity for the polymer nanojunction of $10-100$ S cm^{-1}. When the potential is lowered again below the threshold, the current returns back to very low values. The abrupt decrease in current in the backward scan is observed at a potential that is slightly more negative than that causing the abrupt current increase in the forward scan. In summary, the conductance of this nanoscaled junction switches on and off as a potential input is switched above and below a voltage threshold.

It is interesting to note that the influence of organic bridges on the junction conductance can be exploited for chemical sensing. Nanogaps fabricated following a similar strategy but lacking the polyaniline bridge alter their conduction after exposure to dilute solutions of small organic molecules [2.66]. Indeed, the organic analytes dock into the nanogaps producing a marked decrease in the junction conductance. The magnitude of the conductance drop happens to be proportional to the analyte–nanoelectrode binding strength. Thus the presence of the analyte in solution can be detected probing the current/voltage characteristics of the nanogaps.

Nanogaps between electrodes patterned on silicon/silicon dioxide supports can be bridged also by DNA double strands [2.67, 68]. The device in Fig. 2.18b has a 10.4 nm long poly(G)–poly(C) DNA oligomer suspended between two nanoelectrodes. It can be fabricated patterning a 30 nm wide slit in a silicon nitride overlayer covering a silicon/silicon dioxide support by electron beam evaporation. Underetching the silicon dioxide layer leaves a silicon nitride finger, which can be sputtered with a platinum layer and chopped to leave a nanogap of 8 nm. At this point, a microdroplet of a dilute solution of DNA is deposited on the device and a bias of 5 V is applied between the two electrodes. Electrostatic forces encourage the deposition of a single DNA wire on top of the nanogap. As soon as the nanowire is in position, current starts to flow across the junction. The current/voltage signature of the electrode/DNA/electrode junction shows currents below 1 pA at low voltage biases. Under these conditions, the DNA nanowire is an insulator. Above a certain voltage threshold, however, the nanowire becomes conducting and currents up to 100 nA can flow across the junction through a single nanowire. Assuming that direct tunneling from electrode to electrode is extremely unlikely for a relatively large gap of 8 nm, the intriguing current/voltage behavior has to be a consequence of the participation of the molecular states in the electron transport process. Two possible mechanisms can be envisaged. Sequential hopping of the electrons between states localized in the DNA base pairs can allow the current flow above a certain voltage threshold. But this mechanism would presumably result in a Coulomb blockade voltage gap that is not observed experimentally. More likely, electronic states delocalized across the entire length of the DNA nanowire are producing a molecular conduction band. The off-set between the molecular conduction band and the Fermi levels of the electrodes is responsible for the insulating behavior at low biases. Above a certain voltage threshold, the molecular band and one of the Fermi levels align facilitating the passage of electrons across the junction.

Carbon nanotubes are extremely versatile building blocks for the assembly of nanoscaled electronic devices. They can be used to bridge nanogaps [2.69–72] and assemble nanoscaled cross junctions [2.73–75]. In Fig. 2.19a, a single-wall carbon nanotube crosses over another one in an orthogonal arrangement [2.73]. Both nanotubes have electrical contacts at their ends. The fabrication of this device involves three main steps. First, alignment marks for the electrodes are patterned on a silicon/silicon dioxide support by electron beam lithography. Then the substrate is exposed to a dichloromethane suspension of single-wall SWNT carbon nanotubes. After washing with isopropanol, crosses of carbon nanotubes in an appropriate alignment relative to the electrode marks are identified by tapping mode atomic force microscopy. Finally chromium/gold electrodes are fabricated on top of the nanotube ends, again, by electron beam lithography. The conductance of individual nanotubes can be probed by exploiting the two electric contacts at their ends. These two-terminal measurements reveal that certain nanotubes have metallic behavior, while others are semiconducting. It follows that three distinct types of cross junctions differing in the nature of their constituent nanotubes can be identified on the silicon/silicon dioxide support. Four terminal current/voltage measurements indicate that junctions formed by two metallic nanotubes have high conductance and ohmic behavior.

Similarly, high junction conductance and ohmic behavior is observed when two semiconducting nanotubes cross. The current/voltage signature of junctions formed when a metallic nanotube crosses a semiconducting one are, instead, completely different. The metallic nanotube depletes the semiconducting one at the junction region producing a nanoscaled Schottky barrier with a pronounced rectifying behavior.

Similar fabrication strategies can be exploited to assemble nanoscaled counterparts of conventional transistors. The device in Fig. 2.19b is assembled patterning an aluminum finger on a silicon/silicon dioxide substrate by electron beam lithography [2.75]. After exposure to air, an insulating aluminum oxide layer forms on the aluminum finger. Then a dichloromethane suspension of single-wall carbon nanotubes is deposited on the resulting substrate. Atomic force microscopy can be used to select carbon nanotubes with a diameter of ≈ 1 nm positioned on the aluminum finger. After registering their coordinates relative to alignment markers, gold contacts can be evaporated on their ends by electron beam lithography. The final assembly is a nanoscaled three-terminal device equivalent to a conventional field effect transistor [2.19]. The two gold contacts are the source and drain terminals, while the underlying aluminum finger reproduces the function of the gate. At a source to drain bias of ≈ -1.3 V, the drain current jumps from ≈ 0 to ≈ 50 nA when the gate voltage is lowered from -1.0 to -1.3 V. Thus moderate changes in the gate voltage vary significantly the current flowing through the nanotube-based device in analogy to a conventional enhancement mode p-type field effect transistor [2.19].

The nanoscaled transistor in Fig. 2.18a has a microscaled silicon gate that extends under the entire chip [2.61, 63]. The configuration in Fig. 2.19b, instead, has nanoscaled aluminum gates for every single carbon nanotube transistor fabricated on the same support [2.75]. It follows that multiple nanoscaled transistors can be fabricated on the same chip and operated independently following this strategy. This unique feature offers the possibility of fabricating nanoscaled digital circuits by interconnecting the terminals of independent nanotube transistors. The examples in Fig. 2.19c,d illustrate the configurations of nanoscaled NOT and NOR gates implemented using one or two nanotube transistors. In Fig. 2.19c, an off-chip bias resistor is connected to the drain terminal of a single transistor while the source is grounded. A voltage input applied to the gate modulates the nanotube conductance altering the voltage output probed at the drain termi-

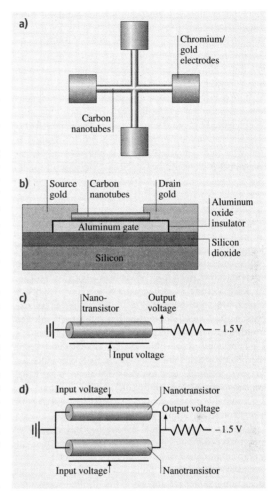

Fig. 2.19 (a) Nanoscaled junctions can be assembled on silicon/silicon dioxide supports crossing pairs of orthogonally arranged single-wall carbon nanotubes with chromium/gold electrical contacts at their ends. (b) Nanotransistors can be fabricated contacting the two ends of a single-wall carbon nanotube deposited on an aluminum/aluminum oxide gate with gold sources and drain. One or two nanotube transistors can be integrated into nanoscaled NOT (c) and NOR (d) logic gates

nal. In particular, a voltage input of -1.5 V lowers the nanotube resistance (26 MΩ) below that of the bias resistor (100 MΩ). As a result, the voltage output drops to 0 V. When the voltage input is raised to 0 V, the nanotube resistance increases above that of the bias resistor and the voltage output becomes -1.5 V. Thus the output of this nanoelectronic device switches from a high (0 V) and to a low (−1.5 V) level as the input shifts from a low (−1.5 V) to a high (0 V) value. The inverse relation between input and output translates into a NOT operation if a negative logic convention (low = 1, high = 0) is applied to both signals.

In Fig. 2.15d, the source terminals of two independent nanotube transistors fabricated on the same chip are connected by a gold wire and grounded [2.75]. Similarly, the two drain terminals are connected by another gold wire and contacted to an off-chip bias resistors. The gate of each nanotube can be stimulated with a voltage input and the voltage output of the device can be probed at their interconnected drain terminals. When the resistance of at least one of the two nanotubes is below that of the resistor, the output is 0 V. When both nanotubes are in a nonconducting mode, the output voltage is −1.5 V. Thus if a low voltage input −1.5 V is applied to one or both transistors, the output is high (0 V). When both voltage inputs are high (0 V), the output is low (−1.5 V). If a negative logic convention (low = 1, high = 0) is applied to all signals, the signal transduction behavior translates in to a NOR operation.

2.4 Conclusions and Outlook

Nature builds nanostructured biomolecules relying on a highly modular approach [2.1]. Small building blocks are connected by robust chemical bonds to generate long strands of repeating units. The synergism of a multitude of attractive supramolecular forces determines the three-dimensional arrangement of the resulting polymeric chains and controls the association of independent strands into single and well-defined entities. Nucleic acids and proteins are two representative classes of biomolecules assembled with subnanometer precision through the subtle interplay of covalent and noncovalent bonds starting from a relatively small pool of nucleotide and amino acid building blocks.

The power of chemical synthesis [2.2] offers the opportunity of mimicking nature's modular approach to nanostructured materials. Following established experimental protocols, small molecular building blocks can be joined together relying on the controlled formation of covalent bonds between designed functional groups. Thus artificial molecules with nanoscaled dimensions can be assembled piece by piece with high structural control. Indeed, helical, tubular, interlocked, and highly branched nanostructures have been all prepared already exploiting this general strategy and the synergism of covalent and noncovalent bonds [2.3].

The chemical construction of nanoscaled molecules from modular building blocks also offers the opportunity for engineering specific properties in the resulting assemblies. In particular, electroactive and photoactive fragments can be integrated into single molecules. The ability of these functional subunits to accept/donate electrons and photons can be exploited to design nanoscaled electronic and photonic devices. Indeed, molecules that respond to electrical and optical stimulations producing detectable outputs have been designed already [2.16]. These chemical systems can be employed to control the interplay of input and output signals at the molecular level. Their conceptual analogy with the signal transduction operated by conventional logic gates in digital circuits is evident. In fact, electroactive and photoactive molecules able to reproduce AND, NOT, and OR operations as well as simple combinational of these basic logic functions are already a reality [2.13, 20, 21].

Most of the molecular switches for digital processing developed so far rely on bulk addressing. In general, relatively large collections of functional molecules are addressed simultaneously in solution. The realization of molecule-based devices with reduced dimensions as well as practical limitations associated with liquid phases in potential applications are encouraging a transition from the solution to the solid state. The general strategy followed so far relies on the deposition of functional molecules on the surfaces of appropriate electrodes following either the Langmuir–Blodgett methodology [2.34] or self-assembly processes [2.35]. The combination of these techniques with the nanofabrication of insulating, metallic, and semiconducting features on appropriate supports has already allowed the realization of fascinating molecule-based devices [2.30–33, 52]. The resulting assemblies integrate inorganic and organic components and, in some instances, even biomolecules to execute specific functions. They can convert optical stimulations into electrical signals. They can execute irreversible and reversible switching operations. They can sense qualitatively and quantitatively specific analytes. They can reproduce the functions of conventional rectifiers and transistors. They can be integrated within functioning nanoelectronic devices capable of simple logic operations.

The remarkable examples of molecule-based materials and devices now available demonstrate the great potential and promise for this research area. At this stage, the only limit left to the design of functional molecules is the imagination of the synthetic chemist. All sort of molecular building blocks with tailored dimensions, shapes, and properties are more or less accessible with the assistance of modern chemical synthesis. Now, the major challenges are (1) to master the operating principles of the molecule-based devices that have been and continue to be assembled and (2) to expand and improve the fabrication strategies available to incorporate molecules into reliable device architectures. As we continue to gather further insights in these directions, design criteria for a wide diversity of molecule-based devices will emerge. It is not unrealistic to foresee the evolution of an entire generation of nanoscaled devices, based on engineered molecular components, that will find applications in a variety of fields ranging from biomedical research to information technology. Perhaps nature can once again illuminate our path, teaching us not only how to synthesize nanostructured molecules but also how to use them. After all, nature is replete with examples of extremely sophisticated molecule-based devices. From tiny bacteria to higher animals, we are all a collection of molecule-based devices.

References

2.1 D. Voet, J.G. Voet: *Biochemistry* (Wiley, New York 1995)

2.2 K.C. Nicolau, E.C. Sorensen: *Classics in Total Synthesis* (VCH, Weinheim 1996)

2.3 J.-M. Lehn: *Supramolecular Chemistry: Concepts and Perspectives* (VCH, Weinheim 1995)

2.4 M.M. Harding, U. Koert, J.-M. Lehn, A. Marquis-Rigault, C. Piguet, J. Siegel: Synthesis of unsubstituted and 4,4′-substituted oligobipyridines as ligand strands for helicate self-assembly, Helv. Chim. Acta **74**, 594–610 (1991)

2.5 J.-M. Lehn, A. Rigault, J. Siegel, B. Harrowfield, B. Chevrier, D. Moras: Spontaneous assembly of double-stranded helicates from oligobipyridine ligands and copper(I) cations: Structure of an inorganic double helix, Proc. Natl. Acad. Sci. USA **84**, 2565–2569 (1987)

2.6 J.-M. Lehn, A. Rigault: Helicates: Tetra- and pentanuclear double helix complexes of Cu(I) and poly(bipyridine) strands, Angew. Chem. Int. Ed. Engl. **27**, 1095–1097 (1988)

2.7 J.D. Hartgerink, J.R. Granja, R.A. Milligan, M.R. Ghadiri: Self-assembling peptide nanotubes, J. Am. Chem. Soc. **118**, 43–50 (1996)

2.8 M.R. Ghadiri, J.R. Granja, R.A. Milligan, D.E. McRee, N. Khazanovich: Self-assembling organic nanotubes based on a cyclic peptide architecture, Nature **366**, 324–327 (1993)

2.9 V. Balzani, A. Credi, F.M. Raymo, J.F. Stoddart: Artificial molecular machines, Angew. Chem. Int. Ed. **39**, 3348–3391 (2000)

2.10 A.J. Bard, L.R. Faulkner: *Electrochemical Methods: Fundamentals and Applications* (Wiley, New York 2000)

2.11 V. Balzani (Ed.): *Electron Transfer in Chemistry* (Wiley-VCH, Weinheim 2001)

2.12 J.D. Coyle: *Principles and Applications of Photochemistry* (Wiley, New York 1988)

2.13 V. Balzani, M. Venturi, A. Credi: *Molecular Devices and Machines* (Wiley-VCH, Weinheim 2003)

2.14 P.R. Ashton, R. Ballardini, V. Balzani, A. Credi, K.R. Dress, E. Ishow, C.J. Kleverlaan, O. Kocian, J.A. Preece, N. Spencer, J.F. Stoddart, M. Venturi, S. Wenger: A photochemically driven molecular-level abacus, Chem. Eur. J. **6**, 3558–3574 (2000)

2.15 M. Irié (Ed.): Photochromism: memories and switches, Chem. Rev. **100**, 1683–1890 (2000)

2.16 B.L. Feringa (Ed.): *Molecular Switches* (Wiley-VCH, Weinheim 2001)

2.17 R.J. Mitchell: *Microprocessor Systems: An Introduction* (Macmillan, London 1995)

2.18 D.R. Smith: *Digital Transmission Systems* (Van Nostrand Reinhold, New York 1993)

2.19 S. Madhu: *Electronics: Circuits and Systems* (SAMS, Indianapolis 1985)

2.20 F.M. Raymo: Digital processing and communication with molecular switches, Adv. Mater. **14**, 401–414 (2002)

2.21 A.P. de Silva: Molecular computation – Molecular logic gets loaded, Nat. Mater. **4**, 15–16 (2005)

2.22 A. Aviram: Molecules for memory, logic and amplification, J. Am. Chem. Soc. **110**, 5687–5692 (1988)

2.23 A.P. de Silva, H.Q.N. Gunaratne, C.P. McCoy: A molecular photoionic AND gate based on fluorescent signaling, Nature **364**, 42–44 (1993)

2.24 M. Asakawa, P.R. Ashton, V. Balzani, A. Credi, G. Mattersteig, O.A. Matthews, M. Montalti, N. Spencer, J.F. Stoddart, M. Venturi: Electrochemically induced molecular motions in pseudorotaxanes: A case of dual-mode (oxidative and reductive) dethreading, Chem. Eur. J. **3**, 1992–1996 (1997)

2.25 F.M. Raymo, S. Giordani, A.J.P. White, D.J. Williams: Digital processing with a three-state molecular switch, J. Org. Chem. **68**, 4158–4169 (2003)

2.26 F.M. Raymo, S. Giordani: Signal communication between molecular switches, Org. Lett. **3**, 3475–3478 (2001)

2.27 F.M. Raymo, S. Giordani: Multichannel digital transmission in an optical network of communicating molecules, J. Am. Chem. Soc. **124**, 2004–2007 (2002)

2.28 F.M. Raymo, S. Giordani: All-optical processing with molecular switches, Proc. Natl. Acad. Sci. USA **99**, 4941–4944 (2002)

2.29 A.J. Bard: *Integrated Chemical Systems: A Chemical Approach to Nanotechnology* (Wiley, New York 1994)

2.30 C. Joachim, J.K. Gimzewski, A. Aviram: Electronics using hybrid-molecular and mono-molecular devices, Nature **408**, 541–548 (2000)

2.31 J.M. Tour: Molecular electronics. Synthesis and testing of components, Acc. Chem. Res. **33**, 791–804 (2000)

2.32 A.R. Pease, J.O. Jeppesen, J.F. Stoddart, Y. Luo, C.P. Collier, J.R. Heath: Switching devices based on interlocked molecules, Acc. Chem. Res. **34**, 433–444 (2001)

2.33 R.M. Metzger: Unimolecular electrical rectifiers, Chem. Rev. **103**, 3803–3834 (2003)

2.34 M.C. Petty: *Langmuir–Blodgett Films: An Introduction* (Cambridge Univ. Press, Cambridge 1996)

2.35 A. Ulman: *An Introduction to Ultrathin Organic Films* (Academic, Boston 1991)

2.36 C. Lee, A.J. Bard: Comparative electrochemical studies of N-methyl-N′-hexadecyl viologen monomolecular films formed by irreversible adsorption and the Langmuir–Blodgett method, J. Electroanal. Chem. **239**, 441–446 (1988)

2.37 C. Lee, A.J. Bard: Cyclic voltammetry and Langmuir film isotherms of mixed monolayers of N-docosoyl-N′-methyl viologen with arachidic acid, Chem. Phys. Lett. **170**, 57–60 (1990)

2.38 M. Fujihira, K. Nishiyama, H. Yamada: Photoelectrochemical responses of optically transparent electrodes modified with Langmuir–Blodgett films consisting of surfactant derivatives of electron donor, acceptor and sensitizer molecules, Thin Solid Films **132**, 77–82 (1985)

2.39 M. Fujihira: Photoelectric conversion with Langmuir–Blodgett films. In: *Nanostructures Based on Molecular Materials*, ed. by W. Göpel, C. Ziegler (VCH, Weinheim 1992) pp. 27–46

2.40 C.P. Collier, E.W. Wong, M. Belohradsky, F.M. Raymo, J.F. Stoddart, P.J. Kuekes, R.S. Williams, J.R. Heath: Electronically configurable molecular-based logic gates, Science **285**, 391–394 (1999)

2.41 E.W. Wong, C.P. Collier, M. Belohradsky, F.M. Raymo, J.F. Stoddart, J.R. Heath: Fabrication and transport properties of single-molecule-thick electrochemical junctions, J. Am. Chem. Soc. **122**, 5831–5840 (2000)

2.42 M. Asakawa, P.R. Ashton, V. Balzani, A. Credi, C. Hamers, G. Mattersteig, M. Montalti, A.N. Shipway, N. Spencer, J.F. Stoddart, M.S. Tolley, M. Venturi, A.J.P. White, D.J. Williams: A chemically and electrochemically switchable [2]catenane incorporating a tetrathiafulvalene unit, Angew. Chem. Int. Ed. **37**, 333–337 (1998)

2.43 V. Balzani, A. Credi, G. Mattersteig, O.A. Matthews, F.M. Raymo, J.F. Stoddart, M. Venturi, A.J.P. White, D.J. Williams: Switching of pseudorotaxanes and catenanes incorporating a tetrathiafulvalene unit by redox and chemical inputs, J. Org. Chem. **65**, 1924–1936 (2000)

2.44 M. Asakawa, M. Higuchi, G. Mattersteig, T. Nakamura, A.R. Pease, F.M. Raymo, T. Shimizu, J.F. Stoddart: Current/voltage characteristics of monolayers of redox-switchable [2]catenanes on gold, Adv. Mater. **12**, 1099–1102 (2000)

2.45 C.P. Collier, G. Mattersteig, E.W. Wong, Y. Luo, K. Beverly, J. Sampaio, F.M. Raymo, J.F. Stoddart, J.R. Heath: A [2]catenane based solid-state electronically reconfigurable switch, Science **289**, 1172–1175 (2000)

2.46 C.P. Collier, J.O. Jeppesen, Y. Luo, J. Perkins, E.W. Wong, J.R. Heath, J.F. Stoddart: Molecular-based electronically switchable tunnel junction devices, J. Am. Chem. Soc. **123**, 12632–12641 (2001)

2.47 J. Chen, M.A. Reed, A.M. Rawlett, J.M. Tour: Large on-off ratios and negative differential resistance in a molecular electronic device, Science **286**, 1550–1552 (1999)

2.48 M.A. Reed, J. Chen, A.M. Rawlett, D.W. Price, J.M. Tour: Molecular random access memory cell, Appl. Phys. Lett. **78**, 3735–3737 (2001)

2.49 D.I. Gittins, D. Bethell, R.J. Nichols, D.J. Schiffrin: Redox-controlled multilayers of discrete gold particles: A novel electroactive nanomaterial, Adv. Mater. **9**, 737–740 (1999)

2.50 D.I. Gittins, D. Bethell, R.J. Nichols, D.J. Schiffrin: Diode-like electron transfer across nanostructured films containing a redox ligand, J. Mater. Chem. **10**, 79–83 (2000)

2.51 D.I. Gittins, D. Bethell, D.J. Schiffrin, R.J. Nichols: A nanometer-scale electronic switch consisting of a metal cluster and redox-addressable groups, Nature **408**, 67–69 (2000)

2.52 A.N. Shipway, M. Lahav, I. Willner: Nanostructured gold colloid electrodes, Adv. Mater. **12**, 993–998 (2000)

2.53 A.N. Shipway, M. Lahav, R. Blonder, I. Willner: Bis-bipyridinium cyclophane receptor–Au nanoparticle superstructure for electrochemical sensing applications, Chem. Mater. **11**, 13–15 (1999)

2.54 M. Lahav, A.N. Shipway, I. Willner, M.B. Nielsen, J.F. Stoddart: An enlarged bis-bipyridinum cyclophane–Au nanoparticle superstructure for selective electrochemical sensing applications, J. Electroanal. Chem. **482**, 217–221 (2000)

2.55 R.E. Gillard, F.M. Raymo, J.F. Stoddart: Controlling self-assembly, Chem. Eur. J. **3**, 1933–1940 (1997)

2.56 F.M. Raymo, J.F. Stoddart: From supramolecular complexes to interlocked molecular compounds, Chemtracts Org. Chem. **11**, 491–511 (1998)

2.57 M. Lahav, T. Gabriel, A.N. Shipway, I. Willner: Assembly of a Zn(II)-porphyrin-bipyridinium dyad and Au-nanoparticle superstructures on conductive surfaces, J. Am. Chem. Soc. **121**, 258–259 (1999)

2.58 M. Lahav, V. Heleg-Shabtai, J. Wasserman, E. Katz, I. Willner, H. Durr, Y. Hu, S.H. Bossmann: Photoelectrochemistry with integrated photosensitizer-electron acceptor Au-nanoparticle arrays, J. Am. Chem. Soc. **122**, 11480–11487 (2000)

2.59 G. Will, S.N. Rao, D. Fitzmaurice: Heterosupramolecular optical write-read-erase device, J. Mater. Chem. **9**, 2297–2299 (1999)

2.60 A. Merrins, C. Kleverlann, G. Will, S.N. Rao, F. Scandola, D. Fitzmaurice: Time-resolved optical spectroscopy of heterosupramolecular assemblies based on nanostructured TiO_2 films modified by chemisorption of covalently linked ruthenium and viologen complex components, J. Phys. Chem. B **105**, 2998–3004 (2001)

2.61 H. Park, J. Park, A.K.L. Lim, E.H. Anderson, A.P. Alivisatos, P.L. McEuen: Nanomechanical oscillations in a single C_{60} transistor, Nature **407**, 57–60 (2000)

2.62 W. Liang, M.P. Shores, M. Bockrath, J.R. Long, H. Park: Kondo resonance in a single-molecule transistor, Nature **417**, 725–729 (2002)

2.63 J. Park, A.N. Pasupathy, J.I. Goldsmith, C. Chang, Y. Yaish, J.R. Petta, M. Rinkoski, J.P. Sethna, H.D. Abruna, P.L. McEuen, D.C. Ralph: Coulomb blockade and the Kondo effect in single-atom transistors, Nature **417**, 722–725 (2002)

2.64 C.Z. Li, H.X. He, N.J. Tao: Quantized tunneling current in the metallic nanogaps formed by electrodeposition and etching, Appl. Phys. Lett. **77**, 3995–3997 (2000)

2.65 H. He, J. Zhu, N.J. Tao, L.A. Nagahara, I. Amlani, R. Tsui: A conducting polymer nanojunction switch, J. Am. Chem. Soc. **123**, 7730–7731 (2001)

2.66 A. Bogozi, O. Lam, H. He, C. Li, N.J. Tao, L.A. Nagahara, I. Amlani, R. Tsui: Molecular adsorption onto metallic quantum wires, J. Am. Chem. Soc. **123**, 4585–4590 (2001)

2.67 A. Bezryadin, C.N. Lau, M. Tinkham: Quantum suppression of superconductivity in ultrathin nanowires, Nature **404**, 971–974 (2000)

2.68 D. Porath, A. Bezryadin, S. de Vries, C. Dekker: Direct measurement of electrical transport through DNA molecules, Nature **403**, 635–638 (2000)

2.69 S.J. Tans, M.H. Devoret, H. Dai, A. Thess, E.E. Smalley, L.J. Geerligs, C. Dekker: Individual single-wall carbon nanotubes as quantum wires, Nature **386**, 474–477 (1997)

2.70 A.F. Morpurgo, J. Kong, C.M. Marcus, H. Dai: Gate-controlled superconducting proximity effect in carbon nanotubes, Nature **286**, 263–265 (1999)

2.71 J. Nygård, D.H. Cobden, P.E. Lindelof: Kondo physics in carbon nanotubes, Nature **408**, 342–346 (2000)

2.72 W. Liang, M. Bockrath, D. Bozovic, J.H. Hafner, M. Tinkham, H. Park: Fabry–Perot interference in a nanotube electron waveguide, Nature **411**, 665–669 (2001)

2.73 M.S. Fuhrer, J. Nygård, L. Shih, M. Forero, Y.-G. Yoon, M.S.C. Mazzoni, H.J. Choi, J. Ihm, S.G. Louie, A. Zettl, P.L. McEuen: Crossed nanotube junctions, Science **288**, 494–497 (2000)

2.74 T. Rueckes, K. Kim, E. Joselevich, G.Y. Tseng, C.-L. Cheung, C.M. Lieber: Carbon nanotube-based nonvolatile random access memory for molecular computing, Science **289**, 94–97 (2000)

2.75 A. Bachtold, P. Hadley, T. Nakanishi, C. Dekker: Logic circuits with carbon nanotube transistors, Science **294**, 1317–1320 (2001)

3. Introduction to Carbon Nanotubes

Marc Monthioux, Philippe Serp, Emmanuel Flahaut, Manitra Razafinimanana, Christophe Laurent, Alain Peigney, Wolfgang Bacsa, Jean-Marc Broto

Carbon nanotubes are remarkable objects that look set to revolutionize the technological landscape in the near future. Tomorrow's society will be shaped by nanotube applications, just as silicon-based technologies dominate society today. Space elevators tethered by the strongest of cables; hydrogen-powered vehicles; artificial muscles: these are just a few of the technological marvels that may be made possible by the emerging science of carbon nanotubes.

Of course, this prediction is still some way from becoming reality; we are still at the stage of evaluating possibilities and potential. Consider the recent example of fullerenes – molecules closely related to nanotubes. The anticipation surrounding these molecules, first reported in 1985, resulted in the bestowment of a Nobel Prize for their discovery in 1996. However, a decade later, few applications of fullerenes have reached the market, suggesting that similarly enthusiastic predictions about nanotubes should be approached with caution.

There is no denying, however, that the expectations surrounding carbon nanotubes are very high. One of the main reasons for this is the anticipated application of nanotubes to electronics. Many believe that current techniques for miniaturizing microchips are about to reach their lowest limits, and that nanotube-based technologies are the best hope for further miniaturization. Carbon nanotubes may therefore provide the building blocks for further technological progress, enhancing our standards of living.

In this chapter, we first describe the structures, syntheses, growth mechanisms and properties of carbon nanotubes. Then we discuss nanotube-related nano-objects, including those formed by reactions and associations of all-carbon nanotubes with foreign atoms, molecules and compounds, which may provide the path to hybrid materials with even better properties than *pristine* nanotubes. Finally, we will describe the most important current and potential applications of carbon nanotubes, which suggest that the future for the carbon nanotube industry looks very promising indeed.

3.1	**Structure of Carbon Nanotubes**	48
	3.1.1 Single-Wall Nanotubes	48
	3.1.2 Multiwall Nanotubes	51
3.2	**Synthesis of Carbon Nanotubes**	53
	3.2.1 Solid Carbon Source-Based Production Techniques for Carbon Nanotubes	53
	3.2.2 Gaseous Carbon Source-Based Production Techniques for Carbon Nanotubes	62
	3.2.3 Miscellaneous Techniques	68
	3.2.4 Synthesis of Carbon Nanotubes with Controlled Orientation	68
3.3	**Growth Mechanisms of Carbon Nanotubes**	70
	3.3.1 Catalyst-Free Growth	71
	3.3.2 Catalytically Activated Growth	71
3.4	**Properties of Carbon Nanotubes**	74
	3.4.1 Overview	74
	3.4.2 General Properties of SWNTs	75
	3.4.3 Adsorption Properties of SWNTs	75
	3.4.4 Electronic and Optical Properties	77
	3.4.5 Mechanical Properties	79
	3.4.6 Reactivity	79
3.5	**Carbon Nanotube-Based Nano-Objects**	80
	3.5.1 Heteronanotubes	80
	3.5.2 Hybrid Carbon Nanotubes	80
	3.5.3 Functionalized Nanotubes	84

3.6 **Applications of Carbon Nanotubes**.......... 85
    3.6.1 Current Applications 86
    3.6.2 Expected Applications
         Related to Adsorption 90
    3.6.3 Expected Applications
         Related to Composite Systems 93

3.7 **Toxicity and Environmental Impact
of Carbon Nanotubes** 99

3.8 **Concluding Remarks** 100

References 101

Carbon nanotubes have long been synthesized as products of the action of a catalyst on the gaseous species originating from the thermal decomposition of hydrocarbons (Sect. 3.2) [3.1]. The first evidence that the nanofilaments produced in this way were actually nanotubes – that they exhibited an inner cavity – can be found in the transmission electron microscope micrographs published by *Radushkevich* and *Lukyanovich* in 1952 [3.2]. This was of course related to and made possible by the progress in transmission electron microscopy. It is then likely that the carbon filaments prepared by *Hughes* and *Chambers* in 1889 [3.3], which is probably the first patent ever deposited in the field, and whose preparation method was also based on the catalytically enhanced thermal cracking of hydrocarbons, were already carbon nanotube-related morphologies. The preparation of vapor-grown carbon fibers was actually reported over a century ago [3.4, 5]. Since then, the interest in carbon nanofilaments/nanotubes has been recurrent, though within a scientific area almost limited to the carbon material scientist community. The reader is invited to consult the review published by *Baker* and *Harris* [3.6] regarding the early works. Worldwide enthusiasm came unexpectedly in 1991, after the catalyst-free formation of nearly perfect concentric multiwall carbon nanotubes (c-MWNTs, Sect. 3.1) was reported [3.7] as by-products of the formation of fullerenes via the electric-arc technique. But the real breakthrough occurred two years later, when attempts to fill the nanotubes in situ with various metals (Sect. 3.5) led to the discovery – again unexpected – of single-wall carbon nanotubes (SWNTs) simultaneously by *Iijima* and *Ichihashi* [3.8] and *Bethune* et al. [3.9]. Single-wall carbon nanotubes were really new nano-objects with properties and behaviors that are often quite specific (Sect. 3.4). They are also beautiful objects for fundamental physics as well as unique molecules for experimental chemistry, although they are still somewhat mysterious since their formation mechanisms are the subject of controversy and are still debated (Sect. 3.3). Potential applications seem countless, although few have reached marketable status so far (Sect. 3.6). Consequently, about five papers a day are currently published by research teams from around the world with carbon nanotubes as the main topic, an illustration of how extraordinarily active – and highly competitive – this field of research is. It is an unusual situation, similar to that for fullerenes, which, by the way, are again carbon nano-objects structurally closely related to nanotubes.

This is not, however, only about scientific exaltation. Economic aspects are leading the game to a greater and greater extent. According to experts, the world market was estimated to be more than 430 million dollars in 2004 and it is predicted to grow to several billion dollars before 2009. That is serious business, and it will be closely related to how scientists and engineers deal with the many challenges found on the path from the beautiful, ideal molecule to the reliable – and it is hoped, cheap – manufactured product.

3.1 Structure of Carbon Nanotubes

It is relatively easy to imagine a single-wall carbon nanotube (SWNT). Ideally, it is enough to consider a perfect graphene sheet (graphene is a polyaromatic monoatomic layer consisting of sp^2-hybridized carbon atoms arranged in hexagons; genuine graphite consists of layers of this graphene) and to roll it into a cylinder (Fig. 3.1), making sure that the hexagonal rings placed in contact join coherently. Then the tips of the tube are sealed by two caps, each cap being a hemi-fullerene of the appropriate diameter (Fig. 3.2a–c).

3.1.1 Single-Wall Nanotubes

Geometrically, there is no restriction on the tube diameter. However, calculations have shown that collapsing the single-wall tube into a flattened two-layer ribbon is energetically more favorable than maintaining the tubular morphology beyond a diameter value of ≈ 2.5 nm [3.10]. On the other hand, it is easy to grasp intuitively that the shorter the radius of curvature, the higher the stress and the energetic cost, although

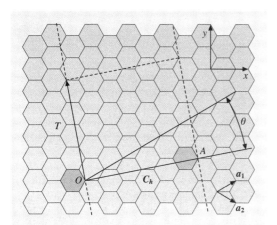

Fig. 3.1 Sketch of the way to make a single-wall carbon nanotube, starting from a graphene sheet (adapted from [3.12])

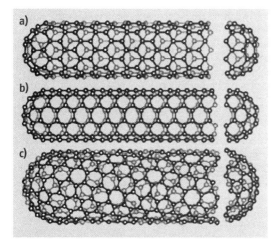

Fig. 3.2a–c Sketches of three different SWNT structures that are examples of (**a**) a zigzag-type nanotube, (**b**) an armchair-type nanotube, (**c**) a helical nanotube (adapted from [3.13])

SWNTs with diameters as low as 0.4 nm have been synthesized successfully [3.11]. A suitable energetic compromise is therefore reached for ≈ 1.4 nm, the most frequent diameter encountered regardless of the synthesis technique (at least for those based on solid carbon sources) when conditions ensuring high SWNT yields are used. There is no such restriction on the nanotube length, which only depends on the limitations of the preparation method and the specific conditions used for the synthesis (thermal gradients, residence time, and so on). Experimental data are consistent with these statements, since SWNTs wider than 2.5 nm are only rarely reported in the literature, whatever the preparation method, while the length of the SWNTs can be in the micrometer or the millimeter range. These features make single-wall carbon nanotubes a unique example of single molecules with huge aspect ratios.

Two important consequences derive from the SWNT structure as described above:

1. All carbon atoms are involved in hexagonal aromatic rings only and are therefore in equivalent positions, except at each nanotube tip, where 6 × 5 = 30 atoms are involved in pentagonal rings (considering that adjacent pentagons are unlikely) – though not more, not less, as a consequence of Euler's rule that also governs the fullerene structure. For ideal SWNTs, chemical reactivity will therefore be highly favored at the tube tips, at the locations of the pentagonal rings.

2. Although carbon atoms are involved in aromatic rings, the C=C bond angles are not planar. This means that the hybridization of carbon atoms is not pure sp^2; it has some degree of the sp^3 character, in a proportion that increases as the tube radius of curvature decreases. The effect is the same as for the C_{60} fullerene molecules, whose radius of curvature is 0.35 nm, and whose bonds therefore have 10% sp^3 character [3.14]. On the one hand, this is believed to make the SWNT surface a bit more reactive than regular, planar graphene, even though it still consists of aromatic ring faces. On the other hand, this somehow induces variable overlapping of energy bands, resulting in unique and versatile electronic behavior (Sect. 3.4).

As illustrated by Fig. 3.2, there are many ways to roll a graphene into a single-wall nanotube, with some of the resulting nanotubes possessing planes of symmetry both parallel and perpendicular to the nanotube axis (such as the SWNTs from Fig. 3.2a,b), while others do not (such as the SWNT from Fig. 3.2c). Similar to the terms used for molecules, the latter are commonly called *chiral* nanotubes, since they are unable to be superimposed on their own image in a mirror. *Helical* is however sometimes preferred (see below). The various ways to roll graphene into tubes are therefore mathematically defined by the vector of helicity C_h, and the angle of helicity θ, as follows (referring to Fig. 3.1)

$$OA = C_h = na_1 + ma_2$$

with

$$a_1 = \frac{a\sqrt{3}}{2}x + \frac{a}{2}y \quad \text{and} \quad a_2 = \frac{a\sqrt{3}}{2}x - \frac{a}{2}y,$$

where $a = 2.46\,\text{Å}$

and

$$\cos\theta = \frac{2n+m}{2\sqrt{n^2+m^2+nm}},$$

where n and m are the integers of the vector OA considering the unit vectors a_1 and a_2.

The vector of helicity $C_h (= OA)$ is perpendicular to the tube axis, while the angle of helicity θ is taken with respect to the so-called zigzag axis: the vector of helicity that results in nanotubes of the *zigzag* type (see below). The diameter D of the corresponding nanotube is related to C_h by the relation

$$D = \frac{|C_h|}{\pi} = \frac{a_{CC}\sqrt{3(n^2+m^2+nm)}}{\pi},$$

where

$$\underset{\text{(graphite)}}{1.41\,\text{Å}} \leq a_{C=C} \leq \underset{(C_{60})}{1.44\,\text{Å}}.$$

The C–C bond length is actually elongated by the curvature imposed by the structure; the average bond length in the C_{60} fullerene molecule is a reasonable upper limit, while the bond length in flat graphene in genuine graphite is the lower limit (corresponding to an infinite radius of curvature). Since C_h, θ, and D are all expressed as a function of the integers n and m, they are sufficient to define any particular SWNT by denoting them (n, m). The values of n and m for a given SWNT can be simply obtained by counting the number of hexagons that separate the extremities of the C_h vector following the unit vector a_1 first and then a_2 [3.12]. In the example of Fig. 3.1, the SWNT that is obtained by rolling the graphene so that the two shaded aromatic cycles can be superimposed exactly is a $(4, 2)$ chiral

Fig. 3.3 Image of two neighboring chiral SWNTs within a SWNT bundle as seen using high-resolution scanning tunneling microscopy (courtesy of Prof. Yazdani, University of Illinois at Urbana, USA)

nanotube. Similarly, SWNTs from Fig. 3.2a–c are $(9, 0)$, $(5, 5)$, and $(10, 5)$ nanotubes respectively, thereby providing examples of zigzag-type SWNT (with an angle of helicity $= 0°$), armchair-type SWNT (with an angle of helicity of $30°$) and a chiral SWNT, respectively. This also illustrates why the term *chiral* is sometimes inappropriate and should preferably be replaced with *helical*. Armchair (n, n) nanotubes, although definitely achiral from the standpoint of symmetry, exhibit a nonzero *chiral angle*. Zigzag and *armchair* qualifications for achiral nanotubes refer to the way that the carbon atoms are displayed at the edge of the nanotube cross section (Fig. 3.2a,b). Generally speaking, it is clear from Figs. 3.1 and 3.2a that having the vector of helicity perpendicular to any of the three overall C=C bond directions will provide zigzag-type SWNTs, denoted $(n, 0)$, while having the vector of helicity parallel to one of the three C=C bond directions will provide armchair-type SWNTs, denoted (n, n). On the other hand, because of the sixfold symmetry of the graphene sheet, the angle of helicity θ for the chiral (n, m) nanotubes is such that $0 < \theta < 30°$. Figure 3.3 provides two examples of what chiral SWNTs look like, as seen via atomic force microscopy.

The graphenes in graphite have π electrons which are accommodated by the stacking of graphenes, al-

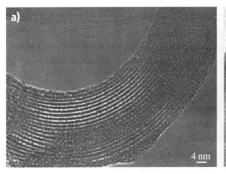

Fig. 3.4a,b High-resolution transmission electron microscopy images of a SWNT rope. (**a**) Longitudinal view. An isolated single SWNT also appears at the top of the image. (**b**) Cross-sectional view (from [3.15])

lowing van der Waals forces to develop. Similar reasons make fullerenes gather and order into fullerite crystals and SWNTs into SWNT ropes (Fig. 3.4a). Provided the SWNT diameter distribution is narrow, the SWNTs in ropes tend to spontaneously arrange into hexagonal arrays, which correspond to the highest compactness achievable (Fig. 3.4b). This feature brings new periodicities with respect to graphite or turbostratic polyaromatic carbon crystals. Turbostratic structure corresponds to graphenes that are stacked with random rotations or translations instead of being piled up following sequential $ABAB$ positions, as in graphite structure. This implies that no lattice atom plane exists other than the graphene planes themselves (corresponding to the (001) atom plane family). These new periodicities give specific diffraction patterns that are quite different to those of other sp^2-carbon-based crystals, although hk reflections, which account for the hexagonal symmetry of the graphene plane, are still present. On the other hand, $00l$ reflections, which account for the stacking sequence of graphenes in regular, *multilayered* polyaromatic crystals (which do not exist in SWNT ropes) are absent. This hexagonal packing of SWNTs within the ropes requires that SWNTs exhibit similar diameters, which is the usual case for SWNTs prepared by electric arc or laser vaporization processes. SWNTs prepared using these methods are actually about 1.35 nm wide (diameter of a (10, 10) tube, among others), for reasons that are still unclear but are related to the growth mechanisms specific to the conditions provided by these techniques (Sect. 3.3).

3.1.2 Multiwall Nanotubes

Building multiwall carbon nanotubes is a little bit more complex, since it involves the various ways graphenes can be displayed and mutually arranged within filamentary morphology. A similar versatility can be expected to the usual textural versatility of polyaromatic solids. Likewise, their diffraction patterns are difficult to differentiate from those of anisotropic polyaromatic solids. The easiest MWNT to imagine is the concentric type (c-MWNT), in which SWNTs with regularly increasing diameters are coaxially arranged (according to a Russian-doll model) into a multiwall nanotube (Fig. 3.5). Such nanotubes are generally formed either by the electric arc technique (without the need for a catalyst), by catalyst-enhanced thermal cracking of gaseous hydrocarbons, or by CO disproportionation (Sect. 3.2). There can be any number of walls (or coaxial tubes), from two upwards. The intertube distance is approximately the same as the intergraphene distance in turbostratic, polyaromatic solids, 0.34 nm (as opposed to 0.335 nm in genuine graphite), since the increasing radius of curvature imposed on the concentric graphenes prevents the carbon atoms from being arranged as in graphite, with each of the carbon atoms from a graphene facing either a ring center or a carbon atom from the neighboring graphene. However, two cases allow a nanotube to reach – totally or partially – the 3-D crystal periodicity of graphite. One is to consider a high number of concentric graphenes: concentric graphenes with a long radius of curvature. In this case, the shift in the relative positions of carbon atoms from superimposed graphenes is so small with respect to that in graphite that some commensurability is possible.

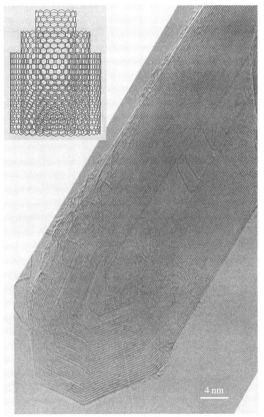

Fig. 3.5 High-resolution transmission electron microscopy image (longitudinal view) of a concentric multiwall carbon nanotube (c-MWNT) prepared using an electric arc. The *insert* shows a sketch of the Russian doll-like arrangement of graphenes

This may result in MWNTs where both structures are associated; in other words they have turbostratic cores and graphitic outer parts [3.16]. The other case occurs for c-MWNTs exhibiting faceted morphologies, originating either from the synthesis process or more likely from subsequent heat treatment at high temperature (such as 2500 °C) in inert atmosphere. Facets allow the graphenes to resume a flat arrangement of atoms (except at the junction between neighboring facets) which allows the specific stacking sequence of graphite to develop.

Another frequent inner texture for multiwall carbon nanotubes is the so-called herringbone texture (h-MWNTs), in which the graphenes make an angle with respect to the nanotube axis (Fig. 3.6). The angle value varies upon the processing conditions (such as the catalyst morphology or the composition of the atmosphere), from 0 (in which case the texture becomes that of a c-MWNT) to 90° (in which case the filament is no longer a tube, see below), and the inner diameter varies so that the tubular arrangement can be

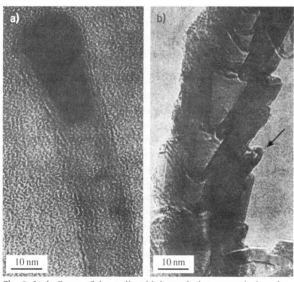

Fig. 3.6a,b Some of the earliest high-resolution transmission electron microscopy images of a herringbone (and bamboo) multiwall nanotube (bh-MWNT, longitudinal view) prepared by CO disproportionation on Fe-Co catalyst. (a) As-grown. The nanotube surface is made of free graphene edges. (b) After 2900 °C heat treatment. Both the herringbone and the bamboo textures have become obvious. Graphene edges from the surface have buckled with their neighbors (*arrow*), closing off access to the intergraphene space (adapted from [3.17])

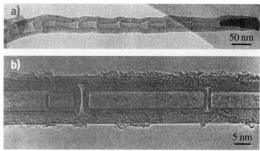

Fig. 3.7a,b Transmission electron microscopy images from bamboo multiwall nanotubes (longitudinal views). (a) Low magnification of a bamboo-herringbone multiwall nanotube (bh-MWNT) showing the nearly periodic nature of the texture, which occurs very frequently. (from [3.18]); (b) high-resolution image of a bamboo-concentric multiwall nanotube (bc-MWNT) (modified from [3.19])

lost [3.20], meaning that the latter are more accurately called nanofibers rather than nanotubes. h-MWNTs are exclusively obtained by processes involving catalysts, generally catalyst-enhanced thermal cracking of hydrocarbons or CO disproportionation. One long-time debated question was whether the herringbone texture, which actually describes the texture projection rather than the overall three-dimensional texture, originates from the scrolllike spiral arrangement of a single graphene ribbon or from the stacking of independent truncated conelike graphenes in what is also called a *cup-stack* texture. It is now demonstrated that both exist [3.21, 22].

Another common feature is the occurrence, to some degree, of a limited amount of graphenes oriented perpendicular to the nanotube axis, thus forming a *bamboo* texture. This is not a texture that can exist on its own; it affect either the c-MWNT (bc-MWNT) or the h-MWNT (bh-MWNT) textures (Figs. 3.6 and 3.7). The question is whether such filaments, although hollow, should still be called nanotubes, since the inner cavity is no longer open all the way along the filament as it is for a genuine tube. These are therefore sometimes referred as *nanofibers* in the literature too.

One nanofilament that definitely cannot be called a nanotube is built from graphenes oriented perpendicular to the filament axis and stacked as piled-up plates. Although these nanofilaments actually correspond to h-MWNTs with a graphene/MWNT axis angle of 90°, an inner cavity is no longer possible, and such filaments are therefore often referred to as *platelet nanofibers* in the literature [3.20].

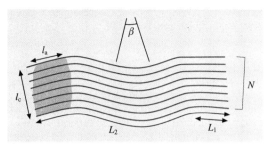

Fig. 3.8 Sketch explaining the various parameters obtained from high-resolution (lattice fringe mode) transmission electron microscopy, used to quantify nanotexture: L_1 is the average length of perfect (distortion-free) graphenes of coherent areas; N is the number of piled-up graphenes in coherent (distortion-free) areas; L_2 is the average length of continuous though distorted graphenes within graphene stacks; β is the average distortion angle. L_1 and N are related to the l_a and l_c values obtained from x-ray diffraction

Unlike SWNTs, whose aspect ratios are so high that it is almost impossible to find the tube tips, the aspect ratios for MWNTs (and carbon nanofibers) are generally lower and often allow one to image tube ends by transmission electron microscopy. Aside from c-MWNTs derived from electric arc (Fig. 3.5), which grow in a catalyst-free process, nanotube tips are frequently found to be associated with the catalyst crystals from which they were formed.

The properties of the MWNT (Sect. 3.4) will obviously largely depend on the perfection and the orientation of the graphenes in the tube (for example, the spiral angles of the nanotubes constituting c-MWNTs has little importance). Graphene orientation is a matter of texture, as described above. Graphene perfection is a matter of nanotexture, which is commonly used to describe other polyaromatic carbon materials, and which is quantified by several parameters preferably obtained from high-resolution transmission electron microscopy (Fig. 3.8). Both texture and nanotexture depend on the processing conditions. While the texture type is a permanent, intrinsic feature which can only be completely altered upon a severe degradation treatment (such as oxidation), the nanotexture can be improved by subsequent thermal treatments at high temperatures (such as $> 2000\,°C$) and potentially degraded by chemical treatments (such as slightly oxidizing conditions).

3.2 Synthesis of Carbon Nanotubes

Producing carbon nanotubes so that the currently planned applications currently planned become marketable will require solving some problems that are more or less restrictive depending on the case. Examples include specifically controlling the configuration (chirality), the purity, or the structural quality of SWNTs, and adapting the production capacity to the application. One objective would be to understand the mechanism of nanotube nucleation and growth perfectly, and this remains a controversial subject despite an intense, worldwide experimental effort. This problem is partly due to our lack of knowledge regarding several parameters controlling the conditions during synthesis. For instance, the exact and accurate role of the catalysts in nanotube growth is often unknown. Given the large number of experimental parameters and considering the large range of conditions that the synthesis techniques correspond to, it is quite legitimate to think of more than one mechanism intervening during nanotube formation.

3.2.1 Solid Carbon Source-Based Production Techniques for Carbon Nanotubes

Among the different SWNT production techniques, the four processes (laser ablation, solar energy, dc electric arc, and three-phase ac arc plasma) presented in this section have at least two points in common: a high-temperature ($1000\,\text{K} < T < 6000\,\text{K}$) medium and the fact that the carbon source originates from the erosion of solid graphite. Despite these common points, the morphologies of the carbon nanostructures and the SWNT yields can differ notably with respect to the experimental conditions.

Before being utilized for carbon nanotube synthesis, these techniques permitted the production of fullerenes. Laser vaporization of graphite was actually the very first method to demonstrate the existence of fullerenes, including the most common one (because it is the most stable and therefore the most abundant), C_{60} [3.23]. On the other hand, the electric arc technique was (and still is) the first method of producing fullerenes in relatively large quantities [3.24–26]. Unlike fullerene formation, which requires the presence of carbon atoms in high-temperature media and the absence of oxygen, the utilization of these techniques for the synthesis of nanotubes (of SWNT type at least) requires an additional condition: the presence of catalysts in either the electrode or the target.

The different mechanisms (such as carbon molecule dissociation and atom recombination processes) involved in these high-temperature techniques take place at different time scales, from nanoseconds to microseconds and even milliseconds. The formation of nanotubes and other graphene-based products occurs afterward with a relatively long delay.

The methods of laser ablation, solar energy, and electric arc are all based on one essential mechanism: the energy transfer resulting from the interaction between either the target material and an external radiation source (a laser beam or radiation emanating from solar energy) or the electrode and the plasma (in case of an electric arc). This interaction causes target or anode erosion, leading to the formation of a plasma: an electrically neutral ionized gas, composed of neutral atoms, charged particles (molecules and ionized species) and electrons. The ionization degree of this plasma, defined by the ratio $(n_e/(n_e + n_o))$, where n_e and n_o are the electron and that of neutral atom densities respectively, highlights the importance of energy transfer between the plasma and the material. The characteristics of this plasma and notably the ranges in temperature and concentrations of the various species present in the plasma thereby depend not only on the nature and composition of the target or the electrode but also on the energy transferred.

One of the advantages of these synthesis techniques is the ability to vary a large number of parameters that modify the composition of the high-temperature medium and consequently allow the most relevant parameters to be determined so that the optimal conditions for the control of carbon nanotube formation can be obtained. However, a major drawback of these techniques – and of any other technique used to produce SWNTs – is that the SWNTs formed are not pure: they are associated with other carbon phases and remnants of the catalyst. Although purification processes have been proposed in the literature and by some commercial companies for removing these undesirable phases, they are all based on oxidation (such as acid-based) processes that are likely to significantly affect the SWNT structure [3.15]. Subsequent thermal treatments at $\approx 1200\,°C$ under inert atmosphere, however, succeed in recovering structural quality somewhat [3.29].

Laser Ablation

After the first laser was built in 1960, physicists immediately made use of it as a means of concentrating a large quantity of energy inside a very small volume within a relatively short time. The consequence of this energy input naturally depends upon the characteristics of the device employed. During the interaction between the laser beam and the material, numerous phenomena occur at the same time and/or follow each other within the a certain time period, and each of these processes are sensitive to different parameters such as the characteristics of the laser beam, the incoming power density (also termed the *fluence*), the nature of the target, and the environment surrounding it. For instance, the solid target can merely heat up, melt or vaporize depending on the power provided.

While this technique was successfully used to synthesize fullerene-related structures for the very first time [3.23], the synthesis of SWNTs by laser ablation took another ten years of research [3.27].

Laser Ablation – Experimental Devices

Two types of laser devices are currently utilized for carbon nanotube production: lasers operating in pulsed mode and lasers operating in continuous mode, with the latter generally providing a smaller fluence.

An example of the layout of a laser ablation device is given in Fig. 3.9. A graphite pellet containing the catalyst is placed in the middle of a quartz tube filled with inert gas and placed in an oven maintained at a temperature of $1200\,°C$ [3.27, 28]. The energy of the laser beam focused on the pellet permits it to vaporize and sublime the graphite by uniformly bombarding its surface. The carbon species, swept along by a flow of neutral gas, are then deposited as soot in different regions: on the con-

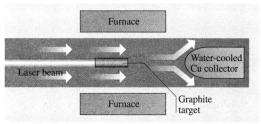

Fig. 3.9 Sketch of an early laser vaporization apparatus (adapted from [3.27, 28])

ical water-cooled copper collector, on the quartz tube walls, and on the backside of the pellet.

Various improvements have been made to this device in order to increase the production efficiency. For example, *Thess* et al. [3.31] employed a second pulsed laser that follows the initial impulsion but at a different frequency in order to ensure a more complete and efficient irradiation of the pellet. This second impulsion vaporizes the coarse aggregates issued from the first ablation, causing them to participate in the active carbon feedstock involved in nanotube growth. Other modifications were suggested by *Rinzler* et al. [3.29], who inserted a second quartz tube of a smaller diameter coaxially inside the first one. This second tube reduces the vaporization zone and so permits an increased amounts of sublimed carbon to be obtained. They also arranged the graphite pellet on a revolving system so that the laser beam uniformly scans its whole surface.

Other groups have realized that, where the target contains both the catalyst and the graphite, the latter evaporates first and the pellet surface becomes more and more metal-rich, resulting in a decrease in the efficiency of nanotube formation during the course of the process. To solve this problem, *Yudasaka* et al. [3.32] utilized two pellets facing each other, one made entirely from the graphite powder and the other from an alloy of transition metals (catalysts), and irradiated them simultaneously.

A sketch of a synthesis reactor based on the vaporization of a target at a fixed temperature by a continuous CO_2 laser beam ($\lambda = 10.6\,\mu m$) is shown in Fig. 3.10 [3.30]. The power can be varied from 100 to 1600 W. The temperature of the target is measured with an optical pyrometer, and these measurements are used to regulate the laser power to maintain a constant vaporization temperature. The gas, heated by contact with the target, acts as a local furnace and creates an extended hot zone, making an external furnace unnecessary. The gas is extracted through a silica pipe, and the solid products formed are carried away by the gas flow through the pipe and then collected on a filter. The synthesis yield is controlled by three parameters: the cooling rate of the medium where the active, secondary catalyst particles are formed, the residence time, and the temperature (in the range 1000–2100 K) at which SWNTs nucleate and grow [3.33].

However, devices equipped with facilities to gather data such as the target temperature in situ are scarce and, generally speaking, this is one of the numerous variables of the laser ablation synthesis technique. The parameters that have been studied the most are the nature of the target, the nature and concentration of the catalyst, the nature of the neutral gas flow, and the temperature of the outer oven.

Laser Ablation – Results

In the absence of catalysts in the target, the soot collected mainly contains multiwall nanotubes (c-MWNTs). Their lengths can reach 300 nm. Their quantity and structural quality are dependent on the oven temperature. The best quality is obtained for an oven temperature set at 1200 °C. At lower oven temperatures, the structural quality decreases, and the nanotubes start presenting many defects [3.27]. As soon as small quantities (a few percent or less) of transition metal (Ni, Co) catalysts are incorporated into the graphite pellet, the products yielded undergo significant modifications, and SWNTs are formed instead of MWNTs. The yield of SWNTs strongly depends on the type of metal catalyst used and is seen to increase with the furnace temperature, among other factors. The SWNTs have remarkably uniform diameters and they self-organize into ropelike crystallites 5–20 nm in diameter and tens to hundreds of micrometers in length (Fig. 3.11). The ends of all of the SWNTs appear to be perfectly closed with hemispherical end-caps that show no evidence of any

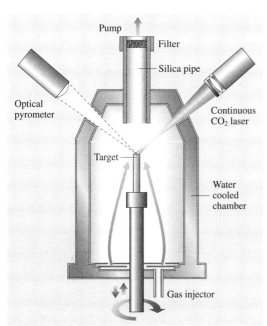

Fig. 3.10 Sketch of a synthesis reactor with a continuous CO_2 laser device (adapted from [3.30])

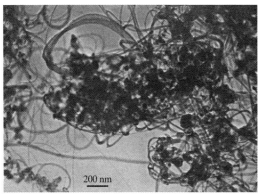

Fig. 3.11 Low-magnification TEM images of a typical raw SWNT material obtained using the laser vaporization technique. The fibrous structures are SWNT bundles, and the dark particles are remnants of the catalyst. Raw SWNT materials obtained from an electric arc exhibit similar features (from [3.15])

associated metal catalyst particle, although, as pointed out in Sect. 3.1, finding the two tips of a SWNT is rather challenging, considering the huge aspect ratio of the nanotube and their entangled nature. Another feature of the SWNTs produced with this technique is that they are supposedly *cleaner* than those produced using other techniques; in other words they associated with smaller amounts of the amorphous carbon that either coats the SWNTs or is gathered into nanoparticles. This advantage, however, only occurs for synthesis conditions designed to ensure high-quality SWNTs. It is not true when high-yield conditions are preferred; in this case SWNTs from an electric arc may appear cleaner than SWNTs from laser vaporization [3.15].

The laser vaporization technique is one of the three methods currently used to prepare SWNTs as commercial products. SWNTs prepared this way were first marketed by Carbon Nanotechnologies Inc. (Houston, USA), with prices as high as 1000 $/g (raw materials) until December 2002. Probably because lowering the amount of impurities in the raw materials using this technique is impossible, they have recently decided to focus on fabricating SWNTs using the HiPCo technique (Sect. 3.2.2). Laser-based methods are generally not considered to be competitive in the long term for the low-cost production of SWNTs compared to CCVD-based methods (Sect. 3.2.2). However, prices as low as 0.03 $/g of raw high concentration have been estimated possible from a pre-industrial project study (Acolt S.A., Yverdon, Switzerland).

Electric Arc Method

Electric arcs between carbon electrodes have been studied as light sources and radiation standards for a very long time. They have however received renewed attention more recently due to their use in the production of new fullerene-related molecular carbon nanostructures, such as genuine fullerenes or nanotubes. This technique was first brought to light by *Krätschmer* et al. [3.24] who utilized it to achieve the production of fullerenes in macroscopic quantities. In the course of investigating other carbon nanostructures formed along with the fullerenes, and more particularly the solid carbon deposit that formed on the cathode, *Iijima* [3.7] discovered the catalyst-free formation of perfect c-MWNT-type carbon nanotubes. Then, as mentioned in the *Introduction*, the catalyst-promoted formation of SWNTs was accidentally discovered after some amounts of transition metals were introduced into the anode in an attempt to fill the c-MWNTs with metals during growth [3.8, 9]. Since then, a lot of work has been carried out by many groups using this technique in order to understand the mechanisms of nanotube growth as well as the role played by the catalysts (if any) in the synthesis of MWNTs and/or SWNTs [3.34–46].

Electric Arc Method – Experimental Devices

The principle of this technique is to vaporize carbon in the presence of catalysts (iron, nickel, cobalt, yttrium, boron, gadolinium, cerium, and so forth) in a reduced atmosphere of inert gas (argon or helium). After triggering an arc between two electrodes, a plasma is formed consisting of the mixture of carbon vapor, the rare gas (helium or argon), and the catalyst vapors. The vaporization is the consequence of energy transfer from the arc to the anode made of graphite doped with catalysts. The importance of the anode erosion rate depends on the power of the arc and also on other experimental conditions. It is worth noting that a high anode erosion does not necessarily lead to a high carbon nanotube production.

An example of a reactor layout is shown in Fig. 3.12. It consists of a cylinder about 30 cm in diameter and about 1 m in height, equipped with diametrically opposed sapphire windows located so that they face the plasma zone, observing the arc. The reactor possesses two valves, one for performing the primary evacuation (0.1 Pa) of the chamber, the other for filling it with a rare gas up to the desired working pressure.

Contrary to the solar energy technique, SWNTs are deposited (provided appropriate catalysts are used) in different regions of the reactor:

1. The collaret, which forms around the cathode
2. The weblike deposits found above the cathode
3. The soot deposited all around the reactor walls and the bottom.

On the other hand, MWNTs are formed in a hard deposit adherent to the cathode whether catalysts are used or not. The cathode deposits form under the cathode. The formation of collaret and web is not systematic and depends on the experimental conditions, as indicated in Table 3.1, as opposed to the cathode deposit and soot, which are obtained consistently.

Two graphite rods of few millimeters in diameter constitute the electrodes between which a potential difference is applied. The dimensions of these electrodes vary according to the authors. In certain cases, the cathode has a greater diameter than the anode in order to facilitate their alignment [3.37, 47]. Other authors utilize electrodes of the same diameter [3.46]. The whole device can be designed horizontally [3.38, 46] or vertically [3.39, 41–43]. The advantage of the latter is the symmetry brought by the verticality with respect to gravity, which facilitates computer modeling (regarding convection flows, for instance).

Two types of anode can be utilized when catalysts are introduced:

1. A graphite anode containing a coaxial hole several centimeters in length into which a mixture of the catalyst and the graphite powder is placed.

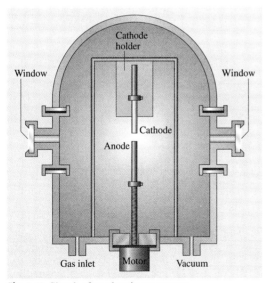

Fig. 3.12 Sketch of an electric arc reactor

2. A graphite anode within which the catalysts are homogeneously dispersed [3.48].

The former are by far the most popular, due to their ease of fabrication.

Optimizing the process in terms of the nanotube yield and quality is achieved by studying the roles of various parameters such as the type of doped anode (homogeneous or heterogeneous catalyst dispersion), the nature as well as the concentration of the catalyst, the nature of the plasmagen gas, the buffer gas pressure, the arc current intensity, and the distance between electrodes. Investigating the influences of these parameters on the type and amount of carbon nanostructures formed is, of course, the preliminary work that has been done. Although electric arc reactors equipped with the facilities to perform such investigations are scarce (Fig. 3.12), investigating the missing link (the effect of varying the parameters on the plasma characteristics – the species concentrations and temperature) is likely to provide a more comprehensive understanding of the phenomena involved during nanotube formation. This has been recently performed using atomic and molecular optical emission spectroscopy [3.39, 41–44, 46].

Finally, we should mention attempts to create an electric arc in liquid media, such as liquid nitrogen [3.49] or water [3.50, 51]. The goal here is to make processing easier, since such systems should not require pumping devices or a closed volume and so they are more likely to allow continuous synthesis. This adaptation has not, however, reached the stage of mass production.

Electric Arc Method – Results
In view of the numerous results obtained with this electric arc technique, it is clear that both the morphology and the production efficiency of nanotubes strongly depends upon the experimental conditions used and, in particular, upon the nature of the catalysts. It is worth noting that the products obtained do not consist solely of carbon nanotubes. Nontubular forms of carbon, such as nanoparticles, fullerenelike structures including C_{60}, poorly organized polyaromatic carbons, nearly amorphous nanofibers, multiwall shells, single-wall nanocapsules, and amorphous carbon have all been obtained, as reported in Table 3.1 [3.40, 42, 43]. In addition, remnants of the catalyst are found all over the place – in the soot, the collaret, the web and the cathode deposit – in various concentrations. Generally, at a helium pressure of about 600 mbar, for an arc current of 80 A and for an electrode gap of 1 mm, the synthesis of

Table 3.1 Different carbon morphologies obtained by changing the type of anode, the type of catalyst and the pressure in a series of arc discharge experiments (electrode gap = 1 mm)

Catalyst (at. %) Arc conditions	0.6Ni + 0.6Co (homogeneous anode) $P \approx 60$ kPa $I \approx 80$ A	0.6Ni + 0.6Co (homogeneous anode) $P \approx 40$ kPa $I \approx 80$ A	0.5Ni + 0.5Co $P \approx 60$ kPa $I \approx 80$ A	4.2Ni + 1Y $P \approx 60$ kPa $I \approx 80$ A
Soot	• **MWNT** + **MWS** + **POPAC** or **Cn** ± catalysts $\phi \approx 3-35$ nm • **NANF** + catalysts • AC particles + catalysts • [*DWNT*], [*SWNT*], ropes or isolated, + POPAC	• **POPAC** and AC particles + **catalysts** $\phi \approx 2-20$ nm • **NANF** + catalysts $\phi \approx 5-20$ nm + MWS • [*SWNT*] $\phi \approx 1-1.4$ nm, distorted or damaged, isolated or ropes + Cn	• AC and POPAC particles + catalysts $\phi \approx 3-35$ nm • **NANF** + catalysts $\phi \approx 4-15$ nm • [*SWNT*] $\phi \approx 1.2$ nm, isolated or ropes	• **POPAC** and AC + particles + catalysts $\phi \leq 30$ nm • SWNT $\phi \approx 1.4$ nm, clean + Cn, short with tips, [*damaged*], isolated or ropes $\phi \leq 25$ nm • [*SWNC*] particles
Web	• [*MWNT*], **DWNT**, ϕ 2.7–**4**–5.7 nm SWNT ϕ 1.2–1.8 nm, isolated or **ropes** $\phi < 15$ nm, + POPAC ± Cn • AC particles + catalysts $\phi \approx 3-40$ nm + MWS • [*NANF*]	None	None	• **SWNT**, $\phi \approx 1.4$ nm, isolated or **ropes** $\phi \leq 20$ nm, + **AC** • POPAC and AC particles + catalysts $\phi \approx 3-\mathbf{10}-40$ nm + MWS
Collaret	• **POPAC** and **SWNC** particles • Catalysts $\phi \approx 3-250$ nm, < **50** nm + MWS • SWNT ϕ 1–1.2 nm, [*opened*], **distorted**, isolated or **ropes** $\phi < 15$ nm, + **Cn** • [*AC*] particles	• AC and **POPAC** particles + catalysts $\phi \approx 3-25$ nm • SWNT $\phi \approx 1-1.4$ nm clean + Cn, [*isolated*] or ropes $\phi < 25$ nm • Catalysts $\phi \approx 5-50$ nm + MWS, • [*SWNC*]	• **Catalysts** $\phi \approx 3-170$ nm + MWS • AC or POPAC particles + **catalysts** $\phi \approx 3-50$ nm • SWNT $\phi \approx 1.4$ nm clean + Cn isolated or ropes $\phi < 20$ nm	• **SWNT** $\phi \approx 1.4-2.5$ nm, clean + Cn, [*damaged*], isolated or **ropes** $\phi < 30$ nm • POPAC or AC particles + catalysts $\phi \approx 3-30$ nm • [*MWS*] + catalysts or catalyst-free
Cathode deposit	• **POPAC** and SWNC particles • **Catalysts** $\phi \approx 5-300$ nm MWS • MWNT $\phi < 50$ nm • [*SWNT*] $\phi \approx 1.6$ nm clean + Cn, isolated or ropes	• **POPAC** and **SWNC** particles + Cn • Catalysts $\phi \approx 20-100$ nm + MWS	• MWS, catalyst-free • MWNT $\phi < 35$ nm • POPAC and PSWNC particles • [*SWNT*], isolated or ropes • [*Catalysts*] $\phi \approx 3-30$ nm	• SWNT $\phi \approx 1.4-4.1$ nm, clean + **Cn**, short with tips, isolated or ropes $\phi \leq 20$ nm. • **POPAC** or AC particles + catalysts $\phi \approx 3-30$ nm • MWS + catalysts $\phi < 40$ nm or catalyst-free • [*MWNT*]

Abundant – Present – [*Rare*]

Glossary: AC: amorphous carbon; POPAC: poorly organized polyaromatic carbon; Cn: fullerenelike structure, including C_{60}; NANF: nearly amorphous nanofiber; MWS: multiwall shell; SWNT: single-wall nanotube; DWNT: double-wall nanotube, MWNT: multiwall nanotube; SWNC: single-wall nanocapsule.

SWNTs is favored by the use of Ni/Y as coupled catalysts [3.8, 38, 52]. In these conditions, which give high SWNT yields, SWNT concentrations are highest in the collaret (50–70%), then in the web ($\approx 50\%$ or less) and then in the soot. On the other hand, c-MWNTs are found in the cathode deposit. SWNT lengths are micrometric and, typical outer diameters are around 1.4 nm. Using the latter conditions (Table 3.1, column 4), Table 3.1 illustrates the consequence of changing the parameters. For instance (Table 3.1, column 3), using Ni/Co instead of Ni/Y as catalysts prevents the formation of SWNTs. But when the Ni/Co catalysts are homogeneously dispersed in the anode (Table 3.1, column 1), the formation of nanotubes is promoted again, but

MWNTs with two or three walls prevail over SWNTs, among which DWNTs (double-wall nanotubes) dominate. However, decreasing the ambient pressure from 60 to 40 kPa (Table 3.1, column 2) again suppresses nanotube formation.

Based on works dealing with the influence of the granulometry of the graphite powders which are mixed with the catalyst powder and placed in hollow-type graphite anodes, recent studies have demonstrated that one of the control keys for growing SWNTs with enhanced purity and yield is for the anode to exhibit a high thermal conductivity with as more limited radial and longitudinal variations as possible [3.53, 54]. This explains why similar results (i.e., enhanced purity and yield) were previously obtained when replacing the graphite powder by diamond powder [3.44, 45] in spite of the low electrical conductivity of diamond, since graphite and diamond powders lead to the same plasma composition once vaporized at high temperatures ($> 4000\,\mathrm{K}$).

A comparison of the plasma characteristics (i.e., radial temperature profiles and CI/NiI concentration ratio) obtained for anodes with different filler material features (i.e., $1/100\,\mu\mathrm{m}$ granulometry and sp^2/sp^3 carbon) is presented in Fig. 3.13a,b respectively. The whole plasma temperature radial profiles obtained using either the Ni/Y/graphite ($\phi \approx 1\,\mu\mathrm{m}$) anode or the Ni/Y/diamond ($\phi \approx 1\,\mu\mathrm{m}$) anode is much smoother than with the standard Ni/Y/graphite ($\phi \approx 100\,\mu\mathrm{m}$) anode, meanwhile exhibiting less extreme temperatures ($\approx 6200\,\mathrm{K}$ for the highest as opposed to $\approx 8000\,\mathrm{K}$ respectively for the standard anode). From 1 mm from the arc axis, temperature is maintained at a constant value at about 4000 K. The absence of large temperature fluctuations is consistent with the fact that the plasma is continuously fed by a rather constant ratio of [carbon]/[catalysts] resulting from the steadier erosion of the anode and a better powder mixture homogenization. In this regard, it might be significant that the smoothest temperature profile over the longest radial distance is obtained for the Ni/Y/graphite ($\varphi \approx 1\,\mu\mathrm{m}$) anode, which has resulted in the highest yield [3.53, 54]. Likewise, the CI/NiI concentration ratio profiles related to either the Ni/Y/graphite ($\phi \approx 1\,\mu\mathrm{m}$) anode or the Ni/Y/diamond anode show a dramatic difference with respect to the Ni/Y/graphite ($\phi \approx 100\,\mu\mathrm{m}$) anode (Fig. 3.13b). They exhibit a fluctuation-free regime along the whole radial profile, with a unique maximum at $\approx 1.3-1.5$ mm from the arc axis. The average ratio is low ($\approx 5 \times 10^5$) due to a relatively low distribution of carbon concentration leading to a higher plasma temperature. This makes a perfect sense, since carbon species are very emissive in the range 4500–6000 K, inducing that radiative losses are more significant when plasmas are enriched in carbon species, leading to colder plasma temperatures, and vice versa. Such a feature is again consistent with the steady erosion of high thermal conductivity anodes. It is also worth noting that, again, area where CI/NiI ratios exhibit

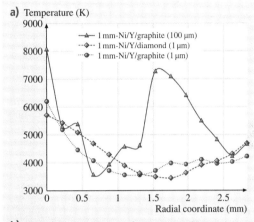

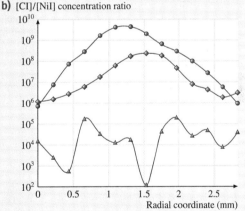

Fig. 3.13a,b Radial temperature profiles (a) and radial [CI]/[NiI] concentration ratio (b) as obtained by emission spectroscopy for hollowed-type anodes with various thermal behaviours. The thermal behaviour was varied by varying the grain size (1 or 100 μm) and the carbon type (sp^2 – graphite, or sp^3 – diamond) of the carbon powder which the hollow core of the anode is filled with (along with yttrium and nickel catalyst powder). Smaller grain size results in better compaction, hence in higher thermal conductivity

maximum values in Fig. 3.13b relate to area of minimum temperature values in Fig. 3.13a. In addition, the Cl/NiI concentration ratio is up to about 3–5 orders of magnitude higher for the fine-grain graphite-containing anode and the diamond-containing anode than for the large-grain graphite-containing anode. Moreover, the Cl/NiI concentration ratio is even higher as ≈ 1.5 orders of magnitude for the fine-grain graphite-containing anode than for the diamond-containing anode.

Highly and homogeneously thermally conductive anodes lead to a steadier anode erosion, hence to steadier plasma characteristics, hence to a more constant variety of the carbon phase formed (SWNTs), finally resulting in an enhanced purity and yield of the latter. Such experiments have revealed, as in the comparison between the results from using homogeneous instead of heterogeneous anodes, that the physical phenomena (charge and heat transfers) that occur in the anode during the arc are of the utmost importance, a factor which was neglected before this.

It is clear that while the use of a rare earth element (such as Y) as a single catalyst does not provide the right conditions to grow SWNTs, associating it with a transition metal (Ni/Y for instance) seems to lead to the best combinations that give the highest SWNT yields [3.47]. On the other hand, using a single rare earth element may lead to unexpected results, such as the closure of graphene edges from a c-MWNT wall with the neighboring graphene edges from the same wall side, leading to the preferred formation of telescopelike and open c-MWNTs that are able to contain nested Gd crystals [3.41, 43]. The effectiveness of bimetallic catalysts is believed to be due to the transitory formation of nickel particles coated with yttrium carbide, which has a lattice constant that is somewhat commensurable with that of graphene [3.55].

Figure 3.14 illustrates other interesting features of the plasma. A common feature is that a huge vertical gradient (≈ 500 K/mm) rapidly establishes (≈ 0.5 mm from the center in the radial direction) from the bottom to the top of the plasma, probably due to convection phenomena (Fig. 3.14a). The zone of actual SWNT formation is beyond the limit of the volume analyzable in the radial direction, corresponding to colder areas. The C_2 concentration increases dramatically from the anode to the cathode and decreases dramatically in the radial direction (Fig. 3.14b). This demonstrates that C_2 moieties are secondary products resulting from the recombination of primary species formed from the anode. It also suggests that C_2 moieties may be the building blocks for MWNTs (formed at the cathode) but not for SWNTs [3.43, 45].

Although many aspects of it still need to be understood, the electric arc method is one of the three methods currently used to produce SWNTs as commercial products. Though not selling bare nanotubes anymore, Nanoledge S.A. (Montpellier, France), for instance, had a current production that reached several tens of kilograms per year (raw SWNTs, in other words unpurified), with a market price of ≈ 65 €/g in 2005, which was much cheaper than any other production method. However, the drop of prices for raw SWNTs

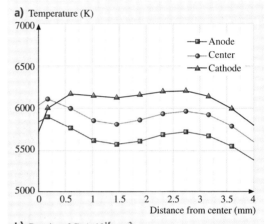

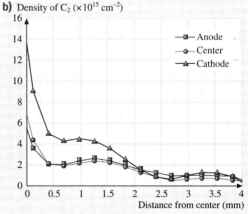

Fig. 3.14a,b Typical temperature (**a**) and C_2 concentration (**b**) profiles for plasma at the anode surface (*squares*), at the center of the plasma (*dots*), and at the cathode surface (*triangles*) at *standard* conditions (see text). Gradients are similar whichever catalyst is used, although absolute values may vary

down to 2–5 €/g which was anticipated for 2007 has not been possible. Actually, Bucky USA (Houston, Texas, USA) are still supplying raw SWNTs derived from electric arcs at a market price of 250 $/g in 2006 (which is, however, a 75% decrease in two years), which is barely lower than the ≈ 350 $/g proposed for 70–90%-purified SWNTs from Nanocarblab (Moscow, Russia).

Three-Phase AC Arc Plasma

An original semi-industrial three-phase AC plasma technology has been developed for the processing of carbon nanomaterials [3.57, 58]. The technology has been specially developed for the treatment of liquid, gaseous or dispersed materials. An electric arc is established between three graphite electrodes. The system is powered by a three-phase AC power supply operated at 600 Hz and at arc currents of 250–400 A. Carbon precursors, gaseous, liquid or solid, are injected at the desired (variable) position into the plasma zone. The reactive mixture can be extracted from the reaction chamber at different predetermined positions. After cooling down to room temperature, the aerosol passes through a filtering system. The main operating parameters, which are freely adjustable, include the arc current,

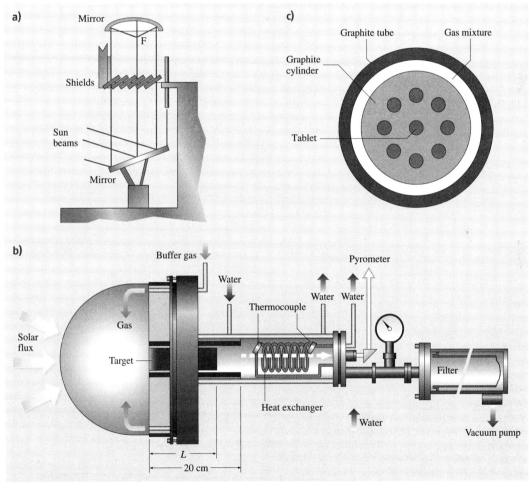

Fig. 3.15a–c Sketch of a solar energy reactor in use in the PROMES-CNRS Laboratory, Odeilho (France). (**a**) Gathering of sun rays, focused at F; (**b**) side view of the experimental set-up at the focus of the 1 MW solar furnace; (**c**) top view of the target graphite rod (adapted from [3.56])

the flow rate and the nature of the plasma gas (N_2, Ar, H_2, He, and so on), the carbon precursor (gaseous, liquid, solid, up to 3 kg/h), the injection and extraction positions, and the quenching rate. This plasma technology has shown very high versatility and it has been demonstrated that it can be used to produce a wide range of carbon nanostructures ranging from carbon blacks to carbon nanotubes over fullerenes with a high product selectivity.

Solar Furnace

Solar furnace devices were originally utilized by several groups to produce fullerenes [3.59–61]. *Heben* et al. [3.62] and *Laplaze* et al. [3.63] later modified their original devices to achieve carbon nanotube production. This modification consisted mainly of using more powerful ovens [3.64, 65].

Solar Furnace – Experimental Devices

The principle of this technique is again based on the sublimation of a mixture of graphite powder and catalysts placed in a crucible in an inert gas. An example of such a device is shown in Fig. 3.15. The solar rays are collected by a plain mirror and reflected toward a parabolic mirror that focuses them directly onto a graphite pellet in a controlled atmosphere (Fig. 3.15a). The high temperature of about 4000 K causes both the carbon and the catalysts to vaporize. The vapors are then dragged by the neutral gas and condense onto the cold walls of the thermal screen. The reactor consists of a brass support cooled by water circulation, upon which Pyrex chambers of various shapes can be fixed (Fig. 3.15b). This support contains a watertight passage permitting the introduction of the neutral gas and a copper rod onto which the target is mounted. The target is a graphite rod that includes pellets containing the catalysts, which is surrounded by a graphite tube (Fig. 3.15c) that acts as both a thermal screen to reduce radiation losses (very important in the case of graphite) and a duct to lead carbon vapors to a filter, which stops soot from being deposited on the Pyrex chamber wall. The graphite rod target replaces the graphite crucible filled with powdered graphite (for fullerene synthesis) or the mixture of graphite and catalysts (for nanotube synthesis) that were used in the techniques we have discussed previously.

These studies primarily investigated the target composition, the type and concentration of catalyst, the flow-rate, the composition and pressure of the plasmagenic gas inside the chamber, and the oven power. The objectives were similar to those of the works associated with the other solid carbon source-based processes. When possible, specific in situ diagnostics (pyrometry, optical emission spectroscopy, and so on) are also performed in order to investigate the roles of various parameters (temperature measurements at the crucible surface, along the graphite tube acting as thermal screen, C_2 radical concentration in the immediate vicinity of the crucible).

Solar Furnace – Results

Some of the results obtained by different groups concerning the influence of the catalyst can be summarized as follows. With Ni/Co, and at low pressure, the sample collected contains mainly MWNTs with bamboo texture, carbon shells, and some bundles of SWNTs [3.64]. At higher pressures, only bundles of SWNTs are obtained, with fewer carbon shells. Relatively long bundles of SWNTs are observed with Ni/Y and at a high pressure. Bundles of SWNTs are obtained in the soot with Co; the diameters of the SWNTs range from 1 to 2 nm. *Laplaze* et al. [3.64] observed very few nanotubes but a large quantity of carbon shells.

In order to proceed to large-scale synthesis of single-wall carbon nanotubes, which is still a challenge for chemical engineers, *Flamant* et al. [3.56] and *Luxembourg* et al. [3.66] recently demonstrated that solar energy-based synthesis is a versatile method for obtaining SWNTs that can be scaled up from 0.1–0.2 to 10 g/h and then to 100 g/h productivity using existing solar furnaces. Experiments performed on a medium scale produced about 10 g/h of SWNT-rich material using various mixtures of catalysts (Ni/Co, Ni/Y, Ni/Ce). A numerical reactor simulation was performed in order to improve the quality of the product, which was subsequently observed to reach 40% SWNT in the soot [3.67].

3.2.2 Gaseous Carbon Source–Based Production Techniques for Carbon Nanotubes

As mentioned in the *Introduction*, the catalysis-enhanced thermal cracking of a gaseous carbon source (hydrocarbons, CO) – commonly referred to as catalytic chemical vapor deposition (CCVD) – has long been known to produce carbon nanofilaments [3.4], so reporting on all of the works published in the field since the beginning of the century is almost impossible. Until the 1990s, however, carbon nanofilaments were mainly produced to act as a core substrate for the subsequent growth of larger (micrometric) carbon fibers – so-called vapor-grown carbon fibers – via thickening in catalyst-free CVD processes [3.68, 69]. We are therefore going to focus instead on more recent attempts to prepare genuine carbon nanotubes.

The synthesis of carbon nanotubes (either single- or multiwalled) by CCVD methods involves the catalytic decomposition of a carbon-containing source on small metallic particles or clusters. This technique involves either an heterogeneous process if a solid substrate is involved or an homogeneous process if everything takes place in the gas phase. The metals generally used for these reactions are transition metals, such as Fe, Co and Ni. It is a rather low-temperature process compared to arc discharge and laser ablation methods, with the formation of carbon nanotubes typically occurring between 600 and 1000 °C. Because of the low temperature, the selectivity of the CCVD method is generally better for the production of MWNTs with respect to graphitic particles and amorphouslike carbon, which remain an important part of the raw arc discharge SWNT samples, for example. Both homogeneous and heterogeneous processes appear very sensitive to the nature and the structure of the catalyst used, as well as to the operating conditions [3.70]. Carbon nanotubes prepared by CCVD methods are generally much longer (a few tens to hundreds of micrometers) than those obtained by arc discharge (a few micrometers). Depending on the experimental conditions, it is possible to grow dense arrays of nanotubes. It is a general statement that MWNTs from CCVD contain more structural defects (exhibit a lower nanotexture) than MWNTs from arc discharge, due to the lower temperature of the reaction, which does not allow any structural rearrangements. These defects can be removed by subsequently applying heat treatments in vacuum or inert atmosphere to the products. Whether such a discrepancy is also true for SWNTs remains questionable. CCVD SWNTs are generally gathered into bundles that are generally of smaller diameter (a few tens of nm) than their arc discharge and laser ablation counterparts (around 100 nm in diameter). Specifically when performed in fluidized-bed reactor [3.71], CCVD provides reasonably good perspectives on large-scale and low-cost processes for the mass production of carbon nanotubes, a key point for their application at the industrial scale.

A final word concerns the nomenclature. Because work in the field started more than a century ago, the names of the carbon objects prepared by this method have changed with time with the authors, research areas, and fashions. These same objects have been called vapor-grown carbon fibers, nanofilaments, nanofibers and nanotubes. For multilayered fibrous morphologies (since single-layered fibrous morphologies can only be SWNT anyway), the exact name should be vapor-grown carbon nanofilaments (VGCNF). Whether or not the filaments are tubular is a matter of textural description, which should go with other textural features such as bamboo, herringbone and concentric (Sect. 3.1.2). In the following, we will therefore use MWNTs for any hollowed nanofilament, whether they contain graphene walls oriented transversally or not. Any other nanofilament will be termed a *nanofiber*.

Heterogeneous Processes

Heterogeneous CCVD processes simply involve passing a gaseous flow containing a given proportion of a hydrocarbon (mainly CH_4, C_2H_2, C_2H_4, or C_6H_6, usually as a mixture with either H_2 or an inert gas such as Ar) over small transition metal particles (Fe, Co, Ni) in a furnace. The particles are deposited onto an inert substrate, by spraying a suspension of the metal particles on it or by another method. The reaction is chemically defined as catalysis-enhanced thermal cracking

$$C_xH_y \rightarrow xC + \tfrac{y}{2}H_2.$$

Catalysis-enhanced thermal cracking was used as long ago as the late nineteenth century. Extensive works on this topic published before the 1990s include those by *Baker* et al. [3.6, 72], or *Endo* et al. [3.73, 74]. Several review papers have been published since then, such as [3.75], in addition to many regular papers.

CO can be used instead of hydrocarbons; the reaction is then chemically defined as catalysis-enhanced disproportionation (the so-called the Boudouard equilibrium)

$$2CO \rightleftharpoons C + CO_2.$$

Heterogeneous Processes – Experimental Devices

The ability of catalysis-enhanced CO disproportionation to make carbon nanofilaments was reported by *Davis* et al. [3.76] as early as 1953, probably for the first time. Extensive follow-up work was performed by *Boehm* [3.77], *Audier* et al. [3.17, 78–80], and *Gadelle* et al. [3.81–84].

Although formation mechanisms for SWNTs and MWNTs can be quite different (Sect. 3.3, or refer to a review article such as [3.85]), many of the catalytic process parameters play similar and important roles in the type of nanotubes formed: the temperature, the duration of the treatment, the gas composition and flow rate, and of course the catalyst nature and size. At a given temperature, depending mainly on the nature of both the catalyst and the carbon-containing gas, the catalytic decomposition will take place at the surfaces of the metal particles, followed by mass transport of the freshly produced carbon by surface or volume diffusion until the carbon concentration reaches the solubility limit, and the precipitation starts.

It is now agreed that CCVD carbon nanotubes form on very small metal particles, typically in the nanometer range [3.85]. These catalytic metal particles are prepared mainly by reducing transition metal compounds (salts, oxides) by H_2 prior to the nanotube formation step (where the carbon containing gas is required). It is possible, however, to produce these catalytic metal particles in situ in the presence of the carbon source, allowing for a one-step process [3.88]. Because controlling the metal particle size is the key issue (they have to be nanosized), coalescence is generally avoided by placing them on an inert support such as an oxide (Al_2O_3, SiO_2, zeolites, $MgAl_2O_4$, MgO) or more rarely on graphite. A low concentration of the catalytic metal precursor is required to limit the coalescence of the metal particles, which can happen during the reduction step. The supported catalysts can be used as a static phase placed within the gas flow, but can also be used as a fine powder suspended into and by the gas phase, in a so-called fluidised bed process. In the latter, the reactor has to be vertical so that to compensate the effect of gravity by the suspending effect of the gas flow.

There are two main ways to prepare the catalyst:

1. The impregnation of a substrate with a solution of a salt of the desired transition metal catalyst
2. The preparation of a solid solution of an oxide of the chosen catalytic metal in a chemically inert and thermally stable host oxide.

The catalyst is then reduced to form the metal particles on which the catalytic decomposition of the carbon source will lead to carbon nanotube growth. In most cases, the nanotubes can then be separated from the catalyst (Fig. 3.16).

Heterogeneous Processes – Results with CCVD Involving Impregnated Catalysts

A lot of work had been done in this area even before the discovery of fullerenes and carbon nanotubes, but although the formation of tubular carbon structures by catalytic processes involving small metal particles was clearly identified, the authors did not focus on the preparation of SWNTs or MWNTs with respect to the other carbon species. Some examples will be given here to illustrate the most striking improvements obtained.

With the impregnation method, the process generally involves four different and successive steps:

1. Impregnation of the support by a solution of a salt (nitrate, chloride) of the chosen metal catalyst
2. Drying and calcination of the supported catalyst to get the oxide of the catalytic metal
3. Reduction in a H_2-containing atmosphere to make the catalytic metal particles
4. The decomposition of a carbon-containing gas over the freshly prepared metal particles that leads to nanotube growth.

For example, *Ivanov* et al. [3.89] prepared nanotubes through the decomposition of C_2H_2 (pure or mixed with H_2) on well-dispersed transition metal par-

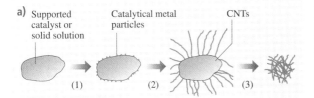

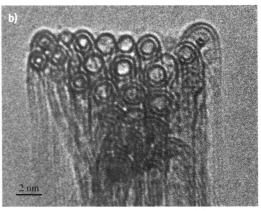

Fig. 3.16 (a) Formation of nanotubes via the CCVD-based impregnation technique. (1) Formation of catalytic metal particles by reduction of a precursor; (2) Catalytic decomposition of a carbon-containing gas, leading to the growth of carbon nanotubes; (3) Removal of the catalyst to recover the nanotubes (from [3.86]). (b) Example of a bundle of double-wall nanotubes (DWNTs) prepared this way (from [3.87])

ticles (Fe, Co, Ni, Cu) supported on graphite or SiO_2. Co-SiO_2 was found to be the best catalyst/support combination for the preparation of MWNTs, but most of the other combinations led to carbon filaments, sometimes covered with amorphouslike carbon. The same authors have developed a precipitation-ion-exchange method that provides a better dispersion of metals on silica compared to the classical impregnation technique. The same group then proposed the use of a zeolite-supported Co catalyst [3.90, 91], resulting in very finely dispersed metal particles (from 1 to 50 nm in diameter). They observed MWNTs with a diameter around 4 nm and only two or three walls only on this catalyst. *Dai* et al. [3.92] have prepared SWNTs by CO disproportionation on nanosized Mo particles. The diameters of the nanotubes obtained are closely related to those of the original particles and range from 1 to 5 nm. The nanotubes obtained by this method are free of an amorphous carbon coating. They also found that a synergetic effect occurs for the alloy instead of the components alone, and one of the most striking examples is the addition of Mo to Fe [3.93] or Co [3.94].

Heterogeneous Processes – Results with CCVD Involving Solid Solution–Based Catalysts

A solid solution of two metal oxides is formed when ions of one metal mix with ions of the other metal. For example, Fe_2O_3 can be prepared in solid solution in Al_2O_3 to give a $Al_{2-2x}Fe_{2x}O_3$ solid solution. The use of a solid solution allows a perfectly homogeneous dispersion of one oxide in the other to be obtained. These solid solutions can be prepared in different ways, but coprecipitation of mixed oxalates and combustion synthesis are the most common methods used to prepare nanotubes. The synthesis of nanotubes by the catalytic decomposition of CH_4 over an $Al_{2-2x}Fe_{2x}O_3$ solid solution was originated by *Peigney* et al. [3.88] and then studied extensively by the same group using different oxides such as spinel-based solid solutions ($Mg_{1-x}M_xAl_2O_4$ with M = Fe, Co, Ni, or a binary alloy [3.86, 95]) or magnesia-based solid solutions [3.86, 96] ($Mg_{1-x}M_xO$, with M = Fe, Co or Ni). Because of the very homogeneous dispersion of the catalytic oxide, it is possible to produce very small catalytic metal particles at the high temperature required for the decomposition of CH_4 (which was chosen for its greater thermal stability compared to other hydrocarbons). The method proposed by these authors involves the heating of the solid solution from room temperature to a temperature of between 850 and 1050 °C in a mixture of H_2 and CH_4, typically containing 18 mol % of CH_4. The nanotubes obtained clearly depend upon the nature of both the transition metal (or alloy) used and the inert oxide (matrix); the latter because the Lewis acidity seems to play an important role [3.97]. For example, in the case of solid solutions containing around 10 wt % of Fe, the amount of carbon nanotubes obtained decreases in the following order depending on the matrix oxide: $MgO > Al_2O_3 > MgAl_2O_4$ [3.86]. In the case of MgO-based solid solutions, the nanotubes can be very easily separated from the catalyst by dissolving it (in diluted HCl for example) [3.96]. The nanotubes obtained are typically gathered into small-diameter bundles (less than 15 nm) with lengths of up to 100 μm. The nanotubes are mainly SWNTs and DWNTs, with diameters of between 1 and 3 nm.

Obtaining pure nanotubes by the CCVD method requires, as for all the other techniques, the removal of the catalyst. When a catalyst supported (impregnated) in a solid solution is used, the supporting – and catalytically inactive – oxide is the main impurity, both in weight and volume. When oxides such as Al_2O_3 or SiO_2 (or even combinations) are used, aggressive treatments involving hot caustic solutions (KOH, NaOH) for Al_2O_3 or the use of HF for SiO_2 are required. These treatments have no effect, however, on other impurities such as other forms of carbon (amorphouslike carbon, graphitized carbon particles and shells, and so on). Oxidizing treatments (air oxidation, use of strong oxidants such as HNO_3, $KMnO_4$, H_2O_2) are thus required and permit the removal of most unwanted forms of carbon, but they result in a low final yield of carbon nanotubes, which are often quite damaged. *Flahaut* et al. [3.96] were the first to use a MgCoO solid solution to prepare SWNTs and DWNTs that could be easily separated without incurring any damage via fast and safe washing with an aqueous HCl solution.

In most cases, only very small quantities of catalyst (typically less than 500 mg) are used, and most claims of *high-yield* productions of nanotubes are based on laboratory experimental data, without taking into account all of the technical problems related to scaling up to a laboratory-scale CCVD reactor. At the present time, although the production of MWNTs is possible on an industrial scale, the production of affordable SWNTs is still a challenge, and controlling the arrangement of and the number of walls in the nanotubes is also problematic. For example, adding small amounts of molybdenum to the catalyst [3.98] can lead to drastic modifications of the nanotube type (from regular nanotubes to carbon nanofibers – Sect. 3.1). *Flahaut* et al. have recently shown that the method used to prepare a particular catalyst can play a very important role [3.99]. Double-walled carbon nanotubes (DWNTs) represent a special case: they are at the frontier between single- (SWNTs) and multiwalled nanotubes (MWNTs). Because they are the MWNTs with the lowest possible number of walls, their structures and properties are very similar to those of SWNTs. Any subsequent functionalization, which is often required to improve the compatibility of nanotubes with their external environment (composites) or to give them new properties (solubility, sensors), will partially damage the external wall, resulting in drastic modifications in terms of both electrical and mechanical properties. This is a serious drawback for SWNTs. In the case of DWNTs, the outer wall can be modified (functionalized) while retaining the structure of the inner tube. DWNTs have been recently synthesised on a gram-scale by CCVD [3.87], with a high purity and a high selectivity (around 80% DWNTs) (Fig. 3.16b).

Homogeneous Processes

The homogenous route, also called the *floating catalyst method*, differs from the other CCVD-based methods because it uses only gaseous species and does not require the presence of any solid phase in the reactor. The basic principle of this technique, similar to the other CCVD processes, is to decompose a carbon source (ethylene, xylene, benzene, carbon monoxide, and so on) on nanosized transition metal (generally Fe, Co, or Ni) particles in order to obtain carbon nanotubes. The catalytic particles are formed directly in the reactor, however, and are not introduced before the reaction, as occurs in supported CCVD for instance.

Homogeneous Processes – Experimental Devices

The typical reactor used in this technique is a quartz tube placed in an oven into which the gaseous feedstock, containing the metal precursor, the carbon source, some hydrogen and a vector gas (N_2, Ar, or He), is passed. The first zone of the reactor is kept at a lower temperature, and the second zone, where the formation of tubes occurs, is heated to 700–1200 °C. The metal precursor is generally a metal-organic compound, such as a zero-valent carbonyl compound like [Fe(CO)$_5$] [3.100], or a metallocene [3.101–103] such as ferrocene, nickelocene or cobaltocene. The use of metal salts, such as cobalt nitrate, has also been reported [3.104]. It may be advantageous to make the reactor vertical, so that gravity acts symmetrically on the gaseous volume inside the furnace.

Homogeneous Processes – Results

The metal-organic compound decomposes in the first zone of the reactor, generating nanosized metallic particles that can catalyze nanotube formation. In the second part of the reactor, the carbon source is decomposed to atomic carbon, which is then responsible for the formation of nanotubes.

This technique is quite flexible and SWNTs [3.105], DWNTs [3.106] and MWNTs [3.107] have been obtained, in proportions depending on the carbon feedstock gas. The technique has also been exploited for some time in the production of vapor-grown carbon nanofibers [3.108].

The main drawback of this type of process is again that it is difficult to control the size of the metal nanoparticles, and thus nanotube formation is often accompanied by the production of undesired carbon forms (amorphous carbon or polyaromatic carbon phases found as various phases or as coatings). In particular, encapsulated forms have been often found as the result of the formation of metal particles that are too large to promote nanotube growth (and so they can end up being totally covered with graphene layers instead).

The same kind of parameters have to be controlled as for heterogeneous processes in order to finely tune this process and selectively obtain the desired morphology and structure of the nanotubes formed, such as: the choice of the carbon source; the reaction temperature; the residence time; the composition of the incoming gaseous feedstock, with particular attention paid to the role played by the proportion of hydrogen, which can influence the orientation of the graphene with respect to the nanotube axis, thus switching from c-MWNT to h-MWNT [3.82]; and the ratio of the metallorganic precursor to the carbon source [3.101]. In an independent study [3.109], it was shown that the general tendency is:

1. To synthesize SWNTs when the ferrocene/benzene molar ratio is high, typically $\approx 15\%$
2. To produce MWNTs when the ferrocene/benzene molar ratio is between ≈ 4 and $\approx 9\%$
3. To synthesize carbon nanofibers when the ferrocene/benzene molar ratio is below $\approx 4\%$.

As recently demonstrated, the overall process can be improved by adding other compounds such as ammonia or sulfur-containing species to the reactive gas phase. The former allows aligned nanotubes and mixed C-N nanotubes [3.110] to be obtained, while the latter results in a significant increase in productivity [3.108, 111].

An interesting result is the increase in yield and purity brought about by a small input of oxygen, as achieved by using alcohol vapors instead of hydrocarbons as feedstock [3.112]. It is assumed that the oxygen preferably burns the poorly organized carbon out into CO_2, thereby enhancing the purity, and prevents the catalyst particles from being encapsulated in the carbon shells too early, making them inactive, thereby enhancing the nanotube yield. Moreover, it was found to promote the formation of SWNTs over MWNTs, since suppressing carbon shell formation suppresses MWNT formation too.

It should be emphasized that only small amounts have been produced so far, and scale-up to industrial levels seems quite difficult due to the large number of parameters that must be considered. A critical one is to be able to increase the quantity of metallorganic compound that is used in the reactor, in order to increase production, without obtaining particles that are too big. This problem has not yet been solved. An additional problem inherent in the process is the possibility of clogging the reactor due to the deposition of metallic nanoparticles on the reactor walls followed by carbon deposition. An interesting alternative could be the injection, into the vertical floating reactor, of a supported catalyst powder instead of an organometallic compound. This approach has allowed the continuous production of single-walled carbon nanotubes with scaling capability up to 220 g/h [3.114].

A significant breakthrough concerning this technique could be the HiPCo process developed at Rice University, which produces SWNTs of very high purity [3.115, 116]. This gas phase catalytic reaction uses carbon monoxide to produce, from [$Fe(CO)_5$], a SWNT material that is claimed to be relatively free of by-products. The temperature and pressure conditions required are applicable to industrial plants. Upon heating, the [$Fe(CO)_5$] decomposes into atoms which condense into larger clusters, and SWNT nucleate and grow on these particles in the gas phase via CO disproportionation (the Boudouard reaction, see *Heterogeneous Processes* in Sect. 3.1.2):

The company Carbon Nanotechnologies Inc. (Houston, USA) currently sells raw SWNT materials prepared in this way, at a market price of 375 $/g, or 500 $/g if purified (2005 data). Other companies that specialize in MWNTs include Applied Sciences Inc. (Cedarville, USA), currently has a production facility of ≈ 40 tons/year of ≈ 100 nm large MWNTs (Pyrograf-III), and Hyperion Catalysis (Cambridge, USA), which makes MWNT-based materials. Though prepared in a similar way by CCVD-related processes, MWNTs remain far less expensive than SWNTs, reaching prices as low as 0.055 $/g (current ASI fares for Pyrograf-III grade).

Templating

Another interesting technique, although one that is definitely not suitable for mass production (and so we only touch on it briefly here), is the templating technique. It is the only other method aside from the electric arc technique that is able to synthesize carbon nanotubes without any catalyst. Any other work reporting the catalyst-free formation of nanotubes is actually likely to have involved the presence of catalytic metallic impurities in the reactor or some other factors that caused a chemical gradient in the system. Another useful aspect of this approach is that it allows aligned nanotubes to be obtained naturally, without the help of any subsequent alignment procedure. However, the template must be removed (dissolved) to recover the nanotubes, in which case the alignment of the nanotubes is lost.

Templating – Experimental Devices

The principle of this technique is to deposit the solid carbon coating obtained from the CVD method onto the walls of a porous substrate whose pores are arranged in parallel channels. The feedstock is again a hydrocarbon, such as a common source of carbon. The substrate can be alumina or zeolite for instance, which present natural channel pores, while the whole system is heated to a temperature that cracks the hydrocarbon selected as the carbon source (Fig. 3.17).

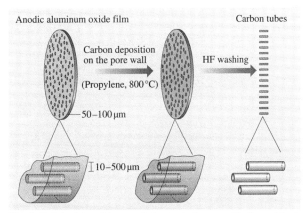

Fig. 3.17 Principle of the templating technique used in the catalyst-free formation of single-walled or concentric-type multiwalled carbon nanotubes (from [3.113])

Templating – Results

Provided the chemical vapor deposition mechanism (which is actually better described as a chemical vapor infiltration mechanism) is well controlled, synthesis results in the channel pore walls being coated with a variable number of graphenes. Both MWNTs (exclusively concentric type) and SWNTs can be obtained. The smallest SWNTs (diameters ≈ 0.4 nm) ever obtained (Sect. 3.1) were actually been synthesized using this technique [3.11]. The nanotube lengths are directly determined by the channel lengths; in other words by the thickness of the substrate plate. One main advantage of the technique is the purity of the tubes (no catalyst remnants, and few other carbon phases). On the other hand, the nanotube structure is not closed at both ends, which can be an advantage or a drawback depending on the application. For instance, the porous matrix must be dissolved using one of the chemical treatments previously cited in order to recover the tubes. The fact that the tubes are open makes them even more sensitive to attack from acids.

3.2.3 Miscellaneous Techniques

In addition to the major techniques described in Sects. 3.2.1 and 3.2.2, many attempts to produce nanotubes in various ways, often with a specific goal in mind, such as looking for a low-cost or a catalyst-free production process, can be found in the literature. As yet, none has been convincing enough to be presented as a serious alternative to the major processes described previously. Some examples are provided in the following.

Hsu et al. [3.117] have succeeded in preparing MWNTs (including coiled MWNTs, a peculiar morphology resembling a spring) by a catalyst-free (although Li was present) electrolytic method, by running a 3–5 A current between two graphite electrodes (the anode was a graphite crucible and the cathode a graphite rod). The graphite crucible was filled with lithium chloride, while the whole system was heated in air or argon at $\approx 600\,°C$. As with many other techniques, by-products such as encapsulated metal particles, carbon shells, amorphous carbon, and so on, are formed.

Cho et al. [3.118] have proposed a pure chemistry route to nanotubes, using the polyesterification of citric acid onto ethylene glycol at $50\,°C$, followed by polymerization at $135\,°C$ and then carbonization at $300\,°C$ under argon, followed by oxidation at $400\,°C$ in air. Despite the latter oxidation step, the solid product contains short MWNTs, although they obviously have poor nanotextures. By-products such as carbon shells and amorphous carbon are also formed.

Li et al. [3.119] have also obtained short MWNTs through a catalyst-free (although Si is present) pyrolytic method which involves heating silicon carbonitride nanograins in a BN crucible to $1200–1900\,°C$ in nitrogen within a graphite furnace. No details are given about the possible occurrence of by-products, but they are likely considering the complexity of the chemical system (Si-C-B-N) and the high temperatures involved.

Terranova et al. [3.120] have investigated the catalyzed reaction between a solid carbon source and atomic hydrogen. Graphite nanoparticles (≈ 20 nm) are sent with a stream of H_2 onto a Ta filament heated at $2200\,°C$. The species produced, whatever they are, then hit a Si polished plate warmed to $900\,°C$ that supports transition metal particles. The whole chamber is kept in a dynamic vacuum of 40 Torr. SWNTs are supposed to form according to the authors, although their images are not very convincing. One major drawback of the method, besides its complexity compared to the others, is that it is difficult to recover the *nanotubes* from the Si substrates to which they seem to be firmly bonded.

The final example is an attempt to prepare nanotubes by diffusion flame synthesis [3.121]. A regular gaseous hydrocarbon source (ethylene, ...) along with ferrocene vapor is passed into a laminar diffusion flame derived from air and CH_4 of temperature $500–1200\,°C$. SWNTs are formed, together with encapsulated metal particles, soot, and so on. In addition to a low yield, the SWNT structure is quite poor.

3.2.4 Synthesis of Carbon Nanotubes with Controlled Orientation

Several applications (such as field emission-based displays Sect. 3.6) require that carbon nanotubes grow as highly aligned bunches, in highly ordered arrays, or that they are located at specific positions. In this case, the purpose of the process is not mass production but controlled growth and purity, with subsequent control of nanotube morphology, texture and structure. Generally speaking, the more promising methods for the synthesis of aligned nanotubes are based on CCVD processes, which involve the use of molecular precursors as carbon sources, and the method of thermal cracking assisted by the catalytic activity of transition metal (Co, Ni, Fe) nanoparticles deposited onto solid supports. Although this approach initially produced mainly MWNTs, DWNT [3.122] and SWNT [3.123] arrays can be selectively obtained today. Generally speaking,

SWNTs and DWNTs nucleate at higher temperatures than MWNTs [3.124].

However, the catalyst-free templating methods related to those described in Sect. 3.2.2 are not considered here, due to the lack of support after the template is removed, which means that the previous alignment is not maintained.

During the CCVD growth, nanotubes can self-assemble into nanotube bunches aligned perpendicular to the substrate if the catalyst film on the substrate has a critical thickness [3.127, 128]. The driving forces for this alignment are the van der Waals interactions between the nanotubes, which allow them to grow perpendicularly to the substrates. If the catalyst nanoparticles are deposited onto a mesoporous substrate, the mesoscopic pores may also have an effect on the alignment when the growth starts, thus controlling the growth direction of the nanotubes. Two kinds of substrates have been used so far for this purpose: mesoporous silica [3.129, 130] and anodic alumina [3.131].

Different methods of depositing metal particles onto substrates have been reported in the literature:

1. Deposition of a thin film on alumina substrates using metallic salt precursor impregnation followed by oxidation/reduction steps [3.132].
2. Embedding catalyst particles in mesoporous silica by sol–gel processes [3.129].
3. Thermal evaporation of Fe, Co, Ni or Co-Ni metal alloys on SiO_2 or quartz substrates under high vacuum [3.133, 134].
4. Photolithographic patterning of metal-containing photoresist polymer using conventional black and white films as a mask [3.135] or photolithography and the inductive plasma deep etching technique [3.136].
5. Electrochemical deposition into pores in anodic aluminum oxide templates [3.131].
6. Deposition of colloidal suspensions of catalyst particles with tailored diameters on a support [3.137–141], by spin-coating for instance.
7. Stamping a catalyst precursor over a patterned silicon wafer is also possible and has been used to grow networks of nanotubes parallel to the substrate (Fig. 3.18a), or more generally to localize the growth of individual CNTs [3.142].

A technique that combines the advantages of electron beam lithography and template methods has also been reported for the large-scale production of ordered MWNTs [3.143] or AFM tips [3.144].

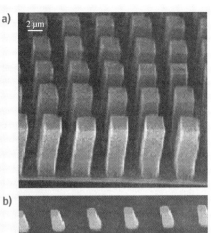

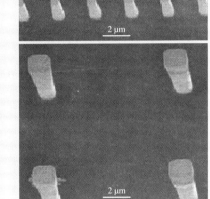

Fig. 3.18 (a) Example of a controlled network of nanotubes grown parallel to the substrate [3.125]; (b) example of a free-standing MWNT array obtained from the pyrolysis of a gaseous carbon source over catalyst nanoparticles previously deposited onto a patterned substrate. Each square-base rod is a bunch of MWNTs aligned perpendicular to the surface (from [3.126])

Depositing the catalyst nanoparticles onto a prepatterned substrate allows one to control the frequency of local occurrence and the arrangement of the nanotube bunches formed. The materials produced mainly consist of arrayed, densely packed, freestanding, aligned MWNTs (Fig. 3.18b), which are quite suitable for field emission-based applications for instance [3.126]. SWNTs have also been produced, and it was reported that the introduction of water vapor during the CVD process allows impurity-free SWNTs to be synthesized [3.145], due to a mechanism related to that

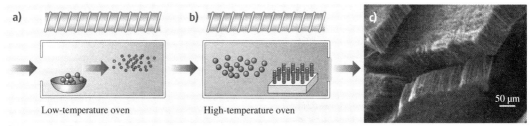

Fig. 3.19a–c Sketch of a double-furnace CCVD device used in the organometallic/hydrocarbon copyrolysis process. (a) Sublimation of the precursor. (b) Decomposition of the precursor and MWNT growth onto the substrate. (c) Example of the densely packed and aligned MWNT material obtained (from [3.146])

previously proposed for the effect of using alcohol instead of hydrocarbon feedstock [3.127].

When a densely packed coating of vertically aligned MWNTs is desired (Fig. 3.19c), another route is the pyrolysis of hydrocarbons in the presence of organometallic precursor molecules like metallocene or iron pentacarbonyl, operating in a dual furnace system (Fig. 3.19a,b). The organometallic precursor (such as ferrocene) is first sublimed at low temperature in the first furnace or injected as a solution along with the hydrocarbon feedstock, and then the whole system is pyrolyzed at higher temperature in the second furnace [3.105, 146–149]. The important parameters here are the heating or feeding rate of ferrocene, the flow rates of the vector gas (Ar or N_2) and the gaseous hydrocarbon, and the temperature of pyrolysis (650–1050 °C). Generally speaking, the codeposition process using [Fe(CO)$_5$] as the catalyst source results in thermal decomposition at elevated temperatures, producing atomic iron that deposits on the substrates in the hot zone of the reactor. Since nanotube growth occurs at the same time as the introduction of [Fe(CO)$_5$], the temperatures chosen for the growth depend on the carbon feedstock utilized; for example, they can vary from 750 °C for acetylene to 1100 °C for methane. Mixtures of [FeCp$_2$] and xylene or [FeCp$_2$] and acetylene have also been successfully used to produce freestanding MWNTs.

The nanotube yield and quality are directly linked to the amount and size of the catalyst particles, and since the planar substrates used do not exhibit high surface areas, the dispersion of the metal can be a key step in the process. It has been observed that an etching pretreatment of the surface of the deposited catalyst thin film with NH_3 may be critical to efficient nanotube growth of nanotubes since it provides the appropriate metal particle size distribution. It may also favor the alignment of MWNTs and prevent the formation of amorphous carbon due to the thermal cracking of acetylene [3.150]. The application of phthalocyanines of Co, Fe and Ni has also been reported, and in this case the pyrolysis of the organometallic precursors also produces the carbon for the vertically aligned MWNTs [3.151].

Densely packed coatings of vertically aligned MWNTs may also be produced over metal-containing deposits, such as iron oxides on aluminum [3.152], in which case MWNT synthesis takes place on small particles that are formed from the iron oxide deposit. Interestingly, it has been recently reported that well aligned MWNT arrays can be produced on a large scale on ceramic spheres using the floating catalyst techniques [3.153]. Finally, it has been recently reported that the Langmuir–Blodgett method can effectively be used to produce monolayers of aligned noncovalently functionalized SWNTs from organic solvent with dense packing [3.154]. This method seems valid for bulk materials with various diameters and offers the advantage that the SWNT monolayers are readily patterned for device integration by microfabrication.

3.3 Growth Mechanisms of Carbon Nanotubes

The growth mechanisms of carbon nanotubes are still the source of much debate. However, researchers have been impressively imaginative, and have come up with a number of hypotheses. One reason for the debate is that the conditions that allow carbon nanofilaments to grow are very diverse, which means that there are many related growth mechanisms. For a given set of conditions, the true mechanism is probably a combination of or a compromise between some of the proposals. Another reason is that the phenomena that occur during growth are pretty rapid and difficult to observe in situ. It is generally agreed, however, that growth occurs such that the number of dangling bonds is minimized, for energetic reasons.

3.3.1 Catalyst-Free Growth

As already mentioned, in addition to the templating technique, which is merely a chemical vapor infiltration mechanism for pyrolytic carbon, the growth of c-MWNT as a deposit on the cathode in the electric arc method is a rare example of catalyst-free carbon nanofilament growth. The driving force is obviously related to the electric field; in other words to charge transfer from one electrode to the other via the particles contained in the plasma. It is not clear how the MWNT nucleus is formed, but once it has, it may include the direct incorporation of C_2 species into the primary graphene structure, as it was previously proposed for fullerenes [3.155]. This is supported by recent C_2 radical concentration measurements that reveal an increasing concentration of C_2 from the anode being consumed at the growing cathode (Fig. 3.14). This indicates that C_2 are only secondary species and that the C_2 species may actually actively participate in the growth of c-MWNTs in the arc method.

3.3.2 Catalytically Activated Growth

Growth mechanisms involving catalysts are more difficult to ascertain, since they are more diverse. Although it involves a more or less extensive contribution from a VLS (vapor–liquid–solid [3.156]) mechanism, it is quite difficult to find comprehensive and plausible explanations that are able to account for both the various conditions used and the various morphologies observed. What follows is an attempt to provide overall explanations of most of the phenomena, while remaining consistent with the experimental data. We do not consider any hypothesis for which there is a lack of experimental evidence, such as the moving nanocatalyst mechanism, which proposes that dangling bonds from a growing SWNT may be temporarily stabilized by a nanosized catalyst located at the SWNT tip [3.28], or the scooter mechanism, which proposes that dangling bonds are temporarily stabilized by a single catalyst atom which moves around the edge of the SWNT, allowing subsequent C atom addition [3.157].

From various results, it appears that the most important parameters are probably the thermodynamic ones (only temperature will be considered here), the catalyst particle size, and the presence of a substrate. Temperature is critical and basically corresponds to the discrepancy between CCVD methods and solid carbon source-based methods.

Low-Temperature Conditions

Low-temperature conditions are typical used in CCVD, where nanotubes are frequently found to grow far below 1000 °C. If the conditions are such that the catalyst is a crystallized solid, the nanofilament is probably formed via a mechanism similar to a VLS mechanism, in which three steps are defined:

1. Adsorption then decomposition of C-containing gaseous moieties at the catalyst surface
2. Dissolution then diffusion of the C species through the catalyst, thus forming a solid solution
3. Back-precipitation of solid carbon as nanotube walls. The texture is then determined by the orientation of the crystal faces relative to the filament axis (Fig. 3.20), as demonstrated beyond doubt by transmission electron microscopy images such as those in *Rodriguez* et al. [3.20].

This mechanism can therefore provide either c-MWNT, h-MWNT, or platelet nanofibers. The latter, however, are mainly formed in large particle sizes (> 100 nm for example). Platelet nanofibers with low diameters (< 40 nm) have never been observed. The reasons for this are related to graphene energetics, such as the need to reach the optimal ratio between the amount of edge carbon atoms (with dangling bonds) and inner carbon atoms (where all of the σ and π orbitals are satisfied).

If conditions are such that the catalyst is a liquid droplet, due to the use of high temperatures or because a catalyst that melts at a low temperature is employed, a mechanism similar to that described above can still occur, which is really VLS (vapor = gaseous C species, liquid = molten catalyst, S = graphenes), but there are obviously no crystal faces to orient preferentially with the rejected graphenes. Energy minimization requirements will therefore tend to make them concentric and parallel to the filament axis.

With large catalyst particles (or in the absence of any substrate), the mechanisms above will generally follow a *tip growth* scheme: the catalyst will move forward while the rejected carbon will form the nanotube behind, whether there is a substrate or not. In this case, there is a good chance that one end will be open. On

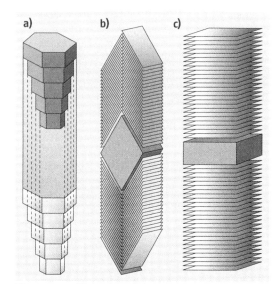

Fig. 3.20a–c Illustration of the possible relationships between the outer morphology of the catalyst crystal and the inner texture of the subsequent carbon nanofilament (adapted from [3.20]). In (**a**), a nanotube with graphenes making the wall arranged concentrically (*concentric MWNT*). In (**b**), a nanofibre with graphenes arranged so that they make an angle with respect to the nanofibre axis (*herringbone nanofibre*). In (**c**), a nanofibre with the graphenes piled up perpendicularly to the nanofibre axis (*platelet nanofibre*). Crystals are drawn with the projected plane perpendicular to the electron beam in a transmission electron microscope; the crystal morphologies and the subsequent graphene arrangements in the out-of-plane dimension are not intended to be accurate representations in these sketches (for example, the graphenes in the herringbone-type nanotubes or nanofibers in (**b**) cannot be arranged like a pile of open books, as sketched here, because it would leave too many dangling bonds)

the other hand, when the catalyst particles deposited onto the substrate are small enough (nanoparticles) to be held in place by interaction forces with the substrate, the growth mechanism will follow a *base growth* scheme, where the carbon nanofilament grows away from the substrate, leaving the catalyst nanoparticle attached to the substrate (Fig. 3.21).

The bamboo texture that affects both the herringbone and the concentric texture may reveal a distinguishing aspect of the dissolution-rejection mechanism: the periodic, discontinuous dynamics of the phenomenon. Once the catalyst has reached the saturation

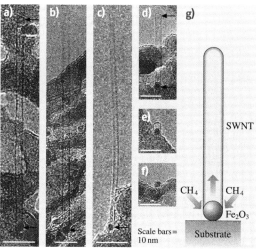

Fig. 3.21a–g High-resolution transmission electron microscopy images of several SWNTs grown from iron-based nanoparticles using the CCVD method, showing that particle sizes determine SWNT diameters in this case (adapted from [3.158]). Yet the catalyst crystal imaged (the *dark spot* at the bottom of each tube) is different for each figure, considering images backward from (**f**) to (**c**) illustrates what could be a sequence of growth of a SWNT from a single nanocrystal, as sketched in (**g**). (**a**) and (**b**) show additional examples of fully grown SWNTs, similar to (**c**)

threshold in terms of its carbon content, it expels it quite suddenly. Then it becomes able to incorporate a given amount of carbon again without having any catalytic activity for a little while. Then over-saturation is reached again, and so on. An exhaustive study of this phenomenon has been carried out by *Jourdain* et al. [3.159].

Therefore, it is clear that 1 catalyst particle = 1 nanofilament in any of the mechanisms above. This explains why, although it is possible to make SWNTs by CCVD methods, controlling the catalyst particle size is critical, since it influences the nanofilament that grows from it. Achieving a really narrow size distribution in CCVD is quite challenging, particularly when nanosizes are required for the growth of SWNTs. Only particles < 2 nm are useful for this (Fig. 3.21), since larger SWNTs are not favored energetically [3.10]. Another distinguishing aspect of the CCVD method and its related growth mechanisms is that the process can occur all along the isothermal zone of the reactor furnace since it is continuously fed with a carbon-rich feedstock, which is generally in excess, with a constant composition at a given species time of flight. Roughly

Table 3.2 Guidelines indicating the relationships between possible carbon nanofilament morphologies and some basic synthesis conditions. Columns (1) and (2) mainly relate to CCVD-based methods; column (3) mainly relates to plasma-based methods

		Increasing temperature... ...and physical state of catalyst			Substrate		Thermal gradient	
		solid (crystallized) (1)	liquid from melting (2)	liquid from condensing atoms (3)	yes	no	low	high
Catalyst particle size	$\lesssim 3$ nm	SWNT	SWNT	?	base-growth		long length	short length
	$\gtrsim 3$ nm	MWNT (c,h,b) platelet nanofiber	c-MWNT	SWNT	tip-growth	tip-growth		
Nanotube diameter		(heterogeneous related to catalyst particle size)		homogeneous (independent from particle size)			(except for SWNTs growing from case (3) catalyst)	
Nanotube/particle		one nanotube/particle		several SWNTs/particle				

speaking, the longer the isothermal zone (in gaseous carbon excess conditions), the longer the nanotubes. This is why the lengths of the nanotubes can be much longer than those obtained using solid carbon source-based methods.

Table 3.2 provides an overview of the relationship between general synthesis conditions and some features of nanotube grown.

High-Temperature Conditions

High-temperature conditions are typical used in solid carbon source-based methods such as the electric arc method, laser vaporization, and the solar furnace method (Sect. 3.2). The huge temperatures involved (several thousands of °C) atomize both the carbon source and the catalyst. Of course, catalyst-based SWNTs do not form in the areas with the highest temperatures (contrary to c-MWNTs in the electric arc method); the medium is a mixture of atoms and radicals, some of which are likely to combine and condense into liquid droplets. At some distance from the atomization zone, the medium is therefore made of carbon metal alloy droplets and of secondary carbon species that range from C_2 to higher order molecules such as corannulene, which is made of a central pentagon surrounded by five hexagons. The preferred formation of such a molecule can be explained by the previous association of carbon atoms into a pentagon, because it is the fastest way to limit dangling bonds at low energetic cost, thereby providing a fixation site for other carbon atoms (or C_2) which also will tend to close into a ring, again to limit dangling bonds. Since adjacent pentagons are not energetically favored, these cycles will be hexagons. Such a molecule is thought to be a probable precursor for fullerenes. Fullerenes are actually always produced, even in conditions that produce SWNTs. The same saturation in C described in Sect. 3.3.2 occurs for the carbon-metal alloy droplet as well, resulting in the precipitation of excess C outside the particle due to the effect of the decreasing thermal gradient in the reactor, which decreases the solubility threshold of C in the metal [3.160]. Once the *inner* carbon atoms reach the surface of the catalyst particle, they meet the *outer* carbon species, including corannulene, that will contribute to capping the merging nanotubes. Once formed and capped, nanotubes can grow both from the inner carbon atoms (Fig. 3.22a), according to the VLS mechanism proposed by *Saito* et al. [3.160], and from the outer carbon atoms, according the adatom mechanism proposed by *Bernholc* et al. [3.161]. In the latter, carbon atoms from the surrounding medium in the reactor are attracted then stabilized by the carbon/catalyst interface

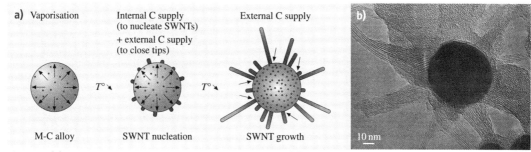

Fig. 3.22 (a) Mechanism proposed for SWNT growth (see text). (b) Transmission electron microscopy image of SWNT growing radially from the surface of a large Ni catalyst particle in an electric arc experiment. (Modified from [3.18])

at the nanotube/catalyst surface contact, promoting their subsequent incorporation at the tube base. The growth mechanism therefore mainly follows the base growth scheme. However, once the nanotubes are capped, any C_2 species that still remains in the medium that meets the growing nanotubes far from the nanotube/catalyst interface may still incorporate the nanotubes from both the side wall or the tip, thereby giving rise to some proportion of Stone–Wales defects [3.45]. The occurrence of a nanometer-thick surface layer of yttrium carbide (onto the main Ni-containing catalyst core), the lattice distance of which is commensurable with that of the C−C distance in graphene (as recently revealed by *Gavillet* et al. [3.55]), could possibly play a beneficial role in stabilizing the nanotube/catalyst interface, which could explain why the SWNT yield is enhanced by bimetallic alloys (as opposed to single metal catalysts).

A major difference from the low-temperature mechanisms described for CCVD methods is that many nanotubes are formed from a single, relatively large (≈ 10–50 nm) catalyst particle (Fig. 3.22b), whose size distribution is therefore not as critical as it is for the low-temperature mechanisms (particles that are too large, however, induce polyaromatic shells instead of nanotubes). This is why the diameters of SWNTs grown at high temperature are much more homogeneous than those associated with CCVD methods. The reason that the most frequent diameter is ≈ 1.4 nm is again a matter of energy balance. Single-wall nanotubes larger than ≈ 2.5 nm are not stable [3.10]. On the other hand, the strain on the C−C bond increases as the radius of curvature decreases. The optimal diameter (1.4 nm) should therefore correspond to the best energetic compromise. Another difference from the low-temperature mechanism for CCVD is that temperature gradients in high temperature methods are huge, and the gas phase composition surrounding the catalysts droplets is also subjected to rapid changes (as opposed to what could happen in a laminar flow of a gaseous feedstock whose carbon source is in excess). This explains why nanotubes from arcs are generally shorter than nanotubes from CCVD, and why mass production by CCVD is favored. In the latter, the metallic particle can act as a catalyst repeatedly as long as the conditions are maintained. In the former, the surrounding conditions change continuously, and the window for efficient catalysis can be very narrow. Decreasing the temperature gradients that occur in solid carbon source-based methods of producing SWNT, such as the electric arc reactor, should therefore increase the SWNT yield and length [3.162]. Amazingly, this is in opposition to what is observed during arc-based fullerene production.

3.4 Properties of Carbon Nanotubes

In previous sections, we noted that the normal planar configuration of graphene can, under certain growth conditions (Sect. 3.3), be changed into a tubular geometry. In this section, we take a closer look at the properties of these carbon nanotubes, which can depend on whether they are arranged as SWNTs or as MWNTs (Sect. 3.1).

3.4.1 Overview

The properties of MWNTs are generally similar to those of regular polyaromatic solids (which may exhibit graphitic, turbostratic or intermediate crystallographic structure). Variations are mainly due to different textural types of the MWNTs considered (concentric, herringbone, bamboo) and the quality of the nanotexture (Sect. 3.1), both of which control the extent of anisotropy. Actually, for polyaromatic solids that consist of stacked graphenes, the bond strength varies significantly depending on whether the in-plane direction is considered (characterized by very strong covalent and therefore very short – 0.142 nm – bonds) or the direction perpendicular to it (characterized by very weak van der Waals and therefore very loose – ≈ 0.34 nm – bonds). Such heterogeneity is not found in single (isolated) SWNTs. However, the heterogeneity returns, along with the related consequences, when SWNTs associate into bundles. Therefore, the properties – and applicability – of SWNTs may also change dramatically depending on whether single SWNT or SWNT ropes are involved.

In the following, we will emphasize the properties of SWNTs, since their unique structures often lead to different properties to regular polyaromatic solids. However, we will also sometimes discuss the properties of MWNTs for comparison.

3.4.2 General Properties of SWNTs

The diameters of SWNT-type carbon nanotubes fall in the nanometer regime, but SWNTs can be hundreds of micrometers long. SWNTs are narrower in diameter than the thinnest line that can be obtained in electron beam lithography. SWNTs are stable up to 750 °C in air (but they are usually damaged before this temperature is reached due to oxidation mechanisms, as demonstrated by the fact that they can be filled with molecules (Sect. 3.5). They are stable up to $\approx 1500-1800\,°\mathrm{C}$ in inert atmosphere, beyond which they transform into regular, polyaromatic solids (phases built with stacked graphenes instead of single graphenes) [3.163]. They have half the mass density of aluminum. The properties of a SWNT, like any molecule, are heavily influenced by the way that its atoms are arranged. The physical and chemical behavior of a SWNT is therefore related to its unique structural features [3.164].

3.4.3 Adsorption Properties of SWNTs

An interesting feature of a SWNT is that it has the highest surface area of any molecule due to the fact that a graphene sheet is probably the only example of a sheetlike molecule that is energetically stable under normal conditions. If we consider an isolated SWNT with one open end (achieved through oxidation treatment for instance), the surface area is equal to that of a single, flat graphene sheet: $\approx 2700\,\mathrm{m^2/g}$ (accounting for both sides).

In reality, nanotubes – specifically SWNTs – are usually associated with other nanotubes in bundles, fibers, films, papers, and so on, rather than as a single entity. Each of these associations has a specific range of porosities that determines its adsorption properties (this topic is also covered in Sect. 3.6.2 on applications). It is therefore more appropriate to discuss adsorption onto the outer or the inner surface of a bundle of SWNTs.

Furthermore, theoretical calculations have predicted that the adsorption of molecules onto the surface or inside of a nanotube bundle is stronger than that onto an individual tube. A similar situation exists for MWNTs, where adsorption can occur on or inside the tubes or between aggregated MWNTs. It has also been shown that the curvature of the graphene sheets constituting the nanotube walls results in a lower heat of adsorption compared to planar graphene (Sect. 3.1.1).

Accessible Specific Surface Area of CNTs

Various studies dealing with the adsorption of nitrogen onto MWNTs [3.165–167] and SWNTs [3.168] have highlighted the porous nature of these two materials. The pores in MWNTs can be divided mainly into hollow inner cavities with small diameters (with narrow size distributions, mainly 3–10 nm) and aggregated pores (with wide size distributions, 20–40 nm), formed by interactions between isolated MWNTs. It is also worth noting that the ultrastrong nitrogen capillarity in the aggregated pores dominates the total adsorption, indicating that the aggregated pores are much more important than the inner cavities of the MWNTs during adsorption. The determination of the space available between a bunch of closed MWNTs has been performed by grand canonical Monte Carlo simulation of nitrogen adsorption, resulting in a satisfactory description of the experimental N_2 adsorption and showing that the distance between nanotubes is in the 4–14 nm range [3.169]. Adsorption of N_2 has been studied on as-prepared and acid-treated SWNTs, and the results obtained highlight the microporous nature of SWNT materials, as opposed to the mesoporous nature of MWNT materials. Also, as opposed to isolated SWNTs (see above), surface areas that are well above $400\,\mathrm{m^2\,g^{-1}}$ have been measured for SWNT-bundle-containing materials, with internal surface areas of $300\,\mathrm{m^2\,g^{-1}}$ or higher.

The theoretical surface area of a carbon nanotube has a broad range, from 50 to $1315\,\mathrm{m^2\,g^{-1}}$ depending on the number of walls, the diameter, and the number of nanotubes in a bundle of SWNTs [3.170]. Experimentally, the surface area of a SWNT is often larger than that of a MWNT. The total surface area of as-grown SWNTs is typically between 400 and 900 m^2 g^{-1} (micropore volume 0.15–0.3 cm^3 g^{-1}), whereas values of 200 and 400 m^2 g^{-1} for as-produced MWNTs are often reported. In the case of SWNTs, the diameters of the tubes and the number of tubes in the bundle will have the most effect on the BET value. It is worth noting that opening or closing the central canal significantly influences the adsorption properties of nanotubes. In the case of MWNTs, chemical treatments such as KOH or NaOH activation are useful for promoting microporosity, and surface areas as high as 1050 m^2 g^{-1} have been reported [3.171, 172]. An efficient two-step treatment (acid +CO$_2$ activation) has been reported to open both ends of MWNTs [3.173]. Therefore, it appears that opening or cutting carbon nanotubes, as well as chemically treating them (using purification steps for example) can considerably affect their surface area and pore structure.

Adsorption Sites and Binding Energy of the Adsorbates

An important problem to solve when considering adsorption onto nanotubes is to identify the adsorption sites. The adsorption of gases into a SWNT bundle can occur inside the tubes (internal sites), in the interstitial triangular channels between the tubes, on the outer surface of the bundle (external sites), or in the grooves formed at the contacts between adjacent tubes on the outside of the bundle (Fig. 3.23). Experimental adsorption studies on SWNT have confirmed the adsorption on internal, external and groove sites [3.175]. Modeling studies have pointed out that the convex surface of the SWNT is more reactive than the concave one and that this difference in reactivity increases as the tube diameter decreases [3.176]. Compared to the highly bent region in fullerenes, SWNTs are only moderately curved and are expected to be much less reactive towards dissociative chemisorption. Models have also predicted enhanced reactivity at the kink sites of bent SWNTs [3.177]. Additionally, it is worth noting that unavoidable imperfections, such as vacancies, Stone–Wales defects, pentagons, heptagons and dopants, are believed to play a role in tailoring the adsorption properties [3.178].

Considering closed-end SWNTs first, simple molecules can be adsorbed onto the walls of the outer nanotubes of the bundle and preferably on the external grooves. In the first stages of adsorption (corresponding to the most attractive sites for adsorption), it seems that adsorption or condensation in the interstitial channels of the SWNT bundles depends on the size of the molecule (and/or on the SWNT diameters) and on their interaction energies [3.179–181]. Opening the tubes favors gas adsorption (including O$_2$, N$_2$ within the inner walls [3.182, 183]). It was found that the adsorption of nitrogen on open-ended SWNT bundles is three times larger than that on closed-ended SWNT bundles [3.184]. The significant influence that the external surface area of the nanotube bundle has on the character of the surface adsorption isotherm of nitrogen (type I, II or even IV of the IUPAC classification) has been demonstrated from theoretical calculations [3.185]. Additionally, it has been shown that the analysis of theoretical adsorption isotherms, determined from a simple model based on the formalism of *Langmuir* and *Fowler*, can help to experimentally determined the ratio of open to closed SWNTs in a sample [3.186]. For hydrogen and other small molecules like CO, computational methods have shown that, for open SWNTs, the pore, interstitial and groove sites are energetically more favorable than surface sites [3.187, 188]. In the case of carbon monoxide, aside from physisorbed CO, CO hydrogen bonds to hydroxyl functionalities created on the SWNTs by acid purification have been identified [3.188]. FTIR and temperature-programmed desorption (TPD) experiments have shown that NH$_3$ or NO$_2$ adsorb molecularly and that NO$_2$ is slightly more strongly bound than NH$_3$ [3.189]. For NO$_2$, the formation of nitrito (O-bonded) complexes is preferred to nitro (N-bonded) ones. For ozone, a strong oxidizing agent, theoretical calculations have shown that physisorption occurs on ideal, defect-free SWNT, whereas strong chemisorption occurs on Stone–Wales defects, highlighting the key role of defective sites in adsorption properties [3.190]. Finally, for acetone, TPD experiments have shown that this molecule chemisorbs on SWNT while physisorption occurs on graphite [3.191].

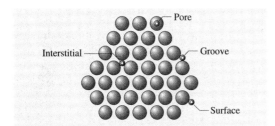

Fig. 3.23 Sketch of a SWNT bundle, illustrating the four different adsorption sites (adapted from [3.174])

For MWNTs, adsorption can occur in the aggregated pores, inside the tube or on the external walls. In the latter case, the presence of defects, as incomplete graphene layers, must be taken into consideration. Although adsorption between the graphenes (intercalation) has been proposed in the case of hydrogen adsorption in h-MWNTs or platelet nanofibers [3.192], it is unlikely to occur for many molecules due to steric effects and should not prevail for small molecules due to the long diffusion paths involved. In the case of inorganic fluorides (BF_3, TiF_4, NbF_5 and WF_6), accommodation of the fluorinated species into the carbon lattice has been shown to result from intercalation and adsorption/condensation phenomena. In this case, doping-induced charge transfer has been demonstrated [3.193].

Only a few studies deal with adsorption sites in MWNTs, but it has been shown that butane adsorbs more onto MWNTs with smaller outside diameters, which is consistent with another statement that the strain on curved graphene surfaces affects sorption. Most of the butane adsorbs to the external surface of the MWNTs while only a small fraction of the gas condenses in the pores [3.194]. Comparative adsorption of krypton or of ethylene onto MWNTs or onto graphite has allowed scientists to determine the dependence of the adsorption and wetting properties of the nanotubes on their specific morphologies. Nanotubes were found to have higher condensation pressures and lower heats of adsorption than graphite [3.195]. These differences are mainly due to decreased lateral interactions between the adsorbed molecules, related to the curvature of the graphene sheets.

A limited number of theoretical as well as experimental studies on the binding energies of gases onto carbon nanotubes exist. While most of these studies report low binding energies on SWNTs, consistent with physisorption, some experimental results, in particular for hydrogen, are still controversial (Sect. 3.6.2). For platelet nanofibers, the initial dissociation of hydrogen on graphite edge sites, which constitute most of the nanofiber surface, has been proposed [3.196]. For carbon nanotubes, a mechanism that involves H_2 dissociation on the residual metal catalyst followed by H spillover and adsorption on the most reactive nanotube sites was envisaged [3.197]. Similarly, simply mixing carbon nanotubes with supported palladium catalysts increased the hydrogen uptake of the carbon by a factor of three, due to hydrogen spillover from the supported catalyst [3.198]. Doping nanotubes with alkali may enhance hydrogen adsorption, due to charge transfer from the alkali metal to the nanotube, which polarizes the H_2 molecule and induces dipole interactions [3.199].

Generally speaking, the adsorbates can be either charge donors or acceptors to the nanotubes. Trends in the binding energies of gases with different van der Waals radii suggest that the groove sites of SWNTs are the preferred low coverage adsorption sites due to their higher binding energies. Finally, several studies have shown that, at low coverage, the binding energy of the adsorbate on SWNT is between 25 and 75% higher than the binding energy on a single graphene. This discrepancy can be attributed to an increase of effective coordination at the binding sites, such as the groove sites, in SWNTs bundles [3.200, 201]. Representative results on the adsorption properties of SWNTs and MWNTs are summarized in Table 3.3.

3.4.4 Electronic and Optical Properties

The electronic states in SWNTs are strongly influenced by their one-dimensional cylindrical structures.

Table 3.3 Adsorption properties and sites of SWNTs and MWNTs. The letters in the *Absorption sites* column refer to Fig. 3.22. The data in the last two columns are from [3.174]

Type of nanotube	Porosity ($cm^3 g^{-1}$)	Surface area (m^2/g)	Binding energy of the adsorbate	Adsorption sites	Attractive potential per site (eV)	Surface area per site (m^2/g)
SWNT (bundle)	Microporous V_{micro}: 0.15–0.3	400–900	Low, mainly physisorption 25–75% > graphite	Surface (A) Groove (B) Pore (C) Interstitial (D)	0.049 0.089 0.062 0.119	483 22 783 45
MWNT	Mesoporous	200–400	Physisorption	Surface Pore Aggregated pores	–	–

One-dimensional subbands are formed that have strong singularities in the density of states (Van Hove singularities) [3.202]. By rolling the graphene sheet to form a tube, new periodic boundary conditions are imposed on the electronic wavefunctions, which give rise to one-dimensional subbands: $CnK = 2q$ where q is an integer. Cn is the roll-up vector $na\mathbf{1} + ma\mathbf{2}$ which defines the helicity (chirality) and the diameter of the tube (Sect. 3.1). Much of the electronic band structure of CNTs can be derived from the electronic band structure of graphene by applying the periodic boundary conditions of the tube under consideration. The conduction and the valence bands of the graphene only touch at six corners (K points) of the Brillouin zone [3.203]. If one of these subbands passes through the K point, the nanotube is metallic; otherwise it is semiconducting. This is a unique property that is not found in any other one-dimensional system, which means that for certain orientations of the honeycomb lattice with respect to the tube axis (chirality), some nanotubes are semiconducting and others are metallic. The band gap for semiconducting tubes is found to be inversely proportional to the tube diameter. As pointed out in Sect. 3.1, knowing (n, m) allows us, in principle, to predict whether the tube is metallic or not. The energy gap decreases for larger tube diameters and MWNTs with larger diameter are found to have properties similar to other forms of regular, polyaromatic solids. It has been shown that electronic conduction mostly occurs through the external tube for MWNTs; even so, interactions with internal tubes often cannot be neglected and they depend upon the helicity of the neighboring tubes [3.204]. The electronic and optical properties of the tubes are considerably influenced by the environment [3.205]. Under externally applied pressure, the small interaction between the tube walls results in the internal tubes experiencing reduced pressure [3.206]. The electronic transition energies are in the infrared and visible spectral range. The one-dimensional Van Hove singularities have a large influence on the optical properties of CNTs. Visible light is selectively and strongly absorbed, which can lead to the spontaneous burning of agglomerated SWNTs in air at room temperature [3.207]. Strong Coulomb interaction in quasi-one dimension leads to the formation of excitons with very large binding energies in CNTs (200–400 meV), and degenerated states at the K, K' points lead to multiple exciton states with dipole allowed (bright) and dipole forbidden transitions (dark) [3.208, 209]. Photoluminescence can be observed in individual SWNT aqueous suspensions stabilized by the addition of surfactants. Detailed photoexcitation maps provide information about the helicity (chirality)-dependent transition energies and the electronic band structures of CNTs [3.210]. Agglomeration of tubes into ropes or bundles influences the electronic states of CNTs. Photoluminescence signals are quenched for agglomerated tubes.

CNTs are model systems for the study of one-dimensional transport in materials. Apart from the singularities in the density of states, electron–electron interactions are expected to show drastic changes at the Fermi edge; the electrons in CNTs are not described by a Fermi liquid, but instead by a Luttinger liquid model [3.211] that describes electronic transport in one-dimensional systems. It is expected that the variation of electronic conductance vs. temperature follows a power law, with zero conductance at low temperatures. Depending on how L_ϕ (the coherence length) on the one hand and L_m (the electronic mean free path) on the other hand compare to L (the length of the nanotube), different conduction modes are observed: ballistic if $L \ll L_\phi, L \ll L_m$, diffusive if $L_\phi \ll L_m < L$ and localization if $L_m \ll L_\phi \ll L$. Fluctuations in the conductance can be seen when $L \approx L_\phi$. For ballistic conduction (a small number of defects) [3.212–214], the predicted electronic conductance is independent of the tube length. The conductance value is twice the fundamental conductance unit $G_0 = 4\,\mathrm{e}/h$ due to the existence of two propagating modes. Due to the reduced electron scattering observed for metallic CNTs and their stability at high temperatures, CNTs can support high current densities (max. 10^9 A/cm^2): about three orders of magnitude higher than Cu. Structural defects can, however, lead to quantum interference of the electronic wave function, which localizes the charge carriers in one-dimensional systems and increases resistivity [3.211, 215, 216]. Localization and quantum interference can be strongly influenced by applying a magnetic field [3.217]. At low temperatures, the discrete energy spectrum leads to a Coulomb blockade resulting in oscillations in the conductance as the gate voltage is increased [3.216]. In order to observe the different conductance regimes, it is important to consider the influence of the electrodes where Schottky barriers are formed. Palladium electrodes have been shown to form excellent junctions with nanotubes [3.218]. The influence of superconducting electrodes or ferromagnetic electrodes on electronic transport in CNTs due to spin polarization has also been explored [3.219, 220].

As a probable consequence of both the small number of defects (at least the kind of defects that oppose phonon transport) and the cylindrical topography,

SWNTs exhibit a large phonon mean free path, which results in a high thermal conductivity. The thermal conductivity of SWNTs is comparable to that of a single, isolated graphene layer or high purity diamond [3.221], or possibly higher ($\approx 6000\,\text{W}/(\text{m K})$).

3.4.5 Mechanical Properties

While tubular nanomorphology is also observed for many two-dimensional solids, carbon nanotubes are unique due to the particularly strong bonding between the carbons (sp^2 hybridization of the atomic orbitals) of the curved graphene sheet, which is stronger than in diamond (sp^3 hybridization), as revealed by the difference in C–C bond lengths (0.142 versus 0.154 nm for graphene and diamond respectively). This makes carbon nanotubes – SWNTs or c-MWNTs – particularly stable against deformations. The tensile strength of SWNTs can be 20 times that of steel [3.222] and has actually been measured as $\approx 45\,\text{GPa}$ [3.223]. Very high tensile strength values are also expected for ideal (defect-free) c-MWNTs, since combining perfect tubes concentrically is not supposed to be detrimental to the overall tube strength, provided the tube ends are well capped (otherwise, concentric tubes could glide relative to each other, inducing high strain). Tensile strength values as high as $\approx 150\,\text{GPa}$ have actually been measured for perfect MWNTs from an electric arc [3.224], although the reason for such a high value compared to that measured for SWNTs is not clear. It probably reveals the difficulties involved in carrying out such measurements in a reliable manner. The flexural modulus of perfect MWNTs should logically be higher than that for SWNTs [3.222], with a flexibility that decreases as the number of walls increases. On the other hand, measurements performed on defective MWNTs obtained from CCVD exhibit a range of 3–30 GPa [3.225]. Values of tensile modulus are also the highest values known, 1 TPa for MWNTs [3.226], and possibly even higher for SWNTs, up to 1.3 TPa [3.227, 228]. Figure 3.24 illustrates how defect-free carbon nanotubes could spectacularly revolutionize the field of high performance fibrous materials.

3.4.6 Reactivity

The chemical reactivities of graphite, fullerenes, and carbon nanotubes are similar in many ways. Like any small object, carbon nanotubes have a large surface to interact with their environment (Sect. 3.4.1). It is worth noting, however, that nanotube chemistry differs from

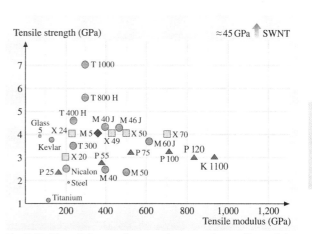

Fig. 3.24 Plot of the tensile strength versus the tensile modulus for various fibrous materials and SWNTs. *Large circles* are PAN-based carbon fibers, which include the fiber with the highest tensile strength available on the market (T1000 from Torayca); *Triangles* are pitch-based carbon fibers, which include the fiber with the highest tensile modulus on the market (K1100 from Amoco)

that observed for regular polyaromatic carbon materials due to the unique shape of the nanotube, its small diameter, and its structural properties. Unlike graphite, perfect SWNTs have no (chemically active) dangling bonds (the reactions of polyaromatic solids is known to occur mainly at graphene edges). Unlike fullerenes, the ratio of *weak* sites (C–C bonds involved in heterocycles) to strong sites (C–C bonds between regular hexagons) is only deviates slightly from 0 for ideal tubes. For C_{60} fullerenes this ratio is $1 - C_{60}$ molecules have 12 pentagons (therefore accounting for $5 \times 12 = 30$ C–C bonds) and 20 hexagons, each of them with three C–C bonds not involved in an adjacent pentagon but shared with a neighboring hexagon (so $20 \times 3 \times 1/2 = 30$ C–C bonds are involved in hexagons only). Although graphene faces are chemically relatively inert, the radius of curvature imposed on the graphene in nanotubes causes the three normally planar C–C bonds caused by sp^2 hybridization to undergo distortions, resulting in bond angles that are closer to three of the four C–C bonds in diamond (characteristic of genuine sp^3 hybridization), as the radius of curvature decreases. Even though it is not enough to make the carbon atoms chemically reactive, one consequence of this is that either nesting sites are created at the concave surface, or strong physisorption sites are created above each carbon atom of the convex surface, both with a bonding efficiency that increases as the nanotube diameter decreases.

As already pointed out in Sect. 3.1, the chemical reactivities of SWNTs (and c-MNWTs) are believed to derive mainly from the caps, since they contain six pentagons each, as opposed to the tube body, which supposedly only contains hexagons. Indeed, applying oxidizing treatments to carbon nanotubes (air oxidation, wet-chemistry oxidation) selectively opens the nanotube tips [3.229]. However, that SWNTs can be opened by oxidation methods and then filled with foreign molecules such as fullerenes (Sect. 3.5) suggests the occurrence of side defects [3.15], whose identity and occurrence were discussed and then proposed to be an average of one Stone–Wales defect every 5 nm along the tube length, involving about 2% of the carbon atoms in a regular (10,10) SWNT [3.230]. A Stone–Wales defect is formed from four adjacent heterocycles, two pentagons and two heptagons, arranged in pairs opposite each other. Such a defect allows localized double bonds to form between the carbon atoms involved in the defect (instead of these electrons participating in the delocalized electron cloud above the graphene as usual, enhancing the chemical reactivity, for example toward chlorocarbenes [3.230]). This means that the overall chemical reactivity of carbon nanotubes should depend strongly on how they are synthesized. For example, SWNTs prepared by the arc-discharge method are believed to contain fewer structural defects than CCVD-synthesized SWNTs, which are more chemically reactive. Of course, the reactivity of h-MWNT-type nanotubes is intrinsically higher, due to the occurrence of accessible graphene edges at the nanotube surface.

3.5 Carbon Nanotube-Based Nano-Objects

3.5.1 Heteronanotubes

It is possible to replace some or all of the carbon atoms in a nanotube with atoms of other elements without damaging the overall honeycomb lattice-based graphene structure. Nanotubes modified in this way are termed here *heteronanotubes*.

The elements used to replace carbon in this case are boron and/or nitrogen. Replacing carbon atoms in this way can result in new behavior (for example, BN nanotubes are electrical insulators), improved properties (resistance to oxidation for instance), or better control over such properties. For instance, one current challenge in carbon SWNT synthesis is to control the processing so that the desired SWNT structure (metallic or semiconductor) is formed selectively. In this regard, it was demonstrated that replacing some C atoms with N or B atoms leads to SWNTs with systematically metallic electrical behavior [3.231, 232].

Some examples of heteronanotubes – mainly MWNTs – can be found in the literature. The heteroatom usually involved is nitrogen, due to the ease with which gaseous or solid nitrogen- and/or boron-containing species (such as N_2, NH_3, BN, HfB_2) can be passed into existing equipment for synthesizing MWNTs [3.231, 233] until complete substitution of carbon occurs [3.234, 235]. An amazing result of such attempts to synthesize hetero-MWNTs is the subsequent formation of *multilayered* c-MWNTs: MWNTs made up of coaxial alternate carbon graphene tubes and boron nitride graphene tubes [3.236]. On the other hand, there are only a few examples of hetero-SWNTs. Syntheses of B- or N-containing SWNTs have recently been reported [3.232, 237], while just one successful synthesis of genuine BN-SWNTs has been reported so far [3.238].

3.5.2 Hybrid Carbon Nanotubes

Hybrid carbon nanotubes are defined here as carbon nanotubes, SWNTs or MWNTs that have inner cavities filled (partially or entirely) with foreign atoms, molecules, compounds or crystals. The terminology X@SWNT (or X@MWNT, if appropriate, where X is the atom, molecule and so on involved) is used for such structures [3.239].

Motivation

But why should we want to fill the cavities of carbon nanotubes [3.230]? The very small inner cavity of nanotubes is an amazing tool for preparing and studying the properties of confined nanostructures of any type, such as salts, metals, oxides, gases, or even discrete molecules like C_{60}, for example. Due to the almost one-dimensional structure of carbon nanotubes (particularly for SWNTs), we might expect that encapsulated material might have different physical and/or chemical properties to the unencapsulated material, and that the hybrid nanotube itself may behave differently to a *pure* nanotube. Indeed, if the volume available inside a carbon nanotube is small enough, the foreign material is largely *surface atoms* of reduced coordination. The original motivation to create such hybrids was to obtain metal nanowires that are likely to of interest in electronics (as quantum wires). In this case, the nanotubes were considered to be nanomolds for the metal filler, and it was probably intended that the nanomold was to be removed afterwards. However, it is likely that this removal of the SWNT *container* to liberate the one-dimensional structure inside it may destroy or at least transform this structure due to the stabilizing effect of interactions with the nanotube wall.

Filling nanotubes while they grow (in situ filling) was one of the pioneering methods of nanotechnology. In most cases, however, the filling step is separate from nanotube synthesis. Three filling methods can then be distinguished: (a) wet chemistry procedures; capillarity-based physical procedures involving (b) a molten material or (c) a sublimated material.

Generally speaking, it is difficult to estimate the filling rate, and this is usually achieved through TEM observation, without obtaining any statistics on the number of tubes observed. Moreover, as far as SWNTs are concerned, the fact that the nanotubes are gathered into bundles makes it difficult to observe the exact number of filled tubes, as well as to estimate the filled length for each tube. It however seems that estimation of filling rates can now be reliably obtained from x-ray studies and Raman spectroscopy.

It is also possible to fill carbon nanotubes with materials that could not have been introduced directly. This is done by first filling the nanotubes with an appropriate precursor (one that is able to sublime, or melt or solubilize) that will later be transformed into the required material by chemical reaction or by a physical interaction, such as electron beam irradiation for example [3.240]. For secondary chemical transformation, reduction by H_2 is often used to obtain nanotubes filled with metals [3.241]. Sulfides can also be obtained if H_2S is used as a reducing agent [3.241].

Because the inner diameters of SWNTs are generally smaller than those of MWNTs, it is more difficult to fill them, and the driving forces involved in this phenomenon are not yet totally understood (see the review paper by *Monthioux* [3.230]). This field is therefore growing fairly rapidly, and so we have chosen to cite the pioneering works and then to focus on more recent works dealing with the more challenging topic of filling SWNTs.

In Situ Filling Method
Initially, most hybrid carbon nanotubes synthesized were based on MWNTs prepared using the electric arc method, and were obtained directly during processing. The filling materials were easily introduced in the system by drilling a central hole in the anode and filling it with the heteroelement. The first hybrid products obtained using this approach were all reported the same year [3.242–246] for heteroelements such as Pb, Bi, YC_2 and TiC. Later on, *Loiseau* and *Pascard* [3.247] showed that MWNTs could also be filled to several μm in length by elements such as Se, Sb, S, and Ge, but only with nanoparticles of elements such as Bi, B, Al and Te. Sulfur was suggested to play an important role during the in situ formation of filled MWNTs using arc discharge [3.248]. This technique is no longer the preferred one because it is difficult to control the filling ratio and yield and to achieve mass production.

Wet Chemistry Filling Method
The wet chemistry method requires that the nanotube tips are opened by chemical oxidation prior to the filling step. This is generally achieved by refluxing the nanotubes in dilute nitric acid [3.249–251], although other oxidizing liquid media may work as well, such as [$HCl + CrO_3$] [3.252] or chlorocarbenes formed from the photolytic dissociation of $CHCl_3$ [3.230], a rare example of a nonacidic liquid route to opening SWNT tips. If a dissolved form (such as a salt or oxide) of the desired metal is introduced during the opening step, some of it will get inside the nanotubes. An annealing treatment (after washing and drying the treated nanotubes) may then lead to the oxide or to the metal, depending on the annealing atmosphere [3.229]. Although the wet chemistry method initially looked promising because a wide variety of materials can be introduced into nanotubes in this way and it operates at temperatures that are not much different from room temperature; however, close attention must be paid to the oxidation method that is used. The damage caused to nanotubes by severe treatments (such by using nitric acid) make them unsuitable for use with SWNTs. Moreover, the filling yield is not very good, probably due to the solvent molecules that also enter the tube cavity: the filled lengths rarely exceed 100 nm. *Mittal* et al. [3.252] have recently filled SWNTs with CrO_3 using wet chemistry with an average yield of $\approx 20\%$.

Molten State Filling Method
The physical filling method involving a liquid (molten) phase is more restrictive, firstly because some materials can decompose when they melt, and secondly because the melting point must be compatible with the nanotubes, so the thermal treatment temperature should

remain below the temperature of transformation or the nanotubes will be damaged. Because the filling occurs due to capillarity, the surface tension threshold of the molten material is $100-200\,\text{N/cm}^2$ [3.253], although this threshold was proposed for MWNTs, whose inner diameters (5–10 nm) are generally larger than those of SWNTs (1–2 nm). In a typical filling experiment, the MWNTs are closely mixed with the desired amount of filler by gentle grinding, and the mixture is then vacuum-sealed in a silica ampoule. The ampoule is then slowly heated to a temperature above the melting point of the filler and slowly cooled. This method does not require that the nanotubes are opened prior to the heat treatment. The mechanism of nanotube opening is yet to be clearly established, but it is certainly related to the chemical reactivities of the molten materials toward carbon, and more precisely toward defects in the tube structure (Sect. 3.4.4).

Most of the works involving the application of this method to SWNTs come from Oxford University [3.254–258], although other groups have followed the same procedure [3.249, 251, 259]. The precursors used to fill the nanotubes were mainly metal halides. Although little is known about the physical properties of halides crystallized within carbon nanotubes, the crystallization of molten salts within small-diameter SWNTs has been studied in detail, and the one-dimensional crystals have been shown to interact strongly with the surrounding graphene wall. For example, *Sloan* et al. [3.256] described two-layer 4 : 4 coordinated KI crystals that formed within SWNTs that were ≈ 1.4 nm in diameter. These two-layer crystals were *all surface* and had no *internal* atoms. Significant lattice distortions occurred compared to the bulk structure of KI, where the normal coordination is 6 : 6 (meaning that each ion is surrounded by six identical close neighbors). Indeed, the distance between two ions across the SWNT capillary is 1.4 times as much as the same distance along the tube axis. This suggests an accommodation of the KI crystal into the confined space provided by the inner nanotube cavity in the constrained crystal direction (across the tube axis). This implies that the interactions between the ions and the surrounding carbon atoms are strong. The volume available within the nanotubes thus somehow controls the crystal structures of inserted materials. For instance, the structures and orientations of encapsulated PbI_2 crystals inside their capillaries were found to differ for SWNTs and DWNTs, depending on the diameter of the confining nanotubes [3.254]. For SWNTs, most of the encapsulated one-dimensional PbI_2 crystals obtained exhibited a strong preferred orientation, with their (110) planes aligning at an angle of around 60° to the SWNT axes, as shown in Fig. 3.25a,b. Due to the extremely small diameters of the nanotube capillaries, individual crystallites are often only a few polyhedral layers thick, as outlined in Fig. 3.25d–h. Due to lattice terminations enforced by capillary confinement, the edging polyhedra must be of reduced coordination, as indicated in Fig. 3.25g,h. Similar crystal growth behavior was generally observed to occur for PbI_2 formed inside DWNTs in narrow nanotubes with diameters comparable to those of SWNTs. As the diameter of the encapsulating capillary increases, however, different preferred orientations are frequently observed (Fig. 3.26). In this example, the PbI_2 crystal is oriented with the [121] direction parallel to the direction of the electron beam (Fig. 3.26a–d). If the PbI_2@DWNT hybrid is viewed *side-on* (as indicated by the arrow in Fig. 3.26e), polyhedral slabs are seen to arrange along the capillary, oriented at an angle of around 45° with respect to the tubule axis. High-yield filling of CNTs by the capillary method is generally difficult but fillings of more than 60% have been reported for different halides, with filling lengths of up to a couple of hundreds of nm [3.260]. Results from the imaging

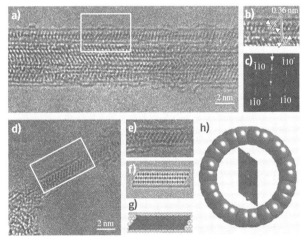

Fig. 3.25a–h HRTEM images and corresponding structural model for PbI_2 filled SWNTs. (**a**) Image of a bundle of SWNTs, all of them being filled with PbI_2. (**b**) Enlargement of the portion framed in (**a**). (**c**) Fourier transform obtained from (**b**) showing the 110 distances at 0.36 nm of a single PbI_2 crystal. (**d**) Image of a single PbI_2-filled SWNT. (**e**) Enlargement of the portion framed in (**b**). (**f**) Simulated HRTEM image, corresponding to (**e**). (**g**) Structural model corresponding to (**f**). (**h**) Structural model of a SWNT filled with a PbI_2 crystal as seen in cross section (from [3.254])

Fig. 3.26a–f HRTEM images (experimental and simulated) and corresponding structural model for a PbI$_2$-filled double-wall carbon nanotube. The larger inner cavity in DWNTs with respect to SWNTs (Fig. 3.25) makes the encapsulated PbI$_2$ crystal orientate differently. (**a**) Image of a single PbI$_2$-filled DWNT, with an insert showing the Fourier transform of the framed portion. (**b**) Enlargement of the portion framed in (**a**). (**c**) Image reconstructed by a second Fourier transform of the inset in (**a**) (= filtered image). (**d**) Structural model corresponding to (**c**). (**e**) and (**f**) Atom and structural models respectively, corresponding to (**d**) (from [3.254]) ▶

and characterization of individual molecules and atomically thin, effectively one-dimensional crystals of rock salt and other halides encapsulated within single-walled carbon nanotubes have recently been reviewed by *Sloan* et al. [3.261].

Sublimation Filling Method

This method is even more restrictive than the previous one, since it is only applicable to a very limited number of compounds due to the need for the filling material to sublimate within the temperature range of thermal stability of the nanotubes. Examples are therefore scarce. Actually, except for a few attempts to fill SWNTs with ZrCl$_4$ [3.257] or selenium [3.262], the first and most successful example published so far is the formation of C$_{60}$@SWNT (nicknamed *peapods*), reported for the first time in 1998 [3.263], where regular $\approx$ 1.4 nm-large SWNTs are filled with C$_{60}$ fullerene molecule chains (Fig. 3.27a). Of course, the process requires that the SWNTs are opened by some method, as discussed previously; typically either acid attack [3.264] or heat treatment in air [3.265]. The opened SWNTs are then inserted into a glass tube together with fullerene powder, which is sealed and placed into a furnace heated above the sublimation temperature for fullerite ($\gtrsim 350\,°\mathrm{C}$). Since there are no filling limitations related to Laplace's law or the presence of solvent (only gaseous molecules are involved), filling efficiencies may actually reach $\approx 100\%$ for this technique [3.265].

C$_{60}$@SWNT has since been shown to possess remarkable behavior traits, such as the ability of the C$_{60}$ molecules to move freely within the SWNT cavity (Fig. 3.27b,c) upon random ionization effects from electron irradiation [3.266], to coalesce into 0.7 nm-wide elongated capsules upon electron irradiation [3.267], or into a 0.7 nm-wide nanotube upon subsequent thermal treatment above 1200 °C under vacuum [3.266, 268]. Annealing C$_{60}$@SWNT material could therefore be

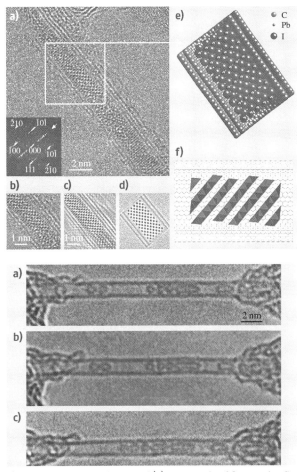

Fig. 3.27a–c HRTEM images of (**a**) an example of five regular C$_{60}$ molecules encapsulated together with two higher fullerenes (C$_{120}$ and C$_{180}$) as distorted capsules (*on the right*) within a regular 1.4 nm-diameter SWNT. (**a–c**) Example of the diffusion of the C$_{60}$-molecules along the SWNT cavity. The time between each image in the sequence is about 10 s. The fact that nothing occurs between (**a**) and (**b**) illustrates the randomness of the ionization events generated by the electron beam that are assumed to be responsible for the molecular displacement

an efficient way to produce DWNTs with constant inner (≈ 0.7 nm) and outer (≈ 1.4 nm) diameters. Using the coalescence of encapsulated fullerenes through both electron irradiation and thermal treatment, it appears to possible to control subsequent DWNT features (inner tube diameter, intertube distance) by varying the electron energy, flow and dose conditions, the temperature, and the outer tube diameter [3.269]. The smallest MWNTs have been obtained in this way.

By synthesizing *endofullerenes* [3.13], it has been possible to use this process to synthesize more complex nanotube-based hybrid materials such as $La_2@C_{80}@SWNTs$ [3.270], $Gd@C_{82}@SWNTs$ [3.271], and $Er_xSc_{3-x}N@C_{80}@SWNT$ [3.272], among other examples. This suggests even more potential applications for peapods, although they are still speculative since the related properties are still being investigated [3.273–275].

The last example discussed here is the successful attempt to produce peapods by a related method, using accelerated fullerene ions (instead of neutral gaseous molecules) to force the fullerenes to enter the SWNT structure [3.276].

3.5.3 Functionalized Nanotubes

Noting the reactivity of carbon nanotubes (Sect. 3.4.6), nanotube functionalization reactions can be divided into two main groups. One is based on the chemical oxidation of the nanotubes (tips, structural defects) leading to carboxylic, carbonyl and/or hydroxyl functions. These functions are then used for additional reactions, to attach oligomeric or polymeric functional entities. The second group is based on direct addition to the graphitic-like surface of the nanotubes (without any intermediate step). Examples of the latter reactions include oxidation or fluorination (an important first step for further functionalization with other organic groups). The properties and applications of functionalized nanotubes have been reviewed in [3.277].

Oxidation of Carbon Nanotubes

Carbon nanotubes are often oxidized and therefore opened before chemical functionalization in order to increase their chemical reactivity (to create dangling bonds). The chemical oxidation of nanotubes is mainly performed using either wet chemistry or gaseous oxidants such as oxygen (typically air) or CO_2. Depending on the synthesis used, the oxidation resistance of nanotubes can vary. When oxidation is achieved using a gas phase, thermogravimetric analysis (TGA) is of great use for determining at which temperature the treatment should be applied. It is important to note that TGA accuracy increases as the heating rate diminishes, while the literature often provides TGA analyses obtained in unoptimized conditions, leading to overestimated oxidation temperatures. Differences in the presence of catalyst remnants (metals or, more rarely, oxides), the type of nanotubes used (SWNTs, c-MWNTs, h-MWNTs), the oxidizing agent used (air, O_2 is an inert gas, CO_2, and so on), as well as the flow rate used make it difficult to compare published results. It is generally agreed, however, that amorphous carbon burns first, followed by SWNTs and then multiwall materials (shells, MWNTs), even if TGA is often unable to separate the different oxidation steps clearly. Air oxidation (static or dynamic conditions) can however be used to prepare samples of very high purity – although the yield is generally low – as monitored by in situ Raman spectroscopy [3.278]. Aqueous solutions of oxidizing reagents are often used for nanotube oxidation. The main reagent is nitric acid, either concentrated or diluted (around 3 mol/l in most cases), but oxidants such as potassium dichromate ($K_2Cr_2O_7$), hydrogen peroxide (H_2O_2) or potassium permanganate ($KMnO_4$) are often used as well. HCl, like HF, does not damage nanotubes because it is not oxidizing.

Functionalization of Oxidized Carbon Nanotubes

The carboxylic groups located at the nanotube tips can be coupled to different chemical groups. Oxidized nanotubes are usually reacted with thionyl chloride ($SOCl_2$) to generate the acyl chloride, even if a direct reaction is theoretically possible with alcohols or amines, for example. The reaction of SWNTs with octadecylamine (ODA) was reported by *Chen* et al. [3.279] after reacting oxidized SWNTs with $SOCl_2$. The functionalized SWNTs are soluble in chloroform ($CHCl_3$), dichloromethane (CH_2Cl_2), aromatic solvents, and carbon bisulfide (CS_2). Many other reactions between functionalized nanotubes (after reaction with $SOCl_2$) and amines have been reported in the literature and will not be reviewed here. Noncovalent reactions between the carboxylic groups of oxidized nanotubes and octadecylammonium ions are possible [3.280], providing solubility in tetrahydrofuran (THF) and CH_2Cl_2. Functionalization by glucosamine using similar procedures [3.281] produced water soluble SWNTs, which is of special interest when considering biological applications of functionalized nanotubes. Functionalization with lipophilic and hydrophilic dendra (with long alkyl chains and oligomeric poly(ethyleneglycol) groups) has been achieved via amination and esterification reactions [3.282], leading to solubility of the functionalized nanotubes in hexane, chloroform, and water. It is interesting to note that, in the latter case, the functional groups could be removed simply by modifying the pH of the solution (base- and acid-catalyzed hydrolysis reaction conditions, [3.283]). One last example is the possible interconnection of nanotubes via chemical functionalization. This has been recently achieved by *Chiu* et al. [3.284] using the acyl chloride method and a bifunctionalized amine to link the nanotubes through the formation of amide bonds. Finally, it has been discovered that imidazolium-ion-functionalized carbon nanotubes are highly dispersible in ionic liquids of analogous chemical structure and that mixtures of functionalized CNT and ionic liquids can form gels upon sonication [3.285] or waxes [3.286] that could find applications as soft composite materials for electrochemistry (sensors, capacitors, or actuators).

Sidewall Functionalization of Carbon Nanotubes

Covalent functionalization of nanotube walls is possible through fluorination reactions. It was first reported by *Mickelson* et al. [3.287], based on F_2 gas (the nanotubes can then be defluorinated, if required, with anhydrous hydrazine). As recently reviewed by *Khabashesku* et al. [3.288], it is then possible to use these fluorinated nanotubes to carry out subsequent derivatization reactions. Thus, sidewall-alkylated nanotubes can be prepared by nucleophilic substitution (Grignard synthesis or reaction with alkyllithium precursors [3.289]). These alkyl sidewall groups can be removed by air oxidation. Electrochemical addition of aryl radicals (from the reduction of aryl diazonium salts) to nanotubes has also been reported by *Bahr* et al. [3.290]. Functionalizations of the external wall of the nanotube by cycloaddition of nitrenes, addition of nuclephilic carbenes or addition of radicals have been described by *Holzinger* et al. [3.291]. Electrophilic addition of dichlorocarbene to SWNTs occurs via a reaction with the deactivated double bonds in the nanotube wall [3.292]. Silanization reactions are another way to functionalize nanotubes, although only tested with MWNTs. *Velasco-Santos* et al. [3.293] have reacted oxidized MWNTs with an organosilane ($RSiR_3$, where R is an organo functional group attached to silicon) and obtained nanotubes with organo functional groups attached via silanol groups.

The noncovalent sidewall functionalization of nanotubes is important because the covalent bonds are associated with changes from sp^2 hybridization to sp^3 carbon hybridization, which corresponds to loss of the graphitelike character. The physical properties of functionalized nanotubes, specifically SWNTs, can therefore be modified. One way to achieve the noncovalent functionalization of nanotubes is to wrap the nanotubes in a polymer [3.294], which permits solubilization (enhancing processing possibilities) while preserving the physical properties of the nanotubes. One reason to functionalize SWNTs is to make them soluble in regular solvents. A promising method to do this was found by *Pénicaud* et al., who made water-soluble by adding charges to SWNTs via the transient and reversible formation of a nanotube salt [3.295].

Finally, it is worth bearing in mind that none of these chemical reactions are specific to nanotubes and so they can affect most of the carbonaceous impurities present in the raw materials as well, making it difficult to characterize the functionalized samples. The experiments must therefore be performed with very pure carbon nanotube samples, which is unfortunately not always the case for the results reported in the literature. On the other hand, purifying the nanotubes to start with may also bias the functionalization experiments, since purification involves chemical treatment. However a demand for such products already exists, and purified then fluorinated SWNTs can be bought for 900 $/g (Carbon Nanotechnologies Inc., 2005).

3.6 Applications of Carbon Nanotubes

A carbon nanotube is inert, has a high aspect ratio and a high tensile strength, has low mass density, high heat conductivity, a large surface area, and a versatile electronic behavior, including high electron conductivity. However, while these are the main characteristics of individual nanotubes, many of them can form secondary structures such as ropes, fibers, papers and thin films with aligned tubes, all with their own specific properties. These properties make them ideal candidates for a large number of applications provided their cost is sufficiently low. The cost of carbon nanotubes depends strongly on both the quality and the production process. High-quality single-shell carbon nanotubes can cost 50–100 times more than gold. However, carbon nanotube synthesis is constantly improving, and sale prices are falling rapidly. The application of carbon nanotubes is therefore a very fast moving field, with new potential applications found every year, even several times per year. Therefore, creating an exhaustive list of these applications is not the aim of this section.

Instead, we will cover the most important applications, and divide them up according to whether they are *current* (Sect. 3.6.1) – they are already on the market, the application is possible in the near future, or because prototypes are currently being developed by profit-based companies – or *expected* applications (Sect. 3.6.2).

3.6.1 Current Applications

Near-Field Microscope Probes

The high mechanical strength of carbon nanotubes makes them almost ideal candidates for use as force sensors in scanning probe microscopy (SPM). They provide higher durability and the ability to image surfaces with a high lateral resolution, the latter being a typical limitation of conventional force sensors (based on ceramic tips). The idea was first proposed and tested by *Dai* et al. [3.92] using c-MWNTs. It was extended to SWNTs by *Hafner* et al. [3.297], since small-diameter SWNTs were believed to give higher resolution than MWNTs due to the extremely short radius of curvature of the tube end. However, commercial nanotube-based tips (such as those made by Piezomax, Middleton, WI, USA) use MWNTs for processing convenience. It is also likely that the flexural modulus of a SWNT is too low, resulting in artifacts that affect the lateral resolution when scanning a rough surface. On the other hand, the flexural modulus of a c-MWNT is believed to increase with the number of walls, although the radius of curvature of the tip increases at the same time. Whether based on SWNT or MWNT, such SPM tips also offer the potential to be functionalized, leading to the prospect of selective imaging based on chemical discrimination in *chemical force microscopy* (CFM). Chemical function imaging using functionalized nanotubes represents a huge step forward in CFM because the tip can be functionalized very specifically (ideally only at the very tip of the nanotube, where the reactivity is the highest), increasing the spatial resolution. The interaction between the chemical species present at the end of the nanotube tip and the surface containing chemical functions can be recorded with great sensitivity, allowing the chemical mapping of molecules [3.298, 299].

Current nanotube-based SPM tips are quite expensive; typically ≈ 450 \$/tip (Nanoscience Co., 2005). This high cost is due to processing difficulties (it is necessary to grow or mount a single MWNT in the appropriate direction at the tip of a regular SPM probe; Fig. 3.28), and the need to individually control the tip quality. The market for nanotube SPM tips has been estimated at ≈ 20 M\$/year.

Field Emission-Based Devices

In a pioneering work by *de Heer* et al. [3.300], carbon nanotubes were shown to be efficient field emitters and this property is currently being used several applications, including flat panel displays for television sets and computers (the first prototype of such a display was exhibited by Samsung in 1999), and devices requiring an electron-producing cathode, such as x-ray sources. The principle of a field emission-based screen is demonstrated in Fig. 3.29a. Briefly, a potential difference is set up between the emitting tips and an extraction grid so that electrons are pulled from the tips onto an electron-sensitive screen layer. Replacing the glass support and protecting the screen using a polymer-based material should even permit the development of flexible screens. Unlike regular (metallic) electron-emitting tips, the structural perfection of carbon nanotubes allows higher electron emission stability, higher mechanical resistance, and longer lifetimes. Most importantly, using them saves energy since the tips operate at a lower heating temperature and require much lower threshold voltage than in other setups. For example, it is possible to produce a current density of $1\,\text{mA/cm}^2$ for a threshold voltage of $3\,\text{V/}\mu\text{m}$ with nanotubes, while it requires $20\,\text{V/}\mu\text{m}$ for graphite powder and $100\,\text{V/}\mu\text{m}$ for regular Mo or Si tips. The subsequent reductions in cost and energy consumption are estimated at 1/3 and 1/10 respectively. Generally speaking, the maximum current density that can be obtained ranges from 10^6 to $10^8\,\text{A/cm}^2$ depending on the nanotubes involved (SWNT or MWNT, opened or capped, aligned or not, and so on) [3.301–303]. Although the side walls of the nanotubes seem to emit as well as the tips, many

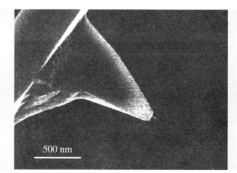

Fig. 3.28 Scanning electron microscopy image of a carbon nanotube (MWNT) mounted onto a regular ceramic tip as a probe for atomic force microscopy (modified from [3.296])

works have investigated the growth of nanotubes perpendicular to the substrate surface as regular arrays (Fig. 3.29b). Besides, it does not appear necessary to use SWNTs instead of MWNTs for many of these applications when they are used as bunches. On the other hand, when considering single, isolated nanotubes, SWNTs are generally less preferable since they permit much lower electron doses than MWNTs, although they often provide a more coherent source (an useful feature for devices such as electron microscopes or x-ray generators).

The market associated with this application is huge. With major companies involved, such as Motorola, NEC, NKK, Samsung, Thales and Toshiba, the first flat TV sets and computers using nanotube-based screens should enter the market in 2007 (Samsung data), once a problem with product lifetime (still only about half that required) is fixed. On the other hand, companies such as Oxford Instruments and Medirad are now commercializing miniature x-ray generators for medical applications that use nanotube-based cold cathodes developed by Applied Nanotech Inc.

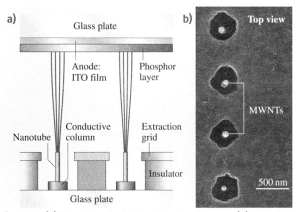

Fig. 3.29 (a) Principle of a field emitter-based screen. (b) Scanning electron microscope image of a nanotube-based emitter system (*top view*). *Round dots* are MWNT tips seen through the holes corresponding to the extraction grid. © *P. Legagneux* (Thales Research & Technology, Orsay, France)

Chemical Sensors

The electrical conductance of semiconductor SWNTs was recently demonstrated to be highly sensitive to changes in the chemical composition of the surrounding atmosphere at room temperature, due to charge transfer between the nanotubes and the molecules from the gases adsorbed onto SWNT surfaces. It has also been shown that there is a linear dependence between the concentration of the adsorbed gas and the change in electrical properties, and that the adsorption is reversible. First tries involved NO_2 or NH_3 [3.304] and O_2 [3.305]. SWNT-based chemical NO_2 and NH_3 sensors are characterized by extremely short response times (Fig. 3.30), unlike conventional sensors [3.304, 306]. The electrical response has been measured by exposing MWNT films to sub-ppm NO_2 concentrations (10–100 ppb in dry air) at different operating temperatures ranging between 25 and 215 °C [3.307]. For SWNTs, the sensor responses are linear for similar concentrations, with detection limits of 44 ppb for NO_2 and 262 ppb for nitrotoluene [3.308]. High sensitivity to water or ammonia vapor has been demonstrated on a SWNT-SiO_2 composite [3.309]. This study indicated the presence of p-type SWNTs dispersed among the predominantly metallic SWNTs, and that the chemisorption of gases on the surface of the semiconductor SWNTs is responsible for the sensing action. Determinations of CO_2 and O_2 concentrations on a SWNT-SiO_2 composite have also been reported [3.310]. By doping nanotubes with palladium

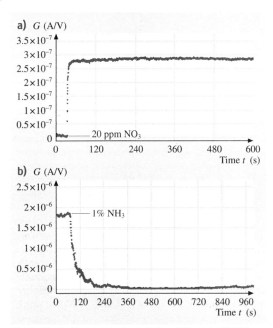

Fig. 3.30a,b Demonstration of the ability of SWNTs to detect trace molecules in inert gases. (a) Increase in the conductance of a single SWNT when 20 ppm of NO_2 are added to an argon gas flow. (b) Same, but with 1% NH_3 added to the argon gas flow (from [3.304])

nanoparticles, *Kong* et al. [3.311] have also shown that the modified material can reveal the presence of hydrogen at levels of up to 400 ppm, whereas the as-grown material was totally ineffective. Miniaturized gas ionization sensors, which work by fingerprinting the ionization characteristics of distinct gases, have also been reported, with detection limits of 25 ppm for NH_3 [3.312].

Generally speaking, the sensitivities of these new nanotube-based sensors are three orders of magnitude higher than those of standard solid state devices. Another reason for using nanotubes instead of current sensors is their simplicity, the facts that they can be placed in very small systems and that they can operate at room temperature, as well as their selectivity. These advantages allow a limited number of sensor device architectures to be built for a variety of industrial purposes, while the current technology requires a large variety of devices based on mixed metal oxides, optomechanics, catalytic beads, electrochemistry, and so on. The market for such devices is expected to be \$ 1.6 billion by 2006, including sensing applications in biological fields and the chemical industry. Nanotube-based sensors are currently being developed by large and small companies, such as Nanomix (Emeryville, USA), for example.

Catalyst Support

Carbon-based materials make good supports in heterogeneous catalytic processes due to their ability to be tailored to a specific need: indeed, activated carbons are already currently employed as catalyst supports due to their high surface areas, their stability at high temperatures (under nonoxidizing atmospheres), and the possibility of controlling both their porous structure and the chemical nature of their surfaces [3.313, 314]. Attention has focused on nanosized fibrous morphologies of carbon have appeared over the last decade, that show great potential for use as supports [3.315]. Carbon nanofibers (also incorrectly called graphite nanofibers) and carbon nanotubes have been successfully used in this area, and have been shown to provide, as catalyst-supporting materials, properties superior to those of such other regular catalyst-supports, such as activated carbon, soot or graphite [3.316–318]. The possibility to use MWNTs as nanoreactors, that means to deposit the active catalytic phase in the inner cavity of the nanotubes and to take advantage of the confinement effect to perform the catalytic reaction, also offers very exciting perspectives [3.319]. Various reactions have been studied [3.316–318]; hydrogenation reactions, Fischer–Tropsch, polymerization and even oxidation reactions, hydrocarbon decomposition and use as fuel cell electrocatalysts are among the most popular domains. The application of graphite nanofibers as direct catalysts for oxidative dehydrogenation [3.320, 321] or methane decomposition [3.322] has also been reported.

The morphology and size of the carbon nanotubes (particularly their aspect ratios), can play a significant role in catalytic applications due to their ability to disperse catalytically active metal particles. Their electronic properties are also of primary importance [3.323], since the conductive support may cause electronic perturbations as well as constraining the geometriies of the dispersed metal particles. A recent comparison between the interactions of transition metal atoms with carbon nanotube walls and their interactions with graphite has shown major differences in bonding sites, magnetic moments, and charge transfer direction [3.324]. Thus the possibility of a strong metal–support interaction must be taken into account. Their mechanical strength is also important, and this makes them resistant to attrition when recycled. Their external and internal surfaces are strongly hydrophobic and adsorb organic molecules strongly. For MWNT-based catalyst-supports, the relatively high surface area and the absence of microporosity (pores < 2 nm), associated with a high meso- and macropore volume (Sect. 3.4.3), result in significant improvements in catalytic activity for liquid phase reactions when compared to catalysts supported on activated carbon. With nanotube supports, the mass transfer of the reactants to the active sites is unlimited, due to the absence of microporosity, and the apparent contact time of the products with the catalyst is diminished, leading to more active and more selective catalytic effects. Finally, as for activated carbon, catalyst-forming is possible and porous granules of carbon nanotubes or electrodes based on carbon nanotubes can be obtained for catalysis or electrocatalysis respectively. Of course, the possibility of shaping these nanomaterials offers interesting perspectives, including for designing structured microreactors [3.325].

The technique usually used to prepare carbon nanotube-supported catalysts is incipient wetness impregnation, in which the purified support is impregnated with a solution of the metal precursor and then dried, calcinated and/or reduced in order to obtain metal particles dispersed on the support. Other techniques such as electrochemical deposition and the use of colloidal chemistry have also been investigated [3.326]. Chemical treatment and/or modification of the carbon nanotube surface were found to be useful ways

Table 3.4 Preparation and catalytic performances of some nanotube-supported catalysts

Catalyst	Preparation route	Catalytic reaction	Comments
Ru/MWNT + SWNT [3.315]	Liquid phase impregnation, no pretreatment of the tubes	Liquid phase cinnamaldehyde hydrogenation	A different kind of metal support interaction compared to activated carbon
Pt/MWNT electrodes [3.331]	Electrodeless plating with prefunctionalization of MWNT	Oxygen reduction for fuel cell applications	High electrocatalytic activity
Rh/MWNT [3.329]	Surface-mediated organometallic synthesis, prefunctionalization of MWNT	Liquid phase hydroformylation and hydrogenation	Higher activity of Rh/MWNT compared to Rh/activated carbon
Ru-alkali/MWNT [3.332]	Liquid phase impregnation, no pretreatment of the tubes	Ammonia synthesis, gas phase reaction	Higher activity with MWNT than with graphite
Rh-phosphine/MWNT [3.333]	Liquid phase grafting from [RhH(CO)(PPh$_3$)$_3$]	Liquid phase hydroformylation	Highly active and regioselective catalyst
Rh/MWNT (confined nanoparticles) [3.319]	Liquid phase impregnation of oxidized MWNTs	Conversion of CO and H$_2$ into ethanol	The overall formation rate of ethanol inside the nanotubes exceeds that on the outside of the nanotubes by more than one order of magnitude

of controlling its hydrophobic or hydrophilic character [3.327]. A strong metal/support interaction can thus be expected from the occurrence of functionalized groups created by the oxidation of the support surface, resulting in smaller particle sizes [3.328]. A more sophisticated technique for achieving the grafting of metal particles onto carbon nanotubes consists of functionalizing the outer surface of the tubes and then performing a chemical reaction with a metal complex, resulting in a good dispersion of the metallic particles (Fig. 3.31) [3.329]. The functionalization of noncovalent carbon nanotubes with polymer multilayers followed by the attachment of gold nanoparticles has also been reported [3.330].

Selected examples of some carbon nanotube-based catalysts together with related preparation routes and catalytic activities are listed in Table 3.4.

The market is important for this application, since it often concerns the heavy chemical industry. It implies and requires mass production of low-cost

Fig. 3.31 Transmission electron microscopy image showing rhodium nanoparticles supported on the surface of an MWNT (from [3.329]) ▶

nanotubes, processed by methods other than those based on solid carbon as the source (Sect. 3.2.1). Such an application also requires some surface reactivity, making the h-MWNT-type nanotubes, with poor nanotextures (Sect. 3.1.2), interesting candidates as starting material for preparing such catalyst supports. Catalysis-enhanced thermal cracking of gaseous carbon precursors is therefore preferred, and pilot plants are already being built by major chemical industrial companies (such as Arkema in France).

3.6.2 Expected Applications Related to Adsorption

Adsorptions of various gases, liquids or metals onto carbon nanotubes, and interactions between them, have attracted much attention recently. The applications resulting from the adsorptive properties of carbon nanotubes can be arbitrarily divided into two groups. The first group is based on the consequences of molecular adsorption on the electronic properties of nanotubes; the main application of this is chemical sensing (Sect. 3.6.1). The second group includes gas storage, gas separation, the use of carbon nanotubes as adsorbants, and results from morphological investigations of carbon nanotubes (surface areas, aspect ratios, and so forth). Among these latter potential applications, the possibility of storing gases – particularly hydrogen – on carbon nanotubes has received most attention.

Gas Storage – Hydrogen

The development of a lightweight and safe system for hydrogen storage is necessary for the widespead use of highly efficient H_2-air fuel cells in transportation vehicles. The US Department of Energy Hydrogen Plan has provided a commercially significant benchmark for the amount of reversible hydrogen adsorption required. This benchmark requires a system weight efficiency (the ratio of H_2 weight to system weight) of 6.5 wt % hydrogen, and a volumetric density of $63\,\text{kg}\,H_2/\text{m}^3$.

The failure to produce a practical storage system for hydrogen has prevented hydrogen from becoming one of the most important transportation fuels. The ideal hydrogen storage system needs to be light, compact, relatively inexpensive, safe, easy to use, and reusable without the need for regeneration. While research and development are continuing into such technologies as liquid hydrogen systems, compressed hydrogen systems, metal hydride systems, and superactivated carbon systems, all have serious disadvantages.

Therefore, there is still a great need for a material that can store hydrogen but is also light, compact, relatively inexpensive, safe, easy to use, and reusable without regeneration. Some recent articles and patents on the very high, reversible adsorption of hydrogen in carbon nanotubes or platelet nanofibers have aroused tremendous interest in the research community, stimulating much experimental and theoretical work. Most of the early works done on hydrogen adsorption on carbon nanotubes have been reviewed in [3.334–340], from the first report about the supposedly highly successful storage of hydrogen in carbon layered nanostructures at room temperature made by a group of Northeastern University [3.192,341], to the multiple yet vain attempts to reproduce this result that followed. Actually, in spite of a worldwide research effort, any work published since then claiming for a hydrogen storage in some nanotextured carbon material with an efficiency better than 1–2% at room temperature or close (and pressure below ≈ 300–500 bar may be regarded as suspicious. Modelling did not help, since it appeared that the calculations are closely constrained by the starting hypotheses. Actually, While considering the same (10,10) SWNT, calculations based on DFT predicted between 14.3 and 1 wt % storage [3.342,343], calculations based on a geometrical model predicted 3.3% [3.334], and calculations based on a quantum mechanical molecular dynamics model predicted 0.47% [3.344]. Therefore, neither experimental results, obviously often biased by procedure problems, nor theoretical results are yet able to demonstrate that an efficient storage of H_2 is possible for carbon nanotubes, whatever the type. However, a definitive statement of failure cannot yet be claimed. Attempts might have failed so far because they were considering by far too simplistic materials, i. e., plain nanotubes. Further efforts have to be made to enhance the adequation of the materials to this specific purpose, in particular:

1. By adjusting the surface properties, which can be modified by mechanical or chemical treatments, e.g. KOH [3.345]
2. By adjusting the texture of the material, such as the pore size [3.346] and possibly the curvature [3.347–349]
3. By complexifying the materials, e.g., by considering nanocomposites combining some host carbon materials and catalyst nanoparticles so as to promote the dissociation of hydrogen molecules to hydrogen atoms that can form bonds with the host [3.340, 350].

In this regard, whether the best carbon material for H$_2$ adsorption will still be nanotube-based is not ascertained.

Gas Storage – Gases Other than Hydrogen

Encouraged by the potential applications related to hydrogen adsorption, several research groups have tried to use carbon nanotubes as a means of stocking and transporting other gases such as oxygen, nitrogen, noble gases (argon and xenon) and hydrocarbons (methane, ethane, and ethylene). These studies have shown that carbon nanotubes could become the world's smallest gas cylinders, combining low weight, easy transportability and safe use with acceptable adsorbed quantities. Nanotubes may also be used in medicine, where it would be extremely useful to physically confine special gases (^{133}Xe for instance) prior to injection.

Kusnetzova et al. [3.351] conducted experiments with xenon and found that the storage capacities of nanotubes can be enhanced by a tremendous amount (a factor of 280, up to a molar ratio of $N_{Xe}/N_C = 0.045$) by opening the SWNT bundles via thermal activation at 800 °C. The gas can be adsorbed inside the nanotubes and the rates of adsorption are also increased using this treatment.

The possibility of storing argon in carbon nanotubes has been studied, with encouraging results, by *Gadd* et al. [3.352]. Their experiments show that large amounts of argon can be trapped in catalytically grown MWNTs (20–150 nm) by hot isostatic pressing (HIPing) for 48 h at 650 °C under an argon pressure of 1700 bar. Energy-dispersive x-ray spectroscopy was used to determine that the gas was located inside the tubes and not on the tube walls. Further studies determined the argon pressure inside the tubes at room temperature. The authors estimated this to be around 600 bar, indicating that equilibrium pressure was attained in the tubes during the HIP-ing and that MWNTs would be a convenient material for storing the gas.

Gas Separation

As SWNTs or MWNTs have regular geometries that can, to some extent, be controlled, they could be used to develop precise separation tools. If the sorption mechanisms are known, it should be possible to control sorption of various gases through particular combinations of temperature, pressure and nanotube morphology. Since the large-scale production of nanotubes is gradually progressing, and this should ultimately result in low costs, accurate separation methods based on carbon nanotubes are now being investigated.

A theoretical study has aimed to determine the effects of different factors such as tube diameter, density and type of the gas used on the flow of molecules inside nanotubes. An atomistic simulation with methane, ethane and ethylene [3.353] has shown that the molecular mobility decreases with decreasing tube for each of the three gases. Ethane and ethylene have smaller mobilities due to the stronger interactions they seem to have with the nanotube walls. In another theoretical study into the possibility of hydrocarbon mixture separation on SWNT bundles, the authors conclude that carbon nanotubes can be used to separate methane/n-butane and methane/isobutene mixtures [3.354] with an efficiency that increases as the average tube diameter decreases. Experimental work was also performed by the same group on the sorption of butane on MWNTs [3.194]. It has been also reported that the Fickian diffusivities of CH$_4$/H$_2$ mixtures in SWNT, like their pure component counterparts, are extraordinarily large when compared with adsorbed gases in other nanoporous materials [3.355].

Grand canonical Monte Carlo simulations of the separation of hydrogen and carbon monoxide by adsorption on SWNTs have also been reported [3.356]. In most of the situations studied, SWNTs were found to adsorb more CO than H$_2$, and excellent separation could again probably be obtained by varying the SWNT average diameter.

Adsorbents

Carbon nanotubes were found to be able to adsorb some toxic gases such as dioxins [3.357], fluoride [3.358], lead [3.359] and alcohols [3.360] better than adsorbent materials in common use, such as activated carbon. These pioneering works opened a new field of applications as cleaning filters for many industrial processes with hazardous by-products. The adsorption of dioxins, which are very common and persistent carcinogenic by-products of many industrial processes, is a good example of the potential of nanotubes in this field. Growing ecological awareness has resulted in the imposition of emission limits on dioxin-generating sources in many countries, but it is difficult to find materials that can act as effective filters, even at extremely low concentrations. *Long* and *Yang* [3.357] found that nanotubes can attract and trap more dioxins than activated carbons or other polyaromatic materials that are currently used as filters. This improvement is probably due to the stronger interaction forces that exist between dioxin molecules and the curved surfaces of nanotubes compared to those for flat graphene sheets.

MWNTs have also been used with success for the adsorption of other pollutants such as volatile organic compounds [3.361], reactive dyes [3.362], or natural organic matter in aqueous solutions [3.363]. MWNTs show also better performances than granular activated carbons for the adsorption of low molecular weight toxins [3.364].

The capacity of Al_2O_3/MWNT to adsorb fluoride from water has been reported to be 13.5 times that of activated carbon and four times that of Al_2O_3 [3.358]. The same group has also reported a capacity of MWNTs to adsorb lead from water that is higher than that for activated carbon [3.359]. The possibility of using graphite nanofibers to purify water from alcohols has also been explored [3.360]. MWNTs were found to be good adsorbents for the removal of dichlorobenzene from wastewaters over a wide range of pH. Typically, the nanotubes adsorb 30 mg of the organic molecule per gram of MWNTs from a 20 mg/l solution [3.365]. It has also been shown that SWNTs act as molecular sponges for molecules such as CCl_4; the nanotubes were in contact with a support surface which also adsorbs molecules, although more weakly than the nanotubes [3.366]. Finally, oxidized carbon nanotubes have been successfully used for the adsorption of heavy metal ions such as Zn(II) [3.367], Cu(II) [3.368], Pb(II) [3.369] or Th(IV) [3.370] from aqueous solutions. While an apolar surface might be more adapted for the adsorption of aromatic organic species, an oxidation of the CNTs that provides a polar and hydrophilic surface is highly desirable for the adsorption of heavy metal ions. These experimental results suggest that carbon nanotubes may be promising adsorbents for removing polluting agents from water.

Biosensors

Attaching molecules of biological interest to carbon nanotubes is an excellent way to produce nanometer-sized biosensors. The electrical conductivities of these functionalized nanotubes would depend on the interaction of the probe with the medium being studied, which would be affected by chemical changes or interactions with the target species. The science of attaching biomolecules to nanotubes is rather recent and was inspired by similar research in the fullerene area. Some results have already been patented, and so such systems may become available in the near future. Using the internal cavities of nanotubes to deliver drugs would be another amazing application, but little work has been carried out so far to investigate the toxicity of nanotubes in the human body. Comparison between the effects of nanotubes and asbestos was investigated by *Huczko* et al. [3.371] and they concluded that the tested samples were innocuous. However, a more recent work has shown that contact with nanotubes may lead to dermal toxicity [3.372] or induce lung lesions characterized by the presence of granulomas [3.373]. *Pantarotto* et al. [3.374] reported the translocation of water-soluble SWNT derivatives across cell membranes and have shown that cell death can be induced by functionalised nanotubes (bioactive peptides), depending upon their concentration in the media. Recent results also indicate that nanotubes may lead to an inflammatory response of the immune system by activating the complement system [3.375].

MWNTs have been used by *Mattson* et al. [3.376] as a substrate for neuronal growth. They have compared the activity of untreated MWNTs with that of MWNTs coated with a bioactive molecule (4-hydroxynonenal) and observed that neurons elaborated multiple neurites on these latter functionalized nanotubes. This is an important result that illustrates the feasibility of using nanotubes as a substrate for nerve cell growth.

Davis et al. [3.377] immobilized different proteins (metallothionein, cytochrome c and c_3, β-lactamase I) in MWNTs and checked whether these molecules were still catalytically active compared to the free ones. They have shown that confining a protein within a nanotube provides some protection for the external environment. Protein immobilization via noncovalent sidewall functionalization was proposed by *Chen* et al. [3.378] using a bifunctional molecule (1-pyrenebutanoic acid, succinimidyl ester). This molecule is tied to the nanotube wall by the pyrenyl group, and amine groups or biological molecules can react with the ester function to form amide bonds. This method was also used to immobilize ferritin and streptavidin onto SWNTs. Its main advantages are that it does not modify the SWNT wall and that it does not perturb the sp^2 structure, so the physical properties of the nanotubes are maintained. *Shim* et al. [3.379] have functionalized SWNTs with biotin and observed specific binding with streptavidin, suggesting biomolecular recognition possibilities. *Dwyer* et al. [3.380] have functionalized SWNTs by covalently coupling DNA strands to them using EDC (1-ethyl-3-(3-dimethylaminopropyl) carbodiimide hydrochloride) but did not test biomolecular recognition; other proteins such as bovine serum albumin (BSA) [3.381] have been attached to nanotubes using the same process (diimide-activated amidation with EDC) and most of the attached proteins remained bioactive. Instead of working with individual nanotubes (or more likely nanotube bundles in the case of SWNTs), *Nguyen* et al. [3.382] have functionalized nanotubes arrayed with a nucleic acid, still using EDC as the coupling agent, in order to realize biosensors based on protein-functionalized nanotubes. *Azamian* et al. [3.383] have immobilized a series of biomolecules (cytochrome c, ferritin, and glucose oxidase) on SWNTs, and they observed that the use of EDC was not always necessary, indicating that the binding was predominantly noncovalent. In the case of glucose oxidase, they tested the catalytic activity of functionalized nanotubes immobilized on a glassy carbon electrode and observed a tenfold greater catalytic response compared to that seen in the absence of modified SWNTs.

Functionalization of nanotubes with biomolecules is still in its infancy, and their use as biosensors may lead to practical applications earlier than expected. For example, functionalized nanotubes can be used as AFM tips (Sect. 3.6.1), allowing single-molecule measurements to be taken using *chemical force microscopy* (CFM). Important improvements in the characterization of biomolecules have even been achieved with unfunctionalized nanotube-based tips (see the review by [3.297]). Nanotube-based biosensors have now been developed. They are based on either field effect transistors [3.384] involving functionalized CNTs (biomolecules) or on electrochemical detection [3.385].

3.6.3 Expected Applications Related to Composite Systems

Because of their exceptional morphological, electrical, thermal, and mechanical characteristics, carbon nanotubes make particularly promising reinforcement materials in composites with metals, ceramics or polymer matrices. Key issues to address include the good dispersion of the nanotubes, the control of the nanotube/matrix bonding, the densification of bulk composites and thin films, and the possibility of aligning the nanotubes. In addition, the nanotube type (SWNT, c-MWNT, h-MWNT, etc.) and origin (arc, laser, CCVD, etc.) are also important variables that control the structural perfection, surface reactivity and aspect ratio of the reinforcement.

The application of carbon nanotubes in this field is expected to lead to major advances in composites. The following sections will give overviews of current work on metal-, ceramic- and polymer-matrix composites containing nanotubes. Nanotubes coated with another material are not considered here. Filled nanotubes are discussed in Sect. 3.5.2.

Metal Matrix Composites

Nanotube-metal matrix composites are still rarely studied. Matrices include Al-, Cu-, Mg-, Ni-, Ni-P-, Ti-, WC-Co- and Zr-based bulk metallic glasses. The materials are generally prepared by standard powder metallurgy techniques, but in this case the nanotube dispersion is not optimal. Other techniques such as plasma spray forming [3.386], the so-called nanoscale-dispersion method [3.387], the rapid solidification technique [3.388] and CCVD [3.389], are being developed. The spark plasma sintering (SPS) technique is sometimes used to densify the composites whilst avoiding matrix-grain growth [3.390, 391]. The room-temperature electrical resistivity of hot-pressed CCVD MWNT-Al composites increases slightly upon increasing the MWNT volume fraction [3.392]. The tensile strengths and elongations of unpurified arc discharge MWNT-Al composites are only slightly affected by annealing at 873 K in contrast to those of pure Al [3.393]. The coefficient of thermal expansion (CTE) of 1 wt % MWNTs-Al composite fabricated by cold isostatic pressing and hot squeeze technique is 11% lower than to that of pure Al or 2024Al matrix, showing some promises as low-CTE materials. Associated to a high thermal conductivity, such materials would be interesting for applications such as packaging and space structures [3.394]. The Young's modulus of nonpurified arc discharge MWNTs-Ti composite is about 1.7 times that of pure Ti [3.395]. The formation of TiC, probably resulting from a reaction between amorphous carbon and the matrix, was observed, but the MWNTs themselves were not damaged. An increase in the Vickers hardness by a factor of 5.5 over that of pure Ti was associated with the suppression of coarsening of the Ti grains, TiC formation, and the addition of MWNTs. Purified nanotube-WC-Co nanocomposites exhibit better hardness-to-toughness relationships than pure nanocrystalline WC-Co [3.391]. Ni-plated MWNTs give better results than unplated MWNTs in strength tests. Indeed, nanotube coating is a promising way to improve the strength of bonding with the matrix [3.396]. Compressive testing of carbon nanotube-reinforced Zr-based bulk metallic glass composites [3.397] shows that the composites display a high fracture strength. In addition, the composites have strong ultrasonic attenuation characteristics and excellent ability to absorb waves. This implies that such composites may also be useful for shielding acoustic sound or environmental noise. CCVD MWNTs-Cu composites [3.398] also show a higher hardness and a lower friction coefficient and wear loss. Fifty to sixty percent deformation of the composites was observed. Carbon nanotube-Cu composite electrodes have been applied to the amperometric detection of carbohydrates, where they show an enhanced sensitivity compared to detectors based on Cu or nanotubes alone [3.399]. Hot-extruded nanotube-Mg nanocomposites showed a simultaneous increase in yield strength, ultimate tensile strength and ductility, until a threshold of 1.3 wt % was reached [3.400]. The yield strength of SWNT-Fe composites showed substantial enhancement relative to that of similarly treated pure iron materials [3.389]. The work hardening coefficient and the Vickers hardness coefficient also significantly increased in these composites. Composite films and coatings deposited by electroless or electrodeposition techniques on various substrates have also been studied. The addition of up to 15 vol. % purified SWNTs to nanocrystalline Al films reduces the coefficient of thermal expansion by as much as 65% and the resulting material could be a promising electronic packaging material [3.401]. Ni-carbon nanotube coatings deposited on carbon steel plate by electroless deposition show significantly increased resistance to corrosion [3.402] and higher Vickers microhardness, higher wear resistance, and lower friction coefficient than SiC-reinforced composite deposits [3.403]. Ni-P-SWNT coatings prepared by electroless plating show not only higher wear resistance but also a lower friction coefficient and a higher corrosion resistance compared to Ni-P coatings [3.404].

Ceramic Matrix Composites

Many different ceramic matrices have been studied over the years, although oxides (in particular alumina), are still the most studied [3.405]. There are three main methods for the preparation of CNT-ceramic nanocomposite powders. One is mechanical milling. It usually involves long times that could damage the nanotubes. Wet-milling is preferred but often requires the addition of organic additives to stabilize both the nanotubes and the ceramic powder. This also true for a second method, i.e., the in-situ synthesis of the matrix on preformed nanotubes. It can lead to a good adhesion between the nanotubes and the ceramic, but can be rather complex to implement. A third method is the in-situ synthesis of the nanotubes within the ceramic powder using procedures closely related to those described in Sect. 3.2.2. The densification of the nanocomposite powders is made difficult by the detrimental influence of the nanotubes. The most common method is hot-pressing (HP). Most of the works [3.406–413] report that increasing the nanotube content inhibits the densification of the material. It has been shown for a series of CNT-MgAl$_2$O$_4$ composites [3.413] that, for a low content (below 9 vol. %), CNTs favor the rearrangement of the grains, which is the first shrinkage step, probably owing to a lubricating role which facilitates the sliding at grain contacts or grain boundaries. By contrast, for higher contents, CNTs form a too rigid weblike structure, therefore inhibiting the rearrangement process. In the second sintering step, at higher temperatures, CNTs inhibit the shrinkage, all the more when their content is increased above 5.0 vol. % only, leading to decreasing densifications. Thus, composites in which the nanotubes are very homogeneously dispersed may be more difficult to densify. The spark-plasma sintering (SPS) technique has been reported as an efficient method to achieve the total densification of CNT-oxide composites without damaging the CNT [3.414–417]. Full densification can be reached with SPS at comparatively lower temperatures with substantial shorter holding time. However, the successful densification by SPS at a lower temperature than for HP supposes that matrix grains are non agglomerated and with size in the range few tens of nanometers.

The influence of the nanotube dispersion onto mechanical properties, in particular on toughness, has been controversial. Indeed, strong increases in toughness derived from the measure of Vickers indentation cracks have been reported [3.33], but they were shown to be probably widely overestimated because such materials are very resistant to contact damage [3.418, 419]. *Xia* et al. [3.420] reported microstructural investigations on MWNTs well-aligned in the pores of an alumina membrane. Different possible reinforcement mechanisms induced by the MWNTs have been evidenced, such as crack deflection, crack bridging, MWNT pulling-out, and MWNT collapsing in shear bands. Indeed, although so far neither SENB nor SEVNB result have evidenced that nanotubes can significantly reinforce alumina ceramics, this could be obtained with ceramic-matrix composites in which the nanotubes would have been properly organized. Enhanced wear resistance of composites has been reported [3.421–423]. The microhardness is found to either increase or decrease, and this depends greatly on the powder preparation route. As noted in [3.419], processing-induced changes in the matrix may have greater effects on the mechanical properties than the actual presence of nanotubes. Regarding the thermal properties, nanotube-ceramic composites often show a lower thermal conductivity than the corresponding ceramics, probably caused by too high thermal contact resistances at nanotube-nanotube and nanotube-ceramic grain junctions [3.424, 425]. By contrast, nanotubes greatly increase the electrical conductivity of insulating ceramic nanocomposites [3.408, 411, 426–428], with a low percolation threshold (less than 1 vol. %) due to their very high aspect ratio [3.427]. The electrical conductivity can be tailored within several orders of magnitude directly by the CNTs quantity and is well fitted by the scaling law of the percolation theory with the exponent close to the theoretical value characteristic of a three-dimensional network [3.427]. An anisotropic conductivity is obtained when the nanotubes are aligned within the composite [3.429]. *Zhan* et al. [3.430] reported an increase of the thermoelectric power with increasing temperature for nanotube-zirconia composites.

Polymer Matrix Composites

Nanotube-polymer composites, first reported by *Ajayan* et al. [3.431], are now being intensively studied; especially epoxy- and polymethylmethacrylate (PMMA)-matrix composites. A review of the mechanical properties can be found in [3.432]. In terms of mechanical characteristics, the three key issues that affect the performance of a fiber-polymer composite are the strength and toughness of the fibrous reinforcement, its orientation, and good interfacial bonding, which is crucial to load transfer [3.433]. The ability of the polymer to form large-diameter helices around individual nanotubes favors the formation of a strong bond with the matrix [3.433]. Isolated SWNTs may be more desirable than MWNTs or bundles for dispersion in a matrix because of the weak frictional interactions between layers of MWNTs and between SWNTs in bundles [3.433]. The main mechanisms of load transfer are micromechanical interlocking, chemical bonding and van der Waals bonding between the nanotubes and the matrix. A high interfacial shear stress between the fiber and the matrix will transfer the applied load to the fiber over a short distance [3.434]. SWNTs longer than 10–100 μm would be needed for significant load-bearing ability in the case of nonbonded SWNT-matrix interactions, whereas the critical length for SWNTs cross-linked to the matrix is only 1 μm [3.435]. Defects are likely to limit the working length of SWNTs, however [3.436].

The load transfer to MWNTs dispersed in an epoxy resin was much higher in compression than in tension [3.434]. It was proposed that all of the walls of the MWNTs are stressed in compression, whereas only the outer walls are stressed in tension because all of the inner tubes are sliding within the outer tube. Mechanical tests performed on 5 wt % SWNT-epoxy composites [3.437] showed that SWNT bundles were pulled out of the matrix during the deformation of the material. The influence of the interfacial nanotube/matrix interaction was demonstrated by *Gong* et al. [3.438]. It was also reported that coating regular carbon fiber with MWNTs prior to their dispersion into an epoxy matrix improves the interfacial load transfer, possibly via local stiffening of the matrix near the interface [3.439]. DWNTs-epoxy composites prepared by a standard calendaring technique were shown to possess higher strength, Young's modulus and strain to failure at a nanotube content of only 0.1 wt % [3.440]. A significantly improved fracture toughness was also observed. The influence of the different types of nanotubes (SWNTs, DWNTs and MWNTs) on the mechanical properties of epoxy-matrix composites is discussed in [3.441]. The stiffness and damping properties of SWNT- and MWNT-epoxy composites were investigated for use in structural vibration applications [3.442]. It was shown that enhancement in damping ratio is more dominant than enhancement in stiffness, MWNTs making a better reinforcement than SWNTs. Indeed, up to 700% increase in damping ratio is observed for MWNT-epoxy beam as compared to the plain epoxy beam. Industrial epoxy loaded with 1 wt % unpurified CCVD-prepared SWNTs showed an increase in thermal conductivity of 70 and 125% at 40 K and at room temperature, respectively [3.443]. Also, the Vickers hardness rose by a factor of 3.5 with the SWNT loading up at 2 wt %. An increase in the amount of MWNTs led to an increase of the glass transition temperature of MWNT-epoxy-composites. The effect is stronger when using samples containing functionalized MWNTs [3.444]. *Pecastaings* et al. [3.445] have investigated the role of interfacial effects in carbon nanotube-epoxy nanocomposite behavior.

As for ceramic matrix composites, the electrical characteristics of SWNT- and MWNT-epoxy composites are described by the percolation theory. Very low percolation thresholds (much below 1 wt %) are often reported [3.446–448]. Thermogravimetric analysis shows that, compared to pure PMMA, the thermal degradation of PMMA films occurs at a slightly higher temperature when 26 wt % of MWNTs are added [3.449]. Improving the wetting between the MWNTs and the PMMA by coating the MWNTs with poly(vinylidene fluoride) prior to melt-blending with PMMA resulted in an increased storage modulus [3.450]. The impact strength in aligned SWNT-PMMA composites increased significantly with only 0.1 wt % of SWNTs, possibly because of weak interfacial adhesion and/or of the high flexibility of the SWNTs and/or the pullout and sliding effects of individual SWNTs within bundles [3.451]. The transport properties of arc discharge SWNT-PMMA composite films (10 μm thick) were studied in great detail [3.452, 453]. The electrical conductivity increases by nine orders of magnitude from 0.1 to 8 wt % SWNTs. The room-temperature conductivity is again well described by the standard percolation theory, confirming the good dispersion of the SWNTs in the matrix. The rheological threshold of SWNT-PMMA composites is about 0.12 wt %, smaller than the percolation threshold of electrical conductivity, about 0.39 wt % [3.454]. This is understood in terms of the smaller nanotube–nanotube distance required for electrical conductivity compared to that required to impede polymer mobility. Furthermore, decreased SWNT alignment, improved SWNT dispersion and/or longer polymer chains increase the elastic response of the nanocomposite. The effects of small quantities of SWNTs (up to 1 wt %) in PMMA on its flammability properties were studied [3.455]. The formation of a continuous SWNTs network layer covering the entire surface without any cracks is critical for obtaining the lowest mass-loss rate of the nanocomposites. One of the most interesting development of nanotube-polymer composites is their use for the production of spun fibers, films and textiles with extraordinary mechanical and electrical properties [3.456–462].

Polymer composites with other matrices include CCVD-prepared MWNT-polyvinyl alcohol [3.463], arc-prepared MWNT-polyhydroxyaminoether [3.464], arc-prepared MWNT-polyurethane acrylate [3.465, 466], SWNT-polyurethane acrylate [3.467], SWNT-polycarbonate [3.468], MWNT-polyaniline [3.469], MWNT-polystyrene [3.470], CCVD double-walled nanotubes-polystyrene-polymethylacrylate [3.471], MWNT-polypropylene [3.472, 473], SWNT-polyethylene [3.474–476], SWNT-poly(vinyl acetate) [3.475,476], CCVD-prepared MWNT-polyacrylonitrile. [3.477], SWNT-polyacrylonitrile [3.478], MWNT-oxotitanium phthalocyanine [3.479], arc-prepared MWNT-poly(3-octylthiophene) [3.480], SWNT-poly(3-octylthiophene) [3.481] and CCVD MWNT-poly(3-hexylthiophene) [3.482]. These works deal mainly with films 100–200 μm thick, and aim to study the glass transition of the polymer, its mechanical and electrical characteristics, as well as the photoconductivity.

A great deal of work has also been devoted to the applications of nanotube-polymer composites as materials for molecular optoelectronics, using primarily poly(m-phenylenevinylene-co-2,5-dioctoxy-p-phenylenevinylene) (PmPV) as the matrix. This conjugated polymer tends to coil, forming a helical structure. The electrical conductivity of the composite films (4–36 wt % MWNTs) is increased by eight orders of magnitude compared to that of PmPV [3.483]. Using the MWNT-PmPV composites as the electron transport layer in light-emitting diodes results in a significant increase in brightness [3.484]. The SWNTs act as a hole-trapping material that blocks the holes in the composites; this is probably induced through long-range interactions within the matrix [3.485]. Similar investigations were carried out on arc discharge SWNT-polyethylene dioxythiophene (PEDOT) composite layers [3.486] and MWNT-polyphenylenevinylene composites [3.487].

To conclude, two critical issues must be considered when using nanotubes as components for advanced composites. One is to choose between SWNTs, DWNTs, and MWNTs. The former seem more beneficial to mechanical strengthening, provided that they are isolated or arranged into cohesive yarns so that the load can be conveniently transferred from one SWNT to another. Unfortunately, despite many advances [3.456–461], this is still a technical challenge. The other issue is to tailor the nanotube/matrix interface with respect to the matrix. In this case, DWNTs and MWNTs may be more useful than SWNTs.

Multifunctional Materials

One of the major benefits expected from incorporating carbon nanotubes into other solid or liquid materials is that they endow the material with some electrical conductivity while leaving other properties or behaviors unaffected. As already mentioned in the previous section, the percolation threshold is reached at very low nanotube loadings. Tailoring the electrical conductivity of a bulk material is then achieved by adjusting the nanotube volume fraction in the formerly insulating material while making sure that this fraction is not too large. As demonstrated by *Maruyama* [3.488], there are three areas of interest regarding the electrical conductivity:

1. Electrostatic discharge (for example, preventing fire or explosion hazards in combustible environments or perturbations in electronics, which requires an electrical resistivity of less than 10^{12} Ω cm)

Table 3.5 Applications of nanotube-based multifunctional materials (from [3.488]), © B. Maruyama (WPAFB, Dayton, Ohio) ([a] For electrostatic painting, to mitigate lightning strikes on aircraft, etc., [b] to increase service temperature rating of product, [c] to reduce operating temperatures of electronic packages, [d] reduces warping, [e] reduces microcracking damage in composites)

Fiber fraction	Applications	Mechanical			Electrical			Thermal		Thermo-mechanical	
		Strength/ stiffness	Specific strength	Through-thickness strength	Static dissipation	Surface conduction[a]	EMI shielding	Service[b] temperature	Conduction/ dissipation[c]	Dimensional stability[d]	CTE reduction[e]
Low volume fraction (fillers)											
Elastomers	Tires				×				×		
Thermoplastics	Chip package					×			×		
	Electronics/ housing	×					×	×	×		
Thermosets	Epoxy products	×	×		×		×			×	×
	Composites				×					×	
High volume fraction											
Structural composites	Space/aircraft components			×	×						
High conduction composites	Radiators	×							×	×	
	Heat exchangers	×							×	×	×
	EMI shield	×				×					

2. Electrostatic painting (which requires the material to be painted to have enough electrical conductivity – an electrical resistivity below $10^6\,\Omega\,\text{cm}$ – to prevent the charged paint droplets from being repelled)
3. Electromagnetic interference shielding (which is achieved for an electrical resistivity of less than $10\,\Omega\,\text{cm}$).

Materials are often required to be multifunctional; for example, to have both high electrical conductivity and high toughness, or high thermal conductivity and high thermal stability. An association of several materials, each of them bringing one of the desired features, generally meets this need. The exceptional features and properties of carbon nanotubes make them likely to be a perfect multifunctional material in many cases. For instance, materials used in satellites are often required to be electrical conductive, mechanically self-supporting, able to transport away excess heat, and often to be robust against electromagnetic interference, while being of minimal weight and volume. All of these properties should be possible with a single nanotube-containing composite material instead of complex multimaterials combining layers of polymers, aluminum, copper, and so on. Table 3.5 provides an overview of various fields in which nanotube-based multifunctional materials should find application.

Nanoelectronics

As reported in Sects. 3.1.1 and 3.4.4, SWNT nanotubes can be either metallic (with an electrical conductivity higher than that of copper), or semiconducting. This has inspired the design of several components for nanoelectronics. First, metallic SWNTs can be used as mere ballistic conductors. Moreover, as early as 1995, realizing a rectifying diode by joining one metallic SWNT to one semiconductor SWNT (hetero-junction) was proposed by *Lambin* et al. [3.489], then later by *Chico* et al. [3.490] and *Yao* et al. [3.491]. Also, field effect transistors (FET) can be built by attaching a semiconductor SWNT across two electrodes (source and drain) deposited on an insulating substrate that serves as a gate electrode [3.492, 493]. The association of two such SWNT-based FETs makes a voltage inverter [3.494].

All of the latter developments are fascinating and provide promising outlets for nanotube-based electronics. However, progress is obviously needed before SWNT-based integrated circuits can be constructed on a routine basis. A key issue is the need to be able to selectively prepare either metallic or semiconductor nanotubes. Although a method of selectively destroying metallic SWNTs in bundles of undifferentiated SWNTs [3.496] has been proposed, the method is not scalable and selective synthesis would be preferable. Also, defect-free nanotubes are required. Generally speaking, this relates to another major challenge, which is to be able to fabricate integrated circuits including nanometer-size components (that only sophisticated imaging methods such as AFM are able to visualize) on an industrial scale. An overview of the issues related to the integration of carbon nanotubes into microelectronics systems has been written by *Graham* et al. [3.497].

Nanotools, Nanodevices and Nanosystems

Due to the ability of graphene to expand slightly when electrically charged, nanotubes have been found to act as actuators. *Kim* and *Lieber* [3.495] demonstrated this by designing *nanotweezers*, which are able to grab, manipulate and release nano-objects (the *nanobead* that was handled for the demonstration was actually closer to a micrometer in size than a nanometer), as well as to measure their electrical properties. This was made possible by simply depositing two noninterconnected gold coatings onto a pulled glass micropipette (Fig. 3.32), and then attaching two MWNTs (or two SWNT-bundles) ≈ 20–50 nm in diameter to each of the gold electrodes. Applying a voltage (0–8.5 V) between the two electrodes then makes the tube tips open and close reversibly in a controlled manner.

A similar experiment, again rather simple, was proposed by *Baughman* et al. the same year (1999) [3.498]. This consisted of mounting two SWNT-based paper strips (*bucky-paper*) on both sides of insulating double-sided tape. The two bucky-paper strips had been previously loaded with Na^+ and Cl^-, respectively. When 1 V was applied between the two paper strips, both of them expanded, but the strip loaded with Na^+ expanded a bit more, forcing the whole system to bend. Though performed in a liquid environment, this behavior has inspired the authors to predict a future use for their system in *artificial muscles*.

Another example of amazing nanotools is the nanothermometer proposed by *Gao* and *Bando* [3.499]. A single MWNT was used, which was partially filled with liquid gallium. Temperature variations in the range 50–500 °C cause the gallium to reversibly move up and down within the nanotube cavity at reproducible levels with respect to the temperature values applied.

Of course, nanotools such as nanotweezers or nanothermometers are hardly commercial enough to justify industrial investment. But such experiments are more than just amazing laboratory curiosities. They demonstrate the ability of carbon nanotubes to provide building blocks for future nanodevices, including nanomechanical systems.

Supercapacitors

Supercapacitors consist of two electrodes immersed in an electrolyte (such as 6 M KOH), separated by an insulating ion-permeable membrane. Charging the capacitors is achieved by applying a potential between the two electrodes, which makes the cations and the anions move toward the oppositely charged electrode. Suitable electrodes should exhibit high electrical conductivities and high surface areas, since the capacitance is proportional to these parameters. Actually, the surface area should consist of an appropriate combination of mesopores (to allow the electrolyte components to circulate well, which is related to the charging speed) and micropores (whose walls provide the attractive surfaces and fixation sites for the ions). Based on early work by *Niu* et al. [3.500], such a combination was found to be provided by the specific architecture offered by packed and entan-

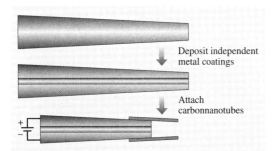

Fig. 3.32 Sketch explaining how the first nanotweezers were designed. The process involves modifying a glass micropipette (dark cone, *top*). Two Au coatings (in gray, *middle*) are deposited so that they are not in contact. Then a voltage is applied to the electrodes (from [3.495])

gled h-MWNTs with poor nanotextures (Sect. 3.1.2). However, activation pretreatments were necessary. For instance, a capacitor made from nanotubes with a surface area of $220\,\mathrm{m^2/g}$ exhibited a capacitance of $20\,\mathrm{F/g}$, which increased to $100\,\mathrm{F/g}$ after an activation treatment was applied to the nanotubes so that their surface area increased to $880\,\mathrm{m^2/g}$ [3.171]. Alternatively, again due to their remarkable architectures derived from their huge aspect ratios, nanotubes can also be used as supports for conductive polymer coatings, such as polypyrrole or polyaniline [3.501], or additives to regular carbon electrodes [3.502], which make the material more open, allowing easier circulation and penetration of ions. Supercapacitors built from such composites can survive more than 2000 charging cycles, with current densities as high as $350\,\mathrm{mA/g}$ [3.503].

Capacitors including nanotubes have already shown capacitances as high as $180-200\,\mathrm{F/g}$, equivalent to those obtained with electrodes built from regular carbon materials, but they have the advantage of faster charging [3.171]. Current work in this area will certainly lead to further optimization of both the nanotube material architecture and the nanotube-supported conductive polymers, meaning that the outlook for the commercial use of nanotubes as components for supercapacitors is positive, and this is ignoring the potential application of second-generation nanotubes (such as nanotube-based nano-objects) in this field. A first attempt to use hybrid nanotubes (Sect. 3.5.2) has already resulted in improved properties with respect to genuine (undoped) nanotube-based systems [3.504].

3.7 Toxicity and Environmental Impact of Carbon Nanotubes

As the number of industrial applications of CNT increases constantly with the production capacity at the worldwide level (estimated to ca. a few hundreds of tons in 2007), it is reasonable to address the issue of their potential impact on both human health and environment. It is important to consider that the large variety of CNTs (SWNT, DWNT, MWNT, hetero-CNTs, hybrid CNTs, etc.) and of synthesis routes (arc-discharge, laser ablation, CCVD, ...) as well as the lack of standardized testing procedures make the investigation of the toxicity of CNTs very difficult, and the comparison of the already published results almost impossible [3.505]. CNTs are mostly found as bundles rather than as individual objects, or more likely as large micrometric agglomerates. All samples contain different levels of residual catalyst(s), depending on the synthesis route and purification steps that they may have undergone. Usual purification treatments involve the combination of acids and oxidising agents, which leads to partial functionalization of the outer wall, making the treated samples more hydrophilic. SWNTs and DWNTs usually form long and flexible bundles (typically hundreds of micrometers long) whereas MWNTs are generally shorter (tens of micrometers) and more rigid. MWNTs also have generally more surface defects, which enhances their chemical reactivity. The specific surface area can range from a few tens of squared metres per gram in the case of densely packed MWNTs to just below $1000\,\mathrm{m^2/g}$ in the case of SWNTs and DWNTs (the theoretical limit being ca. $1300\,\mathrm{m^2/g}$ in the case of individual closed SWNTs).

The main exposure routes for dry CNTs are inhalation and dermal contact (also possible in the case of suspensions). Ingestion is generally not considered (would be accidental), although it is in fact more or less related to inhalation. In the case of suspensions, the main issue concerns their stability. This question has been widely studied worldwide and the general approach is the addition of a surfactant in order to stabilise the CNT in the liquid. The main problem is that all commonly used surfactants are toxic to a certain extent and thus cannot be used in the presence of living cells or animals for in vivo or in vitro investigations, or at such low concentrations that they do not really play anymore the role they are supposed to play. Although a few natural surfactants have been investigated, the stability of the suspensions in the presence of living organisms is often very different (fast destabilisation leading to flocculation). Injection in the bloodstream is envisaged, but would not be accidental (biological applications such as imaging, targeted cell delivery, hyperthermia, etc.). After the CNTs have entered the body, they could travel following different routes depending on the entry point (movements from one organ to another are called translocation) but also mainly on their physicochemical characteristics. Objects recognised as non-self by the immune system usually end up in the liver or the kidneys if they can be transported there, and could possibly be excreted (eliminated) from the body. In the general case, CNTs will just accumulate (biopersistence). They are usually intercepted by macrophages (cells present in all tissues and which role is to phagocyte (engulf and then digest) cellular debris and pathogens as well as to stimulate lymphocytes and other immune cells to respond to the pathogens). Taking into account the small size of macrophages as compared to that of agglomerates, bundles or even individual CNTs, macrophages usually do not manage to get rid of the CNTs by phagocytose. However, they try to do so and thus release reactive oxygen species (ROS), enzymes, cytokines (interferons (IFN)), etc. and agglomerate around them to isolate them from the body. Proteins present in the blood and most biological fluids (complement system – innate immunity) will play a similar role by *labelling* the CNTs (opsonisation) and possibly generating some inflammatory reactions. The complement system strongly interacts with the lymphocytes. These natural phenomena have deleterious consequences on the surrounding tissues: inflammation in a first instance, formation of granuloma (commonly observed in the lungs after exposure to CNTs). Each target organ has its own phagocyte cells (Kupffer cells in the liver, Langerhans cells in the skin, etc.).

Toxicity can be assessed both by in vitro and in vivo experiments. In the case of in vitro assays, cell cultures (usually immortalised cancer cells, but also primary cultures or even stem cells) are exposed to suspensions of CNTs. In the case of in vivo assays, the animals (mice, rats, worms, amphibians, fishes, etc.) are exposed either to aerosols (inhalation) or mainly again to suspensions of CNTs which will be administrated according to different protocols depending on the study (intra-tracheal instillation, injection, contact with the skin, etc.). Extrapolating the toxicity results from animals (or even worse, from cells) to humans is very delicate but the data are however very useful for the sake of comparison in a given system and with given experimental conditions. As soon as CNTs are in contact with a biological fluid, their surface chemistry is likely to be modified very quickly by adsorption of proteins (complement system [3.506], surfactants [3.507], etc.); this adsorption can be very specific [3.506, 507], and is likely to be dynamic and controlled by the affinity of the molecules for the surface of the CNTs (pristine or functionalised). It is thus obvious that the surface chemistry of the CNTs will play a very important role.

The potential use of CNTs in commercial products (Sect. 3.5) begs the question of their fate at the end of their lifecycle. If the impact of CNTs on human health is under investigation for already a few years now, it is noteworthy that the environmental impact has almost not been taken into account. Only a few publications (less than 15) are available to date and the concentration at which ecotoxic effects are evidenced is usually much higher than what could be reasonably found in the environment (unless very local and specific conditions). Due to the potentially very high specific surface area of CNTs, they could act as vectors for pollutants adsorbed on their surface (PAH, polycyclic aromatic hydrocarbons for example), even if themselves do not show any sign of toxicity.

There is currently no consensus about the toxicity of CNTs [3.505], although more than 500 papers have now been published already on this topic within the last 5 years. Despite the worldwide effort devoted to this field of research, the huge variety of CNT types, shapes, composition, etc. will make very difficult to answer this simple question: *are CNT toxic?* The *principle of precaution* should not stop all research in this area but only draw the attention to a more responsible attitude for people working on their synthesis or manipulating them, and industrials willing to include them in consumer products. Gloves should be worn at any time as well as an adapted (FFP3 type) disposable dust mask. Wearing a lab coat is recommended to limit contamination of clothes. CNT wastes should be burnt.

3.8 Concluding Remarks

Carbon nanotubes have been the focus of a lot of research work (and therefore a lot of funding) for nearly two decades now. Considering this investment of time and money, relatively few nanotube applications have reached the market yet. This may remind some of the disappointments associated with fullerene research, originally believed to be so promising, but which has resulted in no significant application after twenty years. However, nanotubes exhibit an extraordinary diversity of morphologies, textures, structures and nanotextures, far beyond that provided by fullerenes. Indeed, the properties of nanotubes are yet to be fully identified, and we should not forget the potential of hybrid nanotubes, heteronanotubes and nanotube-containing composites. The history of nanotubes has only just begun.

References

3.1 M. Monthioux, V.L. Kuznetsov: Who should be given the credit for the discovery of carbon nanotubes?, Carbon **44**, 1621–1623 (2006)

3.2 L.V. Radushkevich, V.M. Lukyanovich: O strukture ugleroda, obrazujucegosja pri termiceskom razlozenii okisi ugleroda na zeleznom kontakte, Zurn. Fis. Chim. **26**, 88–95 (1952), in Russian

3.3 T.V. Hughes, C.R. Chambers: Manufacture of Carbon Filaments, US Patent 405480 (1889)

3.4 P. Schützenberger, L. Schützenberger: Sur quelques faits relatifs à l'histoire du carbone, C. R. Acad. Sci. Paris **111**, 774–778 (1890), in French

3.5 C. Pélabon, H. Pélabon: Sur une variété de carbone filamenteux, C. R. Acad. Sci. Paris **137**, 706–708 (1903), in French

3.6 R.T.K. Baker, P.S. Harris: The formation of filamentous carbon. In: *Chemistry and Physics of Carbon*, Vol. 14, ed. by P.L. Walker Jr., P.A. Thrower (Dekker, New York 1978) pp. 83–165

3.7 S. Iijima: Helical microtubules of graphite carbon, Nature **354**, 56–58 (1991)

3.8 S. Iijima, T. Ichihashi: Single-shell carbon nanotubes of 1 nm diameter, Nature **363**, 603–605 (1993)

3.9 D.S. Bethune, C.H. Kiang, M.S. de Vries, G. Gorman, R. Savoy, J. Vazquez, R. Bayers: Cobalt-catalysed growth of carbon nanotubes with single-atomic-layer walls, Nature **363**, 605–607 (1993)

3.10 J. Tersoff, R.S. Ruoff: Structural properties of a carbon-nanotube crystal, Phys. Rev. Lett. **73**, 676–679 (1994)

3.11 N. Wang, Z.K. Tang, G.D. Li, J.S. Chen: Single-walled 4 Å carbon nanotube arrays, Nature **408**, 50–51 (2000)

3.12 N. Hamada, S.I. Sawada, A. Oshiyama: New one-dimensional conductors, graphite microtubules, Phys. Rev. Lett. **68**, 1579–1581 (1992)

3.13 M.S. Dresselhaus, G. Dresselhaus, P.C. Eklund: *Science of Fullerenes and Carbon Nanotubes* (Academic, San Diego 1995)

3.14 R.C. Haddon: Chemistry of the fullerenes: The manifestation of strain in a class of continuous aromatic molecules, Science **261**, 1545–1550 (1993)

3.15 M. Monthioux, B.W. Smith, B. Burteaux, A. Claye, J. Fisher, D.E. Luzzi: Sensitivity of single-wall nanotubes to chemical processing: An electron microscopy investigation, Carbon **39**, 1261–1272 (2001)

3.16 H. Allouche, M. Monthioux: Chemical vapor deposition of pyrolytic carbon onto carbon nanotubes. Part II – Structure and texture, Carbon **43**, 1265–1278 (2005)

3.17 M. Audier, A. Oberlin, M. Oberlin, M. Coulon, L. Bonnetain: Morphology and crystalline order in catalytic carbons, Carbon **19**, 217–224 (1981)

3.18 Y. Saito: Nanoparticles and filled nanocapsules, Carbon **33**, 979–988 (1995)

3.19 P.J.F. Harris: *Carbon Nanotubes and Related Structures* (Cambridge Univ. Press, Cambridge 1999)

3.20 N.M. Rodriguez, A. Chambers, R.T. Baker: Catalytic engineering of carbon nanostructures, Langmuir **11**, 3862–3866 (1995)

3.21 M. Monthioux, L. Noé, L. Dussault, J.-C. Dupin, N. Latorre, T. Ubieto, E. Romeo, C. Royo, A. Monzón, C. Guimon: Texturising and structurising mechanisms of carbon nanofilament during growth, J. Mater. Chem. **17**, 4611–4618 (2007)

3.22 J. Vera-Agullo, H. Varela-Rizo, J.A. Conesa, C. Almansa, C. Merino, I. Martin-Gullon: Evidence for growth mechanism and helix-spiral cone structure of stacked-cup carbon nanofibers, Carbon **45**, 2751–2758 (2007)

3.23 H.W. Kroto, J.R. Heath, S.C. O'Brien, R.F. Curl, R.E. Smalley: C_{60} Buckminsterfullerene, Nature **318**, 162–163 (1985)

3.24 W. Krätschmer, L.D. Lamb, K. Fostiropoulos, D.R. Huffman: Solid C_{60}: A new form of carbon, Nature **347**, 354–358 (1990)

3.25 L. Fulchieri, Y. Schwob, F. Fabry, G. Flamant, L.F.P. Chibante, D. Laplaze: Fullerene production in a 3-phase AC plasma process, Carbon **38**, 797–803 (2000)

3.26 K. Saidane, M. Razafinimanana, H. Lange, A. Huczko, M. Baltas, A. Gleizes, J.L. Meunier: Fullerene synthesis in the graphite electrode arc process: local plasma characteristics and correlation with yield, J. Phys. D Appl. Phys. **37**, 232–239 (2004)

3.27 T. Guo, P. Nikolaev, A.G. Rinzler, D. Tomanek, D.T. Colbert, R.E. Smalley: Self-assembly of tubular fullerenes, J. Phys. Chem. **99**, 10694–10697 (1995)

3.28 T. Guo, P. Nikolaev, A. Thess, D.T. Colbert, R.E. Smalley: Catalytic growth of single-walled nanotubes by laser vaporisation, Chem. Phys. Lett. **243**, 49–54 (1995)

3.29 A.G. Rinzler, J. Liu, H. Dai, P. Nikolaev, C.B. Huffman, F.J. Rodriguez-Macias, P.J. Boul, A.H. Lu, D. Heymann, D.T. Colbert, R.S. Lee, J.E. Fischer, A.M. Rao, P.C. Eklund, R.E. Smalley: Large scale purification of single wall carbon nanotubes: Process, product and characterization, Appl. Phys. A **67**, 29–37 (1998)

3.30 L.M. Chapelle, J. Gavillet, J.L. Cochon, M. Ory, S. Lefrant, A. Loiseau, D. Pigache: A continuous wave CO_2 laser reactor for nanotube synthesis, Proc. Electron. Prop. Nov. Mater. – XVI Int. Wintersch. – AIP Conf. Proc., ed. by H. Kuzmany, J. Fink, M. Mehring, S. Roth (Springer, Berlin Heidelberg 1999) pp. 237–240

3.31 A. Thess, R. Lee, P. Nikolaev, H. Dai, P. Petit, J. Robert, C. Xu, Y.H. Lee, S.G. Kim, D.T. Colbert, G. Scuseria, D. Tomanek, J.E. Fischer, R.E. Smalley: Crystalline ropes of metallic carbon nanotubes, Science **273**, 487–493 (1996)

3.32 M. Yudasaka, T. Komatsu, T. Ichihashi, S. Iijima: Single wall carbon nanotube formation by laser ablation using double targets of carbon and metal, Chem. Phys. Lett. **278**, 102–106 (1997)

3.33 M. Castignolles, A. Foutel-Richard, A. Mavel, J.L. Cochon, D. Pigache, A. Loiseau, P. Bernier: Combined experimental and numerical study of the parameters controlling the C-SWNT synthesis via laser vaporization, Proc. Electron. Prop. Nov. Mater. – XVI Int. Wintersch. – AIP Conf. Proc., ed. by H. Kuzmany, J. Fink, M. Mehring, S. Roth (Springer, Berlin Heidelberg 2002) pp. 385–389

3.34 T.W. Ebbesen, P.M. Ajayan: Large-scale synthesis of carbon nanotubes, Nature **358**, 220–221 (1992)

3.35 D. Ugarte: Morphology and structure of graphitic soot particles generated in arc-discharge C_{60} production, Chem. Phys. Lett. **198**, 596–602 (1992)

3.36 T.W. Ebbesen: Carbon nanotubes, Ann. Rev. Mater. Sci. **24**, 235–264 (1994)

3.37 T. Beltz, J. Find, D. Herein, N. Pfänder, T. Rühle, H. Werner, M. Wohlers, R. Schlögl: On the production of different carbon forms by electric arc graphite evaporation, Ber. Bunsen. Phys. Chem. **101**, 712–725 (1997)

3.38 C. Journet, W.K. Maser, P. Bernier, A. Loiseau, L.M. de la Chapelle, S. Lefrant, P. Deniard, R. Lee, J.E. Fischer: Large-scale production of single-walled carbon nanotubes by the electric-arc technique, Nature **388**, 756–758 (1997)

3.39 K. Saïdane, M. Razafinimanana, H. Lange, M. Baltas, A. Gleizes, J.J. Gonzalez: Influence of the carbon arc current intensity on fullerene synthesis, Proc. 24th Int. Conf. Phenom. Ioniz. Gases, ed. by P. Pisarczyk, T. Pisarczyk, J. Wotowski (Institute of Plasma Physics and Laser Microfusion, Warsaw 1999) pp. 203–204

3.40 H. Allouche, M. Monthioux, M. Pacheco, M. Razafinimanana, H. Lange, A. Huczko, T.P. Teulet, A. Gleizes, T. Sogabe: Physical characteristics of the graphite-electrode electric-arc as parameters for the formation of single-wall carbon nanotubes, Proc. Eurocarbon, Vol. 2 (Deutsche Keramische Gesellschaft, 2000) pp. 1053–1054

3.41 M. Razafinimanana, M. Pacheco, M. Monthioux, H. Allouche, H. Lange, A. Huczko, A. Gleizes: Spectroscopic study of an electric arc with Gd and Fe doped anodes for the carbon nanotube formation, Proc. 25th Int. Conf. Phenom. Ioniz. Gases, ed. by E. Goto (Nagoya Univ., Nagoya 2001) pp. 297–298

3.42 M. Razafinimanana, M. Pacheco, M. Monthioux, H. Allouche, H. Lange, A. Huczko, P. Teulet, A. Gleizes, C. Goze, P. Bernier, T. Sogabe: Influence of doped graphite electrode in electric arc for the formation of single wall carbon nanotubes, Proc. 6th Eur. Conf. Therm. Plasma Process. – Prog. Plasma Process. Mater., New York 2000, ed. by P. Fauchais (Begell House, New York 2001) pp. 649–654

3.43 M. Pacheco, H. Allouche, M. Monthioux, A. Razafinimanana, A. Gleizes: Correlation between the plasma characteristics and the morphology and structure of the carbon phases synthesised by electric arc discharge, Proc. 25th Bienn. Conf. Carbon, Lexington 2001, ed. by F. Derbyshire (American Carbon Society 2001), Extend. Abstr. (CD-ROM), Novel/14.1

3.44 M. Pacheco, M. Monthioux, M. Razafinimanana, L. Donadieu, H. Allouche, N. Caprais, A. Gleizes: New factors controlling the formation of single-wall carbon nanotubes by arc plasma, Proc. Carbon 2002 Int. Conf., Beijing 2002, ed. by H.-M. Cheng (Shanxi Chunqiu Audio-Visual Press, Beijing 2002), CD-ROM/Oral/I014

3.45 M. Monthioux, M. Pacheco, H. Allouche, M. Razafinimanana, N. Caprais, L. Donadieu, A. Gleizes: New data about the formation of SWNTs by the electric arc method. In: *Electronic Properties of Molecular Nanostructures*, AIP Conf. Proc., ed. by H. Kuzmany, J. Fink, M. Mehring, S. Roth (Springer, Berlin Heidelberg 2002) pp. 182–185

3.46 H. Lange, A. Huczko, M. Sioda, M. Pacheco, M. Razafinimanana, A. Gleizes: Influence of gadolinium on carbon arc plasma and formation of fullerenes and nanotubes, Plasma Chem. Plasma Process **22**, 523–536 (2002)

3.47 C. Journet: La production de nanotubes de carbone. Ph.D. Thesis (University of Montpellier II, Montpellier 1998)

3.48 T. Sogabe, T. Masuda, K. Kuroda, Y. Hirohaya, T. Hino, T. Ymashina: Preparation of B_4C-mixed graphite by pressureless sintering and its air oxidation behavior, Carbon **33**, 1783–1788 (1995)

3.49 M. Ishigami, J. Cumings, A. Zettl, S. Chen: A simple method for the continuous production of carbon nanotubes, Chem. Phys. Lett. **319**, 457–459 (2000)

3.50 Y.L. Hsin, K.C. Hwang, F.R. Chen, J.J. Kai: Production and in-situ metal filling of carbon nanotube in water, Adv. Mater. **13**, 830–833 (2001)

3.51 H.W. Zhu, X.S. Li, B. Jiang, C.L. Xu, C.L. Zhu, Y.F. Zhu, D.H. Wu, X.H. Chen: Formation of carbon nanotubes in water by the electric arc technique, Chem. Phys. Lett. **366**, 664–669 (2002)

3.52 W.K. Maser, P. Bernier, J.M. Lambert, O. Stephan, P.M. Ajayan, C. Colliex, V. Brotons, J.M. Planeix, B. Coq, P. Molinie, S. Lefrant: Elaboration and characterization of various carbon nanostructures, Synth. Met. **81**, 243–250 (1996)

3.53 A. Mansour, M. Razafinimanana, M. Monthioux, M. Pacheco, A. Gleizes: A significant improvement of both yield and purity during SWCNT synthesis via the electric arc process, Carbon **45**, 1651–1661 (2007)

3.54 A. Mansour: Caractérisation expérimentale d'un plasma d'arc électrique en vue du contrôle de la synthèse des nanotubes de carbone monoparois. Ph.D. Thesis (University Paul Sabatier, Toulouse 2007)

3.55 J. Gavillet, A. Loiseau, J. Thibault, A. Maigné, O. Stéphan, P. Bernier: TEM study of the influence of the catalyst composition on the formation and growth of SWNT, Proc. Electron. Prop. Nov. Mater. –

XVI Int. Wintersch. – AIP Conf. Proc., ed. by H. Kuzmany, J. Fink, M. Mehring, S. Roth (Springer, Berlin Heidelberg 2002) pp. 202–206
3.56 G. Flamant, J.F. Robert, S. Marty, J.M. Gineste, J. Giral, B. Rivoire, D. Laplaze: Solar reactor scaling up. The fullerene synthesis case study, Energy **29**, 801–809 (2004)
3.57 T.M. Gruenberger, J. Gonzalez-Aguilar, F. Fabry, L. Fulchieri, E. Grivei, N. Probst, G. Flamant, H. Okuno, J.C. Charlier: Production of carbon nanotubes and other nanostructures via continuous 3-phase AC plasma processing, Fuller. Nanotub. Carbon Nanostruct. **12**, 571–581 (2004)
3.58 H. Okuno, E. Grivel, F. Fabry, T.M. Gruenberger, J.J. Gonzalez-Aguilar, A. Palnichenko, L. Fulchieri, N. Probst, J.C. Chalier: Synthesis of carbon nanotubes and nano-necklaces by thermal plasma process, Carbon **42**, 2543–2549 (2004)
3.59 L.P.F. Chibante, A. Thess, J.M. Alford, M.D. Diener, R.E. Smalley: Solar generation of the fullerenes, J. Phys. Chem. **97**, 8696–8700 (1993)
3.60 C.L. Fields, J.R. Pitts, M.J. Hale, C. Bingham, A. Lewandowski, D.E. King: Formation of fullerenes in highly concentrated solar flux, J. Phys. Chem. **97**, 8701–8702 (1993)
3.61 P. Bernier, D. Laplaze, J. Auriol, L. Barbedette, G. Flamant, M. Lebrun, A. Brunelle, S. Della-Negra: Production of fullerenes from solar energy, Synth. Met. **70**, 1455–1456 (1995)
3.62 M.J. Heben, T.A. Bekkedhal, D.L. Schultz, K.M. Jones, A.C. Dillon, C.J. Curtis, C. Bingham, J.R. Pitts, A. Lewandowski, C.L. Fields: Production of single wall carbon nanotubes using concentrated sunlight, Proc. Symp. Recent Adv. Chem. Phys. Fuller. Rel. Mater., Pennington 1996, ed. by K.M. Kadish, R.S. Ruoff (Electrochemical Society, Pennington 1996) pp. 803–811
3.63 D. Laplaze, P. Bernier, C. Journet, G. Vié, G. Flamant, E. Philippot, M. Lebrun: Evaporation of graphite using a solar furnace, Proc. 8th Int. Symp. Solar Conc. Technol., Köln 1996, ed. by M. Becker, M. Balmer (Müller, Heidelberg 1997) pp. 1653–1656
3.64 D. Laplaze, P. Bernier, W.K. Maser, G. Flamant, T. Guillard, A. Loiseau: Carbon nanotubes: The solar approach, Carbon **36**, 685–688 (1998)
3.65 T. Guillard, S. Cetout, L. Alvarez, J.L. Sauvajol, E. Anglaret, P. Bernier, G. Flamant, D. Laplaze: Production of carbon nanotubes by the solar route, Eur. Phys. J. **5**, 251–256 (1999)
3.66 D. Luxembourg, G. Flamant, A. Guillot, D. Laplaze: Hydrogen storage in solar produced single-walled carbon nanotubes, Mater. Sci. Eng. B **108**, 114–119 (2004)
3.67 G. Flamant, M. Bijeire, D. Luxembourg: Modelling of a solar reactor for single wall nanotubes synthesis, ASME J. Solar Energy Eng. **128**, 1–124 (2006)

3.68 G.G. Tibbetts, M. Endo, C.P. Beetz: Carbon fibers grown from the vapor phase: A novel material, SAMPE Journal **22**, 30 (1989)
3.69 R.T.K. Baker: Catalytic growth of carbon filaments, Carbon **27**, 315–323 (1989)
3.70 E. Lamouroux, P. Serp, P. Kalck: Catalytic chemical vapor deposition routes towards single-walled and double-walled carbon nanotubes, Catal. Rev. Sci. Eng. **49**, 341–405 (2007)
3.71 R. Philippe, A. Morançais, M. Corrias, B. Caussat, Y. Kihn, P. Kalck, D. Plee, P. Gaillard, D. Bernard, P. Serp: Catalytic production of carbon nanotubes by fluidized-bed CVD, Chem. Vap. Depos. **13**, 447–457 (2007)
3.72 R.T.K. Baker, P.S. Harris, R.B. Thomas, R.J. Waite: Formation of filamentous carbon from iron, cobalt, and chromium catalyzed decomposition of acetylene, J. Catal. **30**, 86–95 (1973)
3.73 T. Koyama, M. Endo, Y. Oyuma: Carbon fibers obtained by thermal decomposition of vaporized hydrocarbon, Jpn. J. Appl. Phys. **11**, 445–449 (1972)
3.74 M. Endo, A. Oberlin, T. Koyama: High resolution electron microscopy of graphitizable carbon fiber prepared by benzene decomposition, Jpn. J. Appl. Phys. **16**, 1519–1523 (1977)
3.75 N.M. Rodriguez: A review of catalytically grown carbon nanofibers, J. Mater. Res. **8**, 3233–3250 (1993)
3.76 W.R. Davis, R.J. Slawson, G.R. Rigby: An unusual form of carbon, Nature **171**, 756 (1953)
3.77 H.P. Boehm: Carbon from carbon monoxide disproportionation on nickel and iron catalysts; morphological studies and possible growth mechanisms, Carbon **11**, 583–590 (1973)
3.78 M. Audier, A. Oberlin, M. Coulon: Crystallographic orientations of catalytic particles in filamentous carbon; case of simple conical particles, J. Cryst. Growth **55**, 546–549 (1981)
3.79 M. Audier, M. Coulon: Kinetic and microscopic aspects of catalytic carbon growth, Carbon **23**, 317–323 (1985)
3.80 M. Audier, A. Oberlin, M. Coulon: Study of biconic microcrystals in the middle of carbon tubes obtained by catalytic disproportionation of CO, J. Cryst. Growth **57**, 524–534 (1981)
3.81 A. Thaib, G.A. Martin, P. Pinheiro, M.C. Schouler, P. Gadelle: Formation of carbon nanotubes from the carbon monoxide disproportionation reaction over Co/Al$_2$O$_3$ and Co/SiO$_2$ catalysts, Catal. Lett. **63**, 135–141 (1999)
3.82 P. Pinheiro, M.C. Schouler, P. Gadelle, M. Mermoux, E. Dooryhée: Effect of hydrogen on the orientation of carbon layers in deposits from the carbon monoxide disproportionation reaction over Co/Al$_2$O$_3$ catalysts, Carbon **38**, 1469–1479 (2000)
3.83 P. Pinheiro, P. Gadelle: Chemical state of a supported iron-cobalt catalyst during CO disproportion-

3.84 P. Pinheiro, P. Gadelle, C. Jeandey, J.L. Oddou: Chemical state of a supported iron-cobalt catalyst during CO disproportionation. II. Experimental study, J. Phys. Chem. Solids **62**, 1023–1037 (2001)

ation. I. Thermodynamic study, J. Phys. Chem. Solids **62**, 1015–1021 (2001)

3.85 C. Laurent, E. Flahaut, A. Peigney, A. Rousset: Metal nanoparticles for the catalytic synthesis of carbon nanotubes, New J. Chem. **22**, 1229–1237 (1998)

3.86 E. Flahaut: Synthèse par voir catalytique et caractérisation de composites nanotubes de carbone-metal-oxyde Poudres et matériaux denses. Ph.D. Thesis (Univers. Paul Sabatier, Toulouse 1999)

3.87 E. Flahaut, R. Bacsa, A. Peigney, C. Laurent: Gram-scale CCVD synthesis of double-walled carbon nanotubes, Chem. Commun., 1442–1443 (2003)

3.88 A. Peigney, C. Laurent, F. Dobigeon, A. Rousset: Carbon nanotubes grown in situ by a novel catalytic method, J. Mater. Res. **12**, 613–615 (1997)

3.89 V. Ivanov, J.B. Nagy, P. Lambin, A. Lucas, X.B. Zhang, X.F. Zhang, D. Bernaerts, G. Van Tendeloo, S. Amelinckx, J. Van Landuyt: The study of nanotubules produced by catalytic method, Chem. Phys. Lett. **223**, 329–335 (1994)

3.90 V. Ivanov, A. Fonseca, J.B. Nagy, A. Lucas, P. Lambin, D. Bernaerts, X.B. Zhang: Catalytic production and purification of nanotubules having fullerene-scale diameters, Carbon **33**, 1727–1738 (1995)

3.91 K. Hernadi, A. Fonseca, J.B. Nagy, D. Bernaerts, A. Fudala, A. Lucas: Catalytic synthesis of carbon nanotubes using zeolite support, Zeolites **17**, 416–423 (1996)

3.92 H. Dai, A.G. Rinzler, P. Nikolaev, A. Thess, D.T. Colbert, R.E. Smalley: Single-wall nanotubes produced by metal-catalysed disproportionation of carbon monoxide, Chem. Phys. Lett. **260**, 471–475 (1996)

3.93 A.M. Cassel, J.A. Raymakers, J. Kong, H. Dai: Large scale CVD synthesis of single-walled carbon nanotubes, J. Phys. Chem. B **109**, 6484–6492 (1999)

3.94 B. Kitiyanan, W.E. Alvarez, J.H. Harwell, D.E. Resasco: Controlled production of single-wall carbon nanotubes by catalytic decomposition of CO on bimetallic Co-Mo catalysts, Chem. Phys. Lett. **317**, 497–503 (2000)

3.95 A. Govindaraj, E. Flahaut, C. Laurent, A. Peigney, A. Rousset, C.N.R. Rao: An investigation of carbon nanotubes obtained from the decomposition of methane over reduced $Mg_{1-x}M_xAl_2O_4$ spinel catalysts, J. Mater. Res. **14**, 2567–2576 (1999)

3.96 E. Flahaut, A. Peigney, C. Laurent, A. Rousset: Synthesis of single-walled carbon nanotube-Co-MgO composite powders and extraction of the nanotubes, J. Mater. Chem. **10**, 249–252 (2000)

3.97 J. Kong, A.M. Cassel, H. Dai: Chemical vapor deposition of methane for single-walled carbon nanotubes, Chem. Phys. Lett. **292**, 567–574 (1998)

3.98 E. Flahaut, A. Peigney, W.S. Bacsa, R.R. Bacsa, C. Laurent: CCVD synthesis of carbon nanotubes from (Mg,Co, Mo)O catalysts: Influence of the proportions of cobalt and molybdenum, J. Mater. Chem. **14**, 646–653 (2004)

3.99 E. Flahaut, C. Laurent, A. Peigney: Catalytic CVD synthesis of double and triple-walled carbon nanotubes by the control of the catalyst preparation, Carbon **43**, 375–383 (2005)

3.100 R. Marangoni, P. Serp, R. Feurrer, Y. Kihn, P. Kalck, C. Vahlas: Carbon nanotubes produced by substrate free metalorganic chemical vapor deposition of iron catalyst and ethylene, Carbon **39**, 443–449 (2001)

3.101 R. Sen, A. Govindaraj, C.N.R. Rao: Carbon nanotubes by the metallocene route, Chem. Phys. Lett. **267**, 276–280 (1997)

3.102 Y.Y. Fan, H.M. Cheng, Y.L. Wei, G. Su, S.H. Shen: The influence of preparation parameters on the mass production of vapor grown carbon nanofibers, Carbon **38**, 789–795 (2000)

3.103 L. Ci, J. Wei, B. Wei, J. Liang, C. Xu, D. Wu: Carbon nanofibers and single-walled carbon nanotubes prepared by the floating catalyst method, Carbon **39**, 329–335 (2001)

3.104 M. Glerup, H. Kanzow, R. Almairac, M. Castignolles, P. Bernier: Synthesis of multi-walled carbon nanotubes and nano-fibres using aerosol method with metal-ions as the catalyst precursors, Chem. Phys. Lett. **377**, 293–298 (2003)

3.105 O.A. Nerushev, M. Sveningsson, L.K.L. Falk, F. Rohmund: Carbon nanotube films obtained by thermal vapour deposition, J. Mater. Chem. **11**, 1122–1132 (2001)

3.106 Z. Zhou, L. Ci, L. Song, X. Yan, D. Liu, H. Yuan, Y. Gao, J. Wang, L. Liu, W. Zhou, G. Wang, S. Xie: Producing cleaner double-walled carbon nanotubes in a floating catalyst system, Carbon **41**, 2607–2611 (2003)

3.107 F. Rohmund, L.K.L. Falk, F.E.B. Campbell: A simple method for the production of large arrays of aligned carbon nanotubes, Chem. Phys. Lett. **328**, 369–373 (2000)

3.108 G.G. Tibbetts, C.A. Bernardo, D.W. Gorkiewicz, R.L. Alig: Role of sulfur in the production of carbon fibers in the vapor phase, Carbon **32**, 569–576 (1994)

3.109 S. Bai, F. Li, Q.H. Yang, H.-M. Cheng, J.B. Bai: Influence of ferrocene/benzene mole ratio in the synthesis of carbon nanostructures, Chem. Phys. Lett. **376**, 83–89 (2003)

3.110 W.Q. Han, P. Kholer-Riedlich, T. Seeger, F. Ernst, M. Ruhle, N. Grobert, W.K. Hsu, B.H. Chang, Y.Q. Zhu, H.W. Kroto, M. Terrones, H. Terrones: Aligned CN_x nanotubes by pyrolysis of ferrocene under NH_3 atmosphere, Appl. Phys. Lett. **77**, 1807–1809 (2000)

3.111 L. Ci, Z. Rao, Z. Zhou, D. Tang, X. Yan, Y. Liang, D. Liu, H. Yuan, W. Zhou, G. Wang, W. Liu, S. Xie: Double wall carbon nanotubes promoted by sulfur in a floating iron catalyst CVD system, Chem. Phys. Lett. **359**, 63–67 (2002)

3.112 S. Maruyama, R. Kojima, Y. Miyauchi, S. Chiashi, M. Kohno: Low-temperature synthesis of high-purity single-walled carbon nanotubes from alcohol, Chem. Phys. Lett. **360**, 229–234 (2002)

3.113 T. Kyotani, L.F. Tsai, A. Tomita: Preparation of ultrafine carbon tubes in nanochannels of an anodic aluminum oxide film, Chem. Mater. **8**, 2109–2113 (1996)

3.114 E. Mora, T. Tokune, A.R. Harutyunyan: Continuous production of single-walled carbon nanotubes using a supported floating catalyst, Carbon **45**, 971–977 (2007)

3.115 R.E. Smalley, J.H. Hafner, D.T. Colbert, K. Smith: Catalytic growth of single-wall carbon nanotubes from metal particles, US Patent 19980601010903 (1998)

3.116 P. Nikolaev: Gas-phase production of single-walled carbon nanotubes from carbon monoxide: A review of the HiPco process, J. Nanosci. Nanotechnol. **4**, 307–316 (2004)

3.117 W.K. Hsu, J.P. Hare, M. Terrones, H.W. Kroto, D.R.M. Walton, P.J.F. Harris: Condensed-phase nanotubes, Nature **377**, 687 (1995)

3.118 W.S. Cho, E. Hamada, Y. Kondo, K. Takayanagi: Synthesis of carbon nanotubes from bulk polymer, Appl. Phys. Lett. **69**, 278–279 (1996)

3.119 Y.L. Li, Y.D. Yu, Y. Liang: A novel method for synthesis of carbon nanotubes: Low temperature solid pyrolysis, J. Mater. Res. **12**, 1678–1680 (1997)

3.120 M.L. Terranova, S. Piccirillo, V. Sessa, P. Sbornicchia, M. Rossi, S. Botti, D. Manno: Growth of single-walled carbon nanotubes by a novel technique using nanosized graphite as carbon source, Chem. Phys. Lett. **327**, 284–290 (2000)

3.121 R.L. Vander Wal, T. Ticich, V.E. Curtis: Diffusion flame synthesis of single-walled carbon nanotubes, Chem. Phys. Lett. **323**, 217–223 (2000)

3.122 I. Gunjishima, T. Inoue, S. Yamamuro, K. Sumiyama, A. Okamoto: Synthesis of vertically aligned, double-walled carbon nanotubes from highly active Fe-V-O nanoparticles, Carbon **45**, 1193–1199 (2007)

3.123 G. Zhong, T. Iwasaki, J. Robertson, H. Kawarada: Growth kinetics of 0.5 cm vertically aligned single-walled carbon nanotubes, J. Phys. Chem. B **111**, 1907–1910 (2007)

3.124 H. Cui, G. Eres, J.Y. Howe, A. Puretzki, M. Varela, D.B. Geohegan, D.H. Lowndes: Growth behavior of carbon nanotubes on multilayered metal catalyst film in chemical vapor deposition, Chem. Phys. Lett. **374**, 222–228 (2003)

3.125 A.M. Cassel, N.R. Franklin, T.W. Tombler, E.M. Chan, J. Han, H. Dai: Directed growth of free-standing single-walled carbon nanotubes, J. Am. Chem. Soc. **121**, 7975–7976 (1999)

3.126 S. Fan, M. Chapline, N. Franklin, T. Tombler, A.M. Cassel, H. Dai: Self-oriented regular arrays of carbon nanotubes and their field emission properties, Science **283**, 512–514 (1999)

3.127 Y.Y. Wei, G. Eres, V.I. Merkulov, D.H. Lowdens: Effect of film thickness on carbon nanotube growth by selective area chemical vapor deposition, Appl. Phys. Lett. **78**, 1394–1396 (2001)

3.128 I.T. Han, B.K. Kim, H.J. Kim, M. Yang, Y.W. Jin, S. Jung, N. Lee, S.K. Kim, J.M. Kim: Effect of Al and catalyst thickness on the growth of carbon nanotubes and application to gated field emitter arrays, Chem. Phys. Lett. **400**, 139–144 (2004)

3.129 W.Z. Li, S.S. Xie, L.X. Qian, B.H. Chang, B.S. Zou, W.Y. Zhou, R.A. Zha, G. Wang: Large scale synthesis of aligned carbon nanotubes, Science **274**, 1701–1703 (1996)

3.130 F. Zheng, L. Liang, Y. Gao, J.H. Sukamto, L. Aardahl: Carbon nanotubes synthesis using mesoporous silica templates, Nano Lett. **2**, 729–732 (2002)

3.131 S.H. Jeong, O.-K. Lee, K.H. Lee, S.H. Oh, C.G. Park: Preparation of aligned carbon nanotubes with prescribed dimension: Template synthesis and sonication cutting approach, Chem. Mater. **14**, 1859–1862 (2002)

3.132 N.S. Kim, Y.T. Lee, J. Park, H. Ryu, H.J. Lee, S.Y. Choi, J. Choo: Dependence of vertically aligned growth of carbon nanotubes on catalyst, J. Phys. Chem. B **106**, 9286–9290 (2002)

3.133 C.J. Lee, D.W. Kim, T.J. Lee, Y.C. Choi, Y.S. Park, Y.H. Lee, W.B. Choi, N.S. Lee, G.-S. Park, J.M. Kim: Synthesis of aligned carbon nanotubes using thermal chemical vapor deposition, Chem. Phys. Lett. **312**, 461–468 (1999)

3.134 W.D. Zhang, Y. Wen, S.M. Liu, W.C. Tjiu, G.Q. Xu, L.M. Gan: Synthesis of vertically aligned carbon nanotubes on metal deposited quartz plates, Carbon **40**, 1981–1989 (2002)

3.135 S. Huang, L. Dai, A.W.H. Mau: Controlled fabrication of large scale aligned carbon nanofiber/nanotube patterns by photolithography, Adv. Mater. **14**, 1140–1143 (2002)

3.136 T. Sun, G. Wang, H. Liu, L. Feng, D. Zhu: Control over the wettability of an aligned carbon nanotube film, J. Am. Chem. Soc. **125**, 14996–14997 (2003)

3.137 Y. Huh, J.Y. Lee, J. Cheon, Y.K. Hong, J.Y. Koo, T.J. Lee, C.J. Lee: Controlled growth of carbon nanotubes over cobalt nanoparticles by thermal chemical vapor deposition, J. Mater. Chem. **13**, 2297–2300 (2003)

3.138 Y. Kobayashi, H. Nakashima, D. Takagi, Y. Homma: CVD growth of single-walled carbon nanotubes using size-controlled nanoparticle catalyst, Thin Solid Films **464/465**, 286–289 (2004)

3.139 C.L. Cheung, A. Kurtz, H. Park, C.M. Lieber: Diameter-controlled synthesis of carbon nanotubes, J. Phys. Chem. B **106**, 2429–2433 (2002)

3.140 Y. Huh, J.Y. Lee, J. Cheon, Y.K. Hong, J.Y. Koo, T.J. Lee, C.J. Lee: Controlled growth of carbon nanotubes over cobalt nanoparticles by thermal chemical vapor deposition, J. Mater. Chem. **13**, 2297–2300 (2003)

3.141 M. Paillet, V. Jourdain, P. Poncharal, J.-L. Sauvajol, A. Zahab, J.C. Meyer, S. Roth, N. Cordente, C. Amiens, B. Chaudret: Versatile synthesis of individual single-walled carbon nanotubes from nickel nanoparticles for the study of their physical properties, J. Phys. Chem. B **108**, 17112–17118 (2004)

3.142 S. Casimirius, E. Flahaut, C. Laurent, C. Vieu, F. Carcenac, C. Laberty-Robert: Optimized microcontact printing process for the patterned growth of individual SWNTs, Microelectron. Eng. **73/74**, 564–569 (2004)

3.143 Y. Lei, K.S. Yeong, J.T.L. Thong, W.K. Chim: Large-scale ordered carbon nanotubes arrays initiated from highly ordered catalyst arrays on silicon substrates, Chem. Mater. **16**, 2757–2761 (2004)

3.144 Q. Ye, A.M. Cassel, H. Liu, K.J. Chao, J. Han, M. Meyyappan: Large-scale fabrication of carbon nanotube probe tips for atomic force microscopy critical dimension imaging applications, Nano Lett. **4**, 1301–1308 (2004)

3.145 K. Hata, D.N. Futaba, K. Mizuno, T. Namai, M. Yumara, S. Iijima: Ware-assisted highly efficient synthesis of impurity-free single-walled carbon nanotubes, Science **306**, 1362–1364 (2004)

3.146 R. Andrews, D. Jacques, A.M. Rao, F. Derbyshire, D. Qian, X. Fan, E.C. Dickey, J. Chen: Continous production of aligned carbon nanotubes: A step closer to commercial realization, Chem. Phys. Lett. **303**, 467–474 (1999)

3.147 C.N.R. Rao, R. Sen, B.C. Satishkumar, A. Govindaraj: Large aligned carbon nanotubes bundles from ferrocene pyrolysis, Chem. Commun., 1525–1526 (1998)

3.148 X. Zhang, A. Cao, B. Wei, Y. Li, J. Wei, C. Xu, D. Wu: Rapid growth of well-aligned carbon nanotube arrays, Chem. Phys. Lett. **362**, 285–290 (2002)

3.149 X. Zhang, A. Cao, Y. Li, C. Xu, J. Liang, D. Wu, B. Wei: Self-organized arrays of carbon nanotube ropes, Chem. Phys. Lett. **351**, 183–188 (2002)

3.150 K.S. Choi, Y.S. Cho, S.Y. Hong, J.B. Park, D.J. Kim: Effects of ammonia on the alignment of carbon nanotubes in metal-assisted chemical vapor deposition, J. Eur. Ceram. Soc. **21**, 2095–2098 (2001)

3.151 N.S. Kim, Y.T. Lee, J. Park, J.B. Han, Y.S. Choi, S.Y. Choi, J. Choo, G.H. Lee: Vertically aligned carbon nanotubes grown by pyrolysis of iron, cobalt, and nickel phthalocyanines, J. Phys. Chem. B **107**, 9249–9255 (2003)

3.152 C. Emmeger, J.M. Bonard, P. Mauron, P. Sudan, A. Lepora, B. Grobety, A. Züttel, L. Schlapbach: Synthesis of carbon nanotubes over Fe catalyst on aluminum and suggested growth mechanism, Carbon **41**, 539–547 (2003)

3.153 Q. Zhang, J. Huang, F. Wei, G. Xu, Y. Wang, W. Qian, D. Wang: Large scale production of carbon nanotubes arrays on the sphere surface from liquefied petroleum gas at low cost, Chin. Sci. Bull. **52**, 2896–2902 (2007)

3.154 X. Li, L. Zhang, X. Wang, I. Shimoyama, X. Sun, W.-S. Seo, H. Dai: Assembly of densely aligned single-walled carbon nanotubes from bulk materials Langmuir–Blodgett, J. Am. Chem. Soc. **129**, 4890–4891 (2007)

3.155 M. Endo, H.W. Kroto: Formation of carbon nanofibers, J. Phys. Chem. **96**, 6941–6944 (1992)

3.156 R.S. Wagner: VLS mechanisms of crystal growth. In: *Whisker Technology*, ed. by P.A. Levit (Wiley, New York 1970) pp. 47–72

3.157 Y.H. Lee, S.G. Kim, D. Tomanek: Catalytic growth of single-wall carbon nanotubes: An ab initio study, Phys. Rev. Lett. **78**, 2393–2396 (1997)

3.158 H. Dai: Carbon Nanotubes: Synthesis, integration, and properties, Acc. Chem. Res. **35**, 1035–1044 (2002)

3.159 V. Jourdain, H. Kanzow, M. Castignolles, A. Loiseau, P. Bernier: Sequential catalytic growth of carbon nanotubes, Chem. Phys. Lett. **364**, 27–33 (2002)

3.160 Y. Saito, M. Okuda, N. Fujimoto, T. Yoshikawa, M. Tomita, T. Hayashi: Single-wall carbon nanotubes growing radially from Ni fine particles formed by arc evaporation, Jpn. J. Appl. Phys. **33**, L526–L529 (1994)

3.161 J. Bernholc, C. Brabec, M. Buongiorno Nardelli, A. Malti, C. Roland, B.J. Yakobson: Theory of growth and mechanical properties of nanotubes, Appl. Phys. A **67**, 39–46 (1998)

3.162 M. Pacheco: Synthèse des nanotubes de carbone par arc électrique. Ph.D. Thesis (Université Toulouse III, Toulouse 2003)

3.163 K. Méténier, S. Bonnamy, F. Béguin, C. Journet, P. Bernier, L.M. de la Chapelle, O. Chauvet, S. Lefrant: Coalescence of single walled nanotubes and formation of multi-walled carbon nanotubes under high temperature treatments, Carbon **40**, 1765–1773 (2002)

3.164 P.G. Collins, P. Avouris: Nanotubes for electronics, Sci. Am. **283**, 38–45 (2000)

3.165 Q.-H. Yang, P.X. Hou, S. Bai, M.Z. Wang, H.M. Cheng: Adsorption and capillarity of nitrogen in aggregated multi-walled carbon nanotubes, Chem. Phys. Lett. **345**, 18–24 (2001)

3.166 S. Inoue, N. Ichikuni, T. Suzuki, T. Uematsu, K. Kaneko: Capillary condensation of N_2 on multiwall carbon nanotubes, J. Phys. Chem. **102**, 4689–4692 (1998)

3.167 S. Agnihotri, J.P. Mota, M. Rostam-Abadi, M.J. Rood: Structural characterization of single-walled carbon nanotube bundles by experiment and molecular simulation, Langmuir **21**, 896–904 (2005)

3.168 M. Eswaramoorthy, R. Sen, C.N.R. Rao: A study of micropores in single-walled carbon nanotubes by the adsorption of gases and vapors, Chem. Phys. Lett. **304**, 207–210 (1999)

3.169 S. Furmaniak, A.P. Terzyk, P.A. Gauden, K. Lota, E. Frackowiak, F. Beguin, P. Kowalczyk: Determination of the space between closed multiwalled carbon nanotubes by GCMC simulation of nitrogen adsorption, J. Colloid Interface Sci. **317**, 442–448 (2008)

3.170 A. Peigney, C. Laurent, E. Flahaut, R.R. Bacsa, A. Rousset: Specific surface area of carbon nanotubes and bundles of carbon nanotubes, Carbon **39**, 507–514 (2001)

3.171 E. Frackowiak, S. Delpeux, K. Jurewicz, K. Szostak, D. Cazorla-Amoros, F. Béguin: Enhanced capacitance of carbon nanotubes through chemical activation, Chem. Phys. Lett. **336**, 35–41 (2002)

3.172 E. Raymundo-Piñero, P. Azaïs, T. Cacciaguerra, D. Cazorla-Amorós, A. Linares-Solano, F. Béguin: KOH and NaOH activation mechanisms of multi-walled carbon nanotubes with different structural organisation, Carbon **43**, 786–795 (2005)

3.173 S. Delpeux, K. Szostak, E. Frackowiak, F. Béguin: An efficient two-step process for producing opened multi-walled carbon nanotubes of high purity, Chem. Phys. Lett. **404**, 374–378 (2005)

3.174 K.A. Williams, P.C. Eklund: Monte Carlo simulation of H_2 physisorption in finite diameter carbon nanotube ropes, Chem. Phys. Lett. **320**, 352–358 (2000)

3.175 U. Burghaus, D. Bye, K. Cosert, J. Goering, A. Guerard, E. Kadossov, E. Lee, Y. Nadoyama, N. Richter, E. Schaefer, J. Smith, D. Ulness, B. Wymore: Methanol adsorption in carbon nanotubes, Chem. Phys. Lett. **442**, 344–347 (2007)

3.176 Z. Chen, W. Thiel, A. Hirsch: Reactivity of the convex and concave surfaces of single-walled carbon nanotubes (SWCNTs) towards addition reactions: dependence on the carbon-atom pyramidalization, ChemPhysChem **1**, 93–97 (2003)

3.177 S. Park, D. Srivastava, K. Cho: Generalized reactivity of curved surfaces: carbon nanotubes, Nano Lett. **3**, 1273–1277 (2003)

3.178 X. Lu, Z. Chen, P. Schleyer: Are Stone–Wales defect sites always more reactive than perfect sites in the sidewalls of single-wall carbon nanotubes?, J. Am. Chem. Soc. **127**, 20–21 (2005)

3.179 M. Muris, N. Dupont-Pavlosky, M. Bienfait, P. Zeppenfeld: Where are the molecules adsorbed on single-walled nanotubes?, Surf. Sci. **492**, 67–74 (2001)

3.180 R.B. Hallock, Y.H. Yang: Adsorption of helium and other gases to carbon nanotubes and nanotubes bundles, J. Low Temp. Phys. **134**, 21–30 (2004)

3.181 J. Zhu, Y. Wang, W. Li, F. Wei, Y. Yu: Density functional study of nitrogen adsorption in single-wall carbon nanotubes, Nanotechnology **18**, 095707 (2007)

3.182 A. Fujiwara, K. Ishii, H. Suematsu, H. Kataura, Y. Maniwa, S. Suzuki, Y. Achiba: Gas adsorption in the inside and outside of single-walled carbon nanotubes, Chem. Phys. Lett. **336**, 205–211 (2001)

3.183 C.M. Yang, H. Kanoh, K. Kaneko, M. Yudasaka, S. Iijima: Adsorption behaviors of HiPco single-walled carbon nanotubes aggregates for alcohol vapors, J. Phys. Chem. **106**, 8994–8999 (2002)

3.184 D.H. Yoo, G.H. Rue, M.H.W. Chan, Y.W. Hwang, H.K. Kim: Study of nitrogen adsorbed on open-ended nanotube bundles, J. Phys. Chem. B **107**, 1540–1542 (2003)

3.185 J. Jiang, S.I. Sandler: Nitrogen adsorption on carbon nanotubes bundles: Role of the external surface, Phys. Rev. B **68**, 245412-1–245412-9 (2003)

3.186 M. Arab, F. Picaud, C. Ramseyer, M.R. Babaa, F. Valsaque, E. McRae: Characterization of single wall carbon nanotubes by means of rare gas adsorption, J. Chem. Phys. **126**, 054709 (2007)

3.187 J. Zhao, A. Buldum, J. Han, J.P. Lu: Gas molecule adsorption in carbon nanotubes and nanotube bundles, Nanotechnology **13**, 195–200 (2002)

3.188 C. Matranga, B. Bockrath: Hydrogen-bonded and physisorbed CO in single-walled carbon nanotubes bundles, J. Phys. Chem. B **109**, 4853–4864 (2005)

3.189 M.D. Ellison, M.J. Crotty, D. Koh, R.L. Spray, K.E. Tate: Adsorption of NH_3 and NO_2 on single-walled carbon nanotubes, J. Phys. Chem. B **108**, 7938–7943 (2004)

3.190 S. Picozzi, S. Santucci, L. Lozzi, L. Valentin, B. Delley: Ozone adsorption on carbon nanotubes: The role of Stone–Wales defects, J. Chem. Phys. **120**, 7147–7152 (2004)

3.191 N. Chakrapani, Y.M. Zhang, S.K. Nayak, J.A. Moore, D.L. Carrol, Y.Y. Choi, P.M. Ajayan: Chemisorption of acetone on carbon nanotubes, J. Phys. Chem. B **107**, 9308–9311 (2003)

3.192 A. Chambers, C. Park, R.T.K. Baker, N. Rodriguez: Hydrogen storage in graphite nanofibers, J. Phys. Chem. B **102**, 4253–4256 (1998)

3.193 J. Giraudet, M. Dubois, D. Claves, J.P. Pinheiro, M.C. Schouler, P. Gadelle, A. Hamwi: Modifying the electronic properties of multi-wall carbon nanotubes via charge transfer, by chemical doping with some inorganic fluorides, Chem. Phys. Lett. **381**, 306–314 (2003)

3.194 J. Hilding, E.A. Grulke, S.B. Sinnott, D. Qian, R. Andrews, M. Jagtoyen: Sorption of butane on carbon multiwall nanotubes at room temperature, Langmuir **17**, 7540–7544 (2001)

3.195 K. Masenelli-Varlot, E. McRae, N. Dupont-Pavlosky: Comparative adsorption of simple molecules on carbon nanotubes. Dependence of the adsorption properties on the nanotube morphology, Appl. Surf. Sci. **196**, 209–215 (2002)

3.196 D.J. Browning, M.L. Gerrard, J.B. Lakeman, I.M. Mellor, R.J. Mortimer, M.C. Turpin: Studies into the storage of hydrogen in carbon nanofibers: Proposal of a possible mechanism, Nano Lett. **2**, 201–205 (2002)

3.197 F.H. Yang, R.T. Yang: Ab initio molecular orbital study of adsorption of atomic hydrogen on graphite: insight into hydrogen storage in carbon nanotubes, Carbon **40**, 437–444 (2002)

3.198 A.D. Lueking, R.T. Yang: Hydrogen spillover to enhance hydrogen storage – Study of the effect of carbon physicochemical properties, Appl. Catal. A **265**, 259–268 (2004)

3.199 G.E. Froudakis: Why alkali-metal-doped carbon nanotubes possess high hydrogen uptake, Nano Lett. **1**, 531–533 (2001)

3.200 H. Ulbricht, G. Moos, T. Hertel: Physisorption of molecular oxygen on single-wall carbon nanotube bundles and graphite, Phys. Rev. B **66**, 075404-1–075404-7 (2002)

3.201 H. Ulbricht, J. Kriebel, G. Moos, T. Hertel: Desorption kinetics and interaction of Xe with single-wall carbon nanotube bundles, Chem. Phys. Lett. **363**, 252–260 (2002)

3.202 J.-C. Charlier, X. Blase, S. Roche: Electronic and transport properties of carbon nanotubes, Rev. Mod. Phys. **79**, 677–732 (2007)

3.203 R. Saito, G. Dresselhaus, M.S. Dresselhaus: *Physical Properties of Carbon Nanotubes* (Imperial College Press, London 1998)

3.204 A. Charlier, E. McRae, R. Heyd, M.F. Charlier, D. Moretti: Classification for double-walled carbon nanotubes, Carbon **37**, 1779–1783 (1999)

3.205 A. Charlier, E. McRae, R. Heyd, M.F. Charlier: Metal semi-conductor transitions under uniaxial stress for single- and double-walled carbon nanotubes, J. Phys. Chem. Solids **62**, 439–444 (2001)

3.206 P. Puech, H. Hubel, D. Dunstan, R.R. Bacsa, C. Laurent, W.S. Bacsa: Discontinuous tangential stress in double wall carbon nanotubes, Phys. Rev. Lett. **93**, 095506 (2004)

3.207 P.M. Ajayan, M. Terrrones, A. de la Guardia, V. Hue, N. Grobert, B.Q. Wei, H. Lezec, G. Ramanath, T.W. Ebbesen: Nanotubes in a flash – Ignition and reconstruction, Science **296**, 705 (2002)

3.208 H. Ajiki, T. Ando: Electronic states of carbon nanotubes, J. Phys. Soc. Jpn. **62**, 1255–1266 (1993)

3.209 T. Ando: Excitons in carbon nanotubes, J. Phys. Soc. Jpn. **66**, 1066 (1997)

3.210 S.M. Bachilo, M.S. Strano, C. Kittrell, R.H. Hauge, R.E. Smalley, R.B. Weisman: Structure-assigned optical spectra of single-walled carbon nanotubes, Science **298**, 2361 (2002)

3.211 M. Bockrath, D.H. Cobden, J. Lu, A.G. Rinzler, R.E. Smalley, L. Balents, P.L. McEuen: Luttinger liquid behaviour in carbon nanotubes, Nature **397**, 598–601 (1999)

3.212 C.T. White, T.N. Todorov: Carbon nanotubes as long ballistic conductors, Nature **393**, 240–242 (1998)

3.213 S. Frank, P. Poncharal, Z.L. Wang, W.A. de Heer: Carbon nanotube quantum resistors, Science **280**, 1744–1746 (1998)

3.214 W. Liang, M. Bockrath, D. Bozovic, J.H. Hafner, M. Tinkham, H. Park: Fabry–Perot interference in a nanotube electron waveguide, Nature **411**, 665–669 (2001)

3.215 L. Langer, V. Bayot, E. Grivei, J.-P. Issi, J.-P. Heremans, C.H. Olk, L. Stockman, C. van Haesendonck, Y. Buynseraeder: Quantum transport in a multiwalled carbon nanotubes, Phys. Rev. Lett. **76**, 479–482 (1996)

3.216 K. Liu, S. Roth, G.S. Duesberg, G.T. Kim, D. Popa, K. Mukhopadhyay, R. Doome, J. B'Nagy: Antilocalization in multiwalled carbon nanotubes, Phys. Rev. B **61**, 2375–2379 (2000)

3.217 G. Fedorov, B. Lassagne, M. Sagnes, B. Raquet, J.M. Broto, F. Triozon, S. Roche, E. Flahaut: Gate-dependent magnetoresistance phenomena in carbon nanotubes, Phys. Rev. Lett. **94**, 66801–66804 (2005)

3.218 A. Javey, J. Guo, Q. Wang, M. Lundstrom, H. Dai: Ballistic carbon nanotube field-effect transistors, Nature **424**, 654–657 (2003)

3.219 Y.A. Kasumov, R. Deblock, M. Kociak, B. Reulet, H. Bouchiat, I.I. Khodos, Y.B. Gorbatov, V.T. Volkov, C. Journet, M. Burghard: Supercurrents through single-walled carbon nanotubes, Science **284**, 1508–1511 (1999)

3.220 B.W. Alphenaar, K. Tsukagoshi, M. Wagner: Magnetoresistance of ferromagnetically contacted carbon nanotubes, Phys. Eng. **10**, 499–504 (2001)

3.221 S. Berber, Y. Kwon, D. Tomanek: Unusually high thermal conductivity of carbon nanotubes, Phys. Rev. Lett. **84**, 4613–4616 (2000)

3.222 M.-F. Yu, O. Lourie, M.J. Dyer, K. Moloni, T.F. Kelley, R.S. Ruoff: Strength and breaking mechanism of multiwalled carbon nanotubes under tensile load, Science **287**, 637–640 (2000)

3.223 D.A. Walters, L.M. Ericson, M.J. Casavant, J. Liu, D.T. Colbert, K.A. Smith, R.E. Smalley: Elastic strain of freely suspended single-wall carbon nanotube ropes, Appl. Phys. Lett. **74**, 3803–3805 (1999)

3.224 B.G. Demczyk, Y.M. Wang, J. Cumings, M. Hetman, W. Han, A. Zettl, R.O. Ritchie: Direct mechanical measurement of the tensile strength and elastic modulus of multiwalled carbon nanotubes, Mater. Sci. Eng. A **334**, 173–178 (2002)

3.225 R.P. Gao, Z.L. Wang, Z.G. Bai, W.A. De Heer, L.M. Dai, M. Gao: Nanomechanics of individual carbon nanotubes from pyrolytically grown arrays, Phys. Rev. Lett. **85**, 622–625 (2000)

3.226 M.M.J. Treacy, T.W. Ebbesen, J.M. Gibson: Exceptionally high Young's modulus observed for individual carbon nanotubes, Nature **381**, 678–680 (1996)

3.227 N. Yao, V. Lordie: Young's modulus of single-wall carbon nanotubes, J. Appl. Phys. **84**, 1939–1943 (1998)

3.228 O. Lourie, H.D. Wagner: Transmission electron microscopy observations of fracture of single-wall carbon nanotubes under axial tension, Appl. Phys. Lett. **73**, 3527–3529 (1998)

3.229 S.C. Tsang, Y.K. Chen, P.J.F. Harris, M.L.H. Green: A simple chemical method of opening and filling carbon nanotubes, Nature **372**, 159–162 (1994)

3.230 M. Monthioux: Filling single-wall carbon nanotubes, Carbon **40**, 1809–1823 (2002)

3.231 W.K. Hsu, S.Y. Chu, E. Munoz-Picone, J.L. Boldu, S. Firth, P. Franchi, B.P. Roberts, A. Shilder, H. Terrones, N. Grobert, Y.Q. Zhu, M. Terrones,

M.E. McHenry, H.W. Kroto, D.R.M. Walton: Metallic behaviour of boron-containing carbon nanotubes, Chem. Phys. Lett. **323**, 572–579 (2000)

3.232 R. Czerw, M. Terrones, J.C. Charlier, X. Blasé, B. Foley, R. Kamalakaran, N. Grobert, H. Terrones, D. Tekleab, P.M. Ajayan, W. Blau, M. Rühle, D.L. Caroll: Identification of electron donor states, in N-doped carbon nanotubes, Nano Lett. **1**, 457–460 (2001)

3.233 O. Stephan, P.M. Ajayan, C. Colliex, P. Redlich, J.M. Lambert, P. Bernier, P. Lefin: Doping graphitic and carbon nanotube structures with boron and nitrogen, Science **266**, 1683–1685 (1994)

3.234 A. Loiseau, F. Willaime, N. Demoncy, N. Schramchenko, G. Hug, C. Colliex, H. Pascard: Boron nitride nanotubes, Carbon **36**, 743–752 (1998)

3.235 C.C. Tang, L.M. de la Chapelle, P. Li, Y.M. Liu, H.Y. Dang, S.S. Fan: Catalytic growth of nanotube and nanobamboo structures of boron nitride, Chem. Phys. Lett. **342**, 492–496 (2001)

3.236 K. Suenaga, C. Colliex, N. Demoncy, A. Loiseau, H. Pascard, F. Willaime: Synthesis of nanoparticles and nanotubes with well separated layers of boronnitride and carbon, Science **278**, 653–655 (1997)

3.237 D. Golberg, Y. Bando, L. Bourgeois, K. Kurashima, T. Sato: Large-scale synthesis and HRTEM analysis of single-walled B- and N-doped carbon nanotube bundles, Carbon **38**, 2017–2027 (2000)

3.238 R.S. Lee, J. Gavillet, M. Lamy de la Chapelle, A. Loiseau, J.-L. Cochon, D. Pigache, J. Thibault, F. Willaime: Catalyst-free synthesis of boron nitride single-wall nanotubes with a preferred zig-zag configuration, Phys. Rev. B **64**, 121405.1–121405.4 (2001)

3.239 B. Burteaux, A. Claye, B.W. Smith, M. Monthioux, D.E. Luzzi, J.E. Fischer: Abundance of encapsulated C_{60} in single-wall carbon nanotubes, Chem. Phys. Lett. **310**, 21–24 (1999)

3.240 D. Ugarte, A. Châtelain, W.A. de Heer: Nanocapillarity and chemistry in carbon nanotubes, Science **274**, 1897–1899 (1996)

3.241 J. Cook, J. Sloan, M.L.H. Green: Opening and filling carbon nanotubes, Fuller. Sci. Technol. **5**, 695–704 (1997)

3.242 P.M. Ajayan, S. Iijima: Capillarity-induced filling of carbon nanotubes, Nature **361**, 333–334 (1993)

3.243 P.M. Ajayan, T.W. Ebbesen, T. Ichihashi, S. Iijima, K. Tanigaki, H. Hiura: Opening carbon nanotubes with oxygen and implications for filling, Nature **362**, 522–525 (1993)

3.244 S. Seraphin, D. Zhou, J. Jiao, J.C. Withers, R. Loufty: Yttrium carbide in nanotubes, Nature **362**, 503 (1993)

3.245 S. Seraphin, D. Zhou, J. Jiao, J.C. Withers, R. Loufty: Selective encapsulation of the carbides of yttrium and titanium into carbon nanoclusters, Appl. Phys. Lett. **63**, 2073–2075 (1993)

3.246 R.S. Ruoff, D.C. Lorents, B. Chan, R. Malhotra, S. Subramoney: Single-crystal metals encapsulated in carbon nanoparticles, Science **259**, 346–348 (1993)

3.247 A. Loiseau, H. Pascard: Synthesis of long carbon nanotubes filled with Se, S, Sb, and Ge by the arc method, Chem. Phys. Lett. **256**, 246–252 (1996)

3.248 N. Demoncy, O. Stephan, N. Brun, C. Colliex, A. Loiseau, H. Pascard: Filling carbon nanotubes with metals by the arc discharge method: The key role of sulfur, Eur. Phys. J. B **4**, 147–157 (1998)

3.249 C.H. Kiang, J.S. Choi, T.T. Tran, A.D. Bacher: Molecular nanowires of 1 nm diameter from capillary filling of single-walled carbon nanotubes, J. Phys. Chem. B **103**, 7449–7551 (1999)

3.250 Z.L. Zhang, B. Li, Z.J. Shi, Z.N. Gu, Z.Q. Xue, L.M. Peng: Filling of single-walled carbon nanotubes with silver, J. Mater. Res. **15**, 2658–2661 (2000)

3.251 A. Govindaraj, B.C. Satishkumar, M. Nath, C.N.R. Tao: Metal nanowires and intercalated metal layers in single-walled carbon nanotubes bundles, Chem. Mater. **12**, 202–205 (2000)

3.252 J. Mittal, M. Monthioux, H. Allouche: Room temperature filling of single-wall carbon nanotubes with chromium oxide in open air, Chem. Phys. Lett. **339**, 311–318 (2001)

3.253 E. Dujardin, T.W. Ebbesen, H. Hiura, K. Tanigaki: Capillarity and wetting of carbon nanotubes, Science **265**, 1850–1852 (1994)

3.254 E. Flahaut, J. Sloan, K.S. Coleman, V.C. Williams, S. Friedrichs, N. Hanson, M.L.H. Green: 1D p-block halide crystals confined into single walled carbon nanotubes, Proc. Mater. Res. Soc. Symp., Vol. 633 (2001) pp. A13.15.1–A13.15.6

3.255 J. Sloan, A.I. Kirkland, J.L. Hutchison, M.L.H. Green: Integral atomic layer architectures of 1D crystals inserted into single walled carbon nanotubes, Chem. Commun., 1319–1332 (2002)

3.256 J. Sloan, M.C. Novotny, S.R. Bailey, G. Brown, C. Xu, V.C. Williams, S. Friedrichs, E. Flahaut, R.L. Callender, A.P.E. York, K.S. Coleman, M.L.H. Green, R.E. Dunin-Borkowski, J.L. Hutchison: Two layer 4 : 4 co-ordinated KI crystals grown within single walled carbon nanotubes, Chem. Phys. Lett. **329**, 61–65 (2000)

3.257 G. Brown, S.R. Bailey, J. Sloan, C. Xu, S. Friedrichs, E. Flahaut, K.S. Coleman, J.L. Hutchinson, R.E. Dunin-Borkowski, M.L.H. Green: Electron beam induced in situ clusterisation of 1D $ZrCl_4$ chains within single-walled carbon nanotubes, Chem. Commun., 845–846 (2001)

3.258 J. Sloan, D.M. Wright, H.G. Woo, S. Bailey, G. Brown, A.P.E. York, K.S. Coleman, J.L. Hutchison, M.L.H. Green: Capillarity and silver nanowire formation observed in single walled carbon nanotubes, Chem. Commun., 699–700 (1999)

3.259 X. Fan, E.C. Dickey, P.C. Eklund, K.A. Williams, L. Grigorian, R. Buczko, S.T. Pantelides, S.J. Pennycook: Atomic arrangement of iodine atoms inside single-walled carbon nanotubes, Phys. Rev. Lett. **84**, 4621–4624 (2000)

3.260 G. Brown, S.R. Bailey, M. Novotny, R. Carter, E. Flahaut, K.S. Coleman, J.L. Hutchison, M.L.H. Green, J. Sloan: High yield incorporation and washing properties of halides incorporated into single walled carbon nanotubes, Appl. Phys. A **76**, 457–462 (2003)

3.261 J. Sloan, D.E. Luzzi, A.I. Kirkland, J.L. Hutchison, M.L.H. Green: Imaging and characterization of molecules and one-dimensional crystals formed within carbon nanotubes, Mater. Res. Soc. Bull. **29**, 265–271 (2004)

3.262 J. Chancolon, F. Archaimbault, A. Pineau, S. Bonnamy: Confinement of selenium into carbon nanotubes, Fuller. Nanotub. Carbon Nanostruct. **13**, 189–194 (2005)

3.263 B.W. Smith, M. Monthioux, D.E. Luzzi: Encapsulated C_{60} in carbon nanotubes, Nature **396**, 323–324 (1998)

3.264 B.W. Smith, D.E. Luzzi: Formation mechanism of fullerene peapods and coaxial tubes: A path to large scale synthesis, Chem. Phys. Lett. **321**, 169–174 (2000)

3.265 K. Hirahara, K. Suenaga, S. Bandow, H. Kato, T. Okazaki, H. Shinohara, S. Iijima: One-dimensional metallo-fullerene crystal generated inside single-walled carbon nanotubes, Phys. Rev. Lett. **85**, 5384–5387 (2000)

3.266 B.W. Smith, M. Monthioux, D.E. Luzzi: Carbon nanotube encapsulated fullerenes: A unique class of hybrid material, Chem. Phys. Lett. **315**, 31–36 (1999)

3.267 D.E. Luzzi, B.W. Smith: Carbon cage structures in single wall carbon nanotubes: A new class of materials, Carbon **38**, 1751–1756 (2000)

3.268 S. Bandow, M. Takisawa, K. Hirahara, M. Yudasoka, S. Iijima: Raman scattering study of double-wall carbon nanotubes derived from the chains of fullerenes in single-wall carbon nanotubes, Chem. Phys. Lett. **337**, 48–54 (2001)

3.269 Y. Sakurabayashi, M. Monthioux, K. Kishita, Y. Suzuki, T. Kondo, M. Le Lay: Tayloring double wall carbon nanotubes?. In: *Molecular Nanostructures*, Am. Inst. Phys. Conf. Proc., Vol. 685, ed. by H. Kuzmany, J. Fink, M. Mehring, S. Roth (Springer, Berlin Heidelberg 2003) pp. 302–305

3.270 B.W. Smith, D.E. Luzzi, Y. Achiba: Tumbling atoms and evidence for charge transfer in $La_2@C_{80}@SWNT$, Chem. Phys. Lett. **331**, 137–142 (2000)

3.271 K. Suenaga, M. Tence, C. Mory, C. Colliex, H. Kato, T. Okazaki, H. Shinohara, K. Hirahara, S. Bandow, S. Iijima: Element-selective single atom imaging, Science **290**, 2280–2282 (2000)

3.272 D.E. Luzzi, B.W. Smith, R. Russo, B.C. Satishkumar, F. Stercel, N.R.C. Nemes: Encapsulation of metallo-fullerenes and metallocenes in carbon nanotubes, Proc. Electron. Prop. Nov. Mater. – XVI Intern. Wintersch. – AIP Conf. Proc., ed. by H. Kuzmany, J. Fink, M. Mehring, S. Roth (Springer, Berlin Heidelberg 2001) pp. 622–626

3.273 D.J. Hornbaker, S.-J. Kahng, S. Misra, B.W. Smith, A.T. Johnson, E.J. Mele, D.E. Luzzi, A. Yazdani: Mapping the one-dimensional electronic states of nanotube peapod structures, Science **295**, 828–831 (2002)

3.274 H. Kondo, H. Kino, T. Ohno: Transport properties of carbon nanotubes encapsulating C_{60} and related materials, Phys. Rev. B **71**, 115413 (2005)

3.275 S.H. Jhang, S.W. Lee, D.S. Lee, Y.W. Park, G.H. Jeong, T. Hirata, R. Hatakeyama, U. Dettlaff, S. Roth, M.S. Kabir, E.E.B. Campbell: Random telegraph noise in carbon nanotube peapod transistors, Fuller. Nanotub. Carbon Nanostruct. **13**, 195–198 (2005)

3.276 G.H. Jeong, R. Hatakeyama, T. Hirata, K. Tohji, K. Motomiya, N. Sato, Y. Kawazoe: Structural deformation of single-walled carbon nanotubes and fullerene encapsulation due to magnetized plasma ion irradiation, Appl. Phys. Lett. **79**, 4213–4215 (2001)

3.277 Y.P. Sun, K. Fu, Y. Lin, W. Huang: Functionalized carbon nanotubes: Properties and applications, Acc. Chem. Res. **35**, 1095–1104 (2002)

3.278 S. Osswald, E. Flahaut, H. Ye, Y. Gogotsi: Elimination of D-band in Raman spectra of double-wall carbon nanotubes by oxidation, Chem. Phys. Lett. **402**, 422–427 (2005)

3.279 J. Chen, M.A. Hamon, M. Hui, C. Yongsheng, A.M. Rao, P.C. Eklund, R.C. Haddon: Solution properties of single-walled carbon nanotubes, Science **282**, 95–98 (1998)

3.280 J. Chen, A.M. Rao, S. Lyuksyutov, M.E. Itkis, M.A. Hamon, H. Hu, R.W. Cohn, P.C. Eklund, D.T. Colbert, R.E. Smalley, R.C. Haddon: Dissolution of full-length single-walled carbon nanotubes, J. Phys. Chem. B **105**, 2525–2528 (2001)

3.281 F. Pompeo, D.E. Resasco: Water solubilization of single-walled carbon nanotubes by functionalization with glucosamine, Nano Lett. **2**, 369–373 (2002)

3.282 Y.P. Sun, W. Huang, Y. Lin, K. Fu, A. Kitaygorodskiy, L.A. Riddle, Y.J. Yu, D.L. Carroll: Soluble dendron-functionalized carbon nanotubes: Preparation, characterization, and properties, Chem. Mater. **13**, 2864–2869 (2001)

3.283 K. Fu, W. Huang, Y. Lin, L.A. Riddle, D.L. Carroll, Y.P. Sun: Defunctionalization of functionalized carbon nanotubes, Nano Lett. **1**, 439–441 (2001)

3.284 P.W. Chiu, G.S. Duesberg, U. Dettlaff-Weglikowska, S. Roth: Interconnection of carbon nanotubes by chemical functionalization, Appl. Phys. Lett. **80**, 3811–3813 (2002)

3.285 T. Fukushima, T. Aida: Ionic liquids for soft functional materials with carbon nanotubes, Chem. Eur. J. **13**, 5048–5058 (2007)

3.286 Y. Lei, C. Xiong, L. Dong, H. Guo, X. Su, J. Yao, Y. You, D. Tian, X. Shang: Ionic liquid of ultralong carbon nanotubes, Small **3**, 1889–1893 (2007)

3.287 E.T. Mickelson, C.B. Huffman, A.G. Rinzler, R.E. Smalley, R.H. Hauge, J.L. Margrave: Fluorination of single-wall carbon nanotubes, Chem. Phys. Lett. **296**, 188–194 (1998)

3.288 V.N. Khabashesku, W.E. Billups, J.L. Margrave: Fluorination of single-wall carbon nanotubes and

subsequent derivatization reactions, Acc. Chem. Res. **35**, 1087–1095 (2002)
3.289 P.J. Boul, J. Liu, E.T. Michelson, C.B. Huffman, L.M. Ericson, I.W. Chiang, K.A. Smith, D.T. Colbert, R.H. Hauge, J.L. Margrave, R.E. Smalley: Reversible side-wall functionalization of buckytubes, Chem. Phys. Lett. **310**, 367–372 (1999)
3.290 J.L. Bahr, J. Yang, D.V. Kosynkin, M.J. Bronikowski, R.E. Smalley, J.M. Tour: Functionalization of carbon nanotubes by electrochemical reduction of aryl diazonium salts: A bucky paper electrode, J. Am. Chem. Soc. **123**, 6536–6542 (2001)
3.291 M. Holzinger, O. Vostrowsky, A. Hirsch, F. Hennrich, M. Kappes, R. Weiss, F. Jellen: Sidewall functionalization of carbon nanotubes, Angew. Chem. Int. Ed. **40**, 4002–4005 (2001)
3.292 Y. Chen, R.C. Haddon, S. Fang, A.M. Rao, P.C. Eklund, W.H. Lee, E.C. Dickey, E.A. Grulke, J.C. Pendergrass, A. Chavan, B.E. Haley, R.E. Smalley: Chemical attachment of organic functional groups to single-walled carbon nanotube material, J. Mater. Res. **13**, 2423–2431 (1998)
3.293 C. Velasco-Santos, A.L. Martinez-Hernandez, M. Lozada-Cassou, A. Alvarez-Castillo, V.M. Castano: Chemical functionalization of carbon nanotubes through an organosilane, Nanotechnology **13**, 495–498 (2002)
3.294 A. Star, J.F. Stoddart, D. Steuerman, M. Diehl, A. Boukai, E.W. Wong, X. Yang, S.W. Chung, H. Choi, J.R. Heath: Preparation and properties of polymer-wrapped single-walled carbon nanotubes, Angew. Chem. Int. Ed. **41**, 1721–1725 (2002)
3.295 A. Pénicaud, P. Poulin, A. Derré, E. Anglaret, P. Petit: Spontaneous dissolution of a single-wall carbon nanotube salt, J. Am. Chem. Soc. **127**, 8–9 (2005)
3.296 R. Stevens, C. Nguyen, A. Cassel, L. Delzeit, M. Meyyapan, J. Han: Improved fabrication approach for carbon nanotube probe devices, Appl. Phys. Lett. **77**, 3453–3455 (2000)
3.297 J.H. Hafner, C.L. Cheung, A.T. Wooley, C.M. Lieber: Structural and functional imaging with carbon nanotube AFM probes, Progr. Biophys. Mol. Biol. **77**, 73–110 (2001)
3.298 S.S. Wong, E. Joselevich, A.T. Woodley, C.L. Cheung, C.M. Lieber: Covalently functionalized nanotubes as nanometre-size probes in chemistry and biology, Nature **394**, 52–55 (1998)
3.299 C.L. Cheung, J.H. Hafner, C.M. Lieber: Carbon nanotube atomic force microscopy tips: Direct growth by chemical vapor deposition and application to high-resolution imaging, Proc. Natl. Acad. Sci. USA **97**, 3809–3813 (2000)
3.300 W.A. de Heer, A. Châtelain, D. Ugarte: A carbon nanotube field-emission electron source, Science **270**, 1179–1180 (1995)
3.301 J.M. Bonard, J.P. Salvetat, T. Stockli, W.A. de Heer, L. Forro, A. Chatelâin: Field emission from single-wall carbon nanotube films, Appl. Phys. Lett. **73**, 918–920 (1998)
3.302 W. Zhu, C. Bower, O. Zhou, G. Kochanski, S. Jin: Large curent density from carbon nanotube field emitters, Appl. Phys. Lett. **75**, 873–875 (1999)
3.303 Y. Saito, R. Mizushima, T. Tanaka, K. Tohji, K. Uchida, M. Yumura, S. Uemura: Synthesis, structure, and field emission of carbon nanotubes, Fuller. Sci. Technol. **7**, 653–664 (1999)
3.304 J. Kong, N.R. Franklin, C. Zhou, M.G. Chapline, S. Peng, K. Cho, H. Dai: Nanotube molecular wire as chemical sensors, Science **287**, 622–625 (2000)
3.305 P.G. Collins, K. Bradley, M. Ishigami, A. Zettl: Extreme oxygen sensitivity of electronic properties of carbon nanotubes, Science **287**, 1801–1804 (2000)
3.306 H. Chang, J.D. Lee, S.M. Lee, Y.H. Lee: Adsorption of NH_3 and NO_2 molecules on carbon nanotubes, Appl. Phys. Lett. **79**, 3863–3865 (2001)
3.307 C. Cantalini, L. Valentini, L. Lozzi, I. Armentano, J.M. Kenny, S. Santucci: NO_2 gas sensitivity of carbon nanotubes obtained by plasma enhanced chemical vapor deposition, Sens. Actuators B **93**, 333–337 (2003)
3.308 J. Li, Y. Lu, Q. Ye, M. Cinke, J. Han, M. Meyyappan: Carbon nanotubes sensors for gas and organic vapor detection, Nano Lett. **3**, 929–933 (2003)
3.309 O.K. Varghese, P.D. Kichambre, D. Gong, K.G. Ong, E.C. Dickey, C.A. Grimes: Gas sensing characteristics of multi-wall carbon nanotubes, Sens. Actuators B **81**, 32–41 (2001)
3.310 K.G. Ong, K. Zeng, C.A. Grimes: A wireless, passive carbon nanotube-based gas sensor, IEEE Sens. J. **2**(2), 82–88 (2002)
3.311 J. Kong, M.G. Chapline, H. Dai: Functionalized carbon nanotubes for molecular hydrogen sensors, Adv. Mater. **13**, 1384–1386 (2001)
3.312 A. Modi, N. Koratkar, E. Lass, B. Wei, P.M. Ajayan: Miniaturized gas ionisation sensors using carbon nanotubes, Nature **424**, 171–174 (2003)
3.313 F. Rodriguez-Reinoso: The role of carbon materials in heterogeneous catalysis, Carbon **36**, 159–175 (1998)
3.314 E. Auer, A. Freund, J. Pietsch, T. Tacke: Carbon as support for industrial precious metal catalysts, Appl. Catal. A **173**, 259–271 (1998)
3.315 J.M. Planeix, N. Coustel, B. Coq, B. Botrons, P.S. Kumbhar, R. Dutartre, P. Geneste, P. Bernier, P.M. Ajayan: Application of carbon nanotubes as supports in heterogeneous catalysis, J. Am. Chem. Soc. **116**, 7935–7936 (1994)
3.316 P. Serp, M. Corrias, P. Kalck: Carbon nanotubes and nanofibers in catalysis, Appl. Catal. A **253**, 337–358 (2003)
3.317 K.P. De Jong, J.W. Geus: Carbon nanofibers: catalytic synthesis and applications, Catal. Rev. **42**, 481–510 (2000)

3.318 N.F. Goldshleger: Fullerene and fullerene-based materials in catalysis, Fuller. Sci. Technol. **9**, 255–280 (2001)

3.319 X. Pan, Z. Fan, W. Chen, Y. Ding, H. Luo, X. Bao: Enhanced ethanol production inside carbon-nanotube reactors containing catalytic particles, Nat. Mater. **6**, 507–511 (2007)

3.320 M.F.R. Pereira, J.L. Figueiredo, J.J.M. Órfão, P. Serp, P. Kalck, Y. Kihn: Catalytic activity of carbon nanotubes in the oxidative dehydrogenation of ethylbenzene, Carbon **42**, 2807–2813 (2004)

3.321 G. Mestl, N.I. Maksimova, N. Keller, V.V. Roddatis, R. Schlögl: Carbon nanofilaments in heterogeneous catalysis: An industrial application for new carbon materials?, Angew. Chem. Int. Ed. Engl. **40**, 2066–2068 (2001)

3.322 N. Muradov: Catalysis of methane decomposition over elemental carbon, Catal. Commun. **2**, 89–94 (2001)

3.323 J.E. Fischer, A.T. Johnson: Electronic properties of carbon nanotubes, Curr. Opin. Solid State Mater. Sci. **4**, 28–33 (1999)

3.324 M. Menon, A.N. Andriotis, G.E. Froudakis: Curvature dependence of the metal catalyst atom interaction with carbon nanotubes walls, Chem. Phys. Lett. **320**, 425–434 (2000)

3.325 N. Ishigami, H. Ago, Y. Motoyama, M. Takasaki, M. Shinagawa, K. Takahashi, T. Ikuta, M. Tsuji: Microreactor utilizing a vertically-aligned carbon nanotube array grown inside the channels, Chem. Commun., 1626 (2007)

3.326 G.G. Wildgoose, C.E. Banks, R.G. Compton: Metal nanoparticles and related materials supported on carbon nanotubes: methods and applications, Small **2**, 182–193 (2006)

3.327 T. Kyotani, S. Nakazaki, W.-H. Xu, A. Tomita: Chemical modification of the inner walls of carbon nanotubes by HNO_3 oxidation, Carbon **39**, 782–785 (2001)

3.328 Z.J. Liu, Z.Y. Yuan, W. Zhou, L.M. Peng, Z. Xu: Co/carbon nanotubes monometallic system: The effects of oxidation by nitric acid, PhysChemChemPhys **3**, 2518–2521 (2001)

3.329 R. Giordano, P. Serp, P. Kalck, Y. Kihn, J. Schreiber, C. Marhic, J.-L. Duvail: Preparation of rhodium supported on carbon canotubes catalysts via surface mediated organometallic reaction, Eur. J. Inorg. Chem. **2003**, 610–617 (2003)

3.330 A. Carrillo, J.A. Swartz, J.M. Gamba, R.S. Kane, N. Chakrapani, B. Wei, P.M. Ajayan: Noncovalent functionalization of graphite and carbon nanotubes with polymer multilayers and gold nanoparticles, Nano Lett. **3**, 1437–1440 (2003)

3.331 Z. Liu, X. Lin, J.Y. Lee, W. Zhang, M. Han, L.M. Gan: Preparation and characterization of platinum-based electrocatalysts on multiwalled carbon nanotubes for proton exchange membrane fuel cells, Langmuir **18**, 4054–4060 (2002)

3.332 H.-B. Chen, J.D. Lin, Y. Cai, X.Y. Wang, J. Yi, J. Wang, G. Wei, Y.Z. Lin, D.W. Liao: Novel multi-walled nanotube-supported and alkali-promoted Ru catalysts for ammonia synthesis under atmospheric pressure, Appl. Surf. Sci. **180**, 328–335 (2001)

3.333 Y. Zhang, H.B. Zhang, G.D. Lin, P. Chen, Y.Z. Yuan, K.R. Tsai: Preparation, characterization and catalytic hydroformylation properties of carbon nanotubes-supported Rh-phosphine catalyst, Appl. Catal. A **187**, 213–224 (1999)

3.334 M.S. Dresselhaus, K.A. Williams, P.C. Eklund: Hydrogen adsorption in carbon materials, Mater. Res. Soc. Bull. **24**, 45–50 (1999)

3.335 H.-M. Cheng, Q.-H. Yang, C. Liu: Hydrogen storage in carbon nanotubes, Carbon **39**, 1447–1454 (2001)

3.336 G.G. Tibbetts, G.P. Meisner, C.H. Olk: Hydrogen storage capacity of carbon nanotubes, filaments, and vapor-grown fibers, Carbon **39**, 2291–2301 (2001)

3.337 F.L. Darkrim, P. Malbrunot, G.P. Tartaglia: Review of hydrogen storage adsorption in carbon nanotubes, Int. J. Hydrogen Energy **27**, 193–202 (2002)

3.338 G.E. Froudakis: Hydrogen interaction with carbon nanotubes: a review of ab initio studies, J. Phys. Condens. Matter **14**, R453–R465 (2002)

3.339 M. Hirscher, M. Becher: Hydrogen storage in carbon nanotubes, J. Nanosci. Nanotechnol. **3**(1/2), 3–17 (2003)

3.340 P. Kowalczyk, R. Hołyst, M. Terrones, H. Terrones: Hydrogen storage in nanoporous carbon materials: myth and facts, PhysChemChemPhys **9**(15), 1786–1792 (2007)

3.341 C. Park, P.E. Anderson, C.D. Tan, R. Hidalgo, N. Rodriguez: Further studies of the interaction of hydrogen with graphite nanofibers, J. Phys. Chem. B **103**, 10572–10581 (1999)

3.342 S.M. Lee, H.Y. Lee: Hydrogen storage in single-walled carbon nanotubes, Appl. Phys. Lett. **76**, 2877–2879 (2000)

3.343 X. Zhang, D. Cao, J. Chen: Hydrogen adsorption storage on single-walled carbon nanotube arrays by a combination of classical potential and density functional theory, J. Phys. Chem. B **107**, 4942–4950 (2003)

3.344 H.M. Cheng, G.P. Pez, A.C. Cooper: Mechanism of hydrogen sorption in single-walled carbon nanotubes, J. Am. Chem. Soc. **123**, 5845–5846 (2001)

3.345 C.-H. Chen, C.-C. Huang: Hydrogen storage by KOH-modified multi-walled carbon nanotubes, Int. J. Hydrogen Energy **32**, 237–246 (2007)

3.346 M.A. de la Casa-Lillo, F. Lamari-Darkrim, D. Cazorla-Amoros, A. Linares-Solano: Hydrogen storage in activated carbons and activated carbon fibers, J. Phys. Chem. B **106**, 10930–10934 (2002)

3.347 P. Marinelli, R. Pellenq, J. Conard: H stocké dans les carbones un site légèrement métastable, Natl. Conf. Mater., Tours (2002), AF-14-020

3.348 G. Mpourmpakis, G.E. Froudakis, G.P. Lithoxoos, J. Samios: Effect of curvature and chirality for hydrogen storage in single-walled carbon nanotubes: a combined ab initio and Monte Carlo investigation, J. Chem. Phys. **126**, 144704 (2007)

3.349 C.I. Weng, S.P. Ju, K.C. Fang, F.P. Chang: Atomistic study of the influences of size, VDW distance and arrangement of carbon nanotubes on hydrogen storage, Comput. Mater. Sci. **40**, 300–308 (2007)

3.350 A.L.M. Reddy, S. Ramaprabhu: Hydrogen storage properties of nanocrystalline Pt dispersed multiwalled carbon nanotubes, Int. J. Hydrogen Energy **32**, 3998–4004 (2007)

3.351 A. Kusnetzova, D.B. Mawhinney, V. Naumenko, J.T. Yates, J. Liu, R.E. Smalley: Enhancement of adsorption inside of single-walled nanotubes: Opening the entry ports, Chem. Phys. Lett. **321**, 292–296 (2000)

3.352 G.E. Gadd, M. Blackford, S. Moricca, N. Webb, P.J. Evans, A.M. Smith, G. Jacobsen, S. Leung, A. Day, Q. Hua: The world's smallest gas cylinders?, Science **277**, 933–936 (1997)

3.353 Z. Mao, S.B. Sinnott: A computational study of molecular diffusion and dynamic flow through carbon nanotubes, J. Phys. Chem. B **104**, 4618–4624 (2000)

3.354 Z. Mao, S.B. Sinnott: Separation of organic molecular mixtures in carbon nanotubes and bundles: Molecular dynamics simulations, J. Phys. Chem. B **105**, 6916–6924 (2001)

3.355 H. Chen, D.S. Sholl: Rapid diffusion of CH_4/H_2 mixtures in single-walled carbon nanotubes, J. Am. Chem. Soc. **126**, 7778–7779 (2004)

3.356 C. Gu, G.-H. Gao, Y.X. Yu, T. Nitta: Simulation for separation of hydrogen and carbon monoxide by adsorption on single-walled carbon nanotubes, Fluid Phase Equilib. **194/197**, 297–307 (2002)

3.357 R.Q. Long, R.T. Yang: Carbon nanotubes as superior sorbent for dioxine removal, J. Am. Chem. Soc. **123**, 2058–2059 (2001)

3.358 Y.H. Li, S. Wang, A. Cao, D. Zhao, X. Zhang, C. Xu, Z. Luan, D. Ruan, J. Liang, D. Wu, B. Wei: Adsorption of fluoride from water by amorphous alumina supported on carbon nanotubes, Chem. Phys. Lett. **350**, 412–416 (2001)

3.359 Y.H. Li, S. Wang, J. Wei, X. Zhang, C. Xu, Z. Luan, D. Wu, B. Wei: Lead adsorption on carbon nanotubes, Chem. Phys. Lett. **357**, 263–266 (2002)

3.360 C. Park, E.S. Engel, A. Crowe, T.R. Gilbert, N.M. Rodriguez: Use of carbon nanofibers in the removal of organic solvents from water, Langmuir **16**, 8050–8056 (2000)

3.361 E. Diaz, S. Ordonez, A. Vega: Adsorption of volatile organic compounds onto carbon nanotubes, carbon nanofibers, and high-surface-area graphites, J. Colloid Interf. Sci. **305**, 7–16 (2007)

3.362 C.-H. Wu: Adsorption of reactive dye onto carbon nanotubes: equilibrium, kinetics and thermodynamics, J. Hazard. Mater. **144**, 93–100 (2007)

3.363 C. Lu, F. Su: Adsorption of natural organic matter by carbon nanotubes, Sep. Purif. Technol. **58**, 113–121 (2007)

3.364 C. Ye, Q.-M. Gong, F.-P. Lu: Adsorption of uraemic toxins on carbon nanotubes, Sep. Purif. Technol. **58**, 2–6 (2007)

3.365 X. Peng, Y. Li, Z. Luan, Z. Di, H. Wang, B. Tian, Z. Jia: Adsorption of 1,2-dichlorobenzene from water to carbon nanotubes, Chem. Phys. Lett. **376**, 154–158 (2003)

3.366 P. Kondratyuk, J.T. Yates: Nanotubes as molecular sponges: the adsorption of CCl_4, Chem. Phys. Lett. **383**, 314–316 (2004)

3.367 C. Lu, H. Chiu: Adsorption of zinc(II) from water with purified carbon nanotubes, Chem. Eng. Sci. **61**, 1138 (2006)

3.368 A. Stafiej, K. Pyrzynska: Adsorption of heavy metal ions with carbon nanotubes, Sep. Purif. Technol. **58**, 49–52 (2007)

3.369 H. Wang, A. Zhou, F. Peng, H. Yu, J. Yang: Mechanism study on adsorption of acidified multiwalled carbon nanotubes to Pb(II), J. Colloid Interf. Sci. **316**, 277–283 (2007)

3.370 C. Chen, X. Li, D. Zhao, X. Tan, X. Wang: Adsorption kinetic, thermodynamic and desorption studies of Th(IV) on oxidized multi-wall carbon nanotubes, Colloids Surf. A Physicochem. Eng. Asp. **302**, 449–454 (2007)

3.371 A. Huczko, H. Lange, E. Calko, H. Grubek-Jaworska, P. Droszcz: Physiological testing of carbon nanotubes: Are they asbestos-like?, Fuller. Sci. Technol. **9**, 251–254 (2001)

3.372 A.A. Shvedova, V. Castranova, E.R. Kisin, D. Schwegler-Berry, A.R. Murray, V.Z. Gandelsman, A.M. Maynard, P. Baron: Exposure to carbon nanotube material: assessment of nanotube cytotoxicity using human keratinocyte cells, Toxical Environ. Health A **66**, 1909–1926 (2003)

3.373 C.W. Lam, J.T. James, R. McCluskey, R.L. Hunter: Pulmonary toxicity of single-wall carbon nanotubes in mice 7 and 90 days after intratracheal instillation, Toxicol. Sci. **77**, 126–134 (2004)

3.374 D. Pantarotto, J.P. Briand, M. Prato, A. Bianco: Translocation of bioactive peptides across cell membranes by carbon nanotubes, Chem. Commun., 16–17 (2004)

3.375 C. Salvador-Morales, E. Flahaut, E. Sim, J. Sloan, M.L.H. Green, R.B. Sim: Complement activation and protein adsorption by carbon nanotubes, Mol. Immun. **43**, 193–201 (2006)

3.376 M.P. Mattson, R.C. Haddon, A.M. Rao: Molecular functionalization of carbon nanotubes and use as substrates for neuronal growth, J. Mol. Neurosci. **14**, 175–182 (2000)

3.377 J.J. Davis, M.L.H. Green, H.A.O. Hill, Y.C. Leung, P.J. Sadler, J. Sloan, A.V. Xavier, S.C. Tsang: The immobilization of proteins in carbon nanotubes, Inorg. Chim. Acta **272**, 261–266 (1998)

3.378 R.J. Chen, Y. Zhang, D. Wang, H. Dai: Noncovalent sidewall functionalization of single-walled carbon nanotubes for protein immobilization, J. Am. Chem. Soc. **123**, 3838–3839 (2001)

3.379 M. Shim, N.W.S. Kam, R.J. Chen, Y. Li, H. Dai: Functionalization of carbon nanotubes for biocompatibility and biomolecular recognition, Nano Lett. **2**, 285–288 (2002)

3.380 C. Dwyer, M. Guthold, M. Falvo, S. Washburn, R. Superfine, D. Erie: DNA-functionalized single-walled carbon nanotubes, Nanotechnology **13**, 601–604 (2002)

3.381 H. Huang, S. Taylor, K. Fu, Y. Lin, D. Zhang, T.W. Hanks, A.M. Rao, Y. Sun: Attaching proteins to carbon nanotubes via diimide-activated amidation, Nano Lett. **2**, 311–314 (2002)

3.382 C.V. Nguyen, L. Delzeit, A.M. Cassell, J. Li, J. Han, M. Meyyappan: Preparation of nucleic acid functionalized carbon nanotube arrays, Nano Lett. **2**, 1079–1081 (2002)

3.383 B.R. Azamian, J.J. Davis, K.S. Coleman, C.B. Bagshaw, M.L.H. Green: Bioelectrochemical single-walled carbon nanotubes, J. Am. Chem. Soc. **124**, 12664–12665 (2002)

3.384 E. Katz, I. Willner: Biomolecule-functionalized carbon nanotubes: Applications in nanobioelectronics, ChemPhysChem **5**, 1084–1104 (2004)

3.385 J. Wang: Carbon-nanotube based electrochemical biosensors: a review, Electroanalysis **17**, 7–14 (2005)

3.386 T. Laha, A. Agarwal, T. McKechnie, S. Seal: Synthesis and characterization of plasma spray formed carbon nanotube reinforced aluminum composite, Mater. Sci. Eng. A **381**, 249–258 (2004)

3.387 T. Noguchi, A. Magario, S. Fukazawa, S. Shimizu, J. Beppu, M. Seki: Carbon nanotube/aluminium composites with uniform dispersion, Mater. Trans. **45**, 602–604 (2004)

3.388 Y.B. Li, Q. Ya, B.Q. Wei, J. Liang, D.H. Wu: Processing of a carbon nanotubes-$Fe_{82}P_{18}$ metallic glass composite, J. Mater. Sci. Lett. **17**, 607–609 (1998)

3.389 A. Goyal, D.A. Wiegand, F.J. Owens, Z. Iqbal: Enhanced yield strength in iron nanocomposite with in situ grown single-wall carbon nanotubes, J. Mater. Res. **21**, 522–528 (2006)

3.390 K.T. Kim, K.H. Lee, S.I. Cha, C.-B. Mo, S.H. Hong: Characterization of carbon nanotubes/Cu nanocomposites processed by using nano-sized Cu powders, Mater. Res. Soc. Symp. Proc. **821**, 111–116 (2004)

3.391 F. Zhang, J. Shen, J. Sun: Processing and properties of carbon nanotubes-nano-WC-Co composites, Mater. Sci. Eng. A **381**, 86–91 (2004)

3.392 C.L. Xu, B.Q. Wei, R.Z. Ma, J. Liang, X.K. Ma, D.H. Wu: Fabrication of aluminum-carbon nanotube composites and their electrical properties, Carbon **37**, 855–858 (1999)

3.393 T. Kuzumaki, K. Miyazawa, H. Ichinose, K. Ito: Processing of carbon nanotube reinforced aluminum composite, J. Mater. Res. **13**, 2445–2449 (1998)

3.394 C.F. Deng, Y.X. Ma, P. Zhang, X.X. Zhang, D.Z. Wang: Thermal expansion behaviors of aluminum composite reinforced with carbon nanotubes, Mater. Lett. **62**, 2301–2303 (2008)

3.395 T. Kuzumaki, O. Ujiie, H. Ichinose, K. Ito: Mechanical characteristics and preparation of carbon nanotube fiber-reinforced Ti composite, Adv. Eng. Mater. **2**, 416–418 (2000)

3.396 E. Carreno-Morelli, J. Yang, E. Couteau, K. Hernadi, J.W. Seo, C. Bonjour, L. Forro, R. Schaller: Carbon nanotube/magnesium composites, Phys. Status Solidi (a) **201**, R53–R55 (2004)

3.397 Z. Bian, R.J. Wang, W.H. Wang, T. Zhang, A. Inoue: Carbon-nanotube-reinforced Zr-based bulk metallic glass composites and their properties, Adv. Funct. Mater. **14**, 55–63 (2004)

3.398 S.R. Dong, J.P. Tu, X.B. Zhang: An investigation of the sliding wear behavior of Cu-matrix composite reinforced by carbon nanotubes, Mater. Sci. Eng. A **313**, 83–87 (2001)

3.399 J. Wang, G. Chen, M. Wang, M.P. Chatrathi: Carbon-nanotube/copper composite electrodes for capillary electrophoresis microchip detection of carbohydrates, Analyst (Cambridge) **129**, 512–515 (2004)

3.400 C.S. Goh, J. Wei, L.C. Lee, M. Gupta: Simultaneous enhancement in strength and ductility reinforcing magnesium with carbon nanotubes, Mater. Sci. Eng. A **423**, 153–156 (2006)

3.401 Q. Ngo, B.A. Cruden, A.M. Cassell, M.D. Walker, Q. Ye, J.E. Koehne, M. Meyyappan, J. Li, C.Y. Yang: Thermal conductivity of carbon nanotube composite films, Mater. Res. Soc. Symp. Proc., Vol. 812 (2004) pp. 179–184

3.402 X.H. Chen, C.S. Chen, H.N. Xiao, F.Q. Cheng, G. Zhang, G.J. Yi: Corrosion behavior of carbon nanotubes-Ni composite coating, Surf. Coat. Technol. **191**, 351–356 (2005)

3.403 X.H. Chen, C.S. Chen, H.N. Xiao, H.B. Liu, L.P. Zhou, S.L. Li, G. Zhang: Dry friction and wear characteristics of nickel/carbon nanotube electroless composite deposits, Tribol. Int. **39**, 22–28 (2006)

3.404 Z. Yang, H. Xu, M.-K. Li, Y.-L. Shi, Y. Huang, H.-L. Li: Preparation and properties of Ni-P/single-walled carbon nanotubes composite coatings by means of electroless plating, Thin Solid Films **466**, 86–91 (2004)

3.405 A. Peigney, C. Laurent: Carbon nanotubes ceramic composites. In: *Ceramic Matrix Composites: Microstructure, Properties and Applications*, ed. by I.M. Low (Woodhead, Cambridge 2006) pp. 309–333

3.406 C. Laurent, A. Peigney, O. Dumortier, A. Rousset: Carbon nanotubes-Fe-alumina nanocomposites. Part II: Microstructure and mechanical properties of the hot-pressed composites, J. Eur. Ceram. Soc. **18**, 2005–2013 (1998)

3.407 A. Peigney, C. Laurent, A. Flahaut, A. Rousset: Carbon nanotubes in novel ceramic matrix nanocomposites, Ceram. Int. **26**, 677–683 (2000)

3.408 E. Flahaut, A. Peigney, C. Laurent, C. Marlière, F. Chastel, A. Rousset: Carbon nanotube-metal-oxide nanocomposites: Microstructure, electrical conductivity and mechanical properties, Acta Mater. **48**, 3803–3812 (2000)

3.409 J.W. An, D.H. You, D.S. Lim: Tribological properties of hot-pressed alumina-CNT composites, Wear **255**, 677–681 (2003)

3.410 J. Ning, J. Zhang, Y. Pan, J. Guo: Surfactants assisted processing of carbon nanotube-reinforced SiO_2 matrix composites, Ceram. Int. **30**, 63–67 (2004)

3.411 Q. Huang, L. Gao: Manufacture and electrical properties of multiwalled carbon nanotube/$BaTiO_3$ nanocomposite ceramics, J. Mater. Chem. **14**, 2536–2541 (2004)

3.412 J. Fan, D. Zhao, M. Wu, Z. Xu, J. Song: Preparation and microstructure of multi-wall carbon nanotubes-toughened Al_2O_3 composite, J. Am. Ceram. Soc. **89**, 750–753 (2006)

3.413 A. Peigney, S. Rul, F. Lefevre-Schlick, C. Laurent: Densification during hot-pressing of carbon nanotube metal-ceramic composites, J. Eur. Ceram. Soc. **27**, 2183–2193 (2007)

3.414 J. Sun, L. Gao, W. Li: Colloidal processing of carbon nanotube/alumina composites, Chem. Mater. **14**, 5169–5172 (2002)

3.415 G.D. Zhan, J.D. Kuntz, J. Wan, A.K. Mukherjee: Single-wall carbon nanotubes as attractive toughening agents in alumina-based composites, Nat. Mater. **2**, 38–42 (2003)

3.416 S.I. Cha, K.T. Kim, K.H. Lee, C.B. Mo, S.H. Hong: Strengthening and toughening of carbon nanotube reinforced alumina nanocomposite fabricated by molecular level mixing process, Scr. Mater. **53**, 793–797 (2005)

3.417 C.B. Mo, S.I. Cha, K.T. Kim, K.H. Lee, S.H. Hong: Fabrication of carbon nanotube reinforced alumina matrix nanocomposite by sol-gel process, Mater. Sci. Eng. A **395**, 124–128 (2005)

3.418 X. Wang, N.P. Padture, H. Tanaka: Contact-damage-resistant ceramic/single-wall carbon nanotubes and ceramic/graphite composites, Nat. Mater. **3**, 539–544 (2004)

3.419 W.A. Curtin, B.W. Sheldon: CNT-reinforced ceramics and metals, Mater. Today **7**, 44–49 (2004)

3.420 Z. Xia, L. Riester, W.A. Curtin, H. Li, B.W. Sheldon, J. Liang, B. Chang, J.M. Xu: Direct observation of toughening mechanisms in carbon nanotube ceramic matrix composites, Acta Mater. **52**, 931–944 (2004)

3.421 D.S. Lim, J.W. An, H.J. Lee: Effect of carbon nanotube addition on the tribological behavior of carbon/carbon composites, Wear **252**, 512–517 (2002)

3.422 D.-S. Lim, D.-H. You, H.-J. Choi, S.-H. Lim, H. Jang: Effect of CNT distribution on tribological behavior of alumina-CNT composites, Wear **259**, 539–544 (2005)

3.423 Z.H. Xia, J. Lou, W.A. Curtin: A multiscale experiment on the tribological of aligned carbon nanotube/ceramic composites, Scr. Mater. **58**, 223–226 (2008)

3.424 G.-D. Zhan, J.D. Kuntz, H. Wang, C.-M. Wang, A.K. Mukherjee: Anisotropic thermal properties of single-wall-carbon-nanotube-reinforced nanoceramics, Philos. Mag. Lett. **84**, 419–423 (2004)

3.425 Q. Huang, L. Gao, Y. Liu, J. Sun: Sintering and thermal properties of multiwalled carbon nanotube-$BaTiO_3$ composites, J. Mater. Chem. **15**, 1995–2001 (2005)

3.426 G.-D. Zhan, J.D. Kuntz, J.E. Garay, A.K. Mukherjee: Electrical properties of nanoceramics reinforced with ropes of single-walled carbon nanotubes, Appl. Phys. Lett. **83**, 1228–1230 (2003)

3.427 S. Rul, F. Lefevre-Schlick, E. Capria, C. Laurent, A. Peigney: Percolation of single-walled carbon nanotubes in ceramic matrix nanocomposites, Acta Mater. **52**, 1061–1067 (2004)

3.428 S.-L. Shi, J. Liang: Electronic transport properties of multiwall carbon nanotubes/yttria-stabilized zirconia composites, J. Appl. Phys. **101**, 023708-5 (2007)

3.429 A. Peigney, E. Flahaut, C. Laurent, F. Chastel, A. Rousset: Aligned carbon nanotubes in ceramic-matrix nanocomposites prepared by high-temperature extrusion, Chem. Phys. Lett. **352**, 20–25 (2002)

3.430 G.-D. Zhan, J.D. Kuntz, A.K. Mukherjee, P. Zhu, K. Koumoto: Thermoelectric properties of carbon nanotube/ceramic nanocomposites, Scr. Mater. **54**, 77–82 (2006)

3.431 P.M. Ajayan, O. Stephan, C. Colliex, D. Trauth: Aligned carbon nanotube arrays formed by cutting a polymer resin-nanotube composite, Science **265**, 1212–1214 (1994)

3.432 J.N. Coleman, U. Khan, Y.K. Gun'ko: Mechanical reinforcement of polymers using carbon nanotubes, Adv. Mater. **18**, 689–706 (2006)

3.433 R. Haggenmueller, H.H. Gommans, A.G. Rinzler, J.E. Fischer, K.I. Winey: Aligned single-wall carbon nanotubes in composites by melt processing methods, Chem. Phys. Lett. **330**, 219–225 (2000)

3.434 L.S. Schadler, S.C. Giannaris, P.M. Ajayan: Load transfer in carbon nanotube epoxy composites, Appl. Phys. Lett. **73**, 3842–3844 (1998)

3.435 S.J.V. Frankland, A. Caglar, D.W. Brenner, M. Griebel: Molecular simulation of the influence of chemical cross-links on the shear strength of carbon nanotube-polymer interfaces, J. Phys. Chem. B **106**, 3046–3048 (2002)

3.436 H.D. Wagner: Nanotube-polymer adhesion: A mechanics approach, Chem. Phys. Lett. **361**, 57–61 (2002)

3.437 P.M. Ajayan, L.S. Schadler, C. Giannaris, A. Rubio: Single-walled carbon nanotube-polymer composites: Strength and weakness, Adv. Mater. **12**, 750–753 (2000)

3.438 X. Gong, J. Liu, S. Baskaran, R.D. Voise, J.S. Young: Surfactant-assisted processing of carbon nano-

3.438 tube/polymer composites, Chem. Mater. **12**, 1049–1052 (2000)

3.439 E.T. Thostenson, W.Z. Li, D.Z. Wang, Z.F. Ren, T.W. Chou: Carbon nanotube/carbon fiber hybrid multiscale composites, J. Appl. Phys. **91**, 6034–6037 (2002)

3.440 F.H. Gojny, M.H.G. Wichmann, U. Kopke, B. Fiedler, K. Schulte: Carbon nanotube-reinforced epoxy-composites: enhanced stiffness and fracture toughness at low nanotube content, Compos. Sci. Technol. **64**, 2363–2371 (2004)

3.441 F.H. Gojny, M.H.G. Wichmann, B. Fiedler, K. Schulte: Influence of different carbon nanotubes on the mechanical properties of epoxy matrix composites – A comparative study, Compos. Sci. Tech. **65**, 2300–2313 (2005)

3.442 H. Rajoria, N. Jalili: Passive vibration damping enhancement using carbon nanotube-epoxy reinforced composites, Compos. Sci. Tech. **65**, 2079–2093 (2005)

3.443 M.J. Biercuk, M.C. Llaguno, M. Radosavljevic, J.K. Hyun, A.T. Johnson, J.E. Fischer: Carbon nanotube composites for thermal management, Appl. Phys. Lett. **80**, 2767–2769 (2002)

3.444 F.H. Gojny, K. Schulte: Functionalisation effect on the thermo-mechanical behaviour of multi-wall carbon nanotube/epoxy-composites, Compos. Sci. Technol. **64**, 2303–2308 (2004)

3.445 G. Pecastaings, P. Delhaes, A. Derre, H. Saadaoui, F. Carmona, S. Cui: Role of interfacial effects in carbon nanotube/epoxy nanocomposite behavior, J. Nanosci. Nanotechnol. **4**, 838–843 (2004)

3.446 S. Barrau, P. Demont, A. Peigney, C. Laurent, C. Lacabanne: Effect of palmitic acid on the electrical conductivity of carbon nanotubes-polyepoxy composite, Macromolecules **36**, 9678–9680 (2003)

3.447 S. Barrau, P. Demont, A. Peigney, C. Laurent, C. Lacabanne: DC and AC conductivity of carbon nanotubes-polyepoxy composites, Macromolecules **36**, 5187–5194 (2003)

3.448 J. Sandler, M.S.P. Shaffer, T. Prasse, W. Bauhofer, K. Schulte, A.H. Windle: Development of a dispersion process for carbon nanotubes in an epoxy matrix and the resulting electrical properties, Polymer **40**, 5967–5971 (1999)

3.449 Z. Jin, K.P. Pramoda, G. Xu, S.H. Goh: Dynamic mechanical behavior of melt-processed multi-walled carbon nanotube/poly(methyl methacrylate) composites, Chem. Phys. Lett. **337**, 43–47 (2001)

3.450 Z. Jin, K.P. Pramoda, S.H. Goh, G. Xu: Poly(vinylidene fluoride)-assisted melt-blending of multi-walled carbon nanotube/poly(methyl methacrylate) composites, Mater. Res. Bull. **37**, 271–278 (2002)

3.451 C.A. Cooper, D. Ravich, D. Lips, J. Mayer, H.D. Wagner: Distribution and alignment of carbon nanotubes and nanofibrils in a polymer matrix, Compos. Sci. Technol. **62**, 1105–1112 (2002)

3.452 J.M. Benoit, B. Corraze, S. Lefrant, W.J. Blau, P. Bernier, O. Chauvet: Transport properties of PMMA-carbon nanotubes composites, Synth. Met. **121**, 1215–1216 (2001)

3.453 J.M. Benoit, B. Corraze, O. Chauvet: Localization, Coulomb interactions, and electrical heating in single-wall carbon nanotubes/polymer composites, Phys. Rev. B **65**, 241405/1–241405/4 (2002)

3.454 F. Du, R.C. Scogna, W. Zhou, S. Brand, J.E. Fischer, K.I. Winey: Nanotube networks in polymer nanocomposites: Rheology and electrical conductivity, Macromolecules **37**, 9048–9055 (2004)

3.455 T. Kashiwagi, F. Du, K.I. Winey, K.M. Groth, J.R. Shields, R.H. Harris Jr., J.F. Douglas: Flammability properties of PMMA-single walled carbon nanotube nanocomposites, Polym. Mater. Sci. Eng. **91**, 90–91 (2004)

3.456 B. Vigolo, A. Pénicaud, C. Coulon, C. Sauder, R. Pailler, C. Journet, P. Bernier, P. Poulin: Macroscopic fibers and ribbons of oriented carbon nanotubes, Science **290**, 1331–1334 (2000)

3.457 B. Vigolo, P. Poulin, M. Lucas, P. Launois, P. Bernier: Improved structure and properties of single-wall carbon nanotube spun fibers, Appl. Phys. Lett. **11**, 1210–1212 (2002)

3.458 P. Poulin, B. Vigolo, P. Launois: Films and fibers of oriented single wall nanotubes, Carbon **40**, 1741–1749 (2002)

3.459 K. Jiang, Q. Li, S. Fan: Spinning continuous carbon nanotube yarn, Nature **419**, 801 (2002)

3.460 M. Zhang, K.R. Atkinson, R.H. Baughman: Multifunctional carbon nanotube yarns by downsizing an ancient technology, Science **306**, 1356–1361 (2004)

3.461 J. Steinmetz, M. Glerup, M. Paillet, P. Bernier, M. Holzinger: Production of pure nanotube fibers using a modified wet-spinning method, Carbon **43**, 2397–2400 (2005)

3.462 A.B. Dalton, S. Collins, E. Munoz, J.M. Razal, V.H. Ebron, J.P. Ferraris, J.N. Coleman, B.G. Kim, R.H. Baughman: Super-tough carbon-nanotube fibers, Nature **423**, 703 (2003)

3.463 M.S.P. Shaffer, A.H. Windle: Fabrication and characterization of carbon nanotube/poly(vinyl alcohol) composites, Adv. Mater. **11**, 937–941 (1999)

3.464 L. Jin, C. Bower, O. Zhou: Alignment of carbon nanotubes in a polymer matrix by mechanical stretching, Appl. Phys. Lett. **73**, 1197–1199 (1998)

3.465 H.D. Wagner, O. Lourie, Y. Feldman, R. Tenne: Stress-induced fragmentation of multiwall carbon nanotubes in a polymer matrix, Appl. Phys. Lett. **72**, 188–190 (1998)

3.466 H.D. Wagner, O. Lourie, X.F. Zhou: Macrofragmentation and microfragmentation phenomena in composite materials, Compos. Part A **30**, 59–66 (1998)

3.467 J.R. Wood, Q. Zhao, H.D. Wagner: Orientation of carbon nanotubes in polymers and its detection by

Raman spectroscopy, Compos. Part A **32**, 391–399 (2001)

3.468 Q. Zhao, J.R. Wood, H.D. Wagner: Using carbon nanotubes to detect polymer transitions, J. Polym. Sci. B **39**, 1492–1495 (2001)

3.469 M. Cochet, W.K. Maser, A.M. Benito, M.A. Callejas, M.T. Martinesz, J.M. Benoit, J. Schreiber, O. Chauvet: Synthesis of a new polyaniline/nanotube composite: In-situ polymerisation and charge transfer through site-selective interaction, Chem. Commun., 1450–1451 (2001)

3.470 D. Qian, E.C. Dickey, R. Andrews, T. Rantell: Load transfer and deformation mechanisms in carbon nanotube-polystyrene composites, Appl. Phys. Lett. **76**, 2868–2870 (2000)

3.471 V. Datsyuk, C. Guerret-Piecourt, S. Dagreou, L. Billon, J.-C. Dupin, E. Flahaut, A. Peigney, C. Laurent: Double walled carbon nanotube/polymer composites via in-situ nitroxide mediated polymerisation of amphiphilic block copolymers, Carbon **43**, 873–876 (2005)

3.472 R. Blake, Y.K. Gun'ko, J. Coleman, M. Cadek, A. Fonseca, J.B. Nagy, W.J. Blau: A generic organometallic approach toward ultra-strong carbon nanotube polymer composites, J. Am. Chem. Soc. **126**, 10226–10227 (2004)

3.473 T. Kashiwagi, E. Grulke, J. Hilding, K. Groth, R. Harris, K. Butler, J. Shields, S. Kharchenko, J. Douglas: Thermal and flammability properties of polypropylene/carbon nanotube nanocomposites, Polymer **45**, 4227–4239 (2004)

3.474 C. Wei, D. Srivastava, K. Cho: Thermal expansion and diffusion coefficients of carbon nanotube–polymer composites, Los Alamos Nat. Lab., Preprint Archive, Condensed Matter (archiv:cond-mat/0203349), 1–11 (2002)

3.475 J.C. Grunlan, M.V. Bannon, A.R. Mehrabi: Latex-based, single-walled nanotube composites: processing and electrical conductivity, Polym. Prepr. **45**, 154–155 (2004)

3.476 J.C. Grunlan, A.R. Mehrabi, M.V. Bannon, J.L. Bahr: Water-based single-walled-nanotube-filled polymer composite with an exceptionally low percolation threshold, Adv. Mater. (Weinheim) **16**, 150–153 (2004)

3.477 C. Pirlot, I. Willems, A. Fonseca, J.B. Nagy, J. Delhalle: Preparation and characterization of carbon nanotube/polyacrylonitrile composites, Adv. Eng. Mater. **4**, 109–114 (2002)

3.478 H. Lam, H. Ye, Y. Gogotsi, F. Ko: Structure and properties of electrospun single-walled carbon nanotubes reinforced nanocomposite fibrils by co-electrospinning, Polym. Prepr. **45**, 124–125 (2004)

3.479 L. Cao, H. Chen, M. Wang, J. Sun, X. Zhang, F. Kong: Photoconductivity study of modified carbon nanotube/oxotitanium phthalocyanine composites, J. Phys. Chem. B **106**, 8971–8975 (2002)

3.480 I. Musa, M. Baxendale, G.A.J. Amaratunga, W. Eccleston: Properties of regular poly(3-octyl-thiophene)/multi-wall carbon nanotube composites, Synth. Met. **102**, 1250 (1999)

3.481 E. Kymakis, I. Alexandou, G.A.J. Amaratunga: Single-walled carbon nanotube-polymer composites: Electrical, optical and structural investigation, Synth. Met. **127**, 59–62 (2002)

3.482 K. Yoshino, H. Kajii, H. Araki, T. Sonoda, H. Take, S. Lee: Electrical and optical properties of conducting polymer-fullerene and conducting polymer-carbon nanotube composites, Fuller. Sci. Technol. **7**, 695–711 (1999)

3.483 S.A. Curran, P.M. Ajayan, W.J. Blau, D.L. Carroll, J.N. Coleman, A.B. Dalton, A.P. Davey, A. Drury, B. McCarthy, S. Maier, A. Strevens: A composite from poly(m-phenylenevinylene-co-2,5-dioctoxy-p-phenylenevinylene) and carbon nanotubes. A novel material for molecular optoelectronics, Adv. Mater. **10**, 1091–1093 (1998)

3.484 P. Fournet, D.F. O'Brien, J.N. Coleman, H.H. Horhold, W.J. Blau: A carbon nanotube composite as an electron transport layer for M3EH-PPV based light-emitting diodes, Synth. Met. **121**, 1683–1684 (2001)

3.485 H.S. Woo, R. Czerw, S. Webster, D.L. Carroll, J. Ballato, A.E. Strevens, D. O'Brien, W.J. Blau: Hole blocking in carbon nanotube-polymer composite organic light-emitting diodes based on poly(m-phenylene vinylene-co-2,5-dioctoxy-p-phenylene vinylene), Appl. Phys. Lett. **77**, 1393–1395 (2000)

3.486 H.S. Woo, R. Czerw, S. Webster, D.L. Carroll, J.W. Park, J.H. Lee: Organic light emitting diodes fabricated with single wall carbon nanotubes dispersed in a hole conducting buffer: The role of carbon nanotubes in a hole conducting polymer, Synth. Met. **116**, 369–372 (2001)

3.487 H. Ago, K. Petritsch, M.S.P. Shaffer, A.H. Windle, R.H. Friend: Composites of carbon nanotubes and conjugated polymers for photovoltaic devices, Adv. Mater. **11**, 1281–1285 (1999)

3.488 B. Maruyama, K. Alam: Carbon nanotubes and nanofibers in composite materials, SAMPE Journal **38**, 59–70 (2002)

3.489 P. Lambin, A. Fonseca, J.P. Vigneron, J. B'Nagy, A.A. Lucas: Structural and electronic properties of bent carbon nanotubes, Chem. Phys. Lett. **245**, 85–89 (1995)

3.490 L. Chico, V.H. Crespi, L.X. Benedict, S.G. Louie, M.L. Cohen: Pure carbon nanoscale devices: Nanotube heterojunctions, Phys. Rev. Lett. **76**, 971–974 (1996)

3.491 Z. Yao, H.W.C. Postma, L. Balents, C. Dekker: Carbon nanotube intramolecular junctions, Nature **402**, 273–276 (1999)

3.492 S.J. Tans, A.R.M. Verschueren, C. Dekker: Room temperature transistor based on single carbon nanotube, Nature **393**, 49–52 (1998)

3.493 R. Martel, T. Schmidt, H.R. Shea, T. Hertel, P. Avouris: Single and multi-wall carbon nanotube field effect transistors, Appl. Phys. Lett. **73**, 2447–2449 (1998)

3.494 V. Derycke, R. Martel, J. Appenzeller, P. Avouris: Carbon nanotube inter- and intramolecular logic gates, Nano Lett. **1**, 453–456 (2001)

3.495 P. Kim, C.M. Lieber: Nanotube nanotweezers, Science **286**, 2148–2150 (1999)

3.496 P.G. Collins, M.S. Arnold, P. Avouris: Engineering carbon nanotubes using electrical breakdown, Science **292**, 706–709 (2001)

3.497 A.P. Graham, G.S. Duesberg, W. Hoenlein, F. Kreupl, M. Liebau, R. Martin, B. Rajasekharan, W. Pamler, R. Seidel, W. Steinhoegl, E. Unger: How do carbon nanotubes fit into the semiconductor roadmap?, Appl. Phys. A **80**, 1141–1151 (2005)

3.498 R.H. Baughman, C. Changxing, A.A. Zakhidov, Z. Iqbal, J.N. Barisci, G.M. Spinks, G.G. Wallace, A. Mazzoldi, D. de Rossi, A.G. Rinzler, O. Jaschinki, S. Roth, M. Kertesz: Carbon nanotubes actuators, Science **284**, 1340–1344 (1999)

3.499 Y. Gao, Y. Bando: Carbon nanothermometer containing gallium, Nature **415**, 599 (2002)

3.500 C. Niu, E.K. Sichel, R. Hoch, D. Moy, H. Tennent: High power electro-chemical capacitors based on carbon nanotube electrodes, Appl. Phys. Lett. **70**, 1480–1482 (1997)

3.501 E. Frackowiak, F. Béguin: Electrochemical storage of energy in carbon nanotubes and nanostructured carbons, Carbon **40**, 1775–1787 (2002)

3.502 C. Portet, P.L. Taberna, P. Simon, E. Flahaut: Influence of carbon nanotubes addition on carbon-carbon supercapacitor performances in organic electrolyte, J. Power. Sources **139**, 371–378 (2005)

3.503 E. Frackowiak, K. Jurewicz, K. Szostak, S. Delpeux, F. Béguin: Nanotubular materials as electrodes for supercapacitors, Fuel Process. Technol. **77**, 213–219 (2002)

3.504 G. Lota, E. Frackowiak, J. Mittal, M. Monthioux: High performance supercapacitor from chromium oxide-nanotubes based electrodes, Chem. Phys. Lett. **434**, 73–77 (2007)

3.505 R. Hurt, M. Monthioux, A. Kane (Eds.): Toxicology of carbon nanomaterials, Carbon **44**(6), 1028–1033 (2006), Special issue

3.506 C. Salvador-Morales, E. Flahaut, E. Sim, J. Sloan, M.L.H. Green, R.B. Sim: Complement activation and protein adsorption by carbon nanotubes, Mol. Immun. **43**, 193–201 (2006)

3.507 C. Salvador-Morales, P. Townsend, E. Flahaut, C. Vénien-Bryan, A. Vlandas, M.L.H. Green, R.B. Sim: Binding of pulmonary surfactant proteins to carbon nanotubes; potential for damage to lung immune defence mechanisms, Carbon **45**, 607–617 (2007)

4. Nanowires

Mildred S. Dresselhaus, Yu-Ming Lin, Oded Rabin, Marcie R. Black, Jing Kong, Gene Dresselhaus

This chapter provides an overview of recent research on inorganic nanowires, particularly metallic and semiconducting nanowires. Nanowires are one-dimensional, anisotropic structures, small in diameter, and large in surface-to-volume ratio. Thus, their physical properties are different than those of structures of different scale and dimensionality. While the study of nanowires is particularly challenging, scientists have made immense progress in both developing synthetic methodologies for the fabrication of nanowires, and developing instrumentation for their characterization. The chapter is divided into three main sections: Sect. 4.1 the synthesis, Sect. 4.2 the characterization and physical properties, and Sect. 4.3 the applications of nanowires. Yet, the reader will discover many links that make these aspects of nanoscience intimately interdepent.

4.1	Synthesis	121
	4.1.1 Template-Assisted Synthesis	121
	4.1.2 VLS Method for Nanowire Synthesis	124
	4.1.3 Other Synthesis Methods	126
	4.1.4 Hierarchical Arrangement and Superstructures of Nanowires	128
4.2	Characterization and Physical Properties of Nanowires	130
	4.2.1 Structural Characterization	130
	4.2.2 Mechanical Properties	135
	4.2.3 Transport Properties	136
	4.2.4 Optical Properties	147
4.3	Applications	152
	4.3.1 Electrical Applications	152
	4.3.2 Thermoelectric Applications	154
	4.3.3 Optical Applications	154
	4.3.4 Chemical and Biochemical Sensing Devices	157
	4.3.5 Magnetic Applications	158
4.4	Concluding Remarks	159
References		159

Nanowires are attracting much interest from those seeking to apply nanotechnology and (especially) those investigating nanoscience. Nanowires, unlike other low-dimensional systems, have two quantum-confined directions but one unconfined direction available for electrical conduction. This allows nanowires to be used in applications where electrical conduction, rather than tunneling transport, is required. Because of their unique density of electronic states, in the limit of small diameters nanowires are expected to exhibit significantly different optical, electrical and magnetic properties to their bulk 3-D crystalline counterparts. Increased surface area, very high density of electronic states and joint density of states near the energies of their van Hove singularities, enhanced exciton binding energy, diameter-dependent bandgap, and increased surface scattering for electrons and phonons are just some of the ways in which nanowires differ from their corresponding bulk materials. Yet the sizes of nanowires are typically large enough (> 1 nm in the quantum-confined direction) to result in local crystal structures that are closely related to their parent materials, allowing theoretical predictions about their properties to be made based on knowledge of their bulk properties.

Not only do nanowires exhibit many properties that are similar to, and others that are distinctly different from, those of their bulk counterparts, nanowires also have the advantage from an applications standpoint in that some of the materials parameters critical for certain properties can be independently controlled in nanowires but not in their bulk counterparts. Certain properties can also be enhanced nonlinearly in small-diameter nanowires, by exploiting the singular aspects of the 1-D electronic density of states.

Table 4.1 Selected syntheses of nanowires by material

Material	Growth Technique	Reference	Material	Growth Technique	Reference
ABO_4-type	Template[a]	[4.2]	Ge	High-T, high-P liquid-phase, redox	[4.33]
Ag	DNA-template, redox	[4.3]		VLS[d]	[4.34]
	Template, pulsed ECD[b]	[4.4]		Oxide-assisted	[4.35]
Au	Template, ECD[b]	[4.5, 6]	InAs	VLS[d]	[4.36]
Bi	Stress-induced	[4.7]	MgO	VLS[d]	[4.37]
	Template, vapor-phase	[4.8]	Mo	Step decoration, ECD[b] + redox	[4.38]
	Template, ECD[b]	[4.9–11]	Ni	Template, ECD[b]	[4.11, 39, 40]
	Template, pressure injection	[4.12–14]	Pb	Liquid-phase[f]	[4.41]
BiSb	Pulsed ECD[b]	[4.15]	PbSe	Liquid phase	[4.42]
Bi_2Te_3	Template, dc ECD[b]	[4.16]		Self assembly of nanocrystals[g]	[4.43]
CdS	Liquid-phase (surfactant), recrystallization	[4.17]	Pd	Step decoration, ECD[b]	[4.44]
	Template, ac ECD[b]	[4.18, 19]	Se	Liquid-phase, recrystallization	[4.45]
CdSe	Liquid-phase (surfactant), redox	[4.20]		Template, pressure injection	[4.46]
	Template, ac ECD[b]	[4.21, 22]	Si	VLS[d]	[4.47]
Cu	Vapor deposition	[4.23]		Laser-ablation VLS[d]	[4.48]
	Template, ECD[b]	[4.24]		Oxide-assisted	[4.49]
Fe	Template, ECD[c]	[4.25, 26]		Low-T VLS[d]	[4.50]
	Shadow deposition	[4.27]	W	Vapor transport	[4.51]
GaN	Template, CVD[c]	[4.28, 29]	Zn	Template, vapor-phase	[4.52]
	VLS[d]	[4.30, 31]		Template, ECD[b]	[4.53]
GaAs	Template, liquid/vapor OMCVD[e]	[4.32]	ZnO	VLS[d]	[4.54]
				Template, ECD[b]	[4.53, 55]

[a] Template synthesis
[b] Electrochemical deposition (ECD)
[c] Chemical vapor deposition (CVD)
[d] Vapor–liquid–solid (VLS) growth
[e] Organometallic chemical vapor deposition (OMCVD)
[f] Liquid phase synthesis
[g] Self assembly of nanocrystals (in liquid phase)

Furthermore, nanowires have been shown to provide a promising framework for applying the *bottom-up* approach [4.1] to the design of nanostructures for nanoscience investigations and for potential nanotechnology applications.

Driven by (1) these new research and development opportunities, (2) the smaller and smaller length scales now being used in the semiconductor, optoelectronics and magnetics industries, and (3) the dramatic development of the biotechnology industry where the action is also at the nanoscale, the nanowire research field has developed with exceptional speed in the last few years. Therefore, a review of the current status of nanowire research is of significant broad interest at the present time. It is the aim of this review to focus on nanowire properties that differ from those of their parent crystalline bulk materials, with an eye toward possible applications that might emerge from the unique properties of nanowires and from future discoveries in this field.

For quick reference, examples of typical nanowires that have been synthesized and studied are listed in Table 4.1. Also of use to the reader are review articles that focus on a comparison between nanowire and nanotube properties [4.56] and the many reviews that have been written about carbon nanotubes [4.57–59], which can be considered as a model one-dimensional system.

4.1 Synthesis

In this section we survey the most common synthetic approaches that have successfully afforded high-quality nanowires of a large variety of materials (Table 4.1). In Sect. 4.1.1, we discuss methods which make use of various templates with nanochannels to confine the nanowire growth in two dimensions. In Sect. 4.1.2, we present the synthesis of nanowires by the vapor–liquid–solid mechanism and its many variations. In Sect. 4.1.3, examples of other synthetic methods of general applicability are presented. The last part of this section (Sect. 4.1.4) features several approaches that have been developed to organize nanowires into simple architectures.

4.1.1 Template-Assisted Synthesis

The template-assisted synthesis of nanowires is a conceptually simple and intuitive way to fabricate nanostructures [4.62–64]. These templates contain very small cylindrical pores or voids within the host material, and the empty spaces are filled with the chosen material, which adopts the pore morphology, to form nanowires. In this section, we describe the templates first, and then describe strategies for filling the templates to make nanowires.

Template Synthesis

In template-assisted synthesis of nanostructures, the chemical stability and mechanical properties of the template, as well as the diameter, uniformity and density of the pores are important characteristics to consider. Templates frequently used for nanowire synthesis include anodic alumina (Al_2O_3), nanochannel glass, ion track-etched polymers and mica films.

Porous anodic alumina templates are produced by anodizing pure Al films in selected acids [4.65–67]. Under carefully chosen anodization conditions, the resulting oxide film possesses a regular hexagonal array of parallel and nearly cylindrical channels, as shown in Fig. 4.1a. The self-organization of the pore structure in an anodic alumina template involves two coupled processes: pore formation with uniform diameters and pore ordering. The pores form with uniform diameters because of a delicate balance between electric field-enhanced diffusion which determines the growth rate of the alumina, and dissolution of the alumina into the acidic electrolyte [4.68]. The pores are believed to self-order because of mechanical stress at the aluminum–alumina interface due to expansion during the anodization. This stress produces a repulsive force between the pores, causing them to arrange in a hexagonal lattice [4.69]. Depending on the anodization conditions, the pore diameter can be systematically varied from ≤ 10 up to 200 nm with a pore density in the range of 10^9–10^{11} pores/cm^2 [4.13, 25, 65, 66]. It has been shown by many groups that the pore size distribution and the pore ordering of the anodic alumina templates can be significantly improved by a two-step anodization technique [4.60, 70, 71], where the aluminum oxide layer is dissolved after the first anodization in an acidic solution followed by a second anodization under the same conditions.

Another type of porous template commonly used for nanowire synthesis is the template type fabricated by chemically etching particle tracks originating from ion bombardment [4.72], such as track-etched polycarbonate membranes (Fig. 4.1b) [4.73, 74], and also mica films [4.39].

Other porous materials can be used as host templates for nanowire growth, as discussed by *Ozin* [4.62]. Nanochannel glass (NCG), for example, contains a regular hexagonal array of capillaries similar to the pore structure in anodic alumina with a packing density as high as 3×10^{10} pores/cm^2 [4.63]. Porous Vycor glass that contains an interconnected network of pores less than 10 nm was also employed for the early

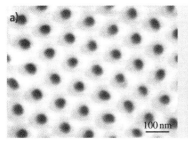
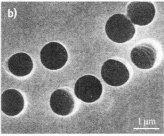

Fig. 4.1 (a) SEM images of the top surfaces of porous anodic alumina templates anodized with an average pore diameter of 44 nm (after [4.60]). (b) SEM image of the particle track-etched polycarbonate membrane, with a pore diameter of 1 μm (after [4.61])

study of nanostructures [4.75]. Mesoporous molecular sieves [4.76], termed MCM-41, possess hexagonally-packed pores with very small channel diameters which can be varied between 2 and 10 nm. Conducting organic filaments have been fabricated in the nanochannels of MCM-41 [4.77]. Recently, the DNA molecule has also been used as a template for growing nanometer-sized wires [4.3].

Diblock copolymers, polymers that consist of two chain segments different properties, have also been utilized as templates for nanowire growth. When two components are immiscible in each other, phase segregation occurs, and depending on their volume ratio, spheres, cylinders and lamellae may self-assemble. To form self-assembled arrays of nanopores, copolymers composed of polystyrene and polymethylmethacrylate [P(S-b-MMA)] [4.79] were used. By applying an electric field while the copolymer was heated above the glass transition temperature of the two constituent polymers, the self-assembled cylinders of PMMA could be aligned with their main axis perpendicular to the film. Selective removal of the PMMA component afforded the preparation of 14 nm diameter ordered pore arrays with a packing density of 1.9×10^{11} cm^{-3}.

Nanowire Template–Assisted Growth by Pressure Injection

The pressure injection technique is often employed for fabricating highly crystalline nanowires from a low-melting point material and when using porous templates with robust mechanical strength. In the high-pressure injection method, the nanowires are formed by pressure-injecting the desired material in liquid form into the evacuated pores of the template. Due to the heating and pressurization processes, the templates used for the pressure injection method must be chemically stable and be able to maintain their structural integrity at high temperatures and at high pressures. Anodic aluminum oxide films and nanochannel glass are two typical materials used as templates in conjunction with the pressure injection filling technique. Metal nanowires (Bi, In, Sn, and Al) and semiconductor nanowires (Se, Te, GaSb, and Bi$_2$Te$_3$) have been fabricated in anodic aluminum oxide templates using this method [4.12, 46, 78].

The pressure P required to overcome the surface tension for the liquid material to fill the pores with a diameter d_W is determined by the Washburn equation [4.80]

$$d_W = -4\gamma \cos\theta / P \,, \qquad (4.1)$$

where γ is the surface tension of the liquid, and θ is the contact angle between the liquid and the template. To reduce the required pressure and to maximize the filling factor, some surfactants are used to decrease the surface tension and the contact angle. For example, the introduction of Cu into the Bi melt can facilitate filling the pores in the anodic alumina template with liquid Bi and can increase the number of nanowires that are formed [4.13]. However, some of the surfactants might cause contamination problems and should therefore be avoided. Nanowires produced by the pressure injection technique usually possess high crystallinity and a preferred crystal orientation along the wire axis. For example, Fig. 4.2 shows the x-ray diffraction (XRD) patterns of Bi nanowire arrays of three different wire diameters with an injection pressure of ≈ 5000 psi [4.78], showing that the major ($> 80\%$) crystal orientation of the wire axes in the 95 and 40 nm diameter Bi nanowire arrays are, respectively, normal to the (202) and (012) lattice planes,

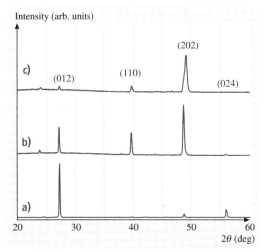

Fig. 4.2a–c XRD patterns of bismuth/anodic alumina nanocomposites with average bismuth wire diameters of (a) 40 nm, (b) 52 nm, and (c) 95 nm [4.78]. The Miller indices corresponding to the lattice planes of bulk Bi are indicated above the individual *peaks*. The majority of the Bi nanowires are oriented along the [10$\bar{1}$1] and [01$\bar{1}$2] directions for $d_W \geq 60$ nm and $d_W \leq 50$ nm, respectively (after [4.13, 78]). The existence of more than one dominant orientation in the 52 nm Bi nanowires is attributed to the transitional behavior of *intermediate*-diameter nanowires as the preferential growth orientation is shifted from [10$\bar{1}$1] to [01$\bar{1}$2] with decreasing d_W

which are denoted by $[10\bar{1}1]$ and $[01\bar{1}2]$ when using a hexagonal unit cell, suggesting a wire diameter-dependent crystal growth direction. On the other hand, 30 nm Bi nanowires produced using a much higher pressure of $> 20\,000$ psi show a different crystal orientation of (001) along the wire axis [4.14], indicating that the preferred crystal orientation may also depend on the applied pressure, with the most dense packing direction along the wire axis for the highest applied pressure.

Electrochemical Deposition

The electrochemical deposition technique has attracted increasing attention as a versatile method for fabricating nanowires in templates. Traditionally, electrochemistry has been used to grow thin films on conducting surfaces. Since electrochemical growth is usually controllable in the direction normal to the substrate surface, this method can be readily extended to fabricate 1-D or 0-D nanostructures, if the deposition is confined within the pores of an appropriate template. In the electrochemical methods, a thin conducting metal film is first coated on one side of the porous membrane to serve as the cathode for electroplating. The length of the deposited nanowires can be controlled by varying the duration of the electroplating process. This method has been used to synthesize a wide variety of nanowires, such as metals (Bi [4.9, 74]; Co [4.81, 82]; Fe [4.25, 83]; Cu [4.73, 84]; Ni [4.39, 81]; Ag [4.85]; Au [4.5]); conducting polymers [4.9, 61]; superconductors (Pb [4.86]); semiconductors (CdS [4.19]); and even superlattice nanowires with A/B constituents (such as Cu/Co [4.73, 84]) have been synthesized electrochemically (Table 4.1).

In the electrochemical deposition process, the chosen template has to be chemically stable in the electrolyte during the electrolysis process. Cracks and defects in the templates are detrimental to the nanowire growth, since the deposition processes primarily occur in the more accessible cracks, leaving most of the nanopores unfilled. Particle track-etched mica films or polymer membranes are typical templates used in simple DC electrolysis. To use anodic aluminum oxide films in the DC electrochemical deposition, the insulating barrier layer which separates the pores from the bottom aluminum substrate has to be removed, and a metal film is then evaporated onto the back of the template membrane [4.87]. Compound nanowire arrays, such as Bi_2Te_3, have been fabricated in alumina templates with a high filling factor using the DC electrochemical deposition [4.16]. Figure 4.3a,b, respectively, shows the top view and the axial cross-sectional SEM images of a Bi_2Te_3 nanowire array [4.16]. The light areas are associated with Bi_2Te_3 nanowires, the dark regions denote empty pores, and the surrounding gray matrix is alumina.

Surfactants are also used with electrochemical deposition when necessary. For example, when using templates derived from PMMA/PS diblock copolymers, a methanol surfactant is used to facilitate pore filling [4.79], thereby achieving a $\approx 100\%$ filling factor.

It is also possible to employ an ac electrodeposition method in anodic alumina templates without the removal of the barrier layer, by utilizing the rectifying properties of the oxide barrier. In ac electrochemical deposition, although the applied voltage is sinusoidal and symmetric, the current is greater during the cathodic half-cycles, making deposition dominant over the stripping, which occurs in the subsequent anodic half-cycles. Since no rectification occurs at defect sites, the deposition and stripping rates are equal, and no material is deposited. Hence, the difficulties associated with cracks are avoided. In this fashion, metals, such as Co [4.82] and Fe [4.25, 83], and semiconductors, such as CdS [4.19], have been deposited into the pores of anodic aluminum oxide templates without removing the barrier layer.

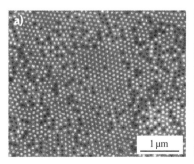 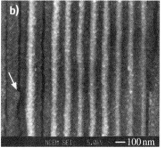

Fig. 4.3 (a) SEM image of a Bi_2Te_3 nanowire array in cross section showing a relatively high pore filling factor. (b) SEM image of a Bi_2Te_3 nanowire array composite along the wire axis (after [4.16])

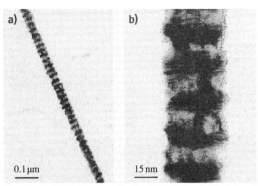

Fig. 4.4 (a) TEM image of a single Co(10 nm)/Cu(10 nm) multilayered nanowire. (b) A selected region of the sample at high magnification (after [4.84])

In contrast to nanowires synthesized by the pressure injection method, nanowires fabricated by the electrochemical process are usually polycrystalline, with no preferred crystal orientations, as observed by XRD studies. However, some exceptions exist. For example, polycrystalline CdS nanowires, fabricated by an ac electrodeposition method in anodic alumina templates [4.19], possibly have a preferred wire growth orientation along the c-axis. In addition, *Xu* et al. have prepared a number of single-crystal II–VI semiconductor nanowires, including CdS, CdSe and CdTe, by DC electrochemical deposition in anodic alumina templates with a nonaqueous electrolyte [4.18, 22]. Furthermore, single-crystal Pb nanowires were formed by pulse electrodeposition under overpotential conditions, but no specific crystal orientation along the wire axis was observed [4.86]. The use of pulse currents is believed to be advantageous for the growth of crystalline wires because the metal ions in the solution can be regenerated between the electrical pulses and therefore uniform deposition conditions can be produced for each deposition pulse. Similarly, single-crystal Ag nanowires were fabricated by pulsed electrodeposition [4.4].

One advantage of the electrochemical deposition technique is the possibility of fabricating multilayered structures within nanowires. By varying the cathodic potentials in the electrolyte, which contains two different kinds of ions, different metal layers can be controllably deposited. Co/Cu multilayered nanowires have been synthesized in this way [4.73, 84]. Figure 4.4 shows TEM images of a single Co/Cu nanowire which is about 40 nm in diameter [4.84]. The light bands represent Co-rich regions and the dark bands represent Cu-rich layers. This electrodeposition method provides a low-cost approach to preparing multilayered 1-D nanostructures.

Vapor Deposition

Vapor deposition of nanowires includes physical vapor deposition (PVD) [4.8], chemical vapor deposition (CVD) [4.29], and metallo-organic chemical vapor deposition (MOCVD) [4.32]. Like electrochemical deposition, vapor deposition is usually capable of preparing smaller-diameter (≤ 20 nm) nanowires than pressure injection methods, since it does not rely on the high pressure and the surface tension involved to insert the material into the pores.

In the physical vapor deposition technique, the material to be filled is first heated to produce a vapor, which is then introduced through the pores of the template and cooled to solidify. Using a specially designed experimental setup [4.8], nearly single-crystal Bi nanowires in anodic aluminum templates with pore diameters as small as 7 nm have been synthesized, and these Bi nanowires were found to possess a preferred crystal growth orientation along the wire axis, similar to the Bi nanowires prepared by pressure injection [4.8, 13].

Compound materials that result from two reacting gases have also be prepared by the chemical vapor deposition (CVD) technique. For example, single-crystal GaN nanowires have been synthesized in anodic alumina templates through a gas reaction of Ga_2O vapor with a flowing ammonia atmosphere [4.28, 29]. A different liquid/gas phase approach has been used to prepare polycrystalline GaAs and InAs nanowires in a nanochannel glass array [4.32]. In this method, the nanochannels are filled with one liquid precursor (such as Me_3Ga or Et_3In) via a capillary effect and the nanowires are formed within the template by reactions between the liquid precursor and the other gas reactant (such as AsH_3).

4.1.2 VLS Method for Nanowire Synthesis

Some of the recent successful syntheses of semiconductor nanowires are based on the so-called vapor–liquid–solid (VLS) mechanism of anisotropic crystal growth. This mechanism was first proposed for the growth of single crystal silicon whiskers 100 nm to hundreds of micrometer in diameter [4.88]. The proposed growth mechanism (Fig. 4.5) involves the absorption of source material from the gas phase into a liquid droplet of catalyst (a molten particle of gold on a silicon substrate in the original work [4.88]). Upon supersaturation of the liquid alloy, a nucleation event generates a solid precipitate of the source material. This seed serves as a preferred site for further deposition of material at the interface of the liquid droplet, promoting the elongation of the seed into a nanowire or a whisker, and suppressing further nucleation events on the same catalyst. Since the liquid droplet catalyzes the incorporation of material from the gas source to the growing crystal, the deposit grows anisotropically as a whisker whose diameter is dictated by the diameter of the liquid alloy droplet. The nanowires thus obtained are of high purity, except for the end containing the solidified catalyst as an alloy particle (Figs. 4.5 and 4.6a). Real-time observations of the alloying, nucleation, and elongation steps in the growth of germanium nanowires from gold nanoclusters by the VLS method were recorded by in situ TEM [4.89].

Reduction of the average wire diameter to the nanometer scale requires the generation of nanosized catalyst droplets. However, due to the balance between the liquid–vapor surface free energy and the free energy of condensation, the size of a liquid droplet, in equilibrium with its vapor, is usually limited to the micrometer range. This obstacle has been overcome in recent years by several new methodologies:

1. Advances in the synthesis of metal nanoclusters have made monodispersed nanoparticles commercially available. These can be dispersed on a solid substrate in high dilution so that when the temperature is raised above the melting point, the liquid clusters do not aggregate [4.47].
2. Alternatively, metal islands of nanoscale sizes can self-form when a strained thin layer is grown or heat-treated on a nonepitaxial substrate [4.34].

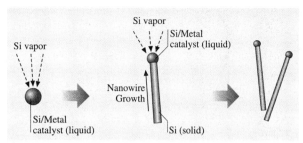

Fig. 4.5 Schematic diagram illustrating the growth of silicon nanowires by the VLS mechanism

3. Laser-assisted catalytic VLS growth is a method used to generate nanowires under nonequilibrium conditions. Using laser ablation of a target containing both the catalyst and the source materials, a plasma is generated from which catalyst nanoclusters nucleate as the plasma cools down. Single crystal nanowires grow as long as the particle remains liquid [4.48].
4. Interestingly, by optimizing the material properties of the catalyst-nanowire system, conditions can be achieved for which nanocrystals nucleate in a liquid catalyst pool supersaturated with the nanowire material, migrate to the surface due to a large surface tension, and continue growing as nanowires perpendicular to the liquid surface [4.50]. In this case, supersaturated nanodroplets are sustained on the outer end of the nanowire due to the low solubility of the nanowire material in the liquid [4.91].

A wide variety of elemental, binary and compound semiconductor nanowires has been synthesized

 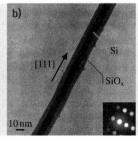 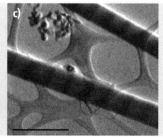

Fig. 4.6 (a) TEM images of Si nanowires produced after laser-ablating a $Si_{0.9}Fe_{0.1}$ target. The *dark spheres* with a slightly larger diameter than the wires are solidified catalyst clusters (after [4.48]). (b) Diffraction contrast TEM image of a Si nanowire. The crystalline Si core appears *darker* than the amorphous oxide surface layer. The *inset* shows the convergent beam electron diffraction pattern recorded perpendicular to the wire axis, confirming the nanowire crystallinity (after [4.48]). (c) STEM image of $Si/Si_{1-x}Ge_x$ superlattice nanowires in the bright field mode. The scale bar is 500 nm (after [4.90])

via the VLS method, and relatively good control over the nanowire diameter and diameter distribution has been achieved. Researchers are currently focusing their attention on the controlled variation of the materials properties along the nanowire axis. In this context, researchers have modified the VLS synthesis apparatus to generate compositionally-modulated nanowires. GaAs/GaP-modulated nanowires have been synthesized by alternately ablating targets of the corresponding materials in the presence of gold nanoparticles [4.92]. p-Si/n-Si nanowires were grown by chemical vapor deposition from alternating gaseous mixtures containing the appropriate dopant [4.92]. $Si/Si_{1-x}Ge_x$ nanowires were grown by combining silicon from a gaseous source with germanium from a periodically ablated target (Fig. 4.6c) [4.90]. NiSi-Si nanowires have been successfully synthesized which directly incorporate a nanowire metal contact into active nanowire devices [4.93]. Finally, using an ultrahigh vacuum chamber and molecular beams, InAs/InP nanowires with atomically sharp interfaces were obtained [4.94]. These compositionally-modulated nanowires are expected to exhibit exciting electronic, photonic, and thermoelectric properties.

Interestingly, silicon and germanium nanowires grown by the VLS method consist of a crystalline core coated with a relatively thick amorphous oxide layer (2–3 nm) (Fig. 4.6b). These layers are too thick to be the result of ambient oxidation, and it has been shown that these oxides play an important role in the nanowire growth process [4.49, 95]. Silicon oxides were found to serve as a special and highly selective catalyst that significantly enhances the yield of Si nanowires without the need for metal catalyst particles [4.49, 95, 96]. A similar yield enhancement was also found in the synthesis of Ge nanowires from the laser ablation of Ge powder mixed with GeO_2 [4.35]. The Si and Ge nanowires produced from these metal-free targets generally grow along the [112] crystal direction [4.97], and have the benefit that no catalyst clusters are found on either ends of the nanowires. Based on these observations and other TEM studies [4.35, 95, 97], an oxide-enhanced nanowire growth mechanism different from the classical VLS mechanism was proposed, where no metal catalyst is required during the laser ablation-assisted synthesis [4.95]. It is postulated that the nanowire growth is dependent on the presence of SiO (or GeO) vapor, which decomposes in the nanowire tip region into both Si (or Ge), which is incorporated into the crystalline phase, and SiO_2 (or GeO_2), which contributes to the outer coating. The initial nucleation

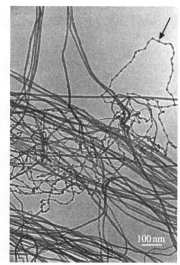

Fig. 4.7 TEM image showing the two major morphologies of Si nanowires prepared by the oxide-assisted growth method (after [4.95]). Notice the absence of metal particles when compared to Fig. 4.6a. The *arrow* points at an oxide-linked chain of Si nanoparticles

events generate oxide-coated spherical nanocrystals. The [112] crystal faces have the fastest growth rate, and therefore the nanocrystals soon begin elongating along this direction to form one-dimensional structures. The Si_mO or Ge_mO ($m > 1$) layer on the nanowire tips may be in or at temperatures near their molten states, catalyzing the incorporation of gas molecules in a directional fashion [4.97]. Besides nanowires with smooth walls, a second morphology of chains of unoriented nanocrystals linked by oxide necks is frequently observed (indicated by an arrow in Fig. 4.7). In addition, it was found by STM studies that about 1% of the wires consist of a regular array of two alternating segments, 10 and 5 nm in length, respectively [4.98]. The segments, whose junctions form an angle of 30°, are probably a result of alternating growth along different crystallographic orientations [4.98]. Branched and hyperbranched Si nanowire structures have also been synthesized by *Whang* et al. [4.99].

4.1.3 Other Synthesis Methods

In this section we review several other general procedures available for the synthesis of a variety of nanowires. We focus on *bottom-up* approaches, which afford many kinds of nanowires in large numbers, and do not require highly sophisticated equipment (such as scanning microscopy or lithography-based methods), and exclude cases for which the nanowires are not self-sustained (such as in the case of atomic rows on the surface of crystals).

A solution-phase synthesis of nanowires with controllable diameters has been demonstrated [4.45, 101], without the use of templates, catalysts, or surfactants. Instead, *Gates* et al. make use of the anisotropy of the crystal structure of trigonal selenium and tellurium, which can be viewed as rows of 1-D helical atomic chains. Their approach is based on the mass transfer of atoms during an aging step from a high free-energy solid phase (e.g., amorphous selenium) to a seed (e.g., trigonal selenium nanocrystal) which grows preferentially along one crystallographic axis. The lateral dimension of the seed, which dictates the diameter of the nanowire, can be controlled by the temperature of the nucleation step. Furthermore, Se/Te alloy nanowires were synthesized by this method, and Ag_2Se compound nanowires were obtained by treating selenium nanowires with $AgNO_3$ [4.102–104]. In a separate work, tellurium nanowires were transformed into Bi_2Te_3 nanowires by their reaction with $BiPh_3$ [4.105].

More often, however, the use of surfactants is necessary to promote the anisotropic 1-D growth of nanocrystals. Solution phase synthetic routes have been optimized to produce monodispersed quantum dots, (zero-dimensional isotropic nanocrystals) [4.106]. Surfactants are necessary in this case to stabilize the interfaces of the nanoparticles and to retard oxidation and aggregation processes. Detailed studies on the effect of growth conditions revealed that they can be manipulated to induce a directional growth of the nanocrystals, usually generating nanorods (aspect ratio of ≈ 10), and in favorable cases, nanowires with high aspect ratios. Heath and LeGoues synthesized germanium nanowires by reducing a mixture of $GeCl_4$ and phenyl-$GeCl_3$ at high temperature and high pressure. The phenyl ligand was essential for the formation of high aspect ratio nanowires [4.33]. In growing CdSe nanorods [4.20], *Alivisatos* et al. used a mixture of two surfactants, whose concentration ratio influenced the structure of the nanocrystal. It is believed that different surfactants have different affinities, and different absorption rates, for the different crystal faces of CdSe, thereby regulating the growth rates of these faces. In the liquid phase synthesis of Bi nanowires, the additive $NaN(SiMe_3)_2$ induces the growth of nanowires oriented along the [110] crystal direction from small bismuth seed clusters, while water solely retarded the growth along the [001] direction, inducing the growth of hexagonal-plate particles [4.105]. A coordinating alkyl-diamine solvent was used to grow polycrystalline PbSe nanowires at low temperatures [4.42]. Here, the surfactant-induced directional growth is believed to occur through to the formation of organometallic complexes in which the bidentate ligand assumes the equatorial positions, thus hindering the ions from approaching each other in this plane. Additionally, the alkyl-diamine molecules coat the external surface of the wire, preventing lateral growth. The aspect ratio of the wires increased as the temperature was lowered in the range $10\,°C < T < 117\,°C$. Ethylenediamine was used to grow CdS nanowires and tetrapods by a solvothermal recrystallization process starting with CdS nanocrystals or amorphous particles [4.17]. While the coordinating solvent was crucial for the nanowire growth, its role in the shape and phase control was not clarified.

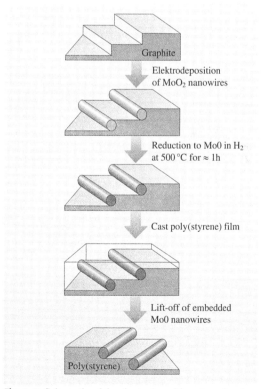

Fig. 4.8 Schematic of the electrodeposition step edge decoration of HOPG (highly oriented pyrolytic graphite) for the synthesis of molybdenum nanowires (after [4.38, 100])

Stress-induced crystalline bismuth nanowires have been grown from sputtered films of layers of Bi and CrN. The nanowires presumably grow from defects and cleavage fractures in the film, and are up to several millimeters in length with diameters ranging from 30 to 200 nm [4.7]. While the exploration of this technique has only begun, stress-induced unidirectional growth should be applicable to a variety of composite films.

Selective electrodeposition along the step edges in highly oriented pyrolytic graphite (HOPG) was used to obtain MoO_2 nanowires as shown in Fig. 4.8. The site-selectivity was achieved by applying a low overpotential to the electrochemical cell in which the HOPG served as cathode, thus minimizing the nucleation events on less favorable sites (plateaux). While these nanowires cannot be removed from the substrate, they can be reduced to metallic molybdenum nanowires, which can then be released as free-standing nanowires. Other metallic nanowires were also obtained by this method [4.38, 100]. In contrast to the template synthesis approaches described above, in this method the substrate only defines the position and orientation of the nanowire, not its diameter. In this context, other surface morphologies, such as self-assembled grooves in etched crystal planes, have been used to generate nanowire arrays via gas-phase shadow deposition (for example: Fe nanowires on (110)NaCl [4.27]). The cross section of artificially prepared superlattice structures has also been used for site-selective deposition of parallel and closely spaced nanowires [4.109]. Nanowires prepared on the above-mentioned substrates would have semicircular, rectangular, or other unconventional cross-sectional shapes.

4.1.4 Hierarchical Arrangement and Superstructures of Nanowires

Ordering nanowires into useful structures is another challenge that needs to be addressed in order to harness the full potential of nanowires for applications. We will first review examples of nanowires with nontrivial structures, and then proceed to describe methods used to create assemblies of nanowires of a predetermined structure.

We mentioned in Sect. 4.1.2 that the preparation of nanowires with a graded composition or with a superlattice structure along their main axis was demonstrated by controlling the gas phase chemistry as a function of time during the growth of the nanowires by the VLS method. Control of the composition along the axial dimension was also demonstrated by a template-assisted method, for example by the consecutive electrochemical deposition of different metals in the pores of an alumina template [4.110]. Alternatively, the composition can be varied along the radial dimension of the nanowire, for example by first growing a nanowire by the VLS method and then switching the synthesis conditions to grow a different material on the surface of the nanowire by CVD. This technique was demonstrated for the synthesis of Si/Ge and Ge/Si coaxial (or core–shell) nanowires [4.111], and it was shown that the outer shell can be formed epitaxially on the inner core by a thermal annealing process. *Han* et al. demonstrated the versatility of MgO nanowire arrays grown by the VLS method as templates for the PLD deposition of oxide coatings to yield MgO/YBCO, MgO/LCMO, MgO/PZT and MgO/Fe_3O_4 core/shell nanowires, all exhibiting epitaxial growth of the shell on the MgO core [4.37]. A different approach was adopted by *Wang* et al. who generated a mixture of coaxial and biaxial SiC-SiO$_x$ nanowires by the catalyst-free high-temperature reaction of amorphous silica and a carbon/graphite mixture [4.112].

A different category of nontrivial nanowires is that of nanowires with a nonlinear structure, resulting from multiple one-dimensional growth steps. Members of this category are tetrapods, which were mentioned in the context of the liquid phase synthesis (Sect. 4.1.3).

   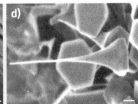

Fig. 4.9a–d SEM images of (**a**) sixfold- (**b**) fourfold- and (**c**) twofold-symmetry nanobrushes made of an In_2O_3 core and ZnO nanowire brushes (after [4.107]), and of (**d**) ZnO nanonails (after [4.108]). The scale bars are (**a**) 1 μm, (**b**) 500 nm, (**c**) 500 nm, and (**d**) 200 nm

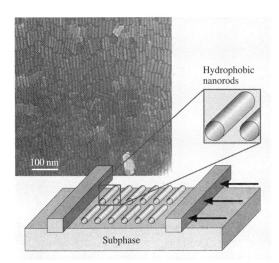

Fig. 4.10 A TEM image of a smectic phase of a BaCrO$_4$ nanorod film (*left inset*) achieved by the Langmuir–Blodgett technique, as depicted by the illustration (after [4.113])

In this process, a tetrahedral quantum dot core is first grown, and then the conditions are modified to induce one-dimensional growth of a nanowire from each one of the facets of the tetrahedron. A similar process produced high-symmetry In$_2$O$_3$/ZnO hierarchical nanostructures. From a mixture of heat-treated In$_2$O$_3$, ZnO, and graphite powders, faceted In$_2$O$_3$ nanowires were first obtained, on which oriented shorter ZnO nanowires were crystallized [4.107]. Brushlike structures were obtained as a mixture of 11 structures of different symmetries. For example, two, four, or six rows of ZnO nanorods could be found on different core nanowires, depending on the crystallographic orientation of the main axis of the core nanowire, as shown in Fig. 4.9. Comblike structures made entirely of ZnO were also reported [4.54].

Controlling the position of a nanowire in the growth process is important for preparing devices or test structures containing nanowires, especially when it involves a large array of nanowires. Post-synthesis methods to align and position nanowires include microfluidic channels [4.114], Langmuir–Blodgett assemblies [4.113], and electric field-assisted assembly [4.115]. The first method involves the orientation of the nanowires by the liquid flow direction when a nanowire solution is injected into a microfluidic channel assembly and by the interaction of the nanowires with the side walls of the channel. The second method involves the alignment of nanowires at a liquid–gas or liquid–liquid interface by the application of compressive forces on the interface (Fig. 4.10). The aligned nanowire films can then be transferred onto a substrate and lithography methods can be used to define interconnects. This allows the nanowires to be organized with a controlled alignment and spacing over large areas. Using this method, centimeter-scale arrays containing thousands of single silicon nanowire field-effect transistors with high performance could be assembled to make large-scale nanowire circuits and devices [4.99, 116]. The third technique is based on dielectrophoretic forces that pull polarizable nanowires toward regions of high field strength. The nanowires align between two isolated electrodes which are capacitatively coupled to a pair of buried electrodes biased with an AC voltage. Once a nanowire shorts the electrodes, the electric field is eliminated, preventing more nanowires from depositing. The above techniques have been successfully used to prepare electronic circuitry and optical devices out of nanowires (Sects. 4.3.1 and 4.3.3). Alternatively, alignment and positioning of the nanowires can be specified and controlled during their growth by the proper design of the synthesis method. For example, ZnO nanowires prepared by the VLS method were grown into an array in which both their position on the substrate and their growth direction and orientation were controlled [4.54]. The nanowire growth region was defined by patterning the gold film, which serves as a catalyst for the ZnO nanowire growth, employing soft-lithography, e-beam lithography, or photolithography. The orientation of the nanowires was achieved by selecting a substrate with a lattice structure matching that of the nanowire material to facilitate the epitaxial growth. These conditions result in an array of nanowire posts at predetermined positions, all vertically aligned with the same crystal growth orientation (Fig. 4.11). Similar rational GaN nanowire arrays have been synthesized epitaxially on (100)LiAlO$_2$ and (111)MgO single-crystal substrates. In addition, control over the crystallographic growth directions of nanowires was achieved by lattice-matching to different substrates. For example, GaN nanowires on (100)LiAlO$_2$ substrates grow oriented along the [110] direction, whereas (111)MgO substrates result in the growth of GaN nanowires with an [001] orientation, due to the different lattice-matching constraints [4.117]. A similar structure could be obtained by the template-mediated electrochemical synthesis of nanowires (Sect. 4.1.1), particularly if anodic alumina with its parallel and ordered channels is used. The control over the location

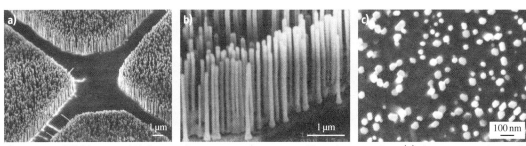

Fig. 4.11a-c SEM images of ZnO nanowire arrays grown on a sapphire substrate, where (**a**) shows patterned growth, (**b**) shows a higher resolution image of the parallel alignment of the nanowires, and (**c**) shows the faceted side-walls and the hexagonal cross section of the nanowires. For nanowire growth, the sapphire substrates were coated with a 1.0–3.5 nm thick patterned layer of Au as the catalyst, using a TEM grid as the shadow mask. These nanowires have been used for nanowire laser applications (after [4.122])

of the nucleation of nanowires in the electrochemical deposition is determined by the pore positions and the back-electrode geometry. The pore positions can be precisely controlled by imprint lithography [4.118].

By growing the template on a patterned conductive substrate that serves as a back-electrode [4.119–121] different materials can be deposited in the pores at different regions of the template.

4.2 Characterization and Physical Properties of Nanowires

In this section we review the structure and properties of nanowires and their interrelationship. The discovery and investigation of nanostructures were spurred on by advances in various characterization and microscopy techniques that enabled material characterization to take place at smaller and smaller length scales, reaching length scales down to individual atoms. For applications, characterizing the structural properties of nanowires is especially important, so that a reproducible relationship between their desired functionality and their geometrical and structural characteristics can be established. Due to the enhanced surface-to-volume ratio in nanowires, their properties may depend sensitively on their surface conditions and geometrical configurations. Even nanowires made of the same material may possess dissimilar properties due to differences in their crystal phase, crystalline size, surface conditions, and aspect ratios, which depend on the synthesis methods and conditions used in their preparation.

4.2.1 Structural Characterization

Structural and geometric factors play an important role in determining the various attributes of nanowires, such as their electrical, optical and magnetic properties. Therefore, various novel tools have been developed and employed to obtain this important structural information at the nanoscale. At the micrometer scale, optical techniques are extensively used for imaging structural features. Since the sizes of nanowires are usually comparable to or, in most cases, much smaller than the wavelength of visible light, traditional optical microscopy techniques are usually limited when characterizing the morphology and surface features of nanowires. Therefore, electron microscopy techniques play a more dominant role at the nanoscale. Since electrons interact more strongly than photons, electron microscopy is particularly sensitive relative to x-rays for the analysis of tiny samples.

In this section we review and give examples of how scanning electron microscopy, transmission electron microscopy, scanning probe spectroscopies, and diffraction techniques are used to characterize the structures of nanowires. To provide the necessary basis for developing reliable structure–property relations, multiple characterization tools are applied to the same samples.

Scanning Electron Microscopy
SEM usually produces images down to length scales of ≈ 10 nm and provides valuable information regarding the structural arrangement, spatial distribution, wire density, and geometrical features of the nanowires. The examples of SEM micrographs shown in Figs. 4.1 and 4.3 indicate that structural features at the 10 nm to

10 μm length scales can be probed, providing information on the size, size distribution, shapes, spatial distributions, density, nanowire alignment, filling factors, granularity, etc.. As another example, Fig. 4.11a shows an SEM image of ZnO nanowire arrays grown on a sapphire substrate [4.122], which provides evidence for the nonuniform spatial distribution of the nanowires on the substrate, which was attained by patterning the catalyst film to define high-density growth regions and nanowire-free regions. Figure 4.11b, showing a higher magnification of the same system, indicates that these ZnO nanowires grow perpendicular to the substrate, are well-aligned with approximately equal wire lengths, and have wire diameters in the range $20 \leq d_W \leq 150$ nm. The SEM micrograph in Fig. 4.11c provides further information about the surface of the nanowires, showing it to be well-faceted, forming a hexagonal cross section, indicative of nanowire growth along the $\langle 0001 \rangle$ direction. Both the uniformity of the nanowire size, their alignment perpendicular to the substrate, and their uniform growth direction, as suggested by the SEM data, are linked to the good epitaxial interface between the (0001) plane of the ZnO nanowire and the (110) plane of the sapphire substrate. (The crystal structures of ZnO and sapphire are essentially incommensurate, with the exception that the a-axis of ZnO and the c-axis of sapphire are related almost exactly by a factor of 4, with a mismatch of less than 0.08% at room temperature [4.122].) The well-faceted nature of these nanowires has important implications for their lasing action (Sect. 4.3.2). Figure 4.12 shows an SEM image of GaN nanowires synthesized by a laser-assisted catalytic growth method [4.30], indicating a random spatial orientation of the nanowire axes and a wide diameter distribution for these nanowires, in contrast to the ZnO wires in Fig. 4.11 and to arrays of well-aligned nanowires prepared by template-assisted growth (Fig. 4.3).

Transmission Electron Microscopy

TEM and high-resolution transmission electron microscopy (HRTEM) are powerful imaging tools for studying nanowires at the atomic scale, and they usually provide more detailed geometrical features than are seen in SEM images. TEM studies also yield information regarding the crystal structure, crystal quality, grain size, and crystal orientation of the nanowire axis. When operating in the diffraction mode, selected area electron diffraction (SAED) patterns can be made to determine the crystal structures of nanowires. As an example, the TEM images in Fig. 4.13 show four different morphologies for Si nanowires prepared by the laser ablation of a Si target [4.123]: (a) spring-shaped; (b) fishbone-shaped (indicated by solid arrow) and frogs egg-shaped (indicated by the hollow arrow), (c) pearl-shaped, while (d) shows the poly-sites of nanowire nucleation. The crystal quality of nanowires is revealed from high-resolution TEM images with atomic resolution, along with selected area electron diffraction (SAED) patterns. For example, Fig. 4.14 shows a TEM image of one of the GaN nanowires from Fig. 4.12, indicating single crystallinity and showing (100) lattice planes, thus indicating the growth direction of the nanowire. This information is supplemented by the corresponding electron diffraction pattern in the upper right. A more comprehensive review of the application of TEM for growth orientation indexing and crystal defect characterization in nanowires is available elsewhere [4.124].

The high resolution of the TEM also permits surface structures of the nanowires to be studied. In many cases, the nanowires are sheathed with a native oxide layer, or an amorphous oxide layer that forms during the growth process. This can be seen in Fig. 4.6b for silicon nanowires and in Fig. 4.15 for germanium nanowires [4.35], showing a mass–thickness contrast TEM image and a selected-area electron diffraction pattern of a Ge nanowire. The main TEM image shows that these Ge nanowires possess an amorphous GeO_2 sheath with a crystalline Ge core that is oriented in the [211] direction.

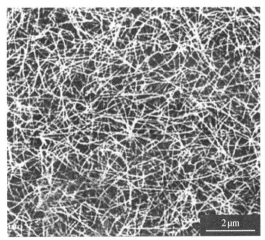

Fig. 4.12 SEM image of GaN nanowires in a mat arrangement synthesized by laser-assisted catalytic growth. The nanowires have diameters and lengths on the order of 10 nm and 10 μm, respectively (after [4.30])

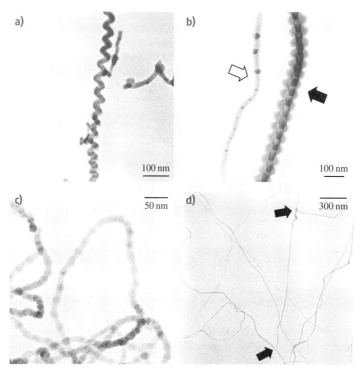

Fig. 4.13a–d TEM morphologies of four special forms of Si nanowires synthesized by the laser ablation of a Si powder target. (**a**) A spring-shaped Si nanowire; (**b**) fishbone-shaped (indicated by a *solid arrow*) and frogs egg-shaped (indicated by a *hollow arrow*) Si nanowires; and (**c**) pearl-shaped nanowires, while (**d**) shows polysites for the nucleation of silicon nanowires (indicated by *arrows*) (after [4.123])

Fig. 4.14 Lattice-resolved high-resolution TEM image of one GaN nanowire (*left*) showing that (100) lattice planes are visible perpendicular to the wire axis. The electron diffraction pattern (*top right*) was recorded along the [001] zone axis. A lattice-resolved TEM image (*lower right*) highlights the continuity of the lattice up to the nanowire edge, where a thin native oxide layer is found. The directions of various crystallographic planes are indicated in the *lower right figure* (after [4.30]) ▶

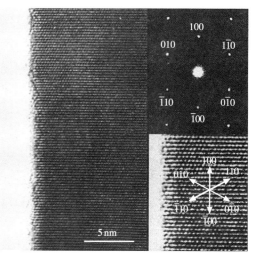

Dynamical processes of the surface layer of nanowires can be studied in-situ using an environmental TEM chamber, which allows TEM observations to be made while different gases are introduced or as the sample is heat-treated at various temperatures, as illustrated in Fig. 4.16. The figure shows high-resolution TEM images of a Bi nanowire with an oxide coating and the effect of a dynamic oxide removal process carried out within the environmental chamber of the TEM [4.125]. The amorphous bismuth-oxide layer coating the nanowire (Fig. 4.16a) is removed by exposure to hydrogen gas within the environmental chamber of the TEM, as indicated in Fig. 4.16b.

By coupling the powerful imaging capabilities of TEM with other characterization tools, such as an electron energy loss spectrometer (EELS) or an energy dispersive x-ray spectrometer (EDS) within the

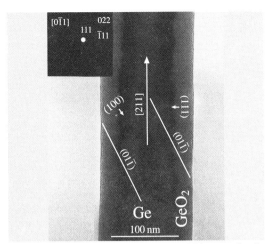

Fig. 4.15 A mass–thickness contrast TEM image of a Ge nanowire taken along the [0$\bar{1}$1] zone axis and a selected-area electron diffraction pattern (*upper left inset*) (after [4.35]). The Ge nanowires were synthesized by laser ablation of a mixture of Ge and GeO$_2$ powder. The core of the Ge nanowire is crystalline, while the surface GeO$_2$ is amorphous

TEM instrument, additional properties of the nanowires can be probed with high spatial resolution. With the EELS technique, the energy and momentum of the incident and scattered electrons are measured in an inelastic electron scattering process to provide information on the energy and momentum of the excitations in the nanowire sample. Figure 4.17 shows the dependence on nanowire diameter of the electron energy loss spectra of Bi nanowires. The spectra were taken from the center of the nanowire, and the shift in the energy of the peak position (Fig. 4.17) indicates the effect of the nanowire diameter on the plasmon frequency in the nanowires. The results show that there are changes in the electronic structure of the Bi nanowires as the wire diameter decreases [4.126]. Such changes in electronic structure as a function of nanowire diameter are also observed in their transport (Sect. 4.2.2) and optical (Sect. 4.2.3) properties, and are related to quantum confinement effects.

EDS measures the energy and intensity distribution of x-rays generated by the impact of the electron beam on the surface of the sample. The elemental composition within the probed area can be determined to a high degree of precision. The technique was particularly useful for the compositional characterization of superlattice

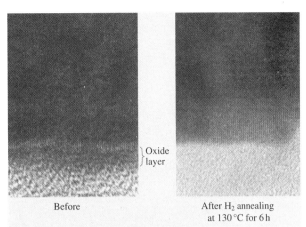

Fig. 4.16 High-resolution transmission electron microscope (HRTEM) image of a Bi nanowire (*left*) before and (*right*) after annealing in hydrogen gas at 130 °C for 6 h within the environmental chamber of the HRTEM instrument to remove the oxide surface layer (after [4.125])

nanowires [4.90] and core–shell nanowires [4.111] (Sect. 4.1.2).

Scanning Tunneling Probes

Several scanning probe techniques, such as scanning tunneling microscopy (STM) [4.127], electric

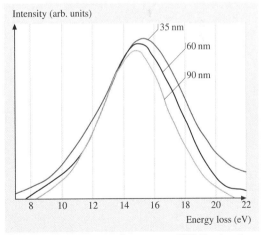

Fig. 4.17 Electron energy loss spectra (EELS) taken from the centers of bismuth nanowires with diameters of 35, 60 and 90 nm. The shift in the volume plasmon peaks is due to the effect of wire diameter on the electronic structure (after [4.126])

field gradient microscopy (EFM) [4.13], magnetic field microscopy (MFM) [4.40], and scanning thermal microscopy (SThM) [4.128], combined with atomic force microscopy (AFM), have been employed to study the structural, electronic, magnetic, and thermal properties of nanowires. A scanning tunneling microscope can be employed to reveal both topographical structural information, such as that illustrated in Fig. 4.18, as well as information on the local electronic density of states of a nanowire, when used in the STS (scanning tunneling spectroscopy) mode. Figure 4.18 shows STM height images (taken in the constant current STM mode) of MoSe molecular wires deposited from a methanol or acetonitrile solution of $Li_2Mo_6Se_6$ onto Au substrates. The STM image of a single MoSe wire (Fig. 4.18a) exhibits a 0.45 nm lattice repeat distance in a MoSe molecular wire. When both STM and STS measurements are made on the same sample, the electronic and structural properties can be correlated, as for example in the joint STM/STS studies on Si nanowires [4.98], showing alternating segments of a single nanowire identified with growth along the [110] and [112] directions, and different I–V characteristics measured for the [110] segments as compared with the [112] segments.

Magnetic field microscopy (MFM) has been employed to study magnetic polarization of magnetic nanowires embedded in an insulating template, such as an anodic alumina template. For example, Fig. 4.19a shows the topographic image of an anodic alumina template filled with Ni nanowires, and Fig. 4.19b demonstrates the corresponding magnetic polarization of each nanowire in the template. This micrograph shows that a magnetic field microscopy probe can distinguish between spin-up and spin-down nanowires in the nanowire array, thereby providing a method for measuring interwire magnetic dipolar interactions [4.40].

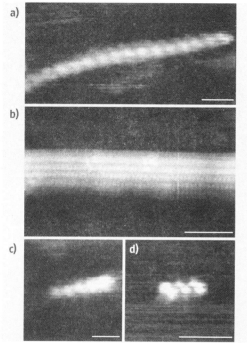

Fig. 4.18a–d STM height images, obtained in the constant current mode, of MoSe chains deposited on an Au(111) substrate. (**a**) A single chain image, and (**b**) a MoSe wire bundle. (**c,d**) Images of MoSe wire fragments containing five and three unit cells, respectively (after [4.127]). The scale bars are all 1 nm

X-ray Analysis

Other characterization techniques that are commonly used to study the crystal structures and chemical compositions of nanowires include x-ray diffraction and x-ray energy dispersion analysis (EDAX). The peak po-

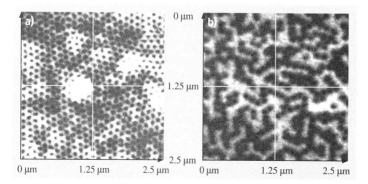

Fig. 4.19 (**a**) Topographic image of a highly-ordered porous alumina template with a period of 100 nm filled with 35 nm diameter nickel nanowires. (**b**) The corresponding MFM (magnetic force microscope) image of the nanomagnet array, showing that the pillars are magnetized alternately *up* (*white*) and *down* (*black*) (after [4.40])

sitions in the x-ray diffraction pattern can be used to determined the chemical composition and the crystal phase structure of the nanowires. For example, Fig. 4.2 shows that Bi nanowires have the same crystal structure and lattice constants as bulk bismuth. Both the x-ray diffraction pattern (XRD) for an array of aligned Bi nanowires (Fig. 4.2) and the SAED pattern for individual Bi nanowires [4.13] suggest that the nanowires have a common axis of nanowire alignment.

As another example of an XRD pattern for an array of aligned nanowires, Fig. 4.20 shows the x-ray diffraction pattern of the ZnO nanowires that are displayed in Fig. 4.11. Only (00ℓ) diffraction peaks are observed for these aligned ZnO nanowires, indicating that their preferred growth direction is (001) along the wire axis. Similarly, XRD was used to confirm the different growth directions of GaN nanowire array grown epitaxially on (100)LiAlO$_2$ and (111)MgO substartes [4.117].

EDAX has been used to determine the chemical compositions and stoichiometries of compound nanowires or impurity contents in nanowires. However, the results from EDAX analysis should be interpreted carefully to avoid systematic errors.

4.2.2 Mechanical Properties

Thermal Stability

Due to the large surface area-to-volume ratio in nanowires and other nanoparticles, the thermal stability of nanowires is anticipated to differ significantly from that of the bulk material. Theoretical studies of materials in confined geometries show that the melting point of the material is reduced in nanostructures, as is the latent heat of fusion, and that large hysteresis can be observed in melting–freezing cycles. These phenomena have been studied experimentally in three types of nanowire systems: porous matrices impregnated with a plurality of nanowires, individual nanowires sheathed by a thin coating, and individual nanowires.

The melting freezing of matrix-supported nanowires can be studied by differential scanning calorimetry (DSC), since large volumes of samples can thus be produced. *Huber* et al. investigated the melting of indium in porous silica glasses with mean pore diameters ranging from 6 to 141 nm [4.129]. The melting point of the pore-confined indium shows a linear dependence on inverse pore diameter, with a maximum melting point depression of 50 K. They also recorded a 6 K difference in the melting temperature and the freezing temperature of 12.8 nm diameter indium. The melting profile

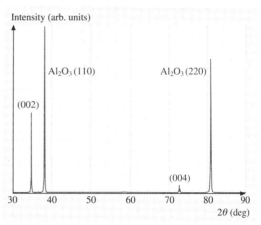

Fig. 4.20 X-ray diffraction pattern of aligned ZnO nanowires (Fig. 4.11) grown on a sapphire substrate. Only $[00\ell]$ diffraction peaks are observed for the nanowires, owing to their well-oriented growth orientation. Strong diffraction peaks for the sapphire substrate are found (after [4.122])

of the pore-confined indium in these samples is broader in temperature than for bulk indium, as expected for the heterogeneity in the pore diameter and in the indium crystal size aspect ratio within the samples.

Sheathed nanowires provide an opportunity to study the melting and recrystallization of individual nanowires. The shell layer surrounding the nanowire provides confinement to keep the liquid phase within the inner cylindrical volume. However, the shell–nanowire surface interaction should be taken into account when analyzing the phase transition thermodynamics and kinetics. *Yang* et al. produced germanium nanowires coated with a thin (1–5 nm) graphite sheath, by pyrolysis of organic molecules over VLS-grown nanowires, and followed the melting and recrystallization of the germanium by variable temperature TEM imaging [4.130]. The melting of the nanowires was followed by the disappearance of the electronic diffraction pattern. It was found that the nanowires began melting from their ends, with the melting front advancing towards the center of the nanowire as the temperature was increased. During the cool-down part of the cycle, the recrystallization of the nanowire occurred instantaneously following significant supercooling. The authors report both the largest melting point suppression recorded thus far for germanium ($\approx 300\,°C$), and a large melting–recrystallization hysteresis of up to $\approx 300\,°C$. Similarly, carbon nanotubes have been filled with various low-temperature metals [4.131]. A nanothermometer has been demonstrated using a 10 nm liquid gallium filled-carbon nanotube, showing an expansion coefficient that is linear in temperature and identical to the bulk value [4.132].

A different behavior was observed in free-standing copper nanowires [4.134]. In this system, there is little interaction between the nanowire surface and the surroundings, and the nanowire is not confined in its diameter, as in the case of the sheathed nanowires. Thermal treatment of the free-standing nanowires leads to their fragmentation into a linear array of metal spheres. Thinner nanowires were more vulnerable than thicker nanowires to the thermal treatment, showing constrictions and segmentation at lower temperatures. Analysis of the temperature response of the nanowires indicates that the nanowire segmentation is a result of the Rayleigh instability, starting with oscillatory perturbations of the nanowire diameter, leading to long cylindrical segments, that become more separated and more spherical at higher temperatures. These observations indicate that annealing and melting are dominated by the surface diffusion of atoms on the entire surface of the nanowire (versus tip-initiated melting).

4.2.3 Transport Properties

The study of electrical transport properties of nanowires is important for nanowire characterization, electronic device applications, and the investigation of unusual transport phenomena arising from one-dimensional quantum effects. Important factors that determine the transport properties of nanowires include the wire diameter, (important for both classical and quantum size effects), material composition, surface conditions, crystal quality, and the crystallographic orientation along the wire axis for materials with anisotropic material parameters, such as the effective mass tensor, the Fermi surface, or the carrier mobility.

Electronic transport phenomena in low-dimensional systems can be roughly divided into two categories: ballistic transport and diffusive transport. Ballistic transport phenomena occur when the electrons can travel across the nanowire without any scattering. In this case, the conduction is mainly determined by the contacts between the nanowire and the external circuit, and the conductance is quantized into an integral number of universal conductance units $G_0 = 2e^2/h$ [4.135, 136]. Ballistic transport phenomena are usually observed in very short quantum wires, such as those produced using mechanically controlled break junctions (MCBJ) [4.137, 138] where the electron mean free path is much longer than the wire length and the conduction is a pure quantum phenomenon. To observe ballistic transport, the thermal energy must also obey the relation $k_B T \ll \varepsilon_j - \varepsilon_{j-1}$, where $\varepsilon_j - \varepsilon_{j-1}$ is the energy separation between subband levels j and $j-1$. On the other hand, for nanowires with lengths much larger than the carrier mean free path, the electrons (or holes) undergo numerous scattering events when they travel along the wire. In this case, the transport is in the diffusive regime, and the conduction is dominated by carrier scattering within the wires, due to phonons (lattice vibrations), boundary scattering, lattice and other structural defects, and impurity atoms.

Conductance Quantization in Metallic Nanowires

The ballistic transport of 1-D systems has been extensively studied since the discovery of quantized conductance in 1-D systems in 1988 [4.135, 136]. The phenomena of conductance quantization occur when the diameter of the nanowire is comparable to the electron Fermi wavelength, which is on the order of 0.5 nm for most metals [4.139]. Most conductance quantization experiments up to the present were performed by bringing together and separating two metal electrodes. As the two metal electrodes are slowly separated, a nanocontact is formed before it breaks completely (Fig. 4.21a), and conductance in integral multiple values of G_0 is observed through these nanocontacts. Figure 4.21b shows the conductance histogram built with 18 000 contact breakage curves between two gold electrodes at room temperature [4.133], with the electrode sep-

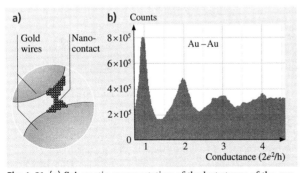

Fig. 4.21 (a) Schematic representation of the last stages of the contact breakage process (after [4.133]). (b) Histogram of conductance values built with 18 000 gold contact breakage experiments in air at room temperature, showing conductance peaks at integral values of G_0. In this experiment the gold electrodes approach and separate at 89 000 Å/s (after [4.133])

aration up to ≈ 1.8 nm. The conductance quantization behavior is found to be independent of the contact material, and has been observed in various metals, such as Au [4.133], Ag, Na, Cu [4.140], and Hg [4.141]. For semimetals such as Bi, conductance quantization has also been observed for electrode separations as long as 100 nm at 4 K because of the long Fermi wavelength (≈ 26 nm) [4.139], indicating that the conductance quantization may be due to the existence of well-defined quantum states localized at a constriction instead of resulting from the atom rearrangement as the electrodes separate. Since conductance quantization is only observed in breaking contacts, or for very narrow and very short nanowires, most nanowires of practical interest (possessing lengths of several micrometer) lie in the diffusive transport regime, where the carrier scattering is significant and should be considered.

I–V Characterization of Semiconducting Nanowires

The electronic transport behavior of nanowires may be categorized based on the relative magnitudes of three length scales: carrier mean free path ℓ_W, the de Broglie wavelength of electrons λ_e, and the wire diameter d_W. For wire diameters much larger than the carrier mean free path ($d_W \gg \ell_W$), the nanowires exhibit transport properties similar to bulk materials, which are independent of the wire diameter, since the scattering due to the wire boundary is negligible compared to other scattering mechanisms. For wire diameters comparable to or smaller than the carrier mean free path ($d_W \approx \ell_W$ or $d_W < \ell_W$), but still much larger than the de Broglie wavelength of the electrons ($d_W \gg \lambda_e$), the transport in nanowires is in the classical finite size regime, where the band structure of the nanowire is still similar to that of bulk, while the scattering events at the wire boundary alter their transport behavior. For wire diameters comparable to the electronic wavelength $d_W \approx \lambda_e$, the electronic density of states is altered dramatically and quantum subbands are formed due to the quantum confinement effect at the wire boundary. In this regime, the transport properties are further influenced by the change in the band structure. Therefore, transport properties for nanowires in the classical finite size and quantum size regimes are highly diameter-dependent.

Researchers have investigated the transport properties of various semiconducting nanowires and have demonstrated their potential for diverse electronic devices, such as for p-n diodes [4.142, 143], field effect transistors [4.142], memory cells, and switches [4.144] (Sect. 4.3.1). So far, the nanowires studied in this context have usually been made from conventional semiconducting materials, such as group IV and III–V compound semiconductors, via the VLS growth method (Sect. 4.1.2), and their nanowire properties have been compared to their well-established bulk properties. Interestingly, the physical principles for describing bulk semiconductor devices also hold for devices based on these semiconducting nanowires with wire diameters of tens of nanometers. For example, Fig. 4.22 shows the current–voltage (I–V) behavior of a 4-by-1 crossed p-Si/n-GaN junction array at room temperature [4.142]. The long horizontal wire in the figure is a p-Si nanowire (10–25 nm in diameter) and the four short vertical wires are n-GaN nanowires (10–30 nm in diameter). Each of the four nanoscale cross points independently forms a p-n junction with current rectification behavior, as shown by the I–V curves in Fig. 4.22, and the junction behavior (for example the turn-on voltage) can be controlled by varying the oxide coating on these nanowires [4.142].

Huang et al. have demonstrated nanowire junction diodes with a high turn-on voltage (≈ 5 V) by increasing the oxide thickness at the junctions. The high turn-on voltage enables the use of the junction in

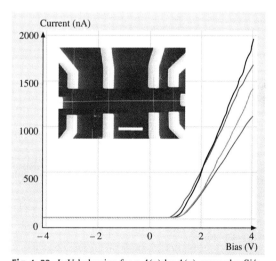

Fig. 4.22 I–V behavior for a 4(p) by 1(n) crossed p-Si/n-GaN junction array shown in the *inset*. The *four curves* represent the I–V response for each of the four junctions, showing similar current rectifying characteristics in each case. The length scale bar between the two *middle junctions* is 2 μm (after [4.142]). The p-Si and n-GaN nanowires are 10–25 and 10–30 nm in diameter, respectively

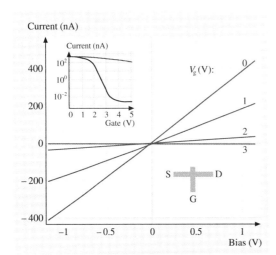

Fig. 4.23 Gate-dependent I–V characteristics of a crossed nanowire field-effect transistor (FET). The n-GaN nanowire is used as the nanogate, with the gate voltage indicated (0, 1, 2, and 3 V). The *inset* shows the current versus V_gate for a nanowire gate (*lower curve*) and for a global back-gate (*top curve*) when the bias voltage is set to 1 V (after [4.142])

a nanoscale FET, as shown in Fig. 4.23 [4.142] where I–V data for a p-Si nanowire are presented, for which the n-GaN nanowire with a thick oxide coating is used as a nanogate. By varying the nanogate voltage, the conductance of the p-Si nanowire can be changed by more than a factor of 10^5 (lower curve in the inset), whereas the conductance changes by only a factor of 10 when a global back-gate is used (top curve in the inset of Fig. 4.23). This behavior may be due to the thin gate dielectric between the crossed nanowires and the better control of the local carrier density through a nanogate. Based on the gate-dependent I–V data from these p-Si nanowires, it is found that the mobility of the holes in the p-Si nanowires may be higher than that for bulk p-Si, although further investigation is required for complete understanding.

Because of the enhanced surface-to-volume ratios of nanowires, their transport behavior may be modified by changing their surface conditions. For example, researchers have found that by coating n-InP nanowires with a layer of redox molecules, such as cobalt phthalocyanine, the conductance of the InP nanowires may change by orders of magnitude upon altering the charge state of the redox molecules to provide bistable nanoscale switches [4.144]. The resistance (or conductance) of some nanowires (such as Pd nanowires) is also very sensitive to the presence of certain gases (e.g., H_2) [4.145, 146], and this property may be utilized for sensor applications to provide improved sensitivity compared to conventional sensors based on bulk material (Sect. 4.3.4).

Although it remains unclear how the size effect may influence the transport properties and device performance of semiconducting nanowires, many of the larger diameter semiconducting nanowires are expected to be described by classical physics, since their quantization energies $\hbar^2/(2m_e d_W^2)$ are usually smaller than the thermal energy $k_B T$. By comparing the quantization energy with the thermal energy, the critical wire diameter below which quantum confinement effects become significant is estimated to be 1 nm for Si nanowires at room temperature, which is much smaller than the sizes of many of the semiconducting nanowires that have been investigated so far. By using material systems with much smaller effective carrier masses m_e (such as bismuth), the critical diameter for which such quantum effects can be observed is increased, thereby facilitating the study of quantum confinement effects. It is for this reason that the bismuth nanowire system has been studied so extensively. Furthermore, since the crystal structure and lattice constants of bismuth nanowires are the same as for 3-D crystalline bismuth, it is possible to carry out detailed model calculations to guide and to interpret transport and optical experiments on bismuth nanowires. For these reasons, bismuth can be considered a model system for studying 1-D effects in nanowires.

Temperature-Dependent Resistance Measurements

Although nanowires with electronic properties similar to their bulk counterparts are promising for constructing nanodevices based on well-established knowledge of their bulk counterparts, it is expected that quantum size effects in nanowires will likely be utilized to generate new phenomena absent in bulk materials, and thus provide enhanced performance and novel functionality for certain applications. In this context, the transport properties of bismuth (Bi) nanowires have been extensively studied, both theoretically [4.147] and experimentally [4.8, 10, 78, 148–150] because of their promise for enhanced thermoelectric performance. Transport studies of ferromagnetic nanowire arrays, such as Ni or Fe, have also received much attention because of their potential for high-density magnetic storage applications [4.151].

The very small electron effective mass components and the long carrier mean free paths in Bi facilitate the study of quantum size effects in the transport properties of nanowires. Quantum size effects are expected to become significant in bismuth nanowires with diameters smaller than 50 nm [4.147], and the fabrication of crystalline nanowires with this diameter range is relatively easy.

Figure 4.24a shows the T dependence of the resistance $R(T)$ for Bi nanowires ($7 \leq d_W < 200$ nm) synthesized by vapor deposition and pressure injection [4.8], illustrating the quantum effects in their temperature-dependent resistance. In Fig. 4.24a, the $R(T)$ behavior of Bi nanowires is dramatically different from that of bulk Bi, and is highly sensitive to the wire diameter. Interestingly, the $R(T)$ curves in Fig. 4.24a show a nonmonotonic trend for large-diameter (70 and 200 nm) nanowires, although $R(T)$ becomes monotonic with T for small-diameter (≤ 48 nm) nanowires. This dramatic change in the behavior of $R(T)$ as a function of d_W is attributed to a unique semimetal–semiconductor transition phenomena in Bi [4.78], induced by quantum size effects. Bi is a semimetal in bulk form, in which the T-point valence band overlaps with the L-point conduction band by 38 meV at 77 K. As the wire diameter decreases, the lowest conduction subband increases in energy and the highest valence subband decreases in energy. Model calculations predict that the band overlap should vanish in Bi nanowires (with their wire axes along the trigonal direction) at a wire diameter ≈ 50 nm [4.147].

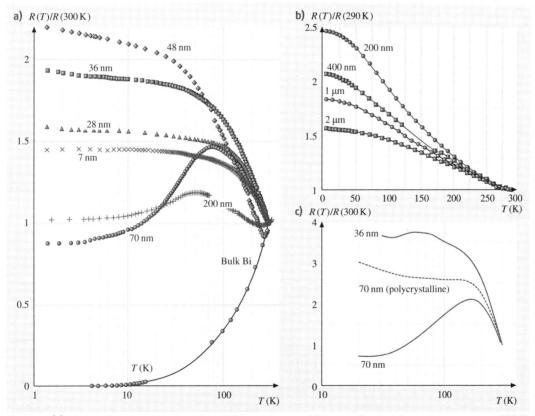

Fig. 4.24 (a) Measured temperature dependence of the resistance $R(T)$ normalized to the room temperature (300 K) resistance for bismuth nanowire arrays of various wire diameters d_W (after [4.8]). (b) $R(T)/R(290\,\text{K})$ for bismuth wires of larger d_W and lower mobility (after [4.10]). (c) Calculated $R(T)/R(300\,\text{K})$ of 36 and 70 nm bismuth nanowires. The *dashed curve* refers to a 70 nm polycrystalline wire with increased boundary scattering (after [4.78])

The resistance of Bi nanowires is determined by two competing factors: the carrier density that increases with T, and the carrier mobility that decreases with T. The nonmonotonic $R(T)$ for large-diameter Bi nanowires is due to a smaller carrier concentration variation at low temperature (≤ 100 K) in semimetals, so that the electrical resistance is dominated by the mobility factor in this temperature range. Based on the semi-classical transport model and the established band structure of Bi nanowires, the calculated $R(T)/R(300\,\text{K})$ for 36 and 70 nm Bi nanowires is shown by the solid curves in Fig. 4.24c to illustrate different $R(T)$ trends for semiconducting and semimetallic nanowires, respectively [4.78]. The curves in Fig. 4.24c exhibit trends consistent with experimental results. The condition for the semimetal–semiconductor transition in Bi nanowires can be experimentally determined, as shown by the measured resistance ratio $R(10\,\text{K})/R(100\,\text{K})$ of Bi nanowires as a function of wire diameter [4.152] in Fig. 4.25. The maximum in the resistance ratio $R(10\,\text{K})/R(100\,\text{K})$ at $d_\text{W} \approx 48$ nm indicates the wire diameter for the transition of Bi nanowires from a semimetallic phase to a semiconducting phase. The semimetal–semiconductor transition and the semiconducting phase in Bi nanowires are examples of new transport phenomena resulting from low dimensionality that are absent in the bulk 3-D phase, and these phenomena further increase the possible benefits from the properties of nanowires for desired applications (Sect. 4.3.2).

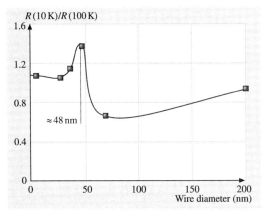

Fig. 4.25 Measured resistance ratio $R(10\,\text{K})/R(100\,\text{K})$ of Bi nanowire array as a function of diameter. The *peak* indicates the transition from a semimetallic phase to a semiconducting phase as the wire diameter decreases (after [4.153])

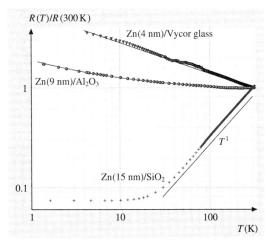

Fig. 4.26 Temperature dependence of the resistance of Zn nanowires synthesized by vapor deposition in various porous templates (after [4.52]). The data are given as *points*, the *full lines* are fits to a T^1 law for 15 nm diameter Zn nanowires in an SiO_2 template, denoted by Zn/SiO_2. Fits to a combined T^1 and $T^{-1/2}$ law were made for the smaller nanowire diameter composite samples denoted by Zn (9 nm)/Al_2O_3 and Zn 4 nm/Vycor glass

It should be noted that good crystal quality is essential for observing the quantum size effect in nanowires, as shown by the $R(T)$ plots in Fig. 4.24a. For example, Fig. 4.24b shows the normalized $R(T)$ measurements of Bi nanowires with larger diameters (200 nm–2 μm) prepared by electrochemical deposition [4.10], and these nanowires possess monotonic $R(T)$ behaviors, quite different from those of the corresponding nanowire diameters shown in Fig. 4.24a. The absence of the resistance maximum in Fig. 4.24b is due to the lower crystalline quality for nanowires prepared by electrochemical deposition, which tends to produce polycrystalline nanowires with a much lower carrier mobility. This monotonic $R(T)$ for semimetallic Bi nanowires with a higher defect level is also confirmed by theoretical calculations, as shown by the dashed curve in Fig. 4.24c for 70 nm wires with increased grain boundary scattering [4.154].

The theoretical model developed for Bi nanowires not only provides good agreement with experimental results, but it also plays an essential role in understanding the influence of the quantum size effect, the boundary scattering, and the crystal quality on their electrical properties. While the electronic density of states may be significantly altered due to quantum confinement effects, various scattering mechanisms related to the transport properties of nanowires can be accounted for by Matthiessen's rule. Furthermore, the transport model has also been generalized to predict the transport properties of Te-doped Bi nanowires [4.78], Sb nanowires [4.155], and BiSb alloy nanowires [4.156], and good agreement between experiment and theory has also been obtained for these cases.

For nanowires with diameters comparable to the phase-breaking length, their transport properties may be further influenced by localization effects. It has been predicted that in disordered systems, the extended electronic wavefunctions become localized near defect sites, resulting in the trapping of carriers and giving rise to different transport behavior. Localization effects are also expected to be more pronounced as the dimensionality and sample size are reduced. Localization effects on the transport properties of nanowire systems have been studied on Bi nanowires [4.158] and, more recently, on Zn nanowires [4.52]. Figure 4.26 shows the measured $R(T)/R(300\,\text{K})$ of Zn nanowires fabricated by vapor deposition in porous silica or alumina [4.52]. While 15 nm Zn nanowires exhibit an $R(T)$ behavior with a T^1 dependence as expected for a metallic wire, the $R(T)$ of 9 and 4 nm Zn nanowires exhibits a temperature dependence of $T^{-1/2}$ at low temperatures, consistent with 1-D localization theory. Thus, due to this localization effect, the use of nanowires with very small diameters for transport applications may be limited.

Magnetoresistance

Magnetoresistance (MR) measurements provide an informative technique for characterizing nanowires, because these measurements yield a great deal of information about the electron scattering with wire boundaries, the effects of doping and annealing on scattering, and localization effects in the nanowires [4.150]. For example, at low fields the MR data show a quadratic dependence on the B field from which carrier mobility estimates can be made (Fig. 4.27 at low B field).

Figure 4.27 shows the longitudinal magnetoresistance (B parallel to the wire axis) for 65 and 109 nm Bi nanowire samples (before thermal annealing) at 2 K. The MR maxima in Fig. 4.27a are due to the classical size effect, where the wire boundary scattering is reduced as the cyclotron radius becomes smaller than the wire radius in the high field limit, resulting in a decrease in the resistivity. This behavior is typical for the longitudinal MR of Bi nanowires in the diameter range of 45

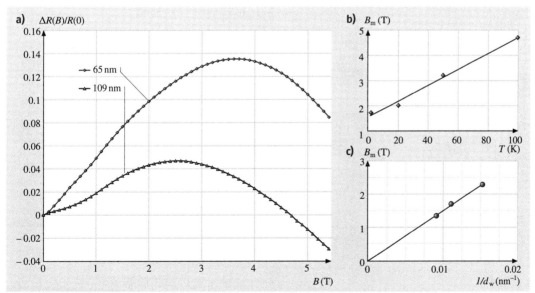

Fig. 4.27 (a) Longitudinal magnetoresistance, $\Delta R(B)/R(0)$, at 2 K as a function of B for Bi nanowire arrays with diameters of 65 and 109 nm before thermal annealing. **(b)** The peak position B_m as a function of temperature for the 109 nm diameter Bi nanowire array after thermal annealing. **(c)** The peak position B_m of the longitudinal MR (after thermal annealing) at 2 K as a function of $1/d_W$, the reciprocal of the nanowire diameter (after [4.157])

to 200 nm [4.8, 149, 150, 157], and the peak position B_m moves to lower B field values as the wire diameter increases, as shown in Fig. 4.27c [4.157], where B_m varies linearly with $1/d_W$. The condition for the occurrence of B_m is approximately given by $B_m \approx 2c\hbar k_F/ed_W$ where k_F is the wave vector at the Fermi energy. The peak position B_m is found to increase linearly with increasing temperature in the range of 2–100 K, as shown in Fig. 4.27b [4.157]. As T is increased, phonon scattering becomes increasingly important, and therefore a higher magnetic field is required to reduce the resistivity associated with boundary scattering sufficiently to change the sign of the MR. Likewise, increasing the grain boundary scattering is also expected to increase the value of B_m at a given T and wire diameter.

The presence of the peak in the longitudinal MR of nanowires requires a high crystal quality with long carrier mean free paths along the nanowire axis, so that most scattering events occur at the wire boundary instead of at a grain boundary, at impurity sites, or at defect sites within the nanowire. Liu et al. have investigated the MR of 400 nm Bi nanowires synthesized by electrochemical deposition [4.74], and no peak in the longitudinal MR is observed. The absence of a magnetoresistance peak may be attributed to a higher defect level in the nanowires produced electrochemically and to a large wire diameter, much longer than the carrier mean free path. The negative MR observed for the Bi nanowire arrays above B_m (Fig. 4.27) shows that wire boundary scattering is a dominant scattering process for the longitudinal magnetoresistance, thereby establishing that the mean free path is larger than the wire diameter and that a ballistic transport behavior is indeed observed in the high field regime.

In addition to the longitudinal magnetoresistance measurements, transverse magnetoresistance measurements ($\boldsymbol{B}$ perpendicular to the wire axis) have also been performed on Bi nanowire array samples [4.8, 150, 157], where a monotonically increasing B^2 dependence over the entire range $0 \leq B \leq 5.5$ T is found for all Bi nanowires studied thus far. This is as expected, since the wire boundary scattering cannot be reduced by a magnetic field perpendicular to the wire axis. The transverse magnetoresistance is also found to be always larger than the longitudinal magnetoresistance in nanowire arrays.

By applying a magnetic field to nanowires at very low temperatures (≤ 5 K), one can induce a transition from a 1-D confined system at low magnetic fields to a 3-D confined system as the field strength increases, as shown in Fig. 4.28 for the longitudinal MR of Bi nanowire arrays of various nanowire diameters

(28–70 nm) for $T < 5$ K [4.150]. In these curves, a subtle steplike feature is seen at low magnetic fields, which is found to depend only on the wire diameter, and is independent of temperature, the orientation of the magnetic field, and even on the nanowire material (see for example Sb nanowires [4.155]). The lack of a dependence of the magnetic field at which the step appears on temperature, field orientation, and material type indicates that the phenomenon is related to the magnetic field length $L_H = (\hbar/eB)^{1/2}$. The characteristic length L_H is the spatial extent of the wave function of electrons in the lowest Landau level, and L_H is independent of the carrier effective masses. Setting $L_H(B_c)$ equal to the diameter d_W of the nanowire defines a critical magnetic field strength B_c below which the wavefunction is con-

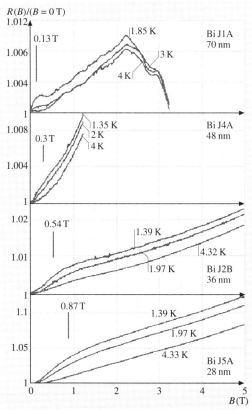

Fig. 4.28 Longitudinal magnetoresistance as a function of magnetic field for Bi nanowires of the diameters indicated. The *vertical bars* indicate the critical magnetic field B_c at which the magnetic length equals the nanowire diameter (after [4.150])

fined by the nanowire boundary (the 1-D regime), and above which the wavefunction is confined by the magnetic field (the 3-D regime). The physical basis for this phenomenon is associated with confinement of a single magnetic flux quantum within the nanowire cross section [4.150]. This phenomenon, though independent of temperature, is observed for $T \leq 5$ K, since the phase breaking length has to be larger than the wire diameter. This calculated field strength B_c indicated in Fig. 4.28 by vertical lines for the appropriate nanowire diameters, provides a good fit to the steplike features in these MR curves.

The Shubnikov–de Haas (SdH) quantum oscillatory effect, which results from the passage of the quantized Landau levels through the Fermi energy as the field strength varies, should, in principle, provide the most direct measurement of the Fermi energy and carrier density. For example, *Heremans* et al. have demonstrated that SdH oscillations can be observed in Bi nanowire samples with diameters down to 200 nm [4.159], and they have demonstrated that Te doping can be used to raise the Fermi energy in Bi nanowires. Such information on the Fermi energy is important because, for certain applications based on nanowires, it is necessary to place the Fermi energy near a subband edge where the density of states has a sharp feature. However, due to the unusual 1-D geometry of nanowires, other characterization techniques that are commonly used in bulk materials to determine the Fermi energy and the carrier concentration (such as Hall measurement) cannot be applied to nanowire systems. The observation of the SdH oscillatory effect requires crystal samples of very high quality which allow carriers to execute a complete cyclotron orbit in the nanowire before they are scattered. For small nanowire diameters, large magnetic fields are required to produce cyclotron radii smaller than the wire radius. For some nanowire systems, all Landau levels may have passed through the Fermi level at such a high field strength, and in such a case, no oscillations can be observed. The localization effect may also prevent the observation of SdH oscillations for very small diameter (≤ 10 nm) nanowires. Observing SdH oscillations in highly doped samples (as may be required for certain applications) may be difficult because impurity scattering reduces the mean free path, requiring high B fields to satisfy the requirement that carriers complete a cyclotron orbit prior to scattering. Therefore, although SdH oscillations provide the most direct method of measuring the Fermi energy and carrier density of nanowire samples, this technique may, however, not work for small-diameter nanowires, nor for nanowires that are heavily doped.

Thermoelectric Properties

Nanowires are predicted to hold great promise for thermoelectric applications [4.147, 160], due to their novel band structure compared to their bulk counterparts and the expected reduction in thermal conductivity associated with enhanced boundary scattering (see below). Due to the sharp density of states at the 1-D subband edges (where the van Hove singularities occur), nanowires are expected to exhibit enhanced Seebeck coefficients compared to their bulk counterparts. Since the Seebeck coefficient measurement is intrinsically independent of the number of nanowires contributing to the signal, the measurements on nanowire arrays of uniform wire diameter are, in principle, as informative as single-wire measurements. The major challenge with measuring the Seebeck coefficients of nanowires lies in the design of tiny temperature probes to accurately determine the temperature difference across the nanowire. Figure 4.29a shows the schematic experimental setup for the Seebeck coefficient measurement of nanowire arrays [4.161], where two thermocouples are placed on both faces of a nanowire array and a heater is attached to one face of the array to generate a temperature gradient along the nanowire axis. Ideally, the size of the thermocouples should be much smaller than the thickness of the nanowire array template (i.e. the nanowire length) to minimize error. However, due to the thinness of most templates ($\leq 50\,\mu$m) and the large size of commercially-available thermocouples ($\approx 12\,\mu$m), the measured Seebeck coefficient values are usually underestimated.

The thermoelectric properties of Bi nanowire systems have been investigated extensively because of their potential as good thermoelectric materials. Figure 4.29b shows the measured Seebeck coefficients $S(T)$ as a function of temperature for nanowire arrays with diameters of 40 and 65 nm and different isoelectronic Sb alloy concentrations [4.154], and $S(T)$ results for bulk Bi are shown (solid curve) for comparison. Thermopower enhancement is observed in Fig. 4.29b as the wire diameter decreases and as the Sb content increases, which is attributed to the semimetal–semiconductor transition induced by quantum confinement and to Sb alloying effects in $Bi_{1-x}Sb_x$ nanowires. *Heremans* et al. have observed a substantial increase in the thermopower of Bi nanowires as the wire diameter decreases further, as shown in Fig. 4.30a for Bi(15 nm)/silica and Bi(9 nm)/alumina nanocom-

Fig. 4.29 (a) Experimental setup for the measurement of the Seebeck coefficient in nanowire arrays (after [4.161]). (b) Measured Seebeck coefficient as a function of temperature for Bi (○, ▽) and $Bi_{0.95}Sb_{0.05}$ (●, ▼) nanowires with different diameters. The *solid curve* denotes the Seebeck coefficient for bulk Bi (after [4.154]) ▶

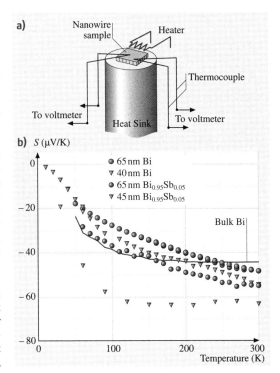

posites [4.52]. The enhancement is due to the sharp density of states near the Fermi energy in a 1-D system. Although the samples in Fig. 4.30a also possess very high electrical resistance ($\sim G\Omega$), the results for the Bi(9 nm)/alumina samples show that the Seebeck coefficient can be enhanced by almost 1000 times relative to bulk material. However, for Bi nanowires with very small diameters (≈ 4 nm), the localization effect becomes dominant, which compromises the thermopower enhancement. Therefore, for Bi nanowires, the optimal wire diameter range for the largest thermopower enhancement is found to be between 4 and 15 nm [4.52].

The effect of the nanowire diameter on the thermopower of nanowires has also been observed in Zn nanowires [4.52]. Figure 4.30b shows the Seebeck coefficient of Zn(9 nm)/alumina and Zn(4 nm)/Vycor glass nanocomposites, also exhibiting enhanced thermopower as the wire diameter decreases. It is found that while 9 nm Zn nanowires still exhibit metallic behavior,

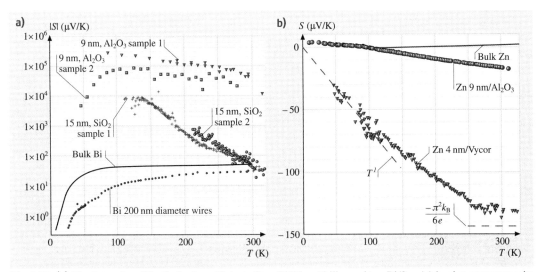

Fig. 4.30 (a) Absolute value of the Seebeck coefficient of two Bi(15 nm)/silica and two Bi(9 nm)/alumina nanocomposite samples, in comparison to bulk Bi and 200 nm Bi nanowires in the pores of alumina templates (after [4.52]). The *full line* on top part of the figure is a fit to a T^{-1} law. The Seebeck coefficient of the Bi(9 nm)/alumina composite is positive; the rest are negative. (b) The Seebeck coefficient of Zn(9 nm)/Al_2O_3 and Zn(4 nm)/Vycor glass nanocomposite samples in comparison to bulk Zn (after [4.52])

the thermopower of 4 nm Zn nanowires shows a different temperature dependence, which may be due to the 1-D localization effect, although further investigation is required for definitive identification of the conduction mechanism in such small nanowires.

Quantum Wire Superlattices

The studies on superlattice nanowires, which possess a periodic modulation in their materials composition along the wire axis, have attracted much attention recently because of their promise in various applications, such as thermoelectrics (Sect. 4.3.2) [4.90, 162], nanobarcodes (Sect. 4.3.3) [4.110], nanolasers (Sect. 4.3.3) [4.92], one-dimensional waveguides, and resonant tunneling diodes [4.94, 163]. Figure 4.31a shows a schematic structure of a superlattice nanowire consisting of interlaced quantum dots of two different materials, as denoted by A and B. Various techniques have been developed to synthesize superlattice nanowire structures with different interface conditions, as mentioned in Sects. 4.1.1 and 4.1.2.

In this superlattice (SL) nanowire structure, the electronic transport along the wire axis is made possible by the tunneling between adjacent quantum dots, while the uniqueness of each quantum dot and its 0-D characteristic behavior is maintained by the energy difference of the conduction or valence bands between quantum dots of different materials (Fig. 4.31b), which provides some amount of quantum confinement. Recently, Björk et al. have observed interesting nonlinear $I-V$ characteristics with a negative differential resistance in one-dimensional heterogeneous structures made of InAs and InP, where InP serves as the potential barrier [4.94, 163]. The nonlinear $I-V$ behavior is associated with the double barrier reso-

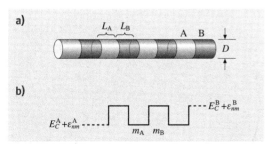

Fig. 4.31 (a) Schematic diagram of superlattice (segmented) nanowires consisting of interlaced nanodots A and B of the indicated length and wire diameter. (b) Schematic potential profile of the subbands in the superlattice nanowire (after [4.162])

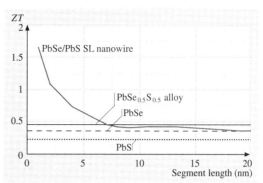

Fig. 4.32 Optimal ZT calculated as a function of segment length for 10 nm diameter PbSe/PbS nanowires at 77 K, where *optimal* refers to the placement of the Fermi level to optimize ZT. The optimal ZT for 10 nm diameter PbSe, PbS, and PbSe$_{0.5}$S$_{0.5}$ nanowires are 0.33, 0.22, and 0.48, respectively (after [4.153])

nant tunneling process in one-dimensional structures, demonstrating that transport phenomena occur in superlattice nanowires via tunneling and the possibility of controlling the electronic band structure of the SL nanowires by carefully selecting the constituent materials. This new kind of structure is especially attractive for thermoelectric applications, because the interfaces between the nanodots can reduce the lattice thermal conductivity by blocking the phonon conduction along the wire axis, while electrical conduction may be sustained and even benefit from the unusual electronic band structures due to the periodic potential perturbation. For example, Fig. 4.32 shows the calculated dimensionless thermoelectric figure of merit $ZT = S^2 \sigma T/\kappa$ (Sect. 4.3.2) where κ is the total thermal conductivity (including both the lattice and electronic contributions) of 10 nm diameter PbS/PbSe superlattice nanowires as a function of the segment length. A higher thermoelectric performance than for PbSe$_{0.5}$S$_{0.5}$ alloy nanowires can be achieved for a 10 nm diameter superlattice nanowire with segment lengths ≤ 7 nm. However, the localization effect, which may become important for very short segment lengths, may jeopardize this enhancement in the ZT of superlattice nanowires [4.153].

Thermal Conductivity of Nanowires

Experimental measurements of the temperature dependence of the thermal conductivity $\kappa(T)$ of individual suspended nanowires have been carried out on study the dependence of $\kappa(T)$ on wire diameter. In this context,

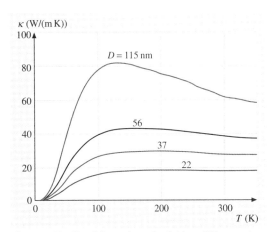

Fig. 4.33 Predicted thermal conductivities of Si nanowires of various diameters (after [4.168])

measurements have been made on nanowires down to only 22 nm in diameter [4.164]. Such measurements are very challenging and are now possible due to technological development in the micro- and nanofabrication of miniature thermal sensors, and the use of nanometer-size thermal scanning probes [4.128, 165, 166]. The experiments show that the thermal conductivity of small homogeneous nanowires may be more than one order of magnitude smaller than in the bulk, due mainly to strong boundary scattering effects [4.167]. Phonon confinement effects may eventually become important in nanowires with even smaller diameters. Measurements on mats of nanowires (Fig. 4.12) do not generally give reliable results because the contact thermal resistance between adjacent nanowires tends to be high, which is in part due to the thin surface oxide coating which most nanowires have. This surface oxide coating may also be important for thermal conductivity measurements on individual suspended nanowires because of the relative importance of phonon scattering at the lateral walls of the nanowire.

The most extensive experimental thermal conductivity measurements have been done on Si nanowires [4.164], where $\kappa(T)$ measurements have been made on nanowires in the diameter range $22 \leq d_W \leq 115$ nm. The results show a large decrease in the peak of $\kappa(T)$, associated with Umklapp processes as d_W decreases, indicating a growing importance of boundary scattering and a corresponding decreasing importance of phonon–phonon scattering. At the smallest wire diameter of 22 nm, a linear $\kappa(T)$ dependence is found experimentally, consistent with a linear T dependence of the specific heat for a 1-D system, and a temperature-independent mean free path and velocity of sound. Further insights are obtained through studies of the thermal conductivity of Si/SiGe superlattice nanowires [4.170].

Model calculations for $\kappa(T)$ based on a radiative heat transfer model have been carried out for Si nanowires [4.168]. These results show that the predicted $\kappa(T)$ behavior for Si nanowires is similar to that observed experimentally in the range of $37 \leq d_W \leq 115$ nm regarding both the functional form of $\kappa(T)$ and the magnitude of the relative decrease in the maximum thermal conductivity κ_{max} as a function of d_W. However, the model calculations predict

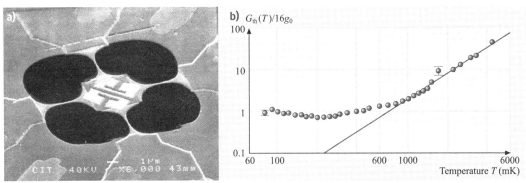

Fig. 4.34 (a) Suspended mesoscopic phonon device used to measure ballistic phonon transport. The device consists of an $4 \times 4\,\mu m^2$ *phonon cavity* (*center*) connected to four Si_3N_4 membranes, 60 nm thick and less than 200 nm wide. The two bright *C*-shaped objects on the phonon cavity are thin film heating and sensing Cr/Au resistors, whereas the *dark regions* are empty space. (b) Log–log plot of the temperature dependence of the thermal conductance G_0 of the structure in (a) normalized to $16g_0$ (see text) (after [4.169])

a substantially larger magnitude for $\kappa(T)$ (by 50% or more) than is observed experimentally. Furthermore, the model calculations (Fig. 4.34) do not reproduce the experimentally observed linear T dependence for the 22 nm nanowires, but rather predict a 3-D behavior for both the density of states and the specific heat in 22 nm nanowires [4.168, 171, 172].

Thermal conductance measurements on GaAs nanowires below 6 K show a power law dependence, but the T dependence becomes somewhat less pronounced below ≈ 2.5 K [4.165]. This deviation from the power law temperature dependence led to a more detailed study of the quantum limit for the thermal conductance. To carry out these more detailed experiments, a mesoscopic phonon resonator and waveguide device were constructed that included four ≈ 200 nm wide and 85 nm thick silicon nitride nanowirelike nanoconstrictions (Fig. 4.33a), and this was used to establish the quantized thermal conductance limit of $g_0 = \pi^2 k_B^2 T/(3h)$ (Fig. 4.33b) for ballistic phonon transport [4.169, 173]. For temperatures above 0.8 K, the thermal conductance in Fig. 4.33b follows a T^3 law, but as T is further reduced, a transition to a linear T dependence is observed, consistent with a phonon mean free path of $\approx 1\,\mu$m, and a thermal conductance value approaching $16g_0$, corresponding to four massless phonon modes per channel and four channels in their phonon waveguide structure (Fig. 4.33a). Ballistic phonon transport occurs when the thermal phonon wavelength (380 nm for the experimental structure) is somewhat greater than the width of the phonon waveguide at the waveguide constriction.

4.2.4 Optical Properties

Optical methods provide an easy and sensitive tool for measuring the electronic structures of nanowires, since optical measurements require minimal sample preparation (for example, contacts are not required) and the measurements are sensitive to quantum effects. Optical spectra of 1-D systems, such as carbon nanotubes, often show intense features at specific energies near singularities in the joint density of states that are formed under strong quantum confinement conditions. A variety of optical techniques have shown that the properties of nanowires are different to those of their bulk counterparts, and this section of the review focuses on these differences in the optical properties of nanowires.

Although optical properties have been shown to provide an extremely important tool for characterizing nanowires, the interpretation of these measurements is not always straightforward. The wavelength of light used to probe the sample is usually smaller than the wire length, but larger than the wire diameter. Hence, the probe light used in an optical measurement cannot be focused solely onto the wire, and the wire and the substrate on which the wire rests (or host material, if the wires are embedded in a template) are probed simultaneously. For measurements, such as photoluminescence (PL), if the substrate does not luminesce or absorb in the frequency range of the measurements, PL measures the luminescence of the nanowires directly and the substrate can be ignored. However, in reflection and transmission measurements, even a nonabsorbing substrate can modify the measured spectra of nanowires.

In this section we discuss the determination of the dielectric function for nanowires in the context of effective medium theories. We then discuss various optical techniques with appropriate examples that sensitively differentiate nanowire properties from those also found in the parent bulk material, placing particular emphasis on electronic quantum confinement effects. Finally, phonon confinement effects are reviewed.

The Dielectric Function

In this subsection, we review the use of effective medium theory as a method to handle the optical properties of nanowires whose diameters are typically smaller than the wavelength of light, noting that observable optical properties of materials can be related to the complex dielectric function [4.174, 175]. Effective medium theories [4.176, 177] can be applied to model the nanowire and substrate as one continuous composite with a single complex dielectric function ($\epsilon_1 + i\epsilon_2$), where the real and imaginary parts of the dielectric function ϵ_1 and ϵ_2 are related to the index of refraction (n) and the absorption coefficient (K) by the relation $\epsilon_1 + i\epsilon_2 = (n + iK)^2$. Since photons at visible or infrared wavelengths *see* a dielectric function for the composite nanowire array/substrate system that is different from that of the nanowire itself, the optical transmission and reflection are different from what they would be if the light were focused only on the nanowire. One commonly observed consequence of effective medium theory is the shift in the plasma frequency in accordance with the percentage of nanowire material that is contained in the composite [4.178]. The plasma resonance occurs when $\epsilon_1(\omega)$ becomes zero, and the plasma frequency of the nanowire composite will shift to lower (higher) energies when the magnitude of the dielectric function of the host materials is larger (smaller) than that of the nanowire.

Although reflection and transmission measurements probe both the nanowire and the substrate, the optical properties of the nanowires can be determined independently. One technique for separating out the dielectric function of the nanowires from the host is to use an effective medium theory in reverse. Since the dielectric function of the host material is often known, and the dielectric function of the composite material can be measured by the standard method of using reflection and transmission measurements in combination with either the Kramer–Kronig relations or Maxwell's equations, the complex dielectric function of the nanowires can be deduced. An example where this approach has been used successfully is for the determination of the frequency dependence of the real and imaginary parts of the dielectric function $\epsilon_1(\omega)$ and $\epsilon_2(\omega)$ for a parallel array of bismuth nanowires filling the pores of an alumina template [4.179].

Characteristic Optical Properties of Nanowires

A wide range of optical techniques are available for the characterization of nanowires, to distinguish their properties from those of their parent bulk materials. Some differences in properties relate to geometric differences, such as the small diameter size and the large length-to-diameter ratio (also called the aspect ratio), while others focus on quantum confinement issues.

Probably the most basic optical technique is to measure the reflection and/or transmission of a nanowire to determine the frequency- dependent real and imaginary parts of the dielectric function. This technique has been used, for example, to study the band gap and its temperature dependence in gallium nitride nanowires in the 10–50 nm range in comparison to bulk values [4.180]. The plasma frequency, free carrier density, and donor impurity concentration as a function of temperature were also determined from the infrared spectra, which is especially useful for nanowire research, since Hall effect measurements cannot be made on nanowires.

Another common method used to study nanowires is photoluminescence (PL) or fluorescence spectroscopy. Emission techniques probe the nanowires directly and the effect of the host material does not have to be considered. This characterization method has been used to study many properties of nanowires, such as the optical gap behavior, oxygen vacancies in ZnO nanowires [4.55], strain in Si nanowires [4.181], and quantum confinement effects in InP nanowires [4.182]. Figure 4.35 shows the photoluminescence of InP nanowires as a function of wire diameter, thereby providing direct information on the effective bandgap. As the wire diameter of an InP nanowire is decreased so that it becomes smaller than the bulk exciton diameter of 19 nm, quantum confinement effects set in, and the band gap is increased. This results in an increase in the PL peak energy. The smaller the effective mass, the larger the quantum confinement effects. When the shift in the peak energy as a function of nanowire diameter Fig. 4.35 is analyzed using an effective mass model, the reduced effective mass of the exciton is deduced to be $0.052\,m_0$, which agrees quite well with the literature value of $0.065\,m_0$ for bulk InP. Although the linewidths of the PL peak for the small-diameter nanowires (10 nm) are smaller at low temperature (7 K), the observation of strong quantum confinement and bandgap tunability effects at room temperature are significant for photonics applications of nanowires (Sect. 4.3.3).

The resolution of photoluminescence (PL) optical imaging of a nanowire is, in general, limited by the wavelength of light. However, when a sample is placed very close to the detector, the light is not given a chance to diffract, and so samples much smaller than the wavelength of light can be resolved. This technique is known as near-field scanning optical microscopy (NSOM) and has been used to successfully image nanowires [4.183]. For example, Fig. 4.36 shows the topographical (a) and (b) NSOM PL images of a single ZnO nanowire.

Magnetooptics can be used to measure the electronic band structure of nanowires. For example, magnetooptics in conjunction with photoconductance has been proposed as a tool to determine band parameters for nanowires, such as the Fermi energy, electron effective masses, and the number of subbands to be considered [4.184]. Since different nanowire subbands have different electrical transmission properties, the electrical conductivity changes when light is used to excite electrons to higher subbands, thereby providing a method for studying the electronic structure of nanowires optically. Magnetooptics can also be used to study the magnetic properties of nanowires in relation to bulk properties [4.27, 185]. For example, the surface magnetooptical Kerr effect has been used to measure the dependence of the magnetic ordering temperature of Fe-Co alloy nanowires on the relative concentration of Fe and Co [4.185], and it was used to find that, unlike in the case of bulk Fe-Co alloys, cobalt in nanowires inhibits magnetic ordering. Nickel nanowires were found to have a strong increase in their magnetooptical activity with respect to bulk nickel. This increase is attributed to the plasmon resonance in the wires [4.186].

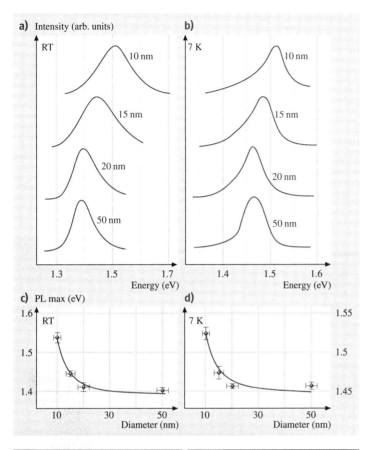

Fig. 4.35a–d Photoluminescence of InP nanowires of varying diameters at 7 K (**b,d**) and room temperature (**a,c**) showing quantum confinement effects of the exciton for wire diameters of less than 20 nm (after [4.182])

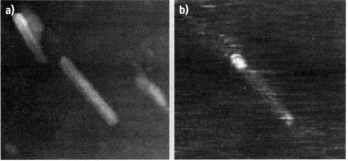

Fig. 4.36 (**a**) Topographical and (**b**) photoluminescence (PL) near-field scanning optical microscopy (NSOM) images of a single ZnO nanowire waveguide (after [4.183])

Nonlinear optical properties of nanowires have received particular attention since the nonlinear behavior is often enhanced compared to bulk materials and the nonlinear effects can be utilized for many applications. One such study measured the second harmonic generation (SHG) and third harmonic generation (THG) in a single nanowire using near-field optical microscopy [4.187]. ZnO nanowires were shown to have strong SHG and THG effects that are highly polarization-sensitive, and this polarization sensitivity can be explained on the basis of optical and geometrical considerations. Some components of the second harmonic polarization tensor are found to be enhanced in nanowires while others are suppressed as the wire diameter is decreased, and such effects could be of interest for device applications. The authors also showed that the second-order nonlinearities are mostly wavelength-independent for $\lambda < 400$ nm, which is in the transparent regime for ZnO, below the onset of band gap absorption, and this observation is also of interest for device applications.

Reflectivity and transmission measurements have also been used to study the effects of quantum confinement and surface effects on the low-energy indirect transition in bismuth nanowires [4.189]. *Black* et al. investigated an intense and sharp absorption peak in bismuth nanowires, which is not observed in bulk bismuth. The energy position E_p of this strong absorption peak increases with decreasing diameter. However, the rate of increase in energy with decreasing diameter $|\partial E_p/\partial d_W|$ is an order of magnitude less than that predicted for either a direct interband transition or for intersubband transitions in bismuth nanowires. On the other hand, the magnitude of $|\partial E_p/\partial d_W|$ agrees well with that predicted for an indirect L-point valence to T-point valence band transition (Fig. 4.37). Since both the initial and final states for the indirect L–T point valence band transition downshift in energy as the wire diameter d_W is decreased, the shift in the absorption peak results from a difference between the effective masses and not from the actual value of either of the masses. Hence the diameter dependence of the absorption peak energy is an order of magnitude less for a valence to valence band indirect transition than for a direct interband L-point transition. Furthermore, the band-tracking effect for the indirect transition gives rise to a large value for the joint density of states, thus accounting for the high intensity of this feature. The enhancement in the absorption resulting from this indirect transition may arise from a gradient in the dielectric function, which is large at the bismuth–air or bismuth–alumina interfaces, or from the relaxation of momentum conservation rules in nanosystems. It should be noted that, in contrast to the surface effect for bulk samples, the whole nanowire contributes to the optical absorption due to the spatial variation in the dielectric function, since the penetration depth is larger than or comparable to the wire diameter. In addition, the intensity can be quite signif-

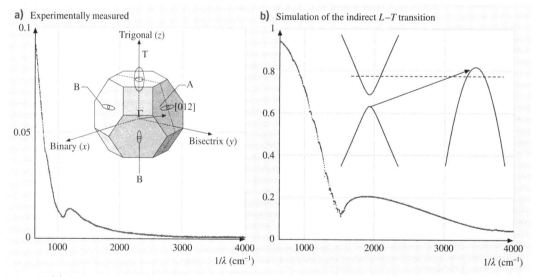

Fig. 4.37 (a) The measured optical transmission spectra as a function of wavenumber $(1/\lambda)$ of a ≈ 45 nm diameter bismuth nanowire array. (b) The simulated optical transmission spectrum resulting from an indirect transition of an L-point electron to a T-point valence subband state. The *insert* in (a) shows the bismuth Brillouin zone, and the locations of the T-point hole and the three L-point electron pockets, including the nondegenerate A, and the doubly-degenerate B pockets. The *insert* in (b) shows the indirect L–T point electronic transition induced by a photon with an energy equal to the energy difference between the initial and final states minus the phonon energy (about $100\,\text{cm}^{-1}$) needed to satisfy conservation of energy in a Stokes process (after [4.188])

icant because there are abundant initial state electrons, final state holes, and appropriate phonons for making an indirect L–T point valence band transition at room temperature. Interestingly, the polarization dependence of this absorption peak is such that the strong absorption is present when the electric field is perpendicular to the wire axis, but is absent when the electric field is parallel to the wire axis, contrary to a traditional polarizer, such as a carbon nanotube where the optical E field is polarized by the nanotube itself and is aligned along the carbon nanotube axis. The observed polarization dependence for bismuth nanowires is consistent with a surface-induced effect that increases the coupling between the L-point and T-point bands throughout the full volume of the nanowire. Figure 4.37 shows the experimentally observed transmission spectrum in bismuth nanowires of ≈ 45 nm diameter (a), and the simulated optical transmission from an indirect transition in bismuth nanowires of ≈ 45 nm diameter is also shown for comparison in (b). The indirect L–T point valence band transition mechanism [4.188] is also consistent with observations of the effect on the optical spectra of a decrease in the nanowire diameter and of n-type doping of bismuth nanowires with Te.

Phonon Confinement Effects

Phonons in nanowires are spatially confined by the nanowire cross-sectional area, crystalline boundaries and surface disorder. These finite size effects give rise to phonon confinement, causing an uncertainty in the phonon wavevector which typically gives rise to a frequency shift and lineshape broadening. Since zone center phonons tend to correspond to maxima in the phonon dispersion curves, the inclusion of contributions from a broader range of phonon wave vectors results in both a downshift in frequency and an asymmetric broadening of the Raman line, which develops a low frequency tail. These phonon confinement effects have been theoretically predicted [4.191, 192] and experimentally observed in GaN [4.190], as shown in Fig. 4.38 for GaN nanowires with diameters in the range 10–50 nm. The application of these theoretical models indicates that broadening effects should be noticeable as the wire diameter in GaN nanowires decreases to ≈ 20 nm. When the wire diameter decreases further to ≈ 10 nm, the frequency downshift and asymmetric Raman line broadening effects should become

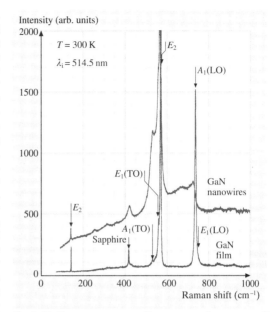

Fig. 4.38 Room-temperature Raman scattering spectra of GaN nanowires and of a 5 μm thick GaN epilayer film with green (514.5 nm) laser excitation. The Raman scattering response was obtained by dividing the measured spectra by the Bose–Einstein thermal factor [4.190]

observable in the Raman spectra for the GaN nanowires but are not found in the corresponding spectra for bulk GaN.

The experimental spectra in Fig. 4.38 show the four $A_1 + E_1 + 2E_2$ modes expected from symmetry considerations for bulk GaN crystals. Two types of quantum confinement effects are observed. The first type is the observation of the downshift and the asymmetric broadening effects discussed above. Observations of such downshifts and asymmetric broadening have also been recently reported in 7 nm diameter Si nanowires [4.193]. A second type of confinement effect found in Fig. 4.38 for GaN nanowires is the appearance of additional Raman features not found in the corresponding bulk spectra and associated with combination modes, and a zone boundary mode. Resonant enhancement effects were also observed for the $A_1(\text{LO})$ phonon at 728 cm^{-1} (Fig. 4.38) at higher laser excitation energies [4.190].

4.3 Applications

In the preceding sections we have reviewed many of the central characteristics that make nanowires in some cases similar to and in some cases very different from their parent materials. We have also shown that some properties are diameter-dependent, and these properties are therefore tunable during synthesis. Thus, it is of great interest to find applications that could benefit in unprecedented ways from both the unique and tunable properties of nanowires and the small sizes of these nanostructures, especially in the miniaturization of conventional devices. As the synthetic methods for the production of nanowires are maturing (Sect. 4.1) and nanowires can be made in reproducible and cost-effective ways, it is only a matter of time before applications will be seriously explored. This is a timely development, as the semiconductor industry will soon be reaching what seems to be its limit in feature size reduction, and approaching a classical-to-quantum size transition. At the same time, the field of biotechnology is expanding through the availability of tremendous genome information and innovative screening assays. Since nanowires are similar in size to the shrinking electronic components and to cellular biomolecules, it is only natural for nanowires to be good candidates for applications in these fields. Commercialization of nanowire devices, however, will require reliable mass production, effective assembly techniques and quality control methods.

In this section, applications of nanowires to electronics (Sect. 4.3.1), thermoelectrics (Sect. 4.3.2), optics (Sect. 4.3.3), chemical and biochemical sensing (Sect. 4.3.4), and magnetic media (Sect. 4.3.5) are discussed.

4.3.1 Electrical Applications

The microelectronics industry continues to face technological (in lithography for example) and economic challenges as the device feature size is decreased, especially below 100 nm. The self-assembly of nanowires might present a way to construct unconventional devices that do not rely on improvements in photolithography and, therefore, do not necessarily imply increasing fabrication costs. Devices made from nanowires have several advantages over those made by photolithography. A variety of approaches have been devised to organize nanowires via self-assembly (Sect. 4.1.4), thus eliminating the need for the expensive lithographic techniques normally required to produce devices the size of typical nanowires that are discussed in this review. In addition, unlike traditional silicon processing, different semiconductors can be used simultaneously in nanowire devices to produce diverse functionalities. Not only can wires of different materials be combined, but a single wire can be made of different materials. For example, junctions of GaAs and GaP show rectifying behavior [4.92], thus demonstrating that good electronic interfaces between two different semiconductors can be achieved in the synthesis of multicomponent nanowires. Transistors made from nanowires could also hold advantages due to their unique morphology. For example, in bulk field effect transistors (FETs), the depletion layer formed below the source and drain region results in a source–drain capacitance which limits the operation speed. However, in nanowires, the conductor is surrounded by an oxide and thus the depletion layer cannot be formed. Thus, depending on the device design, the source–drain capacitance in nanowires could be greatly minimized and possibly eliminated.

Device functionalities common in conventional semiconductor technologies, such as p-n junction diodes [4.142], field-effect transistors [4.144], logic gates [4.142], and light-emitting diodes [4.92, 194], have been recently demonstrated in nanowires, showing their promise as building blocks that could be used to construct complex integrated circuits by employing the *bottom-up* paradigm. Several approaches have been investigated to form nanowire diodes (Sect. 4.2.2). For example, Schottky diodes can be formed by contacting a GaN nanowire with Al electrodes [4.143]. Furthermore, p-n junction diodes can be formed at the crossing of two nanowires, such as the crossing of n- and p-type InP nanowires doped by Te and Zn, respectively [4.194], or Si nanowires doped by phosphorus (n-type) and boron (p-type) [4.195]. In addition to the crossing of two distinctive nanowires, heterogeneous junctions have also been constructed inside a single wire, either along the wire axis in the form of a nanowire superlattice [4.92], or perpendicular to the wire axis by forming a core–shell structure of silicon and germanium [4.111]. These various nanowire junctions not only possess the current rectifying properties (Fig. 4.22) expected of bulk semiconductor devices, but they also exhibit electroluminescence (EL) that may be interesting for optoelectronic applications, as shown in Fig. 4.39 for the electroluminescence of a crossed junction of n- and p-type InP nanowires [4.194] (Sect. 4.3.3).

In addition to the two-terminal nanowire devices, such as the p-n junctions described above, it is found that the conductance of a semiconductor nanowire can be significantly modified by applying voltage at a third gate terminal, implying the utilization of nanowires in field effect transistors (FETs). This gate terminal can either be the substrate [4.30, 196–199], a separate metal contact located close to the nanowire [4.200], or another nanowire with a thick oxide coating in the crossed nanowire junction configuration [4.142]. The operating principles of these nanowire-based FETs are discussed in Sect. 4.2.2. Various logic devices performing basic logic functions have been demonstrated using nanowire junctions [4.142], as shown in Fig. 4.40 for the OR and AND logic gates constructed from 2-by-1 and 1-by-3 nanowire p-n junctions, respectively. By functionalizing nanowires with redox-active molecules to store charge, nanowire FETs were demonstrated with two-level [4.144] and with eight-level [4.201] memory effects, which may be used for nonvolatile memory or as switches. In another advance, In_2O_3 nanowire FETs with high-k dielectric material were demonstrated, and substantially enhanced performance was obtained due to the highly efficient coupling of the gate [4.202]. A vertical FET with a surrounding gate geometry has also been demonstrated, which has the potential for high-density nanoscale memory and logic devices [4.203].

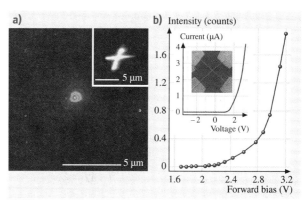

Fig. 4.39a,b Optoelectrical characterization of a crossed nanowire junction formed between 65 nm n-type and 68 nm p-type InP nanowires. (**a**) Electroluminescence (EL) image of the light emitted from a forward-biased nanowire p-n junction at 2.5 V. *Inset*, photoluminescence (PL) image of the junction. (**b**) EL intensity as a function of operation voltage. *Inset*, the SEM image and the I–V characteristics of the junction (after [4.194]). The scale bar in the *inset* is 5 μm

Nanowires have also been proposed for applications associated with electron field emission [4.204], such as flat panel displays, because of their small diameter and large curvature at the nanowire tip, which may reduce the threshold voltage for electron emission [4.205]. In

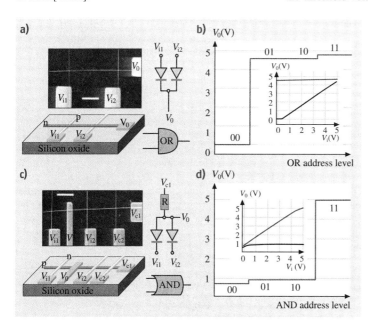

Fig. 4.40a–d Nanowire logic gates: (**a**) Schematic of logic OR gate constructed from a 2 (p-Si) by 1 (n-GaN) crossed nanowire junction. The *inset* shows the SEM image (scale bar: 1 μm) of an assembled OR gate and the symbolic electronic circuit. (**b**) The output voltage of the circuit in (**a**) versus the four possible logic address level inputs: (0,0); (0,1); (1,0); (1,1), where logic 0 input is 0 V and logic 1 is 5 V (same for below). (**c**) Schematic of logic AND gate constructed from a 1 (p-Si) by 3 (n-GaN) crossed nanowire junction. The *inset* shows the SEM image (scale bar: 1 μm) of an assembled AND gate and the symbolic electronic circuit. (**d**) The output voltage of the circuit in (**c**) versus the four possible logic address level inputs (after [4.142])

this regard, the demonstration of very high field emission currents from the sharp tip (≈ 10 nm radius) of a Si cone [4.204], from carbon nanotubes [4.206], from Si nanowires inside a carbon nanotube [4.207], and from Co nanowires [4.208], has stimulated interest in this potential area of application for nanowires.

The concept of constructing electronic devices based on nanowires has already been demonstrated, and the next step for electronic applications would be to devise a feasible method for integration and mass production. We expect that, in order to maintain the growing rate of device density and functionality in the existing electronic industry, new kinds of complementary electronic devices will emerge from this *bottom-up* scheme for nanowire electronics, different from what has been produced by the traditional *top-down* approach pursued by conventional electronics.

4.3.2 Thermoelectric Applications

One proposed application for nanowires is for thermoelectric cooling and for the conversion between thermal and electrical energy [4.171, 209]. The efficiency of a thermoelectric device is measured in terms of a dimensionless figure of merit ZT, where Z is defined as

$$Z = \frac{\sigma S^2}{\kappa}, \tag{4.2}$$

where σ is the electrical conductivity, S is the Seebeck coefficient, κ is the thermal conductivity, and T is the temperature. In order to achieve a high ZT and therefore efficient thermoelectric performance, a high electrical conductivity, a hugh Seebeck coefficient and a low thermal conductivity are required. In 3-D systems, the electronic contribution to κ is proportional to σ in accordance with the Wiedemann–Franz law, and normally materials with high S have a low σ. Hence an increase in the electrical conductivity (for example by electron donor doping) results in an adverse variation in both the Seebeck coefficient (decreasing) and the thermal conductivity (increasing). These two trade-offs set the upper limit for increasing ZT in bulk materials, with the maximum ZT remaining ≈ 1 at room temperature for the 1960–1995 time frame.

The high electronic density of states in quantum-confined structures is proposed as a promising possibility to bypass the Seebeck/electrical conductivity trade-off and to control each thermoelectric-related variable independently, thereby allowing for increased electrical conductivity, relatively low thermal conductivity, and a large Seebeck coefficient simultaneously [4.210].

For example, Figs. 4.29 and 4.30a in Sect. 4.2.3 show an enhanced in S for bismuth and bismuth-antimony nanowires as the wire diameter decreases. In addition to alleviating the undesired connections between σ, S and the electronic contribution to the thermal conductivity, nanowires also have the advantage that the phonon contribution to the thermal conductivity is greatly reduced because of boundary scattering (Sect. 4.2), thereby achieving a high ZT. Figure 4.41a shows the theoretical values for ZT versus sample size for both bismuth thin films (2-D) and nanowires (1-D) in the quantum-confined regime, exhibiting a rapidly increasing ZT as the quantum size effect becomes more and more important [4.210]. In addition, the quantum size effect in nanowires can be combined with other parameters to tailor the band structure and electronic transport behavior (for instance, Sb alloying in Bi) to further optimize ZT. For example, Fig. 4.41b shows the predicted ZT for p-type $Bi_{1-x}Sb_x$ alloy nanowires as a function of wire diameter and Sb content x [4.211]. The occurrence of a local ZT maxima in the vicinity of $x \approx 0.13$ and $d_W \approx 45$ nm is due to the coalescence of ten valence bands in the nanowire and the resulting unusual high density of states for holes, which is a phenomenon absent in bulk $Bi_{1-x}Sb_x$ alloys. For nanowires with very small diameters, it is speculated that localization effects will eventually limit the enhancement of ZT. However, in bismuth nanowires, localization effects are not significant for wires with diameters larger than 9 nm [4.52]. In addition to 1-D nanowires, ZT values as high as ≈ 2 have also been experimentally demonstrated in macroscopic samples containing PbSe quantum dots (0-D) [4.212] and stacked 2-D films [4.167].

Although the application of nanowires to thermoelectrics appears very promising, these materials are still in the research phase of the development cycle and are far from being commercialized. One challenge for thermoelectric devices based on nanowires lies in finding a suitable host material that will not reduce ZT too much due to the unwanted heat conduction through the host material. Therefore, the host material should have a low thermal conductivity and occupy a volume percentage in the composite material that is as low as possible, while still providing the quantum confinement and the support for the nanowires.

4.3.3 Optical Applications

Nanowires also hold promise for optical applications. One-dimensional systems exhibit a singularity in their joint density of states, allowing quantum effects in

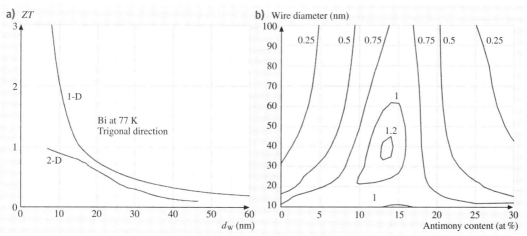

Fig. 4.41 (a) Calculated ZT of 1-D (nanowire) and 2-D (quantum well) bismuth systems at 77 K as a function of d_W, denoting the wire diameter or film thickness. The thermoelectric performance (ZT) is expected to improve greatly when the wire diameter is small enough for the nanowire to become a one-dimensional system. (b) Contour plot of optimal ZT values for p-type $Bi_{1-x}Sb_x$ nanowires versus wire diameter and antimony concentration calculated at 77 K (after [4.211])

nanowires to be optically observable, sometimes even at room temperature. Since the density of states of a nanowire in the quantum limit (small wire diameter) is highly localized in energy, the available states quickly fill up with electrons as the intensity of the incident light is increased. This filling up of the subbands, as well as other effects that are unique to low-dimensional materials, lead to strong optical nonlinearities in quantum wires. Quantum wires may thus yield optical switches with a lower switching energy and increased switching speed compared to currently available optical switches.

Light emission from nanowires can be achieved by photoluminescence (PL) or electroluminescence (EL), distinguished by whether the electronic excitation is achieved by optical illumination or by electrical stimulation across a p-n junction, respectively. PL is often used for optical property characterization, as described in Sect. 4.2.4, but from an applications point of view, EL is a more convenient excitation method. Light-emitting diodes (LEDs) have been achieved in junctions between a p-type and an n-type nanowire (Fig. 4.39) [4.194] and in superlattice nanowires with p-type and n-type segments [4.92]. The light emission was localized to the junction area, and was polarized in the superlattice nanowire. An electrically driven laser was fabricated from CdS nanowires. The wires were assembled by evaporating a metal contact onto an n-type CdS nanowire which resided on a p^+ silicon wafer. The cleaved ends of the wire formed the laser cavity, so that in forward bias, light characteristic of lasing was observed at the end of the wire [4.213]. LEDs have also been achieved with core–shell structured nanowires made of n-GaN/InGaN/p-GaN [4.214].

Light emission from quantum wire p-n junctions is especially interesting for laser applications, because quantum wires can form lasers with lower excitation thresholds than their bulk counterparts and they also exhibit decreased sensitivity of performance to temperature [4.215]. Furthermore, the emission wavelength can be tuned for a given material composition by simply altering the geometry of the wire.

Lasing action has been reported in ZnO nanowires with wire diameters that are much smaller than the wavelength of the light emitted ($\lambda = 385$ nm) [4.122] (Fig. 4.42). Since the edges and lateral surfaces of ZnO nanowires are faceted (Sect. 4.2.1), they form optical cavities that sustain desired cavity modes. Compared to conventional semiconductor lasers, the exciton laser action employed in zinc oxide nanowire lasers exhibits a lower lasing threshold ($\approx 40\,\text{kW/cm}^2$) than their 3-D counterparts ($\approx 300\,\text{kW/cm}^2$). In order to utilize exciton confinement effects in the lasing action, the exciton binding energy (≈ 60 meV in ZnO) must be greater than the thermal energy (≈ 26 meV at 300 K). Decreasing the wire diameter increases the excitation binding energy and lowers the threshold for lasing. PL NSOM imaging confirmed the waveguiding properties of the anisotropic and the well-faceted structure of ZnO

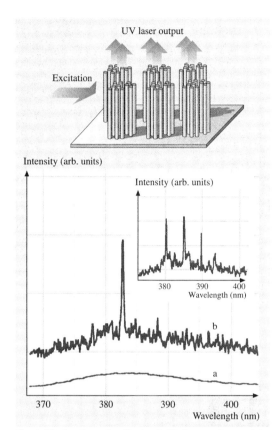

Fig. 4.42 A schematic of lasing in ZnO nanowires and the PL spectra of ZnO nanowires at two excitation intensities. One PL spectrum is taken below the lasing threshold, and the other above it (after [4.122])

sult of the formation of an electron–hole plasma. Heating effects were excluded as the source of the spectral shift. GaN quantum wire UV lasers with a low threshold for lasing action have been achieved using a self-organized GaN(core)/AlGaN(shell) structure [4.219].

Nanowires have also been demonstrated to have good waveguiding properties. Quantitative studies of cadmium sulfide (CdS) nanowire structures show that light propagation takes place with only moderate losses through sharp and even acute angle bends. In addition, active devices made with nanowires have shown that efficient injection into and modulation of light through nanowire waveguides can be achieved [4.220]. By linking ZnO nanowire light sources to SnO_2 waveguides, the possibility of optical integrated circuitry is introduced [4.221].

Nanowire photodetectors were also proposed. ZnO nanowires were found to display a strong photocurrent response to UV light irradiation [4.222]. The conductivity of the nanowire increased by four orders of magnitude compared to the dark state. The response of the nanowire was reversible, and selective to photon energies above the bandgap, suggesting that ZnO nanowires could be a good candidate for optoelectronic switches.

Nanowires have been also proposed for another type of optical switching. Light with its electric field normal to the wire axis excites a transverse free carrier resonance inside the wire, while light with its electric field parallel to the wire axis excites a longitudinal free carrier resonance inside the wire. Since nanowires are highly anisotropic, these two resonances occur at two different wavelengths and thus result in absorption peaks at two different energies. Gold nanowires dispersed in an aqueous solution align along the electric field when a DC voltage is applied. The energy of the absorption peak can be toggled between the transverse and longitudinal resonance energies by changing the alignment of the nanowires under polarized light illumination using an electric field [4.223, 224]. Thus, electro-optical modulation is achieved.

Nanowires may also be used as barcode tags for optical read-out. Nanowires containing gold, silver, nickel, palladium, and platinum were fabricated [4.110] by electrochemical filling of porous anodic alumina, so that each nanowire consisted of segments of various metal constituents. Thus many types of nanowires can be made from a handful of materials, and identified by the order of the metal segments along their main axis, and the length of each segment. Barcode read-out is possible by reflectance optical microscopy. The segment length is limited by the Rayleigh diffraction

nanowires, limiting the emission to the tips of the ZnO nanowires [4.183]. Time-resolved studies have illuminated the dynamics of the emission process [4.216].

Lasing was also observed in ZnS nanowires in anodic aluminum oxide templates [4.217] and in GaN nanowires [4.218]. Unlike ZnO, GaN has a small exciton binding energy, only ≈ 25 meV. Furthermore, since the wire radii used in this study (15–75 nm) [4.218] are larger than the Bohr radius of excitons in GaN (11 nm), the exciton binding energy is not expected to increase in these GaN wires and quantum confinement effects such as those shown in Fig. 4.35 for InP are not expected. However, some tunability of the center of the spectral intensity was achieved by increasing the intensity of the pump power, causing a redshift in the laser emission, which is explained as a bandgap renormalization as a re-

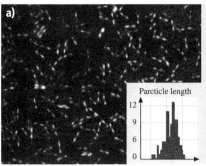
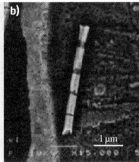

Fig. 4.43 (a) An optical image of many short bar-coded Au-Ag-Au-Au wires and (b) an FE-SEM image of an Au/Ag barcoded wire with multiple strips of varying length. The *insert* in (a) shows a histogram of the particle lengths for 106 particles in this image (after [4.110])

limit, and not by synthesis limitations, and thus can be as small as 145 nm. Figure 4.43a shows an optical image of many Au-Ag-Au-Ag barcoded wires, where the silver segments show higher reflectivity. Figure 4.43b is a backscattering mode FE-SEM image of a single nanowire, highlighting the composition and segment length variations along the nanowire.

Both the large surface area and the high conductivity along the length of a nanowire are favorable for its use in inorganic–organic solar cells [4.225], which offer promise from a manufacturing and cost-effectiveness standpoint. In a hybrid nanocrystal–organic solar cell, the incident light forms bound electron–hole pairs (excitons) in both the inorganic nanocrystal and in the surrounding organic medium. These excitons diffuse to the inorganic–organic interface and disassociate to form an electron and a hole. Since conjugated polymers usually have poor electron mobilities, the inorganic phase is chosen to have a higher electron affinity than the organic phase so that the organic phase carries the holes and the semiconductor carries the electrons. The separated electrons and holes drift to the external electrodes through the inorganic and organic materials, respectively. However, only those excitons formed within an exciton diffusion length from an interface can disassociate before recombining, and therefore the distance between the dissociation sites limits the efficiency of a solar cell. A solar cell prepared from a composite of CdSe nanorods inside poly(3-ethylthiophene) [4.225] yielded monochromatic power efficiencies of 6.9% and power conversion efficiencies of 1.7% under A.M. 1.5 illumination (equal to solar irradiance through 1.5 times the air mass of the Earth at direct normal incidence). The nanorods provide a large surface area with good chemical bonding to the polymer for efficient charge transfer and exciton dissociation. Furthermore, they provide a good conduction path for the electrons to reach the electrode. Their enhanced absorption coefficient and their tunable bandgap are also characteristics that can be used to enhance the energy conversion efficiency of solar cells.

4.3.4 Chemical and Biochemical Sensing Devices

Sensors for chemical and biochemical substances with nanowires as the sensing probe are a very attractive application area. Nanowire sensors will potentially be smaller, more sensitive, demand less power, and react faster than their macroscopic counterparts. Arrays of nanowire sensors could, in principle, achieve nanometer-scale spatial resolution and therefore provide accurate real-time information regarding not only the concentration of a specific analyte but also its spatial distribution. Such arrays could be very useful, for example, for dynamic studies on the effects of chemical gradients on biological cells. The operation of sensors made with nanowires, nanotubes, or nanocontacts is based mostly on the reversible change in the conductance of the nanostructure upon absorption of the agent to be detected, but other detection methods, such as mechanical and optical detection, are conceptually plausible. The increased sensitivity and faster response time of nanowires are a result of the large surface-to-volume ratio and the small cross section available for conduction channels. In the bulk, on the other hand, the abundance of charges can effectively shield external fields, and the abundance of material can afford many alternative conduction channels. Therefore, a stronger chemical stimulus and longer response time are necessary to observe changes in the physical properties of a 3-D sensor in comparison to a nanowire.

It is often necessary to modify the surface of the nanowires to achieve a strong interaction with the analytes that need to be detected. Surface modifications utilize the self-assembly, chemisorption or chemical reactivity of selected organic molecules and polymers towards metal and oxide surfaces. Examples include: thiols on gold, isocyanides on platinum, and siloxanes on silica. These surface coatings regulate the binding and chemical reactivity of other molecules towards the nanowire in a predictable manner [4.226].

Cui et al. placed silicon nanowires made by the VLS method (Sect. 4.1.2) between two metal electrodes and modified the silicon oxide coating of the wire through the addition of molecules that are sensitive to the analyte to be detected [4.227]. For example, a pH sensor was made by covalently linking an amine-containing silane to the surface of the nanowire. Variations in the pH of the solution into which the nanowire was immersed caused protonation and deprotonation of the $-NH_2$ and the $-SiOH$ groups on the surface of the nanowire. The variation in surface charge density regulates the conductance of the nanowire; due to the p-type characteristics of a silicon wire, the conductance increases with the addition of negative surface charge. The combined acid and base behavior of the surface groups results in an approximately linear dependence of the conductance on pH in the pH range 2 to 9, thus leading to a direct readout pH meter. This same type of approach was used for the detection of the binding of biomolecules, such as streptavidin using biotin-modified nanowires (Fig. 4.44). This nanowire-based device has high sensitivity and could detect streptavidin binding down to a concentration of 10 pM (10^{-12} mol). Subsequent results demonstrated the capabilities of these functionalized Si nanowire sensors as DNA sensors down to the femtomolar range [4.228]. The chemical detection devices were made in a field effect transistor geometry, so that the back-gate potential could be used to regulate the conductance in conjugation with the chemical detection and to provide a real-time direct read-out [4.227]. The extension of this device to detect multiple analytes using multiple nanowires, each sensitized to a different analyte, could provide for fast, sensitive, and in situ screening procedures.

A similar approach was used by *Favier* et al., who made a nanosensor for the detection of hydrogen from of an array of palladium nanowires between two metal contacts [4.44]. They demonstrated that nanogaps were present in their nanowire structure, and upon absorption of H_2 and formation of Pd hydride, the nanogap structure would close and improve the electrical contact, thereby increasing the conductance of the nanowire array. The response time of these sensors was 75 ms, and they could operate in the range 0.5–5% H_2 before saturation occurred.

4.3.5 Magnetic Applications

It has been demonstrated that arrays of single-domain magnetic nanowires can be prepared with controlled nanowire diameter and length, aligned along a common direction and arranged in a close-packed ordered array (Sect. 4.1), and that the magnetic properties (coercivity, remanence and dipolar magnetic interwire interaction) can be controlled to achieve a variety of magnetic applications [4.40, 79].

The most interesting of these applications is for magnetic storage, where the large nanowire aspect ratio (length/diameter) is advantageous for preventing the onset of the *superparamagnetic* limit at which the magnetization direction in the magnetic grains can be reversed by the thermal energy $k_B T$, thereby resulting in loss of recorded data in the magnetic recording medium. The magnetic energy in a grain can be increased by increasing either the volume or the anisotropy of the grain. If the volume is increased, the particle size increases, so the resolution is decreased. For spherical magnetized grains, the superparamagnetic limit at room temperature is reached at $70 \, \text{Gbit/in}^2$. In nanowires, the anisotropy is very large and yet the wire diameters are small, so that the magnetostatic switching energy can easily be above the thermal energy while the spatial resolution is large. For magnetic data storage applications, a large aspect ratio is needed for the nanowires in order to maintain a high coercivity, and a sufficient separation between nanowires is needed to suppress interwire magnetic dipolar coupling. Thus

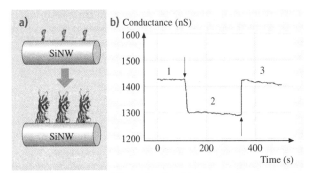

Fig. 4.44 (a) Streptavidin molecules bind to a silicon nanowire functionalized with biotin. The binding of streptavidin to biotin causes the nanowire to change its resistance. (b) The conductance of a biotin-modified silicon nanowire exposed to streptavidin in a buffer solution (regions 1 and 3) and with the introduction of a solution of antibiotin monoclonal antibody (region 2) (after [4.227])

nanowires can form stable and highly dense magnetic memory arrays with packing densities in excess of 10^{11} wires/cm^2.

The onset of superparamagnetism can be prevented in the single-domain magnetic nanowire arrays that have already been fabricated using either porous alumina templates to make Ni nanowires with 35 nm diameters [4.40] or diblock copolymer templates [4.79] to make Co nanowires, with mean diameters of 14 nm and 100% filling of the template pores (Sect. 4.1.1). The ordered magnetic nanowire arrays that have already been demonstrated offer the exciting promise of systems permitting 10^{12} bits/in^2 data storage.

4.4 Concluding Remarks

In this chapter, we reviewed the synthesis, characterization and physical properties of nanowires, placing particular emphasis on nanowire properties that differ from those of the bulk counterparts and potential applications that might result from the special structures and properties of nanowires.

We have shown that the newly emerging field of nanowire research has developed very rapidly over the past few years, driven by the development of a variety of complementary nanowire synthesis methods and effective tools for measuring nanowire structure and properties (Sects. 4.1 and 4.2). At present, much of the progress is at the demonstration-of-concept level, with many gaps in knowledge remaining to be elucidated, theoretical models to be developed, and new nanowire systems to be explored. Having demonstrated that many of the most interesting discoveries to date relate to nanowire properties not present in their bulk material counterparts, we can expect future research emphasis to be increasingly focused on smaller diameter nanowires, where new unexplored physical phenomena related to quantum confinement effects are more likely to be found. We can also expect the development of applications to follow, some coming sooner and others later. Many promising applications are now at the early demonstration stage (Sect. 4.3), but are moving ahead rapidly because of their promise of new functionality, not previously available, in the fields of electronics, optoelectronics, biotechnology, magnetics, and energy conversion and generation, among others. Many exciting challenges remain in advancing both the nanoscience and the nanotechnological promise already demonstrated by the nanowire research described in this review.

References

4.1 R.P. Feynman: There's plenty of room at the bottom, Caltech Eng. Sci. **23**, 22 (1960)

4.2 Y. Mao, S.S. Wong: General, room-temperature method for the synthesis of isolated as well as arrays of single-crystalline ABO$_4$-type nanorods, J. Am. Chem. Soc. **126**, 15245–15252 (2004)

4.3 E. Braun, Y. Eichen, U. Sivan, G. Ben-Yoseph: DNA-templated assembly and electrode attachment of a conducting silver wire, Nature **391**, 775–778 (1998)

4.4 G. Sauer, G. Brehm, S. Schneider, K. Nielsch, R.B. Wehrspohn, J. Choi, H. Hofmeister, U. Gösele: Highly ordered monocrystalline silver nanowire arrays, J. Appl. Phys. **91**, 3243–3247 (2002)

4.5 G.L. Hornyak, C.J. Patrissi, C.M. Martin: Fabrication, characterization and optical properties of gold nanoparticle/porous alumina composites: The non-scattering Maxwell-Garnett limit, J. Phys. Chem. B **101**, 1548–1555 (1997)

4.6 X.Y. Zhang, L.D. Zhang, Y. Lei, L.X. Zhao, Y.Q. Mao: Fabrication and characterization of highly ordered Au nanowire arrays, J. Mater. Chem. **11**, 1732–1734 (2001)

4.7 Y.-T. Cheng, A.M. Weiner, C.A. Wong, M.P. Balogh, M.J. Lukitsch: Stress-induced growth of bismuth nanowires, Appl. Phys. Lett. **81**, 3248–3250 (2002)

4.8 J. Heremans, C.M. Thrush, Y.-M. Lin, S. Cronin, Z. Zhang, M.S. Dresselhaus, J.F. Mansfield: Bismuth nanowire arrays: Synthesis, galvanomagnetic properties, Phys. Rev. B **61**, 2921–2930 (2000)

4.9 L. Piraux, S. Dubois, J.L. Duvail, A. Radulescu, S. Demoustier-Champagne, E. Ferain, R. Legras: Fabrication and properties of organic, metal nanocylinders in nanoporous membranes, J. Mater. Res. **14**, 3042–3050 (1999)

4.10 K. Hong, F.Y. Yang, K. Liu, D.H. Reich, P.C. Searson, C.L. Chien, F.F. Balakirev, G.S. Boebinger: Giant positive magnetoresistance of Bi nanowire arrays in high magnetic fields, J. Appl. Phys. **85**, 6184–6186 (1999)

4.11 A.J. Yin, J. Li, W. Jian, A.J. Bennett, J.M. Xu: Fabrication of highly ordered metallic nanowire arrays by electrodeposition, Appl. Phys. Lett. **79**, 1039–1041 (2001)

4.12 Z. Zhang, J.Y. Ying, M.S. Dresselhaus: Bismuth quantum-wire arrays fabricated by a vacuum melting and pressure injection process, J. Mater. Res. **13**, 1745–1748 (1998)

4.13 Z. Zhang, D. Gekhtman, M.S. Dresselhaus, J.Y. Ying: Processing and characterization of single-crystalline ultrafine bismuth nanowires, Chem. Mater. **11**, 1659–1665 (1999)

4.14 T.E. Huber, M.J. Graf, P. Constant: Processing and characterization of high-conductance bismuth wire array composites, J. Mater. Res. **15**, 1816–1821 (2000)

4.15 L. Li, G. Li, Y. Zhang, Y. Yang, L. Zhang: Pulsed electrodeposition of large-area, ordered $Bi_{1-x}Sb_x$ nanowire arrays from aqueous solutions, J. Phys. Chem. B **108**, 19380–19383 (2004)

4.16 M.S. Sander, A.L. Prieto, R. Gronsky, T. Sands, A.M. Stacy: Fabrication of high-density, high aspect ratio, large-area bismuth telluride nanowire arrays by electrodeposition into porous anodic alumina templates, Adv. Mater. **14**, 665–667 (2002)

4.17 M. Chen, Y. Xie, J. Lu, Y.J. Xiong, S.Y. Zhang, Y.T. Qian, X.M. Liu: Synthesis of rod-, twinrod-, and tetrapod-shaped CdS nanocrystals using a highly oriented solvothermal recrystallization technique, J. Mater. Chem. **12**, 748–753 (2002)

4.18 D. Xu, Y. Xu, D. Chen, G. Guo, L. Gui, Y. Tang: Preparation of CdS single-crystal nanowires by electrochemically induced deposition, Adv. Mater. **12**, 520–522 (2000)

4.19 D. Routkevitch, T. Bigioni, M. Moskovits, J.M. Xu: Electrochemical fabrication of CdS nanowire arrays in porous anodic aluminum oxide templates, J. Phys. Chem. **100**, 14037–14047 (1996)

4.20 L. Manna, E.C. Scher, A.P. Alivisatos: Synthesis of soluble and processable rod-, arrow-, teardrop-, and tetrapod-shaped CdSe nanocrystals, J. Am. Chem. Soc. **122**, 12700–12706 (2000)

4.21 D. Routkevitch, A.A. Tager, J. Haruyama, D. Al-Mawlawi, M. Moskovits, J.M. Xu: Nonlithographic nano-wire arrays: fabrication, physics, and device applications, IEEE Trans. Electron. Dev. **43**, 1646–1658 (1996)

4.22 D.S. Xu, D.P. Chen, Y.J. Xu, X.S. Shi, G.L. Guo, L.L. Gui, Y.Q. Tang: Preparation of II–VI group semiconductor nanowire arrays by dc electrochemical deposition in porous aluminum oxide templates, Pure Appl. Chem. **72**, 127–135 (2000)

4.23 R. Adelung, F. Ernst, A. Scott, M. Tabib-Azar, L. Kipp, M. Skibowski, S. Hollensteiner, E. Spiecker, W. Jäger, S. Gunst, A. Klein, W. Jägermann, V. Zaporojtchenko, F. Faupel: Self-assembled nanowire networks by deposition of copper onto layered-crystal surfaces, Adv. Mater. **14**, 1056–1061 (2002)

4.24 T. Gao, G.W. Meng, J. Zhang, Y.W. Wang, C.H. Liang, J.C. Fan, L.D. Zhang: Template synthesis of single-crystal Cu nanowire arrays by electrodeposition, Appl. Phys. A **73**, 251–254 (2001)

4.25 D. Al-Mawlawi, N. Coombs, M. Moskovits: Magnetic-properties of Fe deposited into anodic aluminum-oxide pores as a function of particle-size, J. Appl. Phys. **70**, 4421–4425 (1991)

4.26 F. Li, R.M. Metzger: Activation volume of α-Fe particles in alumite films, J. Appl. Phys. **81**, 3806–3808 (1997)

4.27 A. Sugawara, T. Coyle, G.G. Hembree, M.R. Scheinfein: Self-organized Fe nanowire arrays prepared by shadow deposition on NaCl(110) templates, Appl. Phys. Lett. **70**, 1043–1045 (1997)

4.28 G.S. Cheng, L.D. Zhang, Y. Zhu, G.T. Fei, L. Li, C.M. Mo, Y.Q. Mao: Large-scale synthesis of single crystalline gallium nitride nanowires, Appl. Phys. Lett. **75**, 2455–2457 (1999)

4.29 G.S. Cheng, L.D. Zhang, S.H. Chen, Y. Li, L. Li, X.G. Zhu, Y. Zhu, G.T. Fei, Y.Q. Mao: Ordered nanostructure of single-crystalline GaN nanowires in a honeycomb structure of anodic alumina, J. Mater. Res. **15**, 347–350 (2000)

4.30 Y. Huang, X. Duan, Y. Cui, C.M. Lieber: Gallium nitride nanowire nanodevices, Nano Lett. **2**, 101–104 (2002)

4.31 X. Duan, C.M. Lieber: Laser-assisted catalytic growth of single crystal GaN nanowires, J. Am. Chem. Soc. **122**, 188–189 (2000)

4.32 A.D. Berry, R.J. Tonucci, M. Fatemi: Fabrication of GaAs, InAs wires in nanochannel glass, Appl. Phys. Lett. **69**, 2846–2848 (1996)

4.33 J.R. Heath, F.K. LeGoues: A liquid solution synthesis of single-crystal germanium quantum wires, Chem. Phys. Lett. **208**, 263–268 (1993)

4.34 Y. Wu, P. Yang: Germanium nanowire growth via simple vapor transport, Chem. Mater. **12**, 605–607 (2000)

4.35 Y.F. Zhang, Y.H. Tang, N. Wang, C.S. Lee, I. Bello, S.T. Lee: Germanium nanowires sheathed with an oxide layer, Phys. Rev. B **61**, 4518–4521 (2000)

4.36 S.J. May, J.-G. Zheng, B.W. Wessels, L.J. Lauhon: Dendritic nanowire growth mediated by a self-assembled catalyst, Adv. Mater. **17**, 598–602 (2005)

4.37 S. Han, C. Li, Z. Liu, B. Lei, D. Zhang, W. Jin, X. Liu, T. Tang, C. Zhou: Transition metal oxide core-shell nanowires: Generic synthesis and transport studies, Nano Lett. **4**, 1241–1246 (2004)

4.38 M.P. Zach, K.H. Ng, R.M. Penner: Molybdenum nanowires by electrodeposition, Science **290**, 2120–2123 (2000)

4.39 L. Sun, P.C. Searson, L. Chien: Electrochemical deposition of nickel nanowire arrays in single-crystal mica films, Appl. Phys. Lett. **74**, 2803–2805 (1999)

4.40 K. Nielsch, R. Wehrspohn, S. Fischer, H. Kronmüller, J. Barthel, J. Kirschner, U. Gösele: Magnetic properties of 100 nm nickel nanowire arrays obtained from

4.41 Y. Wang, X. Jiang, T. Herricks, Y. Xia: Single crystalline nanowires of lead: Large-scale synthesis, mechanistic studies, and transport measurements, J. Phys. Chem. B **108**, 8631–8640 (2004)

ordered porous alumina templates, MRS Symp. Proc. **636**, D1.9-1–D1.9-6 (2001)

4.42 E. Lifshitz, M. Bashouti, V. Kloper, A. Kigel, M.S. Eisen, S. Berger: Synthesis and characterization of PbSe quantum wires, multipods, quantum rods, cubes, Nano Lett. **3**, 857–862 (2003)

4.43 W. Lu, P. Gao, W.B. Jian, Z.L. Wang, J. Fang: Perfect orientation ordered in-situ one-dimensional self-assembly of Mn-doped PbSe nanocrystals, J. Am. Chem. Soc. **126**, 14816–14821 (2004)

4.44 F. Favier, E.C. Walter, M.P. Zach, T. Benter, R.M. Penner: Hydrogen sensors and switches from electrodeposited palladium mesowire arrays, Science **293**, 2227–2231 (2001)

4.45 B. Gates, B. Mayers, B. Cattle, Y. Xia: Synthesis, characterization of uniform nanowires of trigonal selenium, Adv. Funct. Mater. **12**, 219–227 (2002)

4.46 C.A. Huber, T.E. Huber, M. Sadoqi, J.A. Lubin, S. Manalis, C.B. Prater: Nanowire array composites, Science **263**, 800–802 (1994)

4.47 Y. Cui, L.J. Lauhon, M.S. Gudiksen, J. Wang, C.M. Lieber: Diameter-controlled synthesis of single crystal silicon nanowires, Appl. Phys. Lett. **78**, 2214–2216 (2001)

4.48 A.M. Morales, C.M. Lieber: A laser ablation method for the synthesis of crystalline semiconductor nanowires, Science **279**, 208–211 (1998)

4.49 N. Wang, Y.F. Zhang, Y.H. Tang, C.S. Lee, S.T. Lee: SiO_2-enhanced synthesis of Si nanowires by laser ablation, Appl. Phys. Lett. **73**, 3902–3904 (1998)

4.50 M.K. Sunkara, S. Sharma, R. Miranda, G. Lian, E.C. Dickey: Bulk synthesis of silicon nanowires using a low-temperature vapor-liquid-solid method, Appl. Phys. Lett. **79**, 1546–1548 (2001)

4.51 S. Vaddiraju, H. Chandrasekaran, M.K. Sunkara: Vapor phase synthesis of tungsten nanowires, J. Am. Chem. Soc. **125**, 10792–10793 (2003)

4.52 J.P. Heremans, C.M. Thrush, D.T. Morelli, M.-C. Wu: Thermoelectric power of bismuth nanocomposites, Phys. Rev. Lett. **88**, 216801-1–216801-4 (2002)

4.53 Y. Li, G.S. Cheng, L.D. Zhang: Fabrication of highly ordered ZnO nanowire arrays in anodic alumina membranes, J. Mater. Res. **15**, 2305–2308 (2000)

4.54 P. Yang, H. Yan, S. Mao, R. Russo, J. Johnson, R. Saykally, N. Morris, J. Pham, R. He, H.-J. Choi: Controlled growth of ZnO nanowires and their optical properties, Adv. Funct. Mater. **12**, 323–331 (2002)

4.55 M.J. Zheng, L.D. Zhang, G.H. Li, W.Z. Shen: Fabrication and optical properties of large-scale uniform zinc oxide nanowire arrays by one-step electrochemical deposition technique, Chem. Phys. Lett. **363**, 123–128 (2002)

4.56 M.S. Dresselhaus, Y.-M. Lin, O. Rabin, A. Jorio, A.G. Souza Filho, M.A. Pimenta, R. Saito, G.G. Samsonidze, G. Dresselhaus: Nanowires and nanotubes, Mater. Sci. Eng. C **23**, 129–140 (2003), also in: Current trends in nanotechnologies: From materials to systems, Eur. Mater. Res. Soc. Symp. Proc., Vol. 140, ed. by W. Jantsch, H. Grimmeiss, G. Marietta (Elsevier, Amsterdam 2002)

4.57 R. Saito, G. Dresselhaus, M.S. Dresselhaus: *Physical Properties of Carbon Nanotubes* (Imperial College Press, London 1998)

4.58 M.S. Dresselhaus, G. Dresselhaus, P. Avouris: Carbon nanotubes: Synthesis, structure, properties and applications, Springer Ser. Top. Appl. Phys. **80**, 1–447 (2001)

4.59 R.C. Haddon: Special issue on carbon nanotubes, Acc. Chem. Res. **35**, 997–1113 (2002)

4.60 Y.-M. Lin, X. Sun, S. Cronin, Z. Zhang, J.Y. Ying, M.S. Dresselhaus: Fabrication, transport properties of Te-doped bismuth nanowire arrays. In: *Molecular Electronics: MRS Symposium Proceedings*, Vol. 582, ed. by S.T. Pantelides, M.A. Reed, J. Murday, A. Aviran (Materials Research Society Press, Pittsburgh 2000) pp. 1–6, Chap. H10.3

4.61 C.R. Martin: Nanomaterials: A membrane-based synthetic approach, Science **266**, 1961–1966 (1994)

4.62 G.A. Ozin: Nanochemistry: synthesis in diminishing dimensions, Adv. Mater. **4**, 612–649 (1992)

4.63 R.J. Tonucci, B.L. Justus, A.J. Campillo, C.E. Ford: Nanochannel array glass, Science **258**, 783–785 (1992)

4.64 J.Y. Ying: Nanoporous systems and templates, Sci. Spec. **18**, 56–63 (1999)

4.65 J.W. Diggle, T.C. Downie, C.W. Goulding: Anodic oxide films on aluminum, Chem. Rev. **69**, 365–405 (1969)

4.66 J.P. O'Sullivan, G.C. Wood: The morphology and mechanism of formation of porous anodic films on aluminum, Proc. R. Soc. Lond. A **317**, 511–543 (1970)

4.67 A.P. Li, F. Müller, A. Birner, K. Neilsch, U. Gösele: Hexagonal pore arrays with a 50-420nm interpore distance formed by self-organization in anodic alumina, J. Appl. Phys. **84**, 6023–6026 (1998)

4.68 J.P. Sullivan, G.C. Wood: The morphology, mechanism of formation of porous anodic films on aluminum, Proc. R. Soc. Lond. A **317**, 511–543 (1970)

4.69 O. Jessensky, F. Müller, U. Gösele: Self-organized formation of hexagonal pore arrays in anodic alumina, Appl. Phys. Lett. **72**, 1173–1175 (1998)

4.70 F. Li, L. Zhang, R.M. Metzger: On the growth of highly ordered pores in anodized aluminum oxide, Chem. Mater. **10**, 2470–2480 (1998)

4.71 H. Masuda, M. Satoh: Fabrication of gold nanodot array using anodic porous alumina as an evaporation mask, Jpn. J. Appl. Phys. **35**, L126–L129 (1996)

4.72 E. Ferain, R. Legras: Track-etched membrane – dynamics of pore formation, Nucl. Instrum. Methods B **84**, 331–336 (1993)

4.73 A. Blondel, J.P. Meier, B. Doudin, J.-P. Ansermet: Giant magnetoresistance of nanowires of multilayers, Appl. Phys. Lett. **65**, 3019–3021 (1994)

4.74 K. Liu, C.L. Chien, P.C. Searson, Y.Z. Kui: Structural and magneto-transport properties of electrodeposited bismuth nanowires, Appl. Phys. Lett. **73**, 1436–1438 (1998)

4.75 C.A. Huber, T.E. Huber: A novel microstructure: semiconductor-impregnated porous Vycor glass, J. Appl. Phys. **64**, 6588–6590 (1988)

4.76 J.S. Beck, J.C. Vartuli, W.J. Roth, M.E. Leonowicz, C.T. Kresge, K.D. Schmitt, C.T.-W. Chu, D.H. Olson, E.W. Sheppard, S.B. McCullen, J.B. Higgins, J.L. Schlenker: A new family of mesoporous molecular sieves prepared with liquid crystal templates, J. Am. Chem. Soc. **114**, 10834–10843 (1992)

4.77 C.-G. Wu, T. Bein: Conducting polyaniline filaments in a mesoporous channel host, Science **264**, 1757–1759 (1994)

4.78 Y.-M. Lin, S.B. Cronin, J.Y. Ying, M.S. Dresselhaus, J.P. Heremans: Transport properties of Bi nanowire arrays, Appl. Phys. Lett. **76**, 3944–3946 (2000)

4.79 T. Thurn-Albrecht, J. Schotter, G.A. Kästle, N. Emley, T. Shibauchi, L. Krusin-Elbaum, K. Guarini, C.T. Black, M.T. Tuominen, T.P. Russell: Ultrahigh-density nanowire arrays grown in self-assembled diblock copolymer templates, Science **290**, 2126–2129 (2000)

4.80 A.W. Adamson: *Physical Chemistry of Surfaces* (Wiley, New York 1982) p. 338

4.81 R. Ferré, K. Ounadjela, J.M. George, L. Piraux, S. Dubois: Magnetization processes in nickel and cobalt electrodeposited nanowires, Phys. Rev. B **56**, 14066–14075 (1997)

4.82 H. Zeng, M. Zheng, R. Skomski, D.J. Sellmyer, Y. Liu, L. Menon, S. Bandyopadhyay: Magnetic properties of self-assembled Co nanowires of varying length and diameter, J. Appl. Phys. **87**, 4718–4720 (2000)

4.83 Y. Peng, H.L. Zhang, S.-L. Pan, H.-L. Li: Magnetic properties and magnetization reversal of α-Fe nanowires deposited in alumina film, J. Appl. Phys. **87**, 7405–7408 (2000)

4.84 L. Piraux, J.M. George, J.F. Despres, C. Leroy, E. Ferain, R. Legras, K. Ounadjela, A. Fert: Giant magnetoresistance in magnetic multilayered nanowires, Appl. Phys. Lett. **65**, 2484–2486 (1994)

4.85 S. Bhattacharrya, S.K. Saha, D. Chakravorty: Nanowire formation in a polymeric film, Appl. Phys. Lett. **76**, 3896–3898 (2000)

4.86 G. Yi, W. Schwarzacher: Single crystal superconductor nanowires by electrodeposition, Appl. Phys. Lett. **74**, 1746–1748 (1999)

4.87 D. Al-Mawlawi, C.Z. Liu, M. Moskovits: Nanowires formed in anodic oxide nanotemplates, J. Mater. Res. **9**, 1014–1018 (1994)

4.88 R.S. Wagner, W.C. Ellis: Vapor-liquid-solid mechanism of single crystal growth, Appl. Phys. Lett. **4**, 89–90 (1964)

4.89 Y. Wu, P. Yang: Direct observation of vapor-liquid-solid nanowire growth, J. Am. Chem. Soc. **123**, 3165–3166 (2001)

4.90 Y. Wu, R. Fan, P. Yang: Block-by-block growth of single-crystalline Si/SiGe superlattice nanowires, Nano Lett. **2**, 83–86 (2002)

4.91 S. Sharma, M.K. Sunkara, R. Miranda, G. Lian, E.C. Dickey: A novel low temperature synthesis method for semiconductor nanowires. In: *Synthesis, Functional Properties and Applications of Nanostructures: Mat. Res. Soc. Symp. Proc.*, San Francisco, Vol. 676, ed. by H.W. Hahn, D.L. Feldheim, C.P. Kubiak, R. Tannenbaum, R.W. Siegel (Materials Research Society Press, Pittsburgh 2001) p. Y1.6

4.92 M.S. Gudiksen, L.J. Lauhon, J. Wang, D.C. Smith, C.M. Lieber: Growth of nanowire superlattice structures for nanoscale photonics and electronics, Nature **415**, 617–620 (2002)

4.93 Y. Wu, J. Xiang, C. Yang, W. Lu, C.M. Lieber: Single-crystal metallic nanowires and metal/semiconductor nanowire heterostructures, Nature **430**, 61–65 (2004)

4.94 M.T. Björk, B.J. Ohlsson, T. Sass, A.I. Persson, C. Thelander, M.H. Magnusson, K. Deppert, L.R. Wallenberg, L. Samuelson: One-dimensional steeplechase for electrons realized, Nano Lett. **2**, 87–89 (2002)

4.95 N. Wang, Y.H. Tang, Y.F. Zhang, C.S. Lee, S.T. Lee: Nucleation and growth of Si nanowires from silicon oxide, Phys. Rev. B **58**, R16024–R16026 (1998)

4.96 Y.F. Zhang, Y.H. Tang, N. Wang, C.S. Lee, I. Bello, S.T. Lee: One-dimensional growth mechanism of crystalline silicon nanowires, J. Cryst. Growth **197**, 136–140 (1999)

4.97 S.T. Lee, Y.F. Zhang, N. Wang, Y.H. Tang, I. Bello, C.S. Lee, Y.W. Chung: Semiconductor nanowires from oxides, J. Mater. Res. **14**, 4503–4507 (1999)

4.98 D.D.D. Ma, C.S. Lee, Y. Lifshitz, S.T. Lee: Periodic array of intramolecular junctions of silicon nanowires, Appl. Phys. Lett. **81**, 3233–3235 (2002)

4.99 D. Whang, S. Jin, C.M. Lieber: Large-scale hierarchical organization of nanowires for functional nanosystems, Jpn. J. Appl. Phys. **43**, 4465–4470 (2004)

4.100 M.P. Zach, K. Inazu, K.H. Ng, J.C. Hemminger, R.M. Penner: Synthesis of molybdenum nanowires with millimeter-scale lengths using electrochemical step edge decoration, Chem. Mater. **14**, 3206–3216 (2002)

4.101 B. Gates, Y. Yin, Y. Xia: A solution-phase approach to the synthesis of uniform nanowires of crystalline selenium with lateral dimensions in the range of 10-30 nm, J. Am. Chem. Soc. **122**, 12582–12583 (2000)

4.102 B. Mayers, B. Gates, Y. Yin, Y. Xia: Large-scale synthesis of monodisperse nanorods of Se/Te alloys through a homogeneous nucleation and solution growth process, Adv. Mater. **13**, 1380–1384 (2001)

4.103 B. Gates, Y. Wu, Y. Yin, P. Yang, Y. Xia: Single-crystalline nanowires of Ag_2Se can be synthesized by templating against nanowires of trigonal Se, J. Am. Chem. Soc. **123**, 11500–11501 (2001)

4.104 B. Gates, B. Mayers, Y. Wu, Y. Sun, B. Cattle, P. Yang, Y. Xia: Synthesis and characterization of crystalline Ag$_2$Se nanowires through a template-engaged reaction at room temperature, Adv. Funct. Mater. **12**, 679–686 (2002)

4.105 H. Yu, P.C. Gibbons, W.E. Buhro: Bismuth, tellurium and bismuth telluride nanowires, J. Mater. Chem. **14**, 595–602 (2004)

4.106 X. Peng, J. Wickham, A.P. Alivisatos: Kinetics of II–VI, III–V colloidal semiconductor nanocrystal growth: 'Focusing' of size distributions, J. Am. Chem. Soc. **120**, 5343–5344 (1998)

4.107 J.Y. Lao, J.G. Wen, Z.F. Ren: Hierarchical ZnO nanostructures, Nano Lett. **2**, 1287–1291 (2002)

4.108 J.Y. Lao, J.Y. Huang, D.Z. Wang, Z.F. Ren: ZnO nanobridges and nanonails, Nano Lett. **3**, 235–238 (2003)

4.109 N.A. Melosh, A. Boukai, F. Diana, B. Gerardot, A. Badolto, P.M. Petroff, J.R. Heath: Ultrahigh-density nanowire lattices and circuits, Science **300**, 112–115 (2003)

4.110 S.R. Nicewarner-Peña, R.G. Freeman, B.D. Reiss, L. He, D.J. Peña, I.D. Walton, R. Cromer, C.D. Keating, M.J. Natan: Submicrometer metallic barcodes, Science **294**, 137–141 (2001)

4.111 L.J. Lauhon, M.S. Gudiksen, D. Wang, C.M. Lieber: Epitaxial core-shell and core-multishell nanowire heterostructures, Nature **420**, 57–61 (2002)

4.112 Z.L. Wang, Z.R. Dai, R.P. Gao, Z.G. Bai, J.L. Gole: Side-by-side silicon carbide-silica biaxial nanowires: Synthesis, structure and mechanical properties, Appl. Phys. Lett. **77**, 3349–3351 (2000)

4.113 P. Yang, F. Kim: Langmuir–Blodgett assembly of one-dimensional nanostructures, ChemPhysChem **3**, 503–506 (2002)

4.114 B. Messer, J.H. Song, P. Yang: Microchannel networks for nanowire patterning, J. Am. Chem. Soc. **122**, 10232–10233 (2000)

4.115 P.A. Smith, C.D. Nordquist, T.N. Jackson, T.S. Mayer, B.R. Martin, J. Mbindyo, T.E. Mallouk: Electric-field assisted assembly and alignment of metallic nanowires, Appl. Phys. Lett. **77**, 1399–1401 (2000)

4.116 S. Jin, D.M. Whang, M.C. McAlpine, R.S. Friedman, Y. Wu, C.M. Lieber: Scalable interconnection and integration of nanowire devices without registration, Nano Lett. **4**, 915–919 (2004)

4.117 T. Kuykendall, P.J. Pauzauskie, Y.F. Zhang, J. Goldberger, D. Sirbuly, J. Denlinger, P.D. Yang: Crystallographic alignment of high-density gallium nitride nanowire arrays, Nat. Mater. **3**, 524–528 (2004)

4.118 H. Masuda, H. Yamada, M. Satoh, H. Asoh, M. Nakao, T. Tamamura: Highly ordered nanochannel-array architecture in anodic alumina, Appl. Phys. Lett. **71**, 2770–2772 (1997)

4.119 O. Rabin, P.R. Herz, S.B. Cronin, Y.-M. Lin, A.I. Akinwande, M.S. Dresselhaus: Nanofabrication using self-assembled alumina templates. In: *Nonlithographic and Lithographic Methods for Nanofabrication: MRS Symposium Proceedings, Boston*, Vol. 636, ed. by J.A. Rogers, A. Karim, L. Merhari, D. Norris, Y. Xia (Materials Research Society Press, Pittsburgh 2001) pp. D4.7-1–D4.7-6

4.120 O. Rabin, P.R. Herz, Y.-M. Lin, A.I. Akinwande, S.B. Cronin, M.S. Dresselhaus: Formation of thick porous anodic alumina films and nanowire arrays on silicon wafers and glass, Adv. Funct. Mater. **13**, 631–638 (2003)

4.121 O. Rabin, P.R. Herz, Y.-M. Lin, S.B. Cronin, A.I. Akinwande, M.S. Dresselhaus: Arrays of nanowires on silicon wafers, 21st Int. Conf. Thermoelectr. Proc. ICT '02 Long Beach (IEEE, Piscataway 2002) pp. 276–279

4.122 M.H. Huang, S. Mao, H. Feick, H. Yan, Y. Wu, H. Kind, E. Weber, R. Russo, P. Yang: Room-temperature ultraviolet nanowire nanolasers, Science **292**, 1897–1899 (2001)

4.123 Y.H. Tang, Y.F. Zhang, N. Wang, C.S. Lee, X.D. Han, I. Bello, S.T. Lee: Morphology of Si nanowires synthesized by high-temperature laser ablation, J. Appl. Phys. **85**, 7981–7983 (1999)

4.124 Y. Ding, Z.L. Wang: Structure analysis of nanowires and nanobelts by transmission electron microscopy, J. Phys. Chem. B **108**, 12280–12291 (2004)

4.125 S.B. Cronin, Y.-M. Lin, O. Rabin, M.R. Black, G. Dresselhaus, M.S. Dresselhaus, P.L. Gai: Bismuth nanowires for potential applications in nanoscale electronics technology, Microsc. Microanal. **8**, 58–63 (2002)

4.126 M.S. Sander, R. Gronsky, Y.-M. Lin, M.S. Dresselhaus: Plasmon excitation modes in nanowire arrays, J. Appl. Phys. **89**, 2733–2736 (2001)

4.127 L. Venkataraman, C.M. Lieber: Molybdenum selenide molecular wires as one-dimensional conductors, Phys. Rev. Lett. **83**, 5334–5337 (1999)

4.128 A. Majumdar: Scanning thermal microscopy, Annu. Rev. Mater. Sci. **29**, 505–585 (1999)

4.129 K.M. Unruh, T.E. Huber, C.A. Huber: Melting and freezing behavior of indium metal in porous glasses, Phys. Rev. B **48**, 9021–9027 (1993)

4.130 Y.Y. Wu, P.D. Yang: Melting and welding semiconductor nanowires in nanotubes, Adv. Mater. **13**, 520–523 (2001)

4.131 P.M. Ajayan, S. Iijima: Capillarity-induced filling of carbon nanotubes, Nature **361**, 333–334 (1993)

4.132 Y. Gao, Y. Bando: Carbon nanothermometer containing gallium, Nature **415**, 599 (2002)

4.133 J.L. Costa-Krämer, N. Garcia, H. Olin: Conductance quantization histograms of gold nanowires at 4 K, Phys. Rev. B **55**, 12910–12913 (1997)

4.134 M.E.T. Morales, A.G. Balogh, T.W. Cornelius, R. Neumann, C. Trautmann: Fragmentation of nanowires driven by Rayleigh instability, Appl. Phys. Lett. **84**, 5337–5339 (2004)

4.135 D.A. Wharam, T.J. Thornton, R. Newbury, M. Pepper, H. Ahmed, J.E.F. Frost, D.G. Hasko, D.C. Peacock, D.A. Ritchie, G.A.C. Jones: One-dimensional transport and the quantization of the ballistic resistance, J. Phys. C **21**, L209–L214 (1988)

4.136 B.J. van Wees, H. van Houten, C.W.J. Beenakker, J.G. Williamson, L.P. Kouvenhoven, D. van der Marel, C.T. Foxon: Quantized conductance of point contacts in a two-dimensional electron gas, Phys. Rev. Lett. **60**, 848–850 (1988)

4.137 C.J. Muller, J.M. van Ruitenbeek, L.J. de-Jongh: Conductance and supercurrent discontinuities in atomic-scale metallic constrictions of variable width, Phys. Rev. Lett. **69**, 140–143 (1992)

4.138 C.J. Muller, J.M. Krans, T.N. Todorov, M.A. Reed: Quantization effects in the conductance of metallic contacts at room temperature, Phys. Rev. B **53**, 1022–1025 (1996)

4.139 J.L. Costa-Krämer, N. Garcia, H. Olin: Conductance quantization in bismuth nanowires at 4 K, Phys. Rev. Lett. **78**, 4990–4993 (1997)

4.140 C.Z. Li, H.X. He, A. Bogozi, J.S. Bunch, N.J. Tao: Molecular detection based on conductance quantization of nanowires, Appl. Phys. Lett. **76**, 1333–1335 (2000)

4.141 J.L. Costa-Krämer, N. Garcia, P. Garcia-Mochales, P.A. Serena, M.I. Marques, A. Correia: Conductance quantization in nanowires formed between micro and macroscopic metallic electrodes, Phys. Rev. B **55**, 5416–5424 (1997)

4.142 Y. Huang, X. Duan, Y. Cui, L.J. Lauhon, K.-H. Kim, C.M. Lieber: Logic gates and computation from assembled nanowire building blocks, Science **294**, 1313–1317 (2001)

4.143 J.-R. Kim, J. Oh, H.M. So, J.-J. Kim, J. Kim, C.J. Lee, S.C. Lyu: Schottky diodes based on a single GaN nanowire, Nanotechnology **13**, 701–704 (2002)

4.144 X. Duan, Y. Huang, C.M. Lieber: Nonvolatile memory and programmable logic from molecule-gated nanowires, Nano Lett. **2**, 487–490 (2002)

4.145 E.C. Walter, R.M. Penner, H. Liu, K.H. Ng, M.P. Zach, F. Favier: Sensors from electrodeposited metal nanowires, Surf. Interface Anal. **34**, 409–412 (2002)

4.146 E.C. Walter, K.H. Ng, M.P. Zach, R.M. Penner, F. Favier: Electronic devices from electrodeposited metal nanowires, Microelectron. Eng. **61/62**, 555–561 (2002)

4.147 Y.-M. Lin, X. Sun, M.S. Dresselhaus: Theoretical investigation of thermoelectric transport properties of cylindrical Bi nanowires, Phys. Rev. B **62**, 4610–4623 (2000)

4.148 K. Liu, C.L. Chien, P.C. Searson: Finite-size effects in bismuth nanowires, Phys. Rev. B **58**, R14681–R14684 (1998)

4.149 Z. Zhang, X. Sun, M.S. Dresselhaus, J.Y. Ying, J. Heremans: Magnetotransport investigations of ultrafine single-crystalline bismuth nanowire arrays, Appl. Phys. Lett. **73**, 1589–1591 (1998)

4.150 J. Heremans, C.M. Thrush, Z. Zhang, X. Sun, M.S. Dresselhaus, J.Y. Ying, D.T. Morelli: Magnetoresistance of bismuth nanowire arrays: A possible transition from one-dimensional to three-dimensional localization, Phys. Rev. B **58**, R10091–R10095 (1998)

4.151 L. Sun, P.C. Searson, C.L. Chien: Finite-size effects in nickel nanowire arrays, Phys. Rev. B **61**, R6463–R6466 (2000)

4.152 Y.-M. Lin, S.B. Cronin, O. Rabin, J.Y. Ying, M.S. Dresselhaus: Transport properties and observation of semimetal-semiconductor transition in Bi-based nanowires. In: *Quantum Confined Semiconductor Nanostructures: MRS Symposium Proceedings, Boston*, Vol. 737-C, ed. by J.M. Buriak, D.D.M. Wayner, F. Priolo, B. White, V. Klimov, L. Tsybeskov (Materials Research Society Press, Pittsburgh 2003) p. F3.14

4.153 Y.-M. Lin, M.S. Dresselhaus: Transport properties of superlattice nanowires and their potential for thermoelectric applications. In: *Quantum Confined Semiconductor Nanostructures: MRS Symposium Proceedings, Boston*, Vol. 737-C, ed. by J.M. Buriak, D.D.M. Wayner, F. Priolo, B. White, V. Klimov, L. Tsybeskov (Materials Research Society Press, Pittsburgh 2003) p. F8.18

4.154 Y.-M. Lin, O. Rabin, S.B. Cronin, J.Y. Ying, M.S. Dresselhaus: Semimetal-semiconductor transition in $Bi_{1-x}Sb_x$ alloy nanowires and their thermoelectric properties, Appl. Phys. Lett. **81**, 2403–2405 (2002)

4.155 J. Heremans, C.M. Thrush, Y.-M. Lin, S.B. Cronin, M.S. Dresselhaus: Transport properties of antimony nanowires, Phys. Rev. B **63**, 085406-1–085406-8 (2001)

4.156 Y.-M. Lin, S.B. Cronin, O. Rabin, J.Y. Ying, M.S. Dresselhaus: Transport properties of $Bi_{1-x}Sb_x$ alloy nanowires synthesized by pressure injection, Appl. Phys. Lett. **79**, 677–679 (2001)

4.157 Z. Zhang, X. Sun, M.S. Dresselhaus, J.Y. Ying, J. Heremans: Electronic transport properties of single crystal bismuth nanowire arrays, Phys. Rev. B **61**, 4850–4861 (2000)

4.158 D.E. Beutler, N. Giordano: Localization and electron-electron interaction effects in thin Bi wires and films, Phys. Rev. B **38**, 8–19 (1988)

4.159 J. Heremans, C.M. Thrush: Thermoelectric power of bismuth nanowires, Phys. Rev. B **59**, 12579–12583 (1999)

4.160 L.D. Hicks, M.S. Dresselhaus: Thermoelectric figure of merit of a one-dimensional conductor, Phys. Rev. B **47**, 16631–16634 (1993)

4.161 Y.-M. Lin, S.B. Cronin, O. Rabin, J. Heremans, M.S. Dresselhaus, J.Y. Ying: Transport properties of Bi-related nanowire systems. In: *Anisotropic Nanoparticles: Synthesis, Characterization and Applications: MRS Symposium Proceedings, Boston*,

4.161 Vol. 635, ed. by S. Stranick, P.C. Searson, L.A. Lyon, C. Keating (Materials Research Society Press, Pittsburgh 2001) pp. C4301–C4306

4.162 Y.-M. Lin, M.S. Dresselhaus: Thermoelectric properties of superlattice nanowires, Phys. Rev. B **68**, 075304 (2003)

4.163 M.T. Björk, B.J. Ohlsson, C. Thelander, A.I. Persson, K. Deppert, L.R. Wallenberg, L. Samuelson: Nanowire resonant tunneling diodes, Appl. Phys. Lett. **81**, 4458–4460 (2002)

4.164 D. Li, Y. Wu, P. Kim, L. Shi, P. Yang, A. Majumdar: Thermal conductivity of individual silicon nanowires, Appl. Phys. Lett. **83**, 2934–2936 (2003)

4.165 T.S. Tighe, J.M. Worlock, M.L. Roukes: Direct thermal conductance measurements on suspended monocrystalline nanostructures, Appl. Phys. Lett. **70**, 2687–2689 (1997)

4.166 S.T. Huxtable, A.R. Abramson, C.-L. Tien, A. Majumdar, C. LaBounty, X. Fan, G. Zeng, J.E. Bowers, A. Shakouri, E.T. Croke: Thermal conductivity of Si/SiGe and SiGe/SiGe superlattices, Appl. Phys. Lett. **80**, 1737–1739 (2002)

4.167 R. Venkatasubramanian, E. Siivola, T. Colpitts, B. O'Quinn: Thin-film thermoelectric devices with high room-temperature figures of merit, Nature **413**, 597–602 (2001)

4.168 C. Dames, G. Chen: Modeling the thermal conductivity of a SiGe segmented nanowire, 21st Int. Conf. Thermoelectr. Proc. ICT '02, Long Beach (IEEE, Piscataway 2002) pp. 317–320

4.169 K. Schwab, J.L. Arlett, J.M. Worlock, M.L. Roukes: Thermal conductance through discrete quantum channels, Physica E **9**, 60–68 (2001)

4.170 D. Li, Y. Wu, R. Fan, P. Yang, A. Majumdar: Thermal conductivity of Si/SiGe superlattice nanowires, Appl. Phys. Lett. **83**, 3186–3188 (2003)

4.171 G. Chen, M.S. Dresselhaus, G. Dresselhaus, J.-P. Fleurial, T. Caillat: Recent developments in thermoelectric materials, Int. Mater. Rev. **48**, 45–66 (2003)

4.172 C. Dames, G. Chen: Theoretical phonon thermal conductivity of Si-Ge superlattice nanowires, J. Appl. Phys. **95**, 682–693 (2004)

4.173 K. Schwab, E.A. Henriksen, J.M. Worlock, M.L. Roukes: Measurement of the quantum of thermal conductance, Nature **404**, 974–977 (2000)

4.174 M. Cardona: *Light Scattering in Solids* (Springer, Berlin Heidelberg 1982)

4.175 P.Y. Yu, M. Cardona: *Fundamentals of Semiconductors* (Springer, Berlin Heidelberg 1995), Chap. 7

4.176 J.C.M. Garnett: Colours in metal glasses, in metallic films, and in metallic solutions, Philos. Trans. R. Soc. Lond. A **205**, 237–288 (1906)

4.177 D.E. Aspnes: Optical properties of thin films, Thin Solid Films **89**, 249–262 (1982)

4.178 U. Kreibig, L. Genzel: Optical absorption of small metallic particles, Surf. Sci. **156**, 678–700 (1985)

4.179 M.R. Black, Y.-M. Lin, S.B. Cronin, O. Rabin, M.S. Dresselhaus: Infrared absorption in bismuth nanowires resulting from quantum confinement, Phys. Rev. B **65**, 195417-1–195417-9 (2002)

4.180 M.W. Lee, H.Z. Twu, C.-C. Chen, C.H. Chen: Optical characterization of wurtzite gallium nitride nanowires, Appl. Phys. Lett. **79**, 3693–3695 (2001)

4.181 D.M. Lyons, K.M. Ryan, M.A. Morris, J.D. Holmes: Tailoring the optical properties of silicon nanowire arrays through strain, Nano Lett. **2**, 811–816 (2002)

4.182 M.S. Gudiksen, J. Wang, C.M. Lieber: Size-depent photoluminescence from single indium phosphide nanowires, J. Phys. Chem. B **106**, 4036–4039 (2002)

4.183 J.C. Johnson, H. Yan, R.D. Schaller, L.H. Haber, R.J. Saykally, P. Yang: Single nanowire lasers, J. Phys. Chem. B **105**, 11387–11390 (2001)

4.184 S. Blom, L.Y. Gorelik, M. Jonson, R.I. Shekhter, A.G. Scherbakov, E.N. Bogachek, U. Landman: Magneto-optics of electronic transport in nanowires, Phys. Rev. B **58**, 16305–16314 (1998)

4.185 J.P. Pierce, E.W. Plummer, J. Shen: Ferromagnetism in cobalt-iron alloy nanowire arrays on w(110), Appl. Phys. Lett. **81**, 1890–1892 (2002)

4.186 S. Melle, J.L. Menendez, G. Armelles, D. Navas, M. Vazquez, K. Nielsch, R.B. Wehrsphon, U. Gösele: Magneto-optical properties of nickel nanowire arrays, Appl. Phys. Lett. **83**, 4547–4549 (2003)

4.187 J.C. Johnson, H. Yan, R.D. Schaller, P.B. Petersen, P. Yang, R.J. Saykally: Near-field imaging of nonlinear optical mixing in single zinc oxide nanowires, Nano Lett. **2**, 279–283 (2002)

4.188 M.R. Black, Y.-M. Lin, S.B. Cronin, M.S. Dresselhaus: Using optical measurements to improve electronic models of bismuth nanowires, 21st Int. Conf. Thermoelectr. Proc. ICT '02, Long Beach, ed. by T. Caillat, J. Snyder (IEEE, Piscataway 2002) pp. 253–256

4.189 M.R. Black, P.L. Hagelstein, S.B. Cronin, Y.-M. Lin, M.S. Dresselhaus: Optical absorption from an indirect transition in bismuth nanowires, Phys. Rev. B **68**, 235417 (2003)

4.190 H.-L. Liu, C.-C. Chen, C.-T. Chia, C.-C. Yeh, C.-H. Chen, M.-Y. Yu, S. Keller, S.P. DenBaars: Infrared and Raman-scattering studies in single-crystalline GaN nanowires, Chem. Phys. Lett. **345**, 245–251 (2001)

4.191 H. Richter, Z.P. Wang, L. Ley: The one phonon Raman-spectrum in microcrystalline silicon, Solid State Commun. **39**, 625–629 (1981)

4.192 I.H. Campbell, P.M. Fauchet: The effects of microcrystal size and shape on the one phonon Raman-spectra of crystalline semiconductors, Solid State Commun. **58**, 739–741 (1986)

4.193 R. Gupta, Q. Xiong, C.K. Adu, U.J. Kim, P.C. Eklund: Laser-induced Fano resonance scattering in silicon nanowires, Nano Lett. **3**, 627–631 (2003)

4.194 X. Duan, Y. Huang, Y. Cui, J. Wang, C.M. Lieber: Indium phosphide nanowires as building blocks

4.195 Y. Cui, C.M. Lieber: Functional nanoscale electronic devices assembled using silicon nanowire building blocks, Science **291**, 851–853 (2001) [preceded by: for nanoscale electronic and optoelectronic devices, Nature **409**, 66–69 (2001)]

4.196 Y. Cui, X. Duan, J. Hu, C.M. Lieber: Doping and electrical transport in silicon nanowires, J. Phys. Chem. B **104**, 101–104 (2000)

4.197 G.F. Zheng, W. Lu, S. Jin, C.M. Lieber: Synthesis and fabrication of high-performance n-type silicon nanowire transistors, Adv. Mater. **16**, 1890–1891 (2004)

4.198 J. Goldberger, D.J. Sirbuly, M. Law, P. Yang: ZnO nanowire transistors, J. Phys. Chem. B **109**, 9–14 (2005)

4.199 D.H. Kang, J.H. Ko, E. Bae, J. Hyun, W.J. Park, B.K. Kim, J.J. Kim, C.J. Lee: Ambient air effects on electrical characteristics of gap nanowire transistors, J. Appl. Phys. **96**, 7574–7577 (2004)

4.200 S.-W. Chung, J.-Y. Yu, J.R. Heath: Silicon nanowire devices, Appl. Phys. Lett. **76**, 2068–2070 (2000)

4.201 C. Li, W. Fan, B. Lei, D. Zhang, S. Han, T. Tang, X. Liu, Z. Liu, S. Asano, M. Meyyappan, J. Han, C. Zhou: Multilevel memory based on molecular devices, Appl. Phys. Lett. **84**, 1949–1951 (2004)

4.202 B. Lei, C. Li, D.Q. Zhang, Q.F. Zhou, K. Shung, C.W. Zhou: Nanowire transistors with ferroelectric gate dielectrics: Enhanced performance and memory effects, Appl. Phys. Lett. **84**, 4553–4555 (2004)

4.203 H.T. Ng, J. Han, T. Yamada, P. Nguyen, Y.P. Chen, M. Meyyappan: Single crystal nanowire vertical surround-gate field-effect transistor, Nano Lett. **4**, 1247–1252 (2004)

4.204 M. Ding, H. Kim, A.I. Akinwande: Observation of valence band electron emission from n-type silicon field emitter arrays, Appl. Phys. Lett. **75**, 823–825 (1999)

4.205 F.C.K. Au, K.W. Wong, Y.H. Tang, Y.F. Zhang, I. Bello, S.T. Lee: Electron field emission from silicon nanowires, Appl. Phys. Lett. **75**, 1700–1702 (1999)

4.206 P.M. Ajayan, O.Z. Zhou: Applications of carbon nanotubes. In: *Carbon Nanotubes: Synthesis, Structure, Properties and Applications*, Springer Ser. Top. Appl. Phys., Vol. 80, ed. by M.S. Dresselhaus, G. Dresselhaus, P. Avouris (Springer, Berlin Heidelberg 2001) pp. 391–425

4.207 M. Lu, M.K. Li, Z.J. Zhang, H.L. Li: Synthesis of carbon nanotubes/si nanowires core-sheath structure arrays and their field emission properties, Appl. Surf. Sci. **218**, 196–202 (2003)

4.208 L. Vila, P. Vincent, L. Dauginet-DePra, G. Pirio, E. Minoux, L. Gangloff, S. Demoustier-Champagne, N. Sarazin, E. Ferain, R. Legras, L. Piraux, P. Legagneux: Growth and field-emission properties of vertically aligned cobalt nanowire arrays, Nano Lett. **4**, 521–524 (2004)

4.209 G. Dresselhaus, M.S. Dresselhaus, Z. Zhang, X. Sun, J. Ying, G. Chen: Modeling thermoelectric behavior in Bi nano-wires, 17th Int. Conf. Thermoelectr. Proc. ICT'98, Nagoya, ed. by K. Koumoto (IEEE, Piscataway 1998) pp. 43–46

4.210 L.D. Hicks, M.S. Dresselhaus: The effect of quantum well structures on the thermoelectric figure of merit, Phys. Rev. B **47**, 12727–12731 (1993)

4.211 O. Rabin, Y.-M. Lin, M.S. Dresselhaus: Anomalously high thermoelectric figure of merit in $Bi_{1-x}Sb_x$ nanowires by carrier pocket alignment, Appl. Phys. Lett. **79**, 81–83 (2001)

4.212 T.C. Harman, P.J. Taylor, M.P. Walsh, B.E. LaForge: Quantum dot superlattice thermoelectric materials and devices, Science **297**, 2229–2232 (2002)

4.213 X. Duan, Y. Huang, R. Agarwal, C.M. Lieber: Single-nanowire electrically driven lasers, Nature **421**, 241 (2003)

4.214 F. Qian, Y. Li, S. Gradecak, D.L. Wang, C.J. Barrelet, C.M. Lieber: Gallium nitride-based nanowire radial heterostructures for nanophotonics, Nano Lett. **4**, 1975–1979 (2004)

4.215 V. Dneprovskii, E. Zhukov, V. Karavanskii, V. Poborchii, I. Salamatini: Nonlinear optical properties of semiconductor quantum wires, Superlattice. Microst. **23**(6), 1217–1221 (1998)

4.216 J.C. Johnson, K.P. Knutsen, H. Yan, M. Law, Y. Zhang, P. Yang, R.J. Saykally: Ultrafast carrier dynamics in single ZnO nanowire and nanoribbon lasers, Nano Lett. **4**, 197–204 (2004)

4.217 J.X. Ding, J.A. Zapien, W.W. Chen, Y. Lifshitz, S.T. Lee, X.M. Meng: Lasing in ZnS nanowires grown on anodic aluminum oxide templates, Appl. Phys. Lett. **85**, 2361 (2004)

4.218 J.C. Johnson, H.-J. Choi, K.P. Knutsen, R.D. Schaller, P. Yang, R.J. Saykally: Single gallium nitride nanowire lasers, Nat. Mater. **1**, 106–110 (2002)

4.219 H.J. Choi, J.C. Johnson, R. He, S.K. Lee, F. Kim, P. Pauzauskie, J. Goldberger, R.J. Saykally, P. Yang: Self-organized GaN quantum wire UV lasers, J. Phys. Chem. B **107**, 8721–8725 (2003)

4.220 C.J. Barrelet, A.B. Greytak, C.M. Lieber: Nanowire photonic circuit elements, Nano Lett. **4**, 1981–1985 (2004)

4.221 M. Law, D.J. Sirbuly, J.C. Johnson, J. Goldberger, R.J. Saykally, P. Yang: Ultralong nanoribbon waveguides for sub-wavelength photonics integration, Science **305**, 1269–1273 (2004)

4.222 H. Kind, H. Yan, B. Messer, M. Law, P. Yang: Nanowire ultraviolet photodetectors and optical switches, Adv. Mater. **14**, 158–160 (2002)

4.223 B.M.I. van der Zande, M.R. Böhmer, L.G.J. Fokkink, C. Schöneberger: Colloidal dispersions of gold rods: synthesis and optical properties, Langmuir **16**, 451–458 (2000)

4.224 B.M.I. van der Zande, G.J.M. Koper, H.N.W. Lekkerkerker: Alignment of rod-shaped gold particles by electric fields, J. Phys. Chem. B **103**, 5754–5760 (1999)

4.225 W.U. Huynh, J.J. Dittmer, A.P. Alivisatos: Hybrid nanorod-polymer solar cells, Science **295**, 2425–2427 (2002)

4.226 L.A. Bauer, N.S. Birenbaum, G.J. Meyer: Biological applications of high aspect ratio nanoparticles, J. Mater. Chem. **14**, 517–526 (2004)

4.227 Y. Cui, Q. Wei, H. Park, C. Lieber: Nanowire nanosensors for highly sensitive and selective detection of biological and chemical species, Science **293**, 1289–1292 (2001)

4.228 J. Hahm, C. Lieber: Direct ultra-sensitive electrical detection of DNA and DNA sequence variations using nanowire nanosensors, Nano Lett. **4**, 51–54 (2004)

5. Template-Based Synthesis of Nanorod or Nanowire Arrays

Huamei (Mary) Shang, Guozhong Cao

This chapter introduces the fundamentals of and various technical approaches developed for template-based synthesis of nanorod arrays. After a brief introduction to various concepts associated with the growth of nanorods, nanowires and nanobelts, the chapter focuses mainly on the most widely used and well established techniques for the template-based growth of nanorod arrays: electrochemical deposition, electrophoretic deposition, template filling via capillary force and centrifugation, and chemical conversion. In each section, the relevant fundamentals are first introduced, and then examples are given to illustrate the specific details of each technique.

5.1	Template-Based Approach	170
5.2	Electrochemical Deposition	171
	5.2.1 Metals	172
	5.2.2 Semiconductors	173
	5.2.3 Conductive Polymers	174
	5.2.4 Oxides	174
5.3	Electrophoretic Deposition	175
	5.3.1 Polycrystalline Oxides	178
	5.3.2 Single Crystal Oxide Nanorod Arrays Obtained by Changing the Local pH	178
	5.3.3 Single Crystal Oxide Nanorod Arrays Grown by Homoepitaxial Aggregation	179
	5.3.4 Nanowires and Nanotubes of Fullerenes and Metallofullerenes	180
5.4	Template Filling	180
	5.4.1 Colloidal Dispersion (Sol) Filling	180
	5.4.2 Melt and Solution Filling	181
	5.4.3 Centrifugation	181
5.5	Converting from Reactive Templates	182
5.6	Summary and Concluding Remarks	182
References		183

Syntheses, characterizations and applications of nanowires, nanorods, nanotubes and nanobelts (also often referred to as one-dimensional nanostructures) are significant areas of current endeavor in nanotechnology. Many techniques have been developed in these areas, and our understanding of the field has been significantly enhanced [5.1–5]. The field is still evolving rapidly with new synthesis methods and new nanowires or nanorods reported in the literature. Evaporation–condensation growth has been successfully applied to the synthesis of various oxide nanowires and nanorods. Similarly, the dissolution–condensation method has been widely used for the synthesis of various metallic nanowires from solutions. The vapor–liquid–solid (VLS) growth method is a highly versatile approach; various elementary and compound semiconductor nanowires have been synthesized using this method [5.6]. Template-based growth of nanowires or nanorods is an even more versatile method for various materials. Substrate ledge or step-induced growth of nanowires or nanorods has also been investigated intensively [5.7]. Except for VLS and template-based growth, most of the above-mentioned methods result in randomly oriented nanowires or nanorods (commonly in the form of powder). The VLS method provides the ability to grow well oriented nanorods or nanowires directly attached to substrates, and is therefore often advantageous for characterization and applications; however, catalysts are required to form a liquid capsule at the advancing surface during growth at elevated temperatures. In addition, the possible incorporation of catalyst into nanowires and the difficulty removing such capsules from the tips of nanowires or nanorods are two disadvantages of this technique. Template-based growth often suffers from the polycrystalline nature of the resultant nanowires and nanorods, in addition to the difficulties involved in finding appropriate templates with pore channels of a desired diameter, length and surface chemistry and in removing the template completely without compromising the integrity of grown nanowires or nanorods. The discussion in this chapter will focus on nanorod and nanowire arrays, although nanotube arrays are mentioned briefly in conjunction with nanorod and nanowire fabrication. In addition, the terms of *nanorod* and *nanowire* are used interchangeably without special distinction in this chapter; this is commonplace in the literature.

In comparison with nanostructured materials in other forms, nanorod arrays offer several advantages for studying properties and for practical applications. Significant progress has been made in studies of the physical properties of individual nanowires and nanorods performed by directly measuring the properties of individual nanostructures. However, such studies generally require a lot of experimental preparation. For example, for electrical conductivity measurements, patterned electrodes are first created on a substrate, and then nanowires or nanorods are dispersed in an appropriate solvent or solution. This nanowire colloidal dispersion is then cast on the substrate containing pattern electrodes. Measurements are carried out after identifying individual nanowires or nanorods bridging two electrodes. The options for manipulating nanowires or nanorods are limited, and it is difficult to improve the contact between the sample and the electrodes to ensure the desired ohmic contact. For practical applications, the output or signal generated by single nanowire- or nanorod-based devices is small, and the signal-to-noise ratio is small, which means that highly sensitive instrumentation is required to accommodate such devices.

5.1 Template-Based Approach

The template approach to preparing free-standing, non-oriented and oriented nanowires and nanorods has been investigated extensively. The most commonly used and commercially available templates are anodized alumina membrane (AAM) [5.8] and radiation track-etched polycarbonate (PC) membranes [5.9]. Other membranes have also been used, such as nanochannel array on glass [5.10], radiation track-etched mica [5.11], mesoporous materials [5.12], porous silicon obtained via electrochemical etching of silicon wafer [5.13], zeolites [5.14] and carbon nanotubes [5.15,16]. Biotemplates have also been explored for the growth of nanowires [5.17] and nanotubes [5.18], such as Cu [5.19], Ni [5.17], Co [5.17], and Au [5.20] nanowires. Commonly used alumina membranes with uniform and parallel pores are produced by the anodic oxidation of aluminium sheet in solutions of sulfuric, oxalic, or phosphoric acids [5.8, 21]. The pores can be arranged in a regular hexagonal array, and densities as high as 10^{11} pores/cm^2 can be achieved [5.22]. Pore size ranging from 10 nm to 100 μm can be achieved [5.22, 23]. PC membranes are made by bombarding a nonporous polycarbonate sheet, typically 6 to 20 μm in thickness, with nuclear fission fragments to create damage tracks, and then chemically etching these tracks into pores [5.9]. In these radiation track-etched membranes, the pores are of uniform size (as small as 10 nm), but they are randomly distributed. Pore densities can be as high as 10^9 pores/cm^2.

In addition to the desired pore or channel size, morphology, size distribution and density of pores, template materials must meet certain requirements. First, the template materials must be compatible with the processing conditions. For example, an electrical insulator is required when a template is used in electrochemical deposition. Except in the case of template-directed synthesis, the template materials should be chemically and thermally inert during synthesis and the following processing steps. Secondly, the material or solution being deposited must wet the internal pore walls. Thirdly, for the synthesis of nanorods or nanowires, the deposition should start from the bottom or from one end of the template channel and proceed from one side to the other. However, for the growth of nanotubules, deposition should start from the pore wall and proceed inwardly. Inward growth may result in pore blockage, so this should be avoided during the growth of *solid* nanorods or nanowires. Kinetically, the correct amount of surface relaxation permits maximal packing density, so a diffusion-limited process is preferred. Other considerations include the ease of release of the nanowires or nanorods from the templates and the ease of handling during the experiments.

AAM and PC membranes are most commonly used for the synthesis of nanorod or nanowire arrays. Both templates are very convenient for the growth of nanorods by various growth mechanisms, but each type of template also has its disadvantages. The advantages of using PC as the template are its easy handling and easy removal by means of pyrolysis at elevated temperatures, but the flexibility of PC is more prone to distortion during the heating process, and removal of the template occurs before complete densification of the nanorods. These factors result in broken and deformed nanorods. The advantage of using AAM as the template is its rigidity and resistance to high temperatures, which allows the nanorods to densify completely before removal. This results in fairly free-standing and unidirectionally-aligned nanorod arrays with a larger surface area than for PC. The problem with AAM is the complete removal of the template after nanorod growth, which is yet to be achieved when using wet chemical etching.

5.2 Electrochemical Deposition

Electrochemical deposition, also known as electrodeposition, involves the oriented diffusion of charged reactive species through a solution when an external electric field is applied, and the reduction of the charged growth species at the growth or deposition surface (which also serves as an electrode). In industry, electrochemical deposition is widely used when coating metals in a process known as electroplating [5.25]. In general, this method is only applicable to electrically conductive materials such as metals, alloys, semiconductors, and electrically conductive polymers. After the initial deposition, the electrode is separated from the depositing solution by the deposit and so the deposit must conduct in order to allow the deposition process to continue. When the deposition is confined to the pores of template membranes, nanocomposites are produced. If the template membrane is removed, nanorod or nanowire arrays are prepared.

When a solid is immersed in a polar solvent or an electrolyte solution, surface charge will develop. The electrode potential is described by the Nernst equation

$$E = E_0 + \frac{RT}{n_i F} \ln(a_i) \,, \tag{5.1}$$

where E_0 is the standard electrode potential (or the potential difference between the electrode and the solution) when the activity a_i of the ions is unity, F is Faraday's constant, R is the gas constant, and T is the temperature. When the electrode potential is higher than

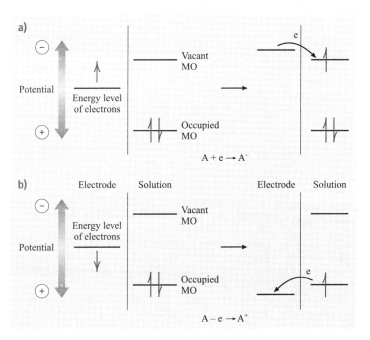

Fig. 5.1a,b Representation of the reduction (a) and oxidation (b) of a species A in solution. The molecular orbitals (MO) shown for species A are the highest occupied MO and the lowest vacant MO. These approximately correspond to the E_0's of the A/A$^-$ and A$^+$/A couples, respectively (after [5.24])

the energy level of a vacant molecular orbital in the electrolyte, electrons will transfer from the electrode to the solution and the electrolyte will be reduced, as shown in Fig. 5.1a [5.24]. On the other hand, if the electrode potential is lower than the energy level of an occupied molecular orbital in the electrolyte, the electrons will transfer from the electrolyte to the electrode, resulting in electrolyte oxidation, as illustrated in Fig. 5.1b [5.24]. These reactions stop when equilibrium is achieved.

When an external electric field is applied between two dissimilar electrodes, charged species flow from one electrode to the other, and electrochemical reactions occur at both electrodes. This process, called electrolysis, converts electrical energy to chemical potential.

The system used to perform electrolysis is called an electrolytic cell. In this cell, the electrode connected to the positive side of the power supply, termed the *anode*, is where an oxidation reaction takes place, whereas the electrode connected to the negative side of the power supply, the *cathode*, is where a reduction reaction proceeds, accompanied by deposition. Therefore, electrolytic deposition is also called cathode deposition, but it is most commonly referred to as electrochemical deposition or electrodeposition.

5.2.1 Metals

The growth of nanowires of conductive materials in an electric field is a self-propagating process [5.27]. Once the small rods form, the electric field and the density of current lines between the tips of nanowires and the opposing electrode are greater than that between two electrodes, due to the shorter distances between the nanowires and the electrodes. This ensures that the species being deposited is constantly attracted preferentially to the nanowire tips, resulting in continued growth. To better control the morphology and size, templates containing channels in the desired shape are used to guide the growth of nanowires. Figure 5.2 illustrates a common setup used for the template-based growth of nanowires [5.26]. The template is attached to the cathode, which is brought into contact with the deposition solution. The anode is placed in the deposition solution, parallel to the cathode. When an electric field is applied, cations diffuse through the channels and deposit on the cathode, resulting in the growth of nanowires inside the template. This figure also shows the current density at different stages of deposition when a constant electric field is applied. The current does not change significantly until the pores are completely filled, at which point the current increases rapidly due to improved contact with the electrolyte solution. The current saturates once the template surface is completely covered. This approach has yielded nanowires made from different metals, including Ni, Co, Cu and Au, with nominal pore diameters of between 10 and 200 nm. The nanowires were found to be true replicas of the pores [5.28]. *Possin* [5.11] prepared various metallic nanowires using radiation track-etched mica. Likewise, *Williams* and *Giordano* [5.29] produced silver nanowires with diameters of less than 10 nm. *Whitney* et al. [5.26] fabricated arrays of nickel and cobalt nanowires, also using PC templates. Single crystal bismuth nanowires have been grown in AAM using pulsed electrodeposition and Fig. 5.3 shows SEM and TEM images

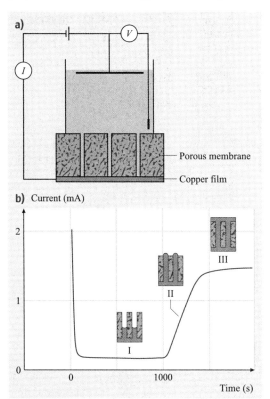

Fig. 5.2a,b Common experimental setup for the template-based growth of nanowires using electrochemical deposition. (**a**) Schematic illustration of the arrangement of the electrodes for nanowire deposition. (**b**) Current–time curve for electrodeposition of Ni into a polycarbonate membrane with 60 nm diameter pores at -1.0 V. Insets depict the different stages of the electrodeposition (after [5.26])

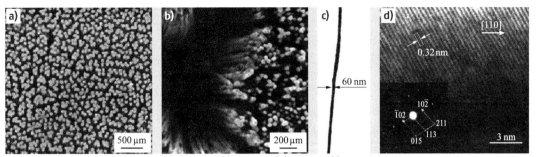

Fig. 5.3a–d SEM images of Bi nanowire arrays: (**a**) top view, (**b**) tilt view. (**c**) TEM image of a typical Bi single nanowire. (**d**) HRTEM image of a typical Bi single nanowire. The *inset* is the corresponding ED pattern (after [5.30])

of the bismuth nanowires [5.30]. Single crystal copper and lead nanowires were prepared by DC electrodeposition and pulse electrodeposition, respectively [5.31, 32]. The growth of single crystal lead nanowires required a greater departure from equilibrium conditions (greater overpotential) compared to the conditions required for polycrystalline ones.

Hollow metal tubules can also be prepared [5.33, 34]. In this case the pore walls of the template are chemically modified by anchoring organic silane molecules so that the metal will preferentially deposit onto the pore walls instead of the bottom electrode. For example, the porous surface of an anodic alumina template was first covered with cyanosilanes; subsequent electrochemical deposition resulted in the growth of gold tubules [5.35]. An electroless electrolysis process has also been investigated for the growth of nanowires and nanorods [5.16, 33, 36]. Electroless deposition is actually a chemical deposition process and it involves the use of a chemical agent to coat a material onto the template surface [5.37]. The main differences between electrochemical deposition and electroless deposition are that the deposition begins at the bottom electrode and the deposited materials must be electrically conductive in the former. The electroless method does not require the deposited materials to be electrically conductive, and the deposition starts from the pore wall and proceeds inwardly. Therefore, in general, electrochemical deposition results in the formation of *solid* nanorods or nanowires of conductive materials, whereas electroless deposition often results in hollow fibrils or nanotubules. For electrochemical deposition, the length of nanowires or nanorods can be controlled by the deposition time, whereas in electroless deposition the length of the nanotubules is solely dependent on the length of the deposition channels or pores. Variation of deposition time would result in a different wall thickness of nanotubules. An increase in deposition time leads to a thick wall, but sometimes the hollow tubule morphology persists even after prolonged deposition.

Although many research groups have reported on the growth of uniformly sized nanorods and nanowires on PC template membranes, *Schönenberger* et al. [5.38] reported that the channels of carbonate membranes were not always uniform in diameter. They grew Ni, Co, Cu, and Au nanowires using polycarbonate membranes with nominal pore diameters of between 10 and 200 nm by an electrolysis method. From both a potentiostatic study of the growth process and a SEM analysis of nanowire morphology, they concluded that the pores were generally not cylindrical with a constant cross section, but instead were rather cigarlike. For pores with a nominal diameter of 80 nm, the middle section of the pores was wider by up to a factor of 3.

5.2.2 Semiconductors

Semiconductor nanowire and nanorod arrays have been synthesized using AAM templates, such as CdSe and CdTe [5.39]. The synthesis of nanowire arrays of bismuth telluride (Bi_2Te_3) provide a good example of the synthesis of compound nanowire arrays by electrochemical deposition. Bi_2Te_3 is of special interest as a thermoelectric material and Bi_2Te_3 nanowire arrays are believed to offer high figures of merit for thermal-electrical energy conversion [5.40, 41]. Both polycrystalline and single crystal Bi_2Te_3 nanowire arrays have been grown by electrochemical deposition inside anodic alumina membranes [5.42, 43]. *Sander* and coworkers [5.42] fabricated Bi_2Te_3 nanowire arrays with diameters as small as ≈ 25 nm from a solution of 0.075 M Bi and 0.1 M Te in 1 M HNO_3 by electrochemical deposition at -0.46 V versus a Hg/Hg_2SO_4

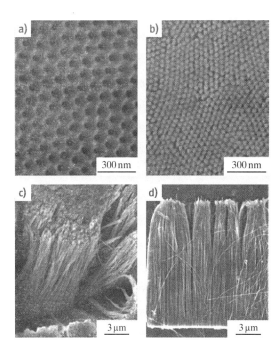

Fig. 5.4a–d SEM photographs of AAM template and Bi_2Te_3 nanowire arrays. (**a**) A typical SEM photograph of AAM. (**b**) Surface view of Bi_2Te_3 nanowire array (eroding time: 5 min). (**c**) Surface view of Bi_2Te_3 nanowire array (eroding time: 15 min). (**d**) Cross-sectional view of Bi_2Te_3 nanowire array (eroding time: 15 min) (after [5.43])

reference electrode. The resultant Bi_2Te_3 nanowire arrays are polycrystalline in nature, and subsequent melting-recrystallization failed to produce single crystal Bi_2Te_3 nanowires. More recently, single crystal Bi_2Te_3 nanowire arrays have been grown from a solution consisted of 0.035 M $Bi(NO_3)_3 \cdot 5H_2O$ and 0.05 M $HTeO_2^+$; the latter was prepared by dissolving Te powder in 5 M HNO_3 by electrochemical deposition. Figure 5.4a shows a typical SEM image of AAM. Both Fig. 5.4b and Fig. 5.4c are surface view of Bi_2Te_3 nanowire array with different eroding time, Fig. 5.4b is 5 min and Fig. 5.4c is 15 min. Figure 5.4d is cross-sectional view of Bi_2Te_3 nanowire array. Figure 5.5 shows TEM image of a cross section of a Bi_2Te_3 nanowire array and an XRD spectrum showing its crystal orientation, respectively. High-resolution TEM and electron diffraction, together with XRD, revealed that [110] is the preferred growth direction of Bi_2Te_3 nanowires. Single crystal nanowire or nanorod arrays can also be made by carefully controlling the initial deposition [5.44]. Simi-larly, large area Sb_2Te_3 nanowire arrays have also been successfully grown by template-based electrochemical deposition, but the nanowires grown are polycrystalline and show no clear preferred growth direction [5.45].

5.2.3 Conductive Polymers

Electrochemical deposition has also been explored for the synthesis of conductive polymer nanowire and nanorod arrays [5.46]. Conductive polymers have great potential for plastic electronics and sensor applications [5.47, 48]. For example, *Schönenberger* et al. [5.38] have made conductive polyporrole nanowires in PC membranes. Nanotubes are commonly observed for polymer materials, as shown in Fig. 5.6 [5.49], in contrast to *solid* metal nanorods or nanowires. It seems that deposition or solidification of polymers inside template pores starts at the surface and proceeds inwardly. *Martin* [5.50] proposed that this phenomenon was caused by the electrostatic attraction between the growing polycationic polymer and the anionic sites along the pore walls of the polycarbonate membrane. In addition, although the monomers are soluble, the polymerized form is insoluble. Hence there is a solvophobic component leading to deposition at the surface of the pores [5.51, 52]. In the final stage, the diffusion of monomers through the inner pores becomes retarded and monomers inside the pores are quickly depleted. The deposition of polymer inside the inner pores stops.

Liang et al. [5.53] reported a direct electrochemical synthesis of oriented nanowires of polyaniline (PANI) – a conducting polymer with a conjugated backbone due to phenyl and amine groups – from solutions using no templates. The experimental design is based on the idea that, in theory, the rate of electropolymerization (or nanowire growth) is related to the current density. Therefore, it is possible to control the nucleation and the polymerization rate simply by adjusting the current density. The synthesis involves electropolymerization of aniline ($C_6H_5NH_2$) and in situ electrodeposition, resulting in nanowire growth.

5.2.4 Oxides

Similar to metals, semiconductors and conductive polymers, some oxide nanorod arrays can be grown directly from solution by electrochemical deposition. For example, V_2O_5 nanorod arrays have been grown on ITO substrate from $VOSO_4$ aqueous solution with VO^{2+} as the growth species [5.54]. At the interface between the

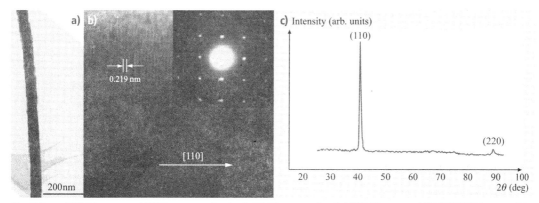

Fig. 5.5a–c TEM images and XRD pattern of a single Bi_2Te_3 nanowire. (**a**) TEM image and (**b**) HRTEM image of the same nanowire. The *inset* is the corresponding ED pattern. (**c**) XRD pattern of Bi_2Te_3 nanowire array (electrodeposition time: 5 min) (after [5.43])

Fig. 5.6a,b SEM images of polymer nanotubes (after [5.49])

electrode (and therefore the subsequent growth surface) and the electrolyte solution, the ionic cluster (VO^{2+}) is oxidized and solid V_2O_5 is deposited through the following reaction

$$2VO^{2+} + 3H_2O \rightarrow V_2O_5 + 6H^+ + 2e^- \,. \quad (5.2)$$

A reduction reaction takes place at the counter electrode

$$2H^+ + 2e^- \rightarrow H_2(g) \,. \quad (5.3)$$

It is obvious that the pH and the concentration of VO^{2+} clusters in the vicinity of the growth surface shift away from that in the bulk solution; both the pH and the VO^{2+} concentration decrease.

ZnO nanowire arrays were fabricated by a one-step electrochemical deposition technique based on an ordered nanoporous alumina membrane [5.55]. The ZnO nanowire array is uniformly assembled into the nanochannels of an anodic alumina membrane and consists of single crystal particles.

5.3 Electrophoretic Deposition

The electrophoretic deposition technique has been widely explored, particularly for the deposition of ceramic and organoceramic materials onto a cathode from colloidal dispersions [5.56–58]. Electrophoretic deposition differs from electrochemical deposition in several aspects. First, the material deposited in the electrophoretic deposition method does not need to be electrically conductive. Second, nanosized particles in colloidal dispersions are typically stabilized by electrostatic or electrosteric mechanisms. As discussed in the previous section, when dispersed in a polar solvent or an electrolyte solution, the surface of a nanoparticle develops an electrical charge via one or more of the following mechanisms: (1) preferential dissolution, (2) deposition of charges or charged species, (3) preferential reduction or oxidation, and (4) physical adsorption of charged

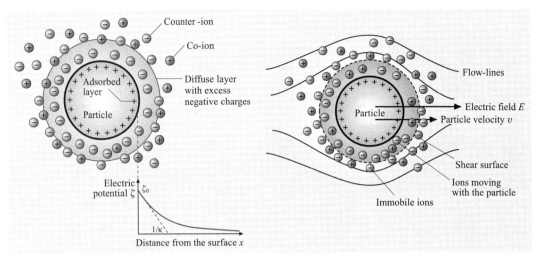

Fig. 5.7 Schematic illustrating electrical double layer structure and the electric potential near the solid surface with both the Stern and Gouy layers indicated. Surface charge is assumed to be positive (after [5.59])

species such as polymers. A combination of electrostatic forces, Brownian motion and osmotic forces results in the formation of a *double layer structure*, schematically illustrated in Fig. 5.7. The figure depicts a positively charged particle surface, the concentration profiles of negative ions (counterions) and positive ions (surface charge-determining ions), and the electric potential profile. The concentration of counterions gradually decreases with distance from the particle surface, whereas that of charge-determining ions increases. As a result, the electric potential decreases with distance. Near the particle surface, the electric potential decreases linearly, in the region known as the Stern layer. Outside of the Stern layer, the decrease follows an exponential relationship. The region between the Stern layer and the point where the electric potential equals zero is called the diffusion layer. Taken together, the Stern layer and diffusion layer is known as the double layer structure in the classical theory of electrostatic stabilization.

Upon the application of an external electric field, charged particles are set in motion, as schematically illustrated in Fig. 5.8 [5.59]. This type of motion is referred to as electrophoresis. When a charged particle moves, some of the solvent or solution surrounding the particle will also move with it, since part of the solvent or solution is tightly bound to the particle. The plane that separates the tightly bound liquid layer from the rest of the liquid is called the *slip plane* (Fig. 5.7). The electric potential at the slip plane is known as the *zeta potential*, which is an important parameter when determining the stability and transport of a colloidal dispersion or a sol. A zeta potential of more than about 25 mV is typically required to stabilize a system [5.60]. The zeta potential ζ around a spherical particle can be described using the relation [5.61]

$$\zeta = \frac{Q}{4\pi\varepsilon_r a(1+\kappa a)}$$

with

$$\kappa = \left(\frac{e^2 \sum n_i z_i^2}{\varepsilon_r \varepsilon_0 k_B T}\right)^{1/2}, \qquad (5.4)$$

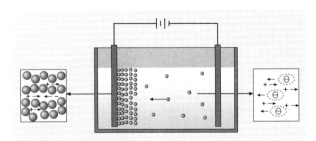

Fig. 5.8 Schematic showing electrophoresis. Upon application of an external electric field to a colloidal system or a sol, the charged nanoparticles or nanoclusters are set in motion (after [5.1])

where Q is the charge on the particle, a is the radius of the particle out to the shear plane, ε_r is the relative dielectric constant of the medium, and n_i and z_i are the bulk concentration and valence of the i-th ion in the system, respectively.

The mobility of a nanoparticle in a colloidal dispersion or a sol μ, is dependent on the dielectric constant of the liquid medium ε_r, the zeta potential of the nanoparticle ζ, and the viscosity η of the fluid. Several forms for this relationship have been proposed, such as the Hückel equation [5.59, 61, 63–65]

$$\mu = \frac{2\varepsilon_r \varepsilon_0 \zeta}{3\pi\eta} \, . \tag{5.5}$$

Electrophoretic deposition simply uses the oriented motion of charged particles in an electrical field to grow films or monoliths by transferring the solid particles from a colloidal dispersion or a sol onto the surface of an electrode. If the particles are positively charged (or more precisely, they have a positive zeta potential), deposition of solid particles will occur at the cathode. Otherwise, deposition will be at the anode. The electrostatic double layers collapse at the electrodes and the particles coagulate, producing porous materials made of compacted nanoparticles. Typical packing densities are far less than the theoretical density of 74 vol.% [5.66]. Many theories have been proposed to explain the processes at the deposition surface during electrophoretic deposition. However, the evolution of structure on the deposition surface is not well understood. The electrochemical processes that take place at the deposition surface and at the electrodes are complex and vary from system to system. The final density is dependent upon

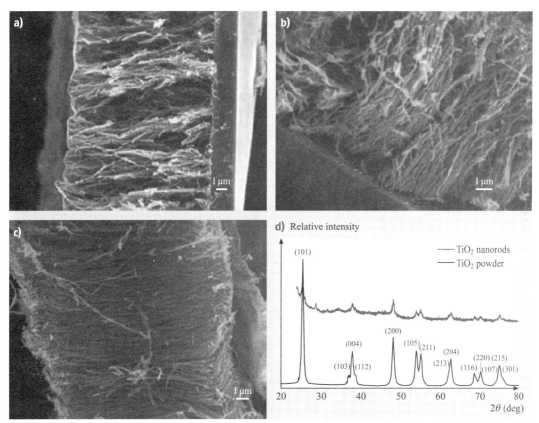

Fig. 5.9a–d SEM micrograph of TiO$_2$ nanorods grown by template-based electrochemically induced sol–gel deposition. The diameters of the nanorods are approximately: (**a**) 180 nm (for the 200 nm polycarbonate membrane); (**b**) 90 nm (for the 100 nm membrane); (**c**) 45 nm (for the 50 nm membrane). (**d**) XRD patterns of both the grown nanorods and a powder derived from the same sol. Both samples consist of the anatase phase only and no peak position shift was observed (after [5.62])

the concentration of particles in sols or colloidal dispersions, the zeta potential, the external electric field, and the reaction kinetics between the surfaces of the particles. A slow reaction and a slow arrival of nanoparticles onto the surface would allow sufficient particle relaxation on the deposition surface, so a high packing density would be expected.

5.3.1 Polycrystalline Oxides

Limmer et al. [5.62, 67–69] combined sol–gel preparation with electrophoretic deposition to prepare nanorods of various complex oxides. One of the advantages of this technique is the ability to synthesize complex oxides and organic–inorganic hybrids with desired stoichiometric compositions. Another advantage is their applicability to a variety of materials. In their approach, conventional sol–gel processing was applied to the synthesis of various sols. By controlling the sol preparation appropriately, nanometer particles of a desired stoichiometric composition were formed, and electrostatically stabilized by pH adjustment. Using radiation-tracked etched polycarbonate membranes with an electric field of ≈ 1.5 V/cm, they have grown nanowires with diameters ranging from 40 to 175 nm and lengths of 10 μm, corresponding to the thickness of the membrane. The materials include anatase TiO_2, amorphous SiO_2, perovskite $BaTiO_3$ and $Pb(Ti, Zr)O_3$, and layered perovskite $Sr_2Nb_2O_7$. Figure 5.9 shows SEM micrographs and XRD patterns of TiO_2 nanorod arrays [5.62].

Wang et al. [5.70] used electrophoretic deposition to make nanorods of ZnO from colloidal sols. ZnO colloidal sol was prepared by hydrolyzing an alcoholic solution of zinc acetate with NaOH, with a small amount of zinc nitrate added as a binder. This solution was then introduced into the pores of anodic alumina membranes at voltages of 10–400 V. It was found that lower voltages led to dense, solid nanorods, while higher voltages caused the formation of hollow tubules. They suggested that the higher voltages cause dielectric breakdown of the anodic alumina, causing it to become as positively charged as the cathode. Electrostatic attraction between the ZnO nanoparticles and the pore walls then leads to tubule formation.

5.3.2 Single Crystal Oxide Nanorod Arrays Obtained by Changing the Local pH

A modified version of sol electrophoretic deposition has been used to grow single crystalline titanium oxide and vanadium pentoxide nanorod arrays from TiO^{2+} and VO_2^+ solutions respectively. *Miao* et al. [5.71] prepared single crystalline TiO_2 nanowires by electrochemically induced sol–gel deposition. Titania electrolyte solution was prepared by dissolving Ti powder into a H_2O_2 and NH_4OH aqueous solution to form TiO^{2+} ionic clusters [5.72]. When an electric field was applied, the TiO^{2+} ionic clusters diffused to the cathode and underwent hydrolysis and condensation reactions, resulting in the deposition of nanorods of amorphous TiO_2 gel. After heating at 240 °C for 24 h in air, single crystal anatase nanorods with diameters of 10, 20, and 40 nm and lengths ranging from 2 to 10 μm were synthesized. The formation of single crystal TiO_2 nanorods here is different to that reported by *Martin*'s group [5.73]. It is suggested that the nanoscale crystallites generated during heating assembled epitaxially to form single crystal nanorods.

During typical sol–gel processing, nanoclusters are formed through homogeneous nucleation and subsequent growth through sequential yet parallel hydrolysis and condensation reactions. Sol electrophoretic deposition enriches and deposits these formed nanoclusters at an appropriate electrode surface under an external electric field. The modified process is to limit and induce the condensation reaction at the growth surface by changing local pH value, which is a result of partial water hydrolysis at the electrode or growth surface

$$2H_2O + 2e^- \rightarrow H_2 + 2OH^- \,, \tag{5.6}$$

$$2VO_2^+ + 2OH^- \rightarrow V_2O_5 + H_2O \,. \tag{5.7}$$

Reaction (5.6), or the electrolysis of water, plays a very important role here. As the reaction proceeds, hydroxyl groups are produced, resulting in increased pH near to the deposition surface. This increase in pH value near to the growth surface initiated and promotes the precipitation of V_2O_5, or reaction (5.7). The initial pH of the VO_2^+ solution is ≈ 1.0, meaning that VO_2^+ is stable. However, when the pH increases to ≈ 1.8, VO_2^+ is no longer stable and solid V_2O_5 forms. Since the change in pH occurs near to the growth surface, reaction (5.7) or deposition is likely to occur on the surface of the electrode through heterogeneous nucleation and subsequent growth. It should be noted that the hydrolysis of water has another effect on the deposition of solid V_2O_5. Reaction (5.6) produces hydrogen on the growth surface. These molecules may poison the growth surface before dissolving into the electrolyte or by forming a gas bubble, which may cause the formation of porous nanorods.

The formation of single crystal nanorods from solutions by pH change-induced surface condensation has been proven by TEM analyses, including high-resolution imaging showing the lattice fringes and electron diffraction. The growth of single crystal nanorods by pH change-induced surface condensation is attributed to evolution selection growth, which is briefly summarized below. The initial heterogeneous nucleation or deposition onto the substrate surface results in the formation of nuclei with random orientations. The subsequent growth of various facets of a nucleus is dependent on the surface energy, and varies significantly from one facet to another [5.74]. For one-dimensional growth, such as film growth, only the highest growth rate with a direction perpendicular to the growth surface will be able to continue to grow. The nuclei with the fastest growth direction perpendicular to the growth surface will grow larger, while nuclei with slower growth rates will eventually cease to grow. Such a growth mechanism results in the formation of columnar structured films where all of the grains have the same crystal orientation (known as textured films) [5.75, 76]. In the case of nanorod growth inside a pore channel, such evolution selection growth is likely to lead to the formation of a single crystal nanorod or a bundle of single crystal nanorods per pore channel. Figure 5.10 shows typical TEM micrographs and selected-area electron diffraction patterns of V_2O_5 nanorods. It is well known that [010] (the b-axis) is the fastest growth direction for a V_2O_5 crystal [5.77, 78], which would explain why single crystal vanadium nanorods or a bundle of single crystal nanorods grow along the b-axis.

5.3.3 Single Crystal Oxide Nanorod Arrays Grown by Homoepitaxial Aggregation

Single crystal nanorods can also be grown directly by conventional electrophoretic deposition. However, several requirements must be met for such growth. First, the nanoclusters or particles in the sol must have a crystalline structure extended to the surface. Second, the deposition of nanoclusters on the growth surface must have a certain degree of reversibility so that the nanoclusters can rotate or reposition prior to their irreversible incorporation into the growth surface. Thirdly, the deposition rate must be slow enough to permit sufficient time for the nanoclusters to rotate or reposition. Lastly, the surfaces of the nanoclusters must be free of strongly attached alien chemical species. Although precise control of all these parameters remains a challenge, the growth of single crystal nanorods through homoepitaxial aggregation of nanocrystals has been demonstrated [5.79, 80]. The formation of single crystalline vanadium pentoxide nanorods by template-based sol electrophoretic deposition can be attributed to homoepitaxial aggregation of crystalline nanoparticles. Thermodynamically it is favorable for the crystalline nanoparticles to aggregate epitaxially; this growth behavior and mechanism is well documented in the literature [5.81, 82]. In this growth mechanism, an initial weak interaction between two nanoparticles allows rotation and migration relative to each other. Obviously, homoepitaxial aggregation is a competitive process and porous structure is expected to form through this homoepitaxial aggregation (as schematically illustrated in Fig. 5.11). Vanadium oxide particles present in a typ-

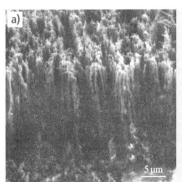
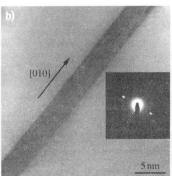
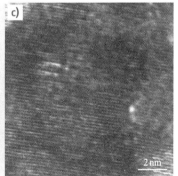

Fig. 5.10 (a) SEM image of V_2O_5 nanorod arrays on an ITO substrate grown in a 200 nm carbonate membrane by sol electrophoretic deposition; (b) TEM image of a V_2O_5 nanorod with its electron diffraction pattern; (c) high-resolution TEM image of the V_2O_5 nanorod showing the lattice fringes (after [5.54])

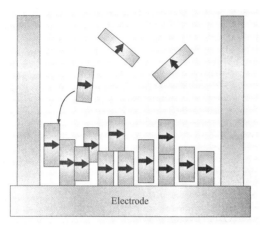

Fig. 5.11 Schematic illustration of the homoepitaxial aggregation growth mechanism of single-crystalline nanorods (after [5.54])

ical sol are known to easily form ordered crystalline structure [5.83], so it is reasonable to expect that homoepitaxial aggregation of vanadium nanocrystals from sol results in the formation of single crystal nanorods. Single crystal nanorods formed in this way are likely to undergo significant shrinkage when fired at high temperatures due to its original porous nature; 50% lateral shrinkage has been observed in vanadium pentoxide nanorods formed by this method. In addition, it might be possible that the electric field and the internal surfaces of the pore channels play significant roles in the orientation of the nanorods, as suggested in the literature [5.84, 85].

5.3.4 Nanowires and Nanotubes of Fullerenes and Metallofullerenes

Electrophoretic deposition in combination with template-based growth has also been successfully explored in the formation of nanowires and nanotubes of carbon fullerenes, such as C_{60} [5.86], or metallofullerenes, such as $Sc@C_{82}(I)$ [5.87]. Typical experiments include the purification or isolation of the fullerenes or metallofullerenes required using multiple-step liquid chromatography and dispersion of the fullerenes in a mixed solvent of acetonitrile/toluene in a ratio of 7 : 1. The electrolyte solution has a relatively low concentration of fullerenes (35 μM) and metallofullerenes (40 μM), and the electrophoretic deposition takes place with an externally applied electric field of 100–150 V with a distance of 5 mm between the two electrodes. Both nanorods and nanotubes of fullerenes or metallofullerenes can form and it is believed that initial deposition occurs along the pore surface. A short deposition time results in the formation of nanotubes, whereas extended deposition leads to the formation of solid nanorods. These nanorods possess either crystalline or amorphous structure.

5.4 Template Filling

Directly filling a template with a liquid mixture precursor is the most straightforward and versatile method for preparing nanowire or nanorod arrays. The drawback of this approach is that it is difficult to ensure complete filling of the template pores. Both nanorods and nanotubules can be obtained depending on the interfacial adhesion and the solidification modes. If the adhesion between the pore walls and the filling material is weak, or if solidification starts at the center (or from one end of the pore, or uniformly throughout the rods), solid nanorods are likely to form. If the adhesion is strong, or if the solidification starts at the interfaces and proceeds inwardly, hollow nanotubules are likely to form.

5.4.1 Colloidal Dispersion (Sol) Filling

Martin and coworkers [5.73, 88] have studied the formation of various oxide nanorods and nanotubules by simply filling the templates with colloidal dispersions (Fig. 5.12). Nanorod arrays of a mesoporous material (SBA-15) were recently synthesized by filling an ordered porous alumina membrane with sol containing surfactant (Pluronic P123) [5.89]. Colloidal dispersions were prepared using appropriate sol–gel processing techniques. The template was placed in a stable sol for various periods of time. The capillary force drives the sol into the pores if the sol has good wettability for the template. After the pores were filled with sol, the template was withdrawn from the sol and dried. The sample was fired at elevated temperatures to remove the template and to densify the sol–gel-derived green nanorods.

A sol typically consists of a large volume fraction of solvent, up to 90% or higher. Although the capillary force may ensure complete filling of the pores with the suspension, the amount of solid occupying the pore space is small. Upon drying and subsequent fir-

Fig. 5.12a–c SEM micrographs of oxide nanorods created by filling the templates with sol–gels: (**a**) ZnO, (**b**) TiO$_2$ and (**c**) hollow nanotube (after [5.73])

ing processes, significant shrinkage would be expected. However, the actual shrinkage observed is small when compared with the pore size. These results indicate that an (unknown) mechanism is acting to enrich the concentration of solid inside the pores. One possible mechanism could be the diffusion of solvent through the membrane, leading to the enrichment of solid on the internal surfaces of the template pores, similar to what happens during ceramic slip casting [5.90]. Figure 5.12a is a top view of ZnO nanotubules array, Fig. 5.12b is TiO$_2$ nanotubules array, Fig. 5.12c is hollow nanotube array. The observed formation of nanotubules (in Fig. 5.12 [5.73]) may imply that this process is indeed present. However, considering the fact that the templates were typically emerged into sol for just a few minutes, diffusion through the membrane and enrichment of the solid inside the pores must be rather rapid processes. It was also noticed that the nanorods made by template filling are commonly polycrystalline or amorphous, although single crystal TiO$_2$ nanorods were sometimes observed for nanorods smaller than 20 nm [5.73].

5.4.2 Melt and Solution Filling

Metallic nanowires can also be synthesized by filling a template with molten metals [5.91]. One example is the preparation of bismuth nanowires using pressure injection of molten bismuth into the nanochannels of an anodic alumina template [5.92]. The anodic alumina template was degassed and immersed in the liquid bismuth at 325 °C ($T_m = 271.5$ °C for Bi), and then high pressure Ar gas of ≈ 300 bar was applied in order to inject liquid Bi into the nanochannels of the template for 5 h. Bi nanowires with diameters of 13–110 nm and large aspect ratios (of up to several hundred) have been obtained. Individual nanowires are believed to be single-crystal. When exposed to air, bismuth nanowires are readily oxidized. An amorphous oxide layer ≈ 4 nm in thickness was observed after 48 h. After 4 weeks, the bismuth nanowires were completely oxidized. Nanowires of other metals, such as In, Sn and Al, and the semiconductors Se, Te, GaSb, and Bi$_2$Te$_3$, were also prepared by injecting molten liquid into anodic alumina templates [5.93].

Polymeric fibrils have been made by filling the template pores with a monomer solution containing the desired monomer and a polymerization reagent, followed by in situ polymerization [5.14, 94–97]. The polymer preferentially nucleates and grows on the pore walls, resulting in tubules at short deposition times.

Metal, oxide and semiconductor nanowires have recently been synthesized using self-assembled mesoporous silica as the template. For example, *Han* et al. [5.98] have synthesized Au, Ag and Pt nanowires in mesoporous silica templates. The mesoporous templates were first filled with aqueous solutions of the corresponding metal salts (such as HAuCl$_4$). After drying and treatment with CH$_2$Cl$_2$, the samples were reduced under H$_2$ flow to form metallic nanowires. *Liu* et al. [5.99] carefully studied the interface between these nanowires and the matrix using high-resolution electron microscopy and electron energy loss spectroscopy techniques. A sharp interface only exists between noble metal nanowires and the matrix. For magnetic nickel oxide, a core–shell nanorod structure containing a nickel oxide core and a thin nickel silicate shell was observed. The magnetic properties of the templated nickel oxide were found to be significantly different from nickel oxide nanopowders due to the alignment of the nanorods. In another study, *Chen* et al. filled the pores of a mesoporous silica template with an aqueous solution of Cd and Mn salts, dried the sample, and reacted it with H$_2$S gas to convert it to (Cd,Mn)S [5.100].

5.4.3 Centrifugation

Filling the template with nanoclusters via centrifugation forces is another inexpensive method for mass producing nanorod arrays. Figure 5.13 shows SEM images of lead zirconate titanate (PZT) nanorod arrays with uniform sizes and unidirectional alignment [5.101]. These nanorod arrays were grown in polycarbonate membrane from PZT sol by centrifugation at 1500 rpm for 60 min. The samples were attached to silica glass

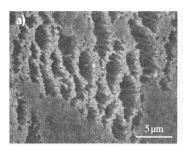

Fig. 5.13a,b SEM images of the top view (*left*) and side view (*right*) of lead zirconate titanate (PZT) nanorod arrays grown in polycarbonate membrane from PZT sol by centrifugation at 1500 rpm for 60 min. Samples were attached to silica glass and fired at 650 °C in air for 60 min (after [5.101])

and fired at 650 °C in air for 60 min. Nanorod arrays of other oxides (silica and titania) were prepared. The advantages of centrifugation include its applicability to any colloidal dispersion system, including those consisting of electrolyte-sensitive nanoclusters or molecules.

5.5 Converting from Reactive Templates

Nanorods or nanowires can also be synthesized using consumable templates, although the resultant nanowires and nanorods are generally not ordered to form aligned arrays. Nanowires of compounds can be prepared using a template-directed reaction. First nanowires or nanorods of one constituent element are prepared, and then these are reacted with chemicals containing the other element desired in order to form the final product. *Gates* et al. [5.102] converted single crystalline trigonal selenium nanowires into single crystalline nanowires of Ag_2Se by reacting Se nanowires with aqueous $AgNO_3$ solutions at room temperature. Nanorods can also be synthesized by reacting volatile metal halides or oxide species with carbon nanotubes to form solid carbide nanorods with diameters of between 2 and 30 nm and lengths of up to 20 μm [5.103]. ZnO nanowires were prepared by oxidizing metallic zinc nanowires [5.104]. Hollow nanotubules of $MoS_2 \approx 30$ μm long and 50 nm in external diameter with wall thicknesses of 10 nm were prepared by filling a solution mixture of the molecular precursors, $(NH_4)_2MoS_4$ and $(NH_4)_2Mo_3S_{13}$, into the pores of alumina membrane templates. Then the template filled with the molecular precursors was heated to an elevated temperature and the molecular precursors were thermally decomposed into MoS_2 [5.105]. Certain polymers and proteins were also used to direct the growth of nanowires of metals or semiconductors. For example, *Braun* et al. [5.106] reported a two-step procedure using DNA as a template for the vectorial growth of a silver nanorods 12 μm in length and 100 nm in diameter. CdS nanowires were prepared by polymer-controlled growth [5.107]. For the synthesis of CdS nanowires, cadmium ions were well distributed in a polyacrylamide matrix. The Cd^{2+}-containing polymer was treated with thiourea (NH_2CSNH_2) solvothermally in ethylenediamine at 170 °C, resulting in degradation of polyacrylamide. Single crystal CdS nanowires 40 nm in diameter and up to 100 μm in length were obtained with preferential [001] orientations.

5.6 Summary and Concluding Remarks

This chapter provides a brief summary of the fundamentals of and techniques used for the template-based synthesis of nanowire or nanorod arrays. Examples were used to illustrate the growth of each nanorod material made with each technique. The literature associated with this field is overwhelming and is expanding very rapidly. This chapter is by no means comprehensive in its coverage of the relevant literature. Four groups of template-based synthesis methods have been reviewed and discussed in detail. Electrochemical deposition or electrodeposition is the method used to grow electrically conductive or semiconductive materials, such as metals, semiconductors, and conductive polymers and oxides. Electrophoretic deposition from colloidal dispersion is the method used to synthesize dielectric nanorods and nanowires. Template filling is conceptually straightforward, although complete filling is often very difficult. Converting reactive templates is a method used to achieve both nanorod arrays and randomly oriented nanowires or nanorods, and it is often combined with other synthetic methods.

This chapter has focused on the growth of solid nanorod and nanowire arrays by template-based synthesis; however, the use of template-based synthesis to synthesize nanotubes, and in particular nanotube arrays, has received increasing attention [5.108]. One of the greatest advantages using template-based synthesis to grow of nanotubes and nanotube arrays is the independent control of the lengths, diameters, and the wall thicknesses of the nanotubes available. While the lengths and the diameters of the resultant nanotubes are dependent on the templates used for the synthesis, the wall thicknesses of the nanotubes can be readily controlled through the duration of growth. Another great advantage of the template-based synthesis of nanotubes is the possibility of multilayered hollow nanotube or solid nanocable structures. For example, Ni@$V_2O_5 \cdot nH_2O$ nanocable arrays have been synthesized by a two-step approach [5.109]. First, Ni nanorod arrays were grown in a PC template by electrochemical deposition, and then the PC template was removed by pyrolysis, followed by sol electrophoretic deposition of $V_2O_5 \cdot nH_2O$ on the surfaces of the Ni nanorod arrays. It is obvious that there is a lot of scope for more research into template-based syntheses of nanorod, nanotube and nanocable arrays, and their applications.

References

5.1 G.Z. Cao: *Nanostructures and Nanomaterials: Synthesis, Properties and Applications* (Imperial College, London 2004)

5.2 Z.L. Wang: *Nanowires and Nanobelts: Materials, Properties and Devices, Nanowires and Nonobelts of Functional Materials*, Vol. 2 (Kluwer, Boston 2003)

5.3 Y. Xia, P. Yang, Y. Sun, Y. Wu, Y. Yin, F. Kim, H. Yan: One-dimensional nanostructures: Synthesis, characterization and applications, Adv. Mater. **15**, 353–389 (2003)

5.4 A. Huczko: Template-based synthesis of nanomaterials, Appl. Phys. A **70**, 365–376 (2000)

5.5 C. Burda, X. Chen, R. Narayanan, M.A. El-Sayed: Chemistry and properties of nanocrystals of different shapes, Chem. Rev. **105**, 1025–1102 (2005)

5.6 X. Duan, C.M. Lieber: General synthesis of compound semiconductor nanowires, Adv. Mater. **12**, 298–302 (2000)

5.7 M.P. Zach, K.H. Ng, R.M. Penner: Molybdenum nanowires by electrodeposition, Science **290**, 2120–2123 (2000)

5.8 R.C. Furneaux, W.R. Rigby, A.P. Davidson: The formation of controlled-porosity membranes from anodically oxidized aluminium, Nature **337**, 147–149 (1989)

5.9 R.L. Fleisher, P.B. Price, R.M. Walker: *Nuclear Tracks in Solids* (Univ. of California Press, Berkeley 1975)

5.10 R.J. Tonucci, B.L. Justus, A.J. Campillo, C.E. Ford: Nanochannel array glass, Science **258**, 783–787 (1992)

5.11 G.E. Possin: A method for forming very small diameter wires, Rev. Sci. Instrum. **41**, 772–774 (1970)

5.12 C. Wu, T. Bein: Conducting polyaniline filaments in a mesoporous channel host, Science **264**, 1757–1759 (1994)

5.13 S. Fan, M.G. Chapline, N.R. Franklin, T.W. Tombler, A.M. Cassell, H. Dai: Self-oriented regular arrays of carbon nanotubes and their field emission properties, Science **283**, 512–514 (1999)

5.14 P. Enzel, J.J. Zoller, T. Bein: Intrazeolite assembly and pyrolysis of polyacrylonitrile, J. Chem. Soc. Chem. Commun. **8**, 633–635 (1992)

5.15 C. Guerret-Piecourt, Y. Le Bouar, A. Loiseau, H. Pascard: Relation between metal electronic structure and morphology of metal compounds inside carbon nanotubes, Nature **372**, 761–765 (1994)

5.16 P.M. Ajayan, O. Stephan, P. Redlich, C. Colliex: Carbon nanotubes as removable templates for metal oxide nanocomposites, nanostructures, Nature **375**, 564–567 (1995)

5.17 M. Knez, A.M. Bittner, F. Boes, C. Wege, H. Jeske, E. Maiâ, K. Kern: Biotemplate synthesis of 3-nm nickel and cobalt nanowires, Nano Lett. **3**, 1079–1082 (2003)

5.18 R. Gasparac, P. Kohli, M.O.M.L. Trofin, C.R. Martin: Template synthesis of nano test tubes, Nano Lett. **4**, 513–516 (2004)

5.19 C.F. Monson, A.T. Woolley: DNA-templated construction of copper nanowires, Nano Lett. **3**, 359–363 (2003)

5.20 Y. Weizmann, F. Patolsky, I. Popov, I. Willner: Telomerase-generated templates for the growing of metal nanowires, Nano Lett. **4**, 787–792 (2004)

5.21 A. Despic, V.P. Parkhuitik: *Modern Aspects of Electrochemistry*, Vol. 20 (Plenum, New York 1989)

5.22 D. Al Mawiawi, N. Coombs, M. Moskovits: Magnetic properties of Fe deposited into anodic aluminum oxide pores as a function of particle size, J. Appl. Phys. **70**, 4421–4425 (1991)

5.23 C.A. Foss, M.J. Tierney, C.R. Martin: Template-synthesis of infrared-transparent metal microcylin-

5.24 A.J. Bard, L.R. Faulkner: *Electrochemical Methods* (Wiley, New York 1980)

5.25 J.B. Mohler, H.J. Sedusky: *Electroplating for the Metallurgist, Engineer and Chemist* (Chemical Publishing, New York 1951)

5.26 T.M. Whitney, J.S. Jiang, P.C. Searson, C.L. Chien: Fabrication and magnetic properties of arrays of metallic nanowires, Science **261**, 1316–1319 (1993)

5.27 F.R.N. Nabarro, P.J. Jackson: Growth of crystal whiskers – A review. In: *Growth and Perfection of Crystals*, ed. by R.H. Doremus, B.W. Roberts, D. Turnbull (Wiley, New York 1958) pp. 11–102

5.28 B.Z. Tang, H. Xu: Preparation, alignment and optical properties of soluble poly(phenylacetylene)-wrapped carbon nanotubes, Macromolecules **32**, 2567–2569 (1999)

5.29 W.D. Williams, N. Giordano: Fabrication of 80 Å metal wires, Rev. Sci. Instrum. **55**, 410–412 (1984)

5.30 C.G. Jin, G.W. Jiang, W.F. Liu, W.L. Cai, L.Z. Yao, Z. Yao, X.G. Li: Fabrication of large-area single crystal bismuth nanowire arrays, J. Mater. Chem. **13**, 1743–1746 (2003)

5.31 M.E.T. Molares, V. Buschmann, D. Dobrev, R. Neumann, R. Scholz, I.U. Schuchert, J. Vetter: Single-crystalline copper nanowires produced by electrochemical deposition in polymeric ion track membranes, Adv. Mater. **13**, 62–65 (2001)

5.32 G. Yi, W. Schwarzacher: Single crystal superconductor nanowires by electrodeposition, Appl. Phys. Lett. **74**, 1746–1748 (1999)

5.33 C.J. Brumlik, V.P. Menon, C.R. Martin: Synthesis of metal microtubule ensembles utilizing chemical, electrochemical and vacuum deposition techniques, J. Mater. Res. **268**, 1174–1183 (1994)

5.34 C.J. Brumlik, C.R. Martin: Template synthesis of metal microtubules, J. Am. Chem. Soc. **113**, 3174–3175 (1991)

5.35 C.J. Miller, C.A. Widrig, D.H. Charych, M. Majda: Microporous aluminum oxide films at electrodes. 4. Lateral charge transport in self-organized bilayer assemblies, J. Phys. Chem. **92**, 1928–1936 (1988)

5.36 W. Han, S. Fan, Q. Li, Y. Hu: Synthesis of gallium nitride nanorods through a carbon nanotube-confined reaction, Science **277**, 1287–1289 (1997)

5.37 G.O. Mallory, J.B. Hajdu (Eds.): *Electroless Plating: Fundamentals and Applications* (AESF, Orlando 1990)

5.38 C. Schönenberger, B.M.I. van der Zande, L.G.J. Fokkink, M. Henny, C. Schmid, M. Krüger, A. Bachtold, R. Huber, H. Birk, U. Staufer: Template synthesis of nanowires in porous polycarbonate membranes: Electrochemistry and morphology, J. Phys. Chem. B **101**, 5497–5505 (1997)

5.39 J.D. Klein, R.D. Herrick II, D. Palmer, M.J. Sailor, C.J. Brumlik, C.R. Martin: Electrochemical fabrication of cadmium chalcogenide microdiode arrays, Chem. Mater. **5**, 902–904 (1993)

5.40 L.D. Hicks, M.S. Dresselhaus: Thermoelectric figure of merit of a one-dimensional conductor, Phys. Rev. B **47**, 679–682 (1993)

5.41 M.S. Dresselhaus, G. Dresselhaus, X. Sun, Z. Zhang, S.B. Cronin, T. Koga: Low-dimensional thermoelectric materials, Phys. Solid State **41**, 679–682 (1999)

5.42 M.S. Sander, R. Gronsky, T. Sands, A.M. Stacy: Structure of bismuth telluride nanowire arrays fabricated by electrodeposition into porous anodic alumina templates, Chem. Mater. **15**, 335–339 (2003)

5.43 C. Lin, X. Xiang, C. Jia, W. Liu, W. Cai, L. Yao, X. Li: Electrochemical fabrication of large-area, ordered Bi_2Te_3 nanowire arrays, J. Phys. Chem. B **108**, 1844–1847 (2004)

5.44 D.S. Xu, Y.J. Xu, D.P. Chen, G.L. Guo, L.L. Gui, Y.Q. Tang: Preparation of CdS single-crystal nanowires by electrochemically induced deposition, Adv. Mater. **12**, 520–522 (2000)

5.45 C. Lin, G. Zhang, T. Qian, X. Li, Z. Yao: Large-area Sb_2Te_3 nanowire arrays, J. Phys. Chem. B **109**, 1430–1432 (2005)

5.46 C. Jérôme, R. Jérôme: Electrochemical synthesis of polypyrrole nanowires, Angew. Chem. Int. Ed. **37**, 2488–2490 (1998)

5.47 A.G. MacDiarmid: Nobel lecture: "Synthetic metals": A novel role for organic polymers, Rev. Mod. Phys. **73**, 701–712 (2001)

5.48 K. Doblhofer, K. Rajeshwar: *Handbook of Conducting Polymers* (Marcel Dekker, New York 1998), Chap. 20

5.49 L. Dauginet, A.-S. Duwez, R. Legras, S. Demoustier-Champagne: Surface modification of polycarbonate and poly(ethylene terephthalate) films and membranes by polyelectrolyte, Langmuir **17**, 3952–3957 (2001)

5.50 C.R. Martin: Membrane-based synthesis of nanomaterials, Chem. Mater. **8**, 1739–1746 (1996)

5.51 C.R. Martin: Template synthesis of polymeric and metal microtubules, Adv. Mater. **3**, 457–459 (1991)

5.52 J.C. Hulteen, C.R. Martin: A general template-based method for the preparation of nanomaterials, J. Mater. Chem. **7**, 1075–1087 (1997)

5.53 L. Liang, J. Liu, C.F. Windisch Jr., G.J. Exarhos, Y. Lin: Assembly of large arrays of oriented conducting polymer nanowires, Angew. Chem. Int. Ed. **41**, 3665–3668 (2002)

5.54 K. Takahashi, S.J. Limmer, Y. Wang, G.Z. Cao: Growth and electrochemical properties of single-crystalline V_2O_5 nanorod arrays, Jpn. J. Appl. Phys. B **44**, 662–668 (2005)

5.55 M.J. Zheng, L.D. Zhang, G.H. Li, W.Z. Shen: Fabrication and optical properties of large-scale uniform zinc oxide nanowire arrays by one-step electrochemical deposition technique, Chem. Phys. Lett. **363**, 123–128 (2002)

5.56 I. Zhitomirsky: Cathodic electrodeposition of ceramic and organoceramic materials. Fundamental aspects, Adv. Colloid Interf. Sci. **97**, 279–317 (2002)

5.57 O.O. Van der Biest, L.J. Vandeperre: Electrophoretic deposition of materials, Annu. Rev. Mater. Sci. **29**, 327–352 (1999)

5.58 P. Sarkar, P.S. Nicholson: Electrophoretic deposition (EPD): Mechanism, kinetics, and application to ceramics, J. Am. Ceram. Soc. **79**, 1987–2002 (1996)

5.59 A.C. Pierre: *Introduction to Sol-Gel Processing* (Kluwer, Norwell 1998)

5.60 J.S. Reed: *Introduction to the Principles of Ceramic Processing* (Wiley, New York 1988)

5.61 R.J. Hunter: *Zeta Potential in Colloid Science: Principles and Applications* (Academic, London 1981)

5.62 S.J. Limmer, T.P. Chou, G.Z. Cao: A study on the growth of TiO_2 using sol electrophoresis, J. Mater. Sci. **39**, 895–901 (2004)

5.63 C.J. Brinker, G.W. Scherer: *Sol-Gel Science: the Physics and Chemistry of Sol-Gel Processing* (Academic, San Diego 1990)

5.64 J.D. Wright, N.A.J.M. Sommerdijk: *Sol-Gel Materials: Chemistry and Applications* (Gordon and Breach, Amsterdam 2001)

5.65 D.H. Everett: *Basic Principles of Colloid Science* (The Royal Society of Chemistry, London 1988)

5.66 W.D. Callister: *Materials Science and Engineering: An Introduction* (Wiley, New York 1997)

5.67 S.J. Limmer, S. Seraji, M.J. Forbess, Y. Wu, T.P. Chou, C. Nguyen, G.Z. Cao: Electrophoretic growth of lead zirconate titanate nanorods, Adv. Mater. **13**, 1269–1272 (2001)

5.68 S.J. Limmer, S. Seraji, M.J. Forbess, Y. Wu, T.P. Chou, C. Nguyen, G.Z. Cao: Template-based growth of various oxide nanorods by sol-gel electrophoresis, Adv. Funct. Mater. **12**, 59–64 (2002)

5.69 S.J. Limmer, G.Z. Cao: Sol-gel electrophoretic deposition for the growth of oxide nanorods, Adv. Mater. **15**, 427–431 (2003)

5.70 Y.C. Wang, I.C. Leu, M.N. Hon: Effect of colloid characteristics on the fabrication of ZnO nanowire arrays by electrophoretic deposition, J. Mater. Chem. **12**, 2439–2444 (2002)

5.71 Z. Miao, D. Xu, J. Ouyang, G. Guo, Z. Zhao, Y. Tang: Electrochemically induced sol-gel preparation of single-crystalline TiO_2 nanowires, Nano Lett. **2**, 717–720 (2002)

5.72 C. Natarajan, G. Nogami: Cathodic electrodeposition of nanocrystalline titanium dioxide thin films, J. Electrochem. Soc. **143**, 1547–1550 (1996)

5.73 B.B. Lakshmi, P.K. Dorhout, C.R. Martin: Sol-gel template synthesis of semiconductor nanostructures, Chem. Mater. **9**, 857–863 (1997)

5.74 A. van der Drift: Evolutionary selection, a principle governing growth orientation in vapor-deposited layers, Philips Res. Rep. **22**, 267–288 (1968)

5.75 G.Z. Cao, J.J. Schermer, W.J.P. van Enckevort, W.A.L.M. Elst, L.J. Giling: Growth of {100} textured diamond films by the addition of nitrogen, J. Appl. Phys. **79**, 1357–1364 (1996)

5.76 M. Ohring: *Materials Science of Thin Films* (Academic, San Diego 2001)

5.77 D. Pan, Z. Shuyuan, Y. Chen, J.G. Hou: Hydrothermal preparation of long nanowires of vanadium oxide, J. Mater. Res. **17**, 1981–1984 (2002)

5.78 V. Petkov, P.N. Trikalitis, E.S. Bozin, S.J.L. Billinge, T. Vogt, M.G. Kanatzidis: Structure of $V_2O_5 \cdot nH_2O$ xerogel solved by the atomic pair distribution function technique, J. Am. Chem. Soc. **124**, 10157–10162 (2002)

5.79 K. Takahashi, S.J. Limmer, Y. Wang, G.Z. Cao: Synthesis, electrochemical properties of single crystal V_2O_5 nanorod arrays by template-based electrodeposition, J. Phys. Chem. B **108**, 9795–9800 (2004)

5.80 G.Z. Cao: Growth of oxide nanorod arrays through sol electrophoretic deposition, J. Phys. Chem. B **108**, 19921–19931 (2004)

5.81 R.L. Penn, J.F. Banfield: Morphology development and crystal growth in nanocrystalline aggregates under hydrothermal conditions: Insights from titania, Geochim. Cosmochim. Acta **63**, 1549–1557 (1999)

5.82 C.M. Chun, A. Navrotsky, I.A. Aksay: Aggregation growth of nanometer-sized $BaTiO_3$ particles, Proc. Microsc. Microanal. (1995) pp. 188–189

5.83 J. Livage: Synthesis of polyoxovanadates via chimie douce, Coord. Chem. Rev. **178–180**, 999–1018 (1998)

5.84 K.V. Saban, J. Thomas, P.A. Varughese, G. Varghese: Thermodynamics of crystal nucleation in an external electric field, Cryst. Res. Technol. **37**, 1188–1199 (2002)

5.85 D. Grier, E. Ben-Jacob, R. Clarke, L.M. Sander: Morphology and microstructure in electrochemical deposition of zinc, Phys. Rev. Lett. **56**, 1264–1267 (1986)

5.86 C.J. Li, Y.G. Guo, B.S. Li, C.R. Wang, L.J. Wan, C.L. Bai: Template synthesis of $Sc@C_{82}$ (I) nanowires and nanotubes at room temperature, Adv. Mater. **17**, 71–73 (2005)

5.87 Y.G. Guo, C.J. Li, L.J. Wan, D.M. Chen, C.R. Wang, C.L. Bai, Y.G. Wang: Well-defined fullerene nanowire arrays, Adv. Funct. Mater. **13**, 626–630 (2003)

5.88 B.B. Lakshmi, C.J. Patrissi, C.R. Martin: Sol-gel template synthesis of semiconductor oxide micro- and nanostructures, Chem. Mater. **9**, 2544–2550 (1997)

5.89 Q. Lu, F. Gao, S. Komarneni, T.E. Mallouk: Ordered SBA-15 nanorod arrays inside a porous alumina membrane, J. Am. Chem. Soc. **126**, 8650–8651 (2004)

5.90 J.S. Reed: *Introduction to Principles of Ceramic Processing* (Wiley, New York 1988)

5.91 C.A. Huber, T.E. Huber, M. Sadoqi, J.A. Lubin, S. Manalis, C.B. Prater: Nanowire array composite, Science **263**, 800–802 (1994)

5.92 Z. Zhang, D. Gekhtman, M.S. Dresselhaus, J.Y. Ying: Processing and characterization of single-crystalline ultrafine bismuth nanowires, Chem. Mater. **11**, 1659–1665 (1999)

5.93 E.G. Wolff, T.D. Coskren: Growth, morphology of magnesium oxide whiskers, J. Am. Ceram. Soc. **48**, 279–285 (1965)

5.94 W. Liang, C.R. Martin: Template-synthesized polyacetylene fibrils show enhanced supermolecular order, J. Am. Chem. Soc. **112**, 9666–9668 (1990)

5.95 S.M. Marinakos, L.C. Brousseau III, A. Jones, D.L. Feldheim: Template synthesis of one-dimensional Au, Au-poly(pyrrole) and poly(pyrrole) nanoparticle arrays, Chem. Mater. **10**, 1214–1219 (1998)

5.96 H.D. Sun, Z.K. Tang, J. Chen, G. Li: Polarized Raman spectra of single-wall carbon nanotubes monodispersed in channels of $AlPO_4$-5 single crystals, Solid State Commun. **109**, 365–369 (1999)

5.97 Z. Cai, J. Lei, W. Liang, V. Menon, C.R. Martin: Molecular and supermolecular origins of enhanced electronic conductivity in template-synthesized polyheterocyclic fibrils. 1. Supermolecular effects, Chem. Mater. **3**, 960–967 (1991)

5.98 Y.J. Han, J.M. Kim, G.D. Stucky: Preparation of noble metal nanowires using hexagonal mesoporous silica SBA-15, Chem. Mater. **12**, 2068–2069 (2000)

5.99 J. Liu, G.E. Fryxell, M. Qian, L.-Q. Wang, Y. Wang: Interfacial chemistry in self-assembled nanoscale materials with structural ordering, Pure Appl. Chem. **72**, 269–279 (2000)

5.100 L. Chen, P.J. Klar, W. Heimbrodt, F. Brieler, M. Fröba: Towards ordered arrays of magnetic semiconductor quantum wires, Appl. Phys. Lett. **76**, 3531–3533 (2000)

5.101 T. Wen, J. Zhang, T.P. Chou, S.J. Limmer, G.Z. Cao: Template-based growth of oxide nanorod arrays by centrifugation, J. Sol-Gel Sci. Tech. **33**, 193–200 (2005)

5.102 B. Gates, Y. Wu, Y. Yin, P. Yang, Y. Xia: Single-crystalline nanowires of Ag_2Se can be synthesized by templating against nanowires of trigonal Se, J. Am. Chem. Soc. **123**, 11500–11501 (2001)

5.103 E.W. Wong, B.W. Maynor, L.D. Burns, C.M. Lieber: Growth of metal carbide nanotubes, nanorods, Chem. Mater. **8**, 2041–2046 (1996)

5.104 Y. Li, G.S. Cheng, L.D. Zhang: Fabrication of highly ordered ZnO nanowire arrays in anodic alumina membranes, J. Mater. Res. **15**, 2305–2308 (2000)

5.105 C.M. Zelenski, P.K. Dorhout: The template synthesis of monodisperse microscale nanofibers, nanotubules of MoS_2, J. Am. Chem. Soc. **120**, 734–742 (1998)

5.106 E. Braun, Y. Eichen, U. Sivan, G. Ben-Yoseph: DNA-templated assembly and electrode attachment of a conducting silver wire, Nature **391**, 775–778 (1998)

5.107 J. Zhan, X. Yang, D. Wang, S. Li, Y. Xie, Y. Xia, Y. Qian: Polymer-controlled growth of CdS nanowires, Adv. Mater. **12**, 1348–1351 (2000)

5.108 Y. Wang, K. Takahashi, H.M. Shang, G.Z. Cao: Synthesis, electrochemical properties of vanadium pentoxide nanotube arrays, J. Phys. Chem. **B109**, 3085–3088 (2005)

5.109 K. Takahashi, Y. Wang, G.Z. Cao: Ni-$V_2O_5 \cdot n H_2O$ core-shell nanocable arrays for enhanced electrochemical intercalation, J. Phys. Chem. B **109**, 48–51 (2005)

6. Templated Self-Assembly of Particles

Tobias Kraus, Heiko Wolf

Nanoparticles are frequently immobilized on substrates to use them as functional elements. In the resulting layer, the particles are accessible, so that their useful properties can be exploited, but their positions are fixed, so that their behavior is stable and reproducible. Frequently, the particles' positions have to be well defined. Templated assembly can position particles even in the low-nanometer size regime, and it can do so efficiently for many particles in parallel. Thus, nanoparticles become building blocks, capable of forming complex superstructures.

Templated assembly is based on a simple idea: particles are brought to a surface that has binding sites which strongly interact with the particles. Ideally, the particles adsorb solely at the predefined binding sites, thus creating the desired arrangement. In reality, it is often a challenge to reach good yields, high precision, and good specificity, in particular for very small particles. Since the method is very general, particles of various materials such as oxides, metals, semiconductors, and polymers can be arranged for applications ranging from microelectronics to optics and biochemistry.

6.1	The Assembly Process	189
	6.1.1 Energy and Length Scales	189
	6.1.2 Mobility, Stability, and Yield	191
	6.1.3 Large Binding Sites	193
	6.1.4 Thermodynamics, Kinetics, and Statistics	193
6.2	Classes of Directed Particle Assembly	194
	6.2.1 Assembly from the Gas Phase	194
	6.2.2 Assembly in the Liquid Phase	195
	6.2.3 Assembly at Gas–Liquid Interfaces	200
6.3	Templates	202
	6.3.1 Chemical Templates	203
	6.3.2 Charges and Electrodes	204
	6.3.3 Topographical Templates	204
	6.3.4 Advanced Templates	204
6.4	Processes and Setups	205
	6.4.1 Setups for Particle Assembly	205
	6.4.2 Particle Printing and Processing	206
6.5	Conclusions	206
References		207

Solid particles with sub-µm diameters are intriguing objects. They have a well-defined surface which is large compared with their volume, so that they interact strongly with their environment. At the same time, particles are clearly defined entities which can be mixed, purified, modified, and arranged into larger structures. This combination has made them popular in fields ranging from biology (where they carry analyte-binding molecules) to semiconductor fabrication (where they confine electrons) [6.1].

It is tempting to try and use such particles as nanoscale building blocks to create functional devices, be their function electronic, mechanical or chemical.

There are two prerequisites: first, particles with narrow size distribution and well-defined structures and surfaces have to be available from different materials in sufficient quantities. Second, these particles have to be arranged such that they provide the desired functionality. Templated particle assembly is one way to do so. A template defines the particle arrangement in advance according to the designer's wishes.

Producing particles of sufficient quality to be used as building blocks is not necessarily simple, but it can be done efficiently. Chemical methods are known to produce particles from very small clusters (with diameters in the low nanometer regime), various shapes of single crystals with diameters from 10 to about 100 nm, and larger particles with diameters up to micrometers. Some syntheses produce particles that are rather monodisperse, the best methods reaching coefficients of variation below 3%. This is still worse than, say, the relative size distribution of bricks in most buildings, but good enough for the particle to arrange spontaneously into ordered supercrystals [6.2]. The particles can be simple crystals or complex structures with a shell that differs from the core, for example, to protect the surface of the core [6.3]. Chemical methods readily produce such core–shell structures which would be exceedingly complicated to make using conventional methods.

As for their arrangement, particles down to about 100 μm in diameter are routinely handled using conventional pick-and-place techniques, a method widely used in industrial processes. Such serial methods become very time consuming at smaller scales, and they fail in the sub-μm regime, where adhesion forces render the simple maneuver of *putting down* a particle very challenging [6.4]. In this size range, particles are dominated by Brownian motion. They move randomly in their suspensions, and alternative assembly methods become necessary for their placement. Templated particle assembly is such an alternative strategy, based on a predefined surface that carries the information on the final particle placement. It can produce a variety of particle arrangements in parallel and over large areas (with typical lateral dimensions up to 10^6 particle diameters).

Templated assembly utilizes the strong interactions of particles with interfaces and their tendency to produce dense packings to create predictable arrangements on a patterned surface. Since the desirable arrangement depends on the desired material properties, it is an advantage of templated assembly to give the user great flexibility in attainable particle arrangements.

There are rather different motivations for the use of well-defined particle arrangements. If single-particle properties are to be exploited (for example, their small size, large surface-to-volume ratio or optical properties), it is often critical to know in advance the exact particle positions. Particles are then commonly arranged into spaced arrays, possibly with alignment marks. In a biological assay, for example, a fluorescence reader can find the individual particles in a regular array according to their position and record their optical properties to gain information on an analyte that had come into contact with the particles [6.5]. Similarly, if particles are used as memory elements [6.6], they need to be electrically addressed – a task that is greatly simplified if their positions are well known in advance.

Interacting particles can exhibit collective properties that depend on their relative arrangement. In the field of metamaterials, for example, the activity of many particles with sizes well below the wavelength of an incident electromagnetic wave leads to unusual far-field behavior [6.7]. From afar, the bulk metamaterial appears to have, for example, a negative refractive index. Optical metamaterials also include photonic crystals, which exhibit a photonic bandgap much like the electronic bandgap of semiconductors due to a periodic potential caused by regular crystals of spherical, diffracting particles. Templated assembly can create such dense structures with well-defined boundaries, and it can influence the packing itself by imposing a desired geometry on the first layer.

More complex structures, possibly including more than one particle type, offer even more complex functionalities. One popular target is *smart materials*, which react to a stimulus in a coherent and useful way. Much like the electronic properties of a semiconductor microchip lead to extremely complex electronic behavior, patterned materials formed from arranged particles might exhibit useful mechanical, thermal or other properties. Another application of such complex structures (which are hard to produce) is anticounterfeiting, where an object is protected by a small particle structure with a unique property that can be detected.

Templated assembly is, of course, competing with more traditional means of micro- and nanofabrication, as covered in other chapters of this Handbook. Templated assembly is advantageous in that it takes advantage of the chemically produced small dimensions of nanoparticles, and it is more general than traditional methods in that it can process a wide variety of available colloids. The actual assembly process can be rather simple and compatible with continuous processing, even under ambient conditions. The most challenging prerequisite is usually the template, which has to be fabricated to provide sufficient definition of the assembled structure.

A process that arranges particles into a regular structure without any template is often called *self-assembly*. Here, the information on the arrangement is not contained in a template but in the properties of the particles themselves. The problem of *programming* the assembly process is thus shifted to the particles, which have to be chosen (or modified) such that they assemble into a certain structure. This is not an easy task, and there are few examples so far of rational materials design using engineered particles. A template, on the other hand, can be defined using classical top-down methods, which provide great flexibility.

Still, templates become hard to fabricate if the particles are small and high patterning resolution is required. A combination of self-assembly and templated assembly is then useful: boundaries are defined by the template, but additional effects such as particle–particle, particle–surface or particle–solvent interactions lead to a predictable particle arrangement inside the boundaries.

We will limit ourselves here to processes with surface-bound templates and disregard supramolecular assembly, although molecular cages might also be regarded as templates. Likewise, biomineralization processes which can be templated using certain surfaces will not be covered here. The main focus is on sub-µm particles that are hard to place using any other method but can be assembled with high quality by means of templated assembly processes.

Even today, larger particles (between ≈ 1 and $100\,\mu m$) are assembled using templated assembly methods, mostly from slurries in an approach called *fluidic assembly* [6.8]. Illumina, Inc. arranges 3 µm-diameter glass beads functionalized with short DNA strands into a regular grid, which can then be used for DNA sequencing. Alien Technology Corporation holds several patents covering the integration of semiconductor pieces into polymers and other carriers, which today it mainly uses for the production of radiofrequency identification (RFID) chips, in which small electronic radiofrequency components are mounted on a paper or polymer label which is attached to an item for wireless identification. Similar methods for much smaller particles are currently being developed, but have not yet been applied industrially.

The challenges that occur when going down in particle size are mostly due to the greater influence of Brownian motion, which disturbs any order formed; strong adhesion to surfaces, which increases unspecific adsorption and makes pick-and-place difficult; and the problem of process control as the particles become harder to resolve with conventional optical methods.

In addition, the dimensions of the targeted nanostructured materials are often comparable to those produced with larger particles, but the number of particles involved is now very much higher (scaling inversely with the particle volume). Even assembly methods with very high yields are therefore bound to produce defects, which might hinder the function of the material. In some interesting applications (such as optical metamaterials), the absolute placement accuracies required to create a discernible optical effect are strict. Templated assembly is in principle able to provide such accuracies – even for many particles – and we will discuss its prerequisites in the next section.

6.1 The Assembly Process

Templated particle assembly involves particle adsorption on surfaces, and the well-developed ideas from adsorption theory (treated in many monographs and reviews) also hold for the case of templated assembly. While in many classical adsorption processes adsorption occurs at unpredictable positions, often until the entire surface is covered, the goal of a templated assembly process is the arrangement of particles with great precision and specificity. In this chapter, we will review some concepts that are less prominent in the adsorption literature. A useful metaphor of the directed assembly process is the energy landscape, which we will introduce here and frequently use to illustrate effects of interaction lengths, particle mobility, time scales, and other features of assembly methods.

6.1.1 Energy and Length Scales

A driving force that brings a colloidal object to a defined position and holds it there has to overcome Brownian motion. This constitutes the minimum requirement for the design of a templated assembly process. In the absence of a driving force, the particle will deviate from its original position r_0 according to

$$\frac{1}{3}\langle(r-r_0)^2\rangle = \frac{k_B T}{6\pi a \eta} t = 2Dt , \qquad (6.1)$$

depending on the temperature T, the particle diameter a, the viscosity of the surrounding fluid η, the time t, and Boltzmann's constant k_B [6.9]. Thus, when averaging over a very large number of particles, a 10 nm-diameter particle in water would move about 51 µm in 60 s.

Table 6.1 Interactions that can drive particle assembly processes

Interaction	Typical range (order of magnitude)
Covalent	0.1 nm
van der Waals	1 nm
Coulomb (electrostatic)	1 nm (polar)–100 nm (apolar)
Hydrophobic	1 nm
Capillary	1 mm

The goal of an assembly process is to overcome this random, diffusional motion (with an energy scale of $k_B T$ and characterized by the diffusion coefficient D) by a bias that induces drift so that the probability of finding a particle at the desired position is markedly increased. Particles are then held in place until the system is quenched in some way, for example, by exchanging its environment.

In order to arrange the particles, templated assembly processes use potentials with minima at the particles' target positions. Such potential wells can be defined using various particle–surface interactions, some of which are listed in Table 6.1. These interactions act over different lengths, have different strengths, and form minima with different geometries, all of which can influence the assembly process.

Let us consider a particle that is moving in a fluid in the vicinity of a surface with *binding sites*, that is, features that interact with the particle more strongly than does the rest of the surface. The particle is mobile and moves randomly due to thermal excitation. Figure 6.1 illustrates this situation: depending on its position, the free energy of the particle will change as the interaction with the binding sites changes. If there is a gradient present, a directing force will act on the particle and bias its random motion towards an energy minimum. This *energy landscape*, formed by the superposition of the interaction, governs the particle's motion.

Some interactions are strong but short-ranged, for example, covalent bonds. In the energy landscape picture shown in Fig. 6.1, they will resemble a steep well into which the particle falls and from which it can hardly escape. On the other hand, the particle can be in close proximity to such steep wells and still not feel their presence. More precisely, the probability distribution of its presence will only be affected locally. When the particle is trapped inside the well, and if the entrapment can be reasonably modeled using a harmonic oscillator, its deviations from the minimum at $x = 0$ equals [6.10]

$$\left\langle x^2 \right\rangle = \frac{k_B T}{m \omega_0^2} \qquad (6.2)$$

for a particle with mass m that is bound as in a harmonic oscillator with a frequency $\omega = \sqrt{k/m}$, the square root of the spring constant over the particle mass. Thus, a steep potential minimum can trap a particle with high accuracy: if the oscillator has a frequency of 1 GHz, a 10 nm particle of gold will deviate by less than a nanometer. The prototypical example of such a strong binding site is a topographical hole from which the particle cannot escape. The walls provide very steep exclusion potentials. Much less steep, but affecting a larger volume, is the well formed by an electrostatic field. In practice, even if the theoretical assembly accuracy of such an electrostatic binding site is limited, it often provides very good results. Other factors turn out to be critical as well – in particular the minimal achievable size of the binding sites and the yield of assembly. An assembled particle can simply block a binding site geometrically by not letting any other particle sufficiently close to the site, but it can also neutralize its charge (at least partially) and therefore hinder the adsorption of additional particles. Such changes in the energy landscape due to adsorption are often critical for the specificity and kinetics of the assembly.

Some of the most relevant interactions in directed assembly processes are summarized in Table 6.1. The exact shapes of the energy landscape caused by a particular interaction potential depend critically on the binding-site geometry, while the interaction lengths depend mainly on the used materials, solvents, and surfactants. Electrostatic interactions in suspensions are

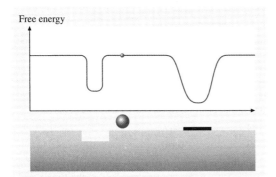

Fig. 6.1 A particle moving in an energy landscape during templated assembly. Its trajectory depends on the shape of the potential wells created by the binding sites, which also influence yield and accuracy of assembly

subject to shielding by ions from the solvent; their strength can also depend on the hydrodynamic situation. van der Waals interactions depend on the dielectric properties of the solvent: their interaction length is generally so short that they do not *funnel* particles from the bulk but trap particles that randomly hit the surface or were attracted by other forces. Supramolecular interactions are not included here because they are too diverse; in general, such interactions tend to be similarly short-ranged as van der Waals interactions. In three-phase systems, capillary forces can occur and exert very long-ranged forces even on small particles.

An important practical limit of the assembly accuracy is the template. The template has to be fabricated, often using top-down methods, to define the final particle positions. It may have binding sites that are large enough for many particles to be trapped inside, either in ordered arrangements or in disordered layers. On the other hand, it may have binding sites that are small enough to accommodate only a single particle. If so, the area of a binding site usually has to be on the order of the particle's projected area. A particle that comes into contact with the binding site might be irreversibly adsorbed immediately. In the energy landscape picture, this would correspond to a well with steep walls and a flat base. On the other hand, if the well has slanted walls and a small base, the particle can align with the binding site with better placement accuracy (Fig. 6.2). If it is not possible to pattern the template with very small binding sites, one either has to accept limited placement accuracy or employ an additional *focusing* mechanism so that the particle will be deposited at a well-defined position inside the binding site. One example is the combination of electrostatic and capillary interactions [6.11].

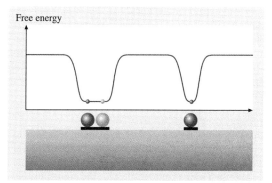

Fig. 6.2 Placement accuracy and yield depend on the geometry and potential shape of the binding sites

Various forces can occur in combination or subsequently during an assembly process. For example, in the classical example of the *convective assembly* of particles in a thin wetting film, hydrodynamic drag and capillary forces act in different stages of the assembly, yielding two-dimensional crystals of particles [6.12]. In templated assembly, one can use such combined effects to cause additional confinement. Aizenberg and her group have shown that the combination of capillary and electrostatic forces produces a *focusing* effect when particles are assembled on larger patches [6.11]. The energy landscape changes with time: its minimum becomes narrower as the liquid evaporates and centers the assembled particle on the binding site.

The formation of a potential *funnel* that guides the particle to its desired position is desirable for successful templated particle assembly. A properly chosen energy landscape ensures high placement accuracy, as discussed above. It also increases the yield of assembly by attracting particles from a larger volume towards the binding site. Assembly is more rapid if particles are guided from a larger volume instead of randomly diffusing until they accidentally arrive at the binding site. On the other hand, unspecific deposition is avoided if secondary minima on the energy landscape are kept shallow and are connected to the global minima (the binding sites) via low-energy pathways.

6.1.2 Mobility, Stability, and Yield

If the energy landscape is appropriate, a particle with sufficient mobility can explore it and assemble. With increasing mobility, it will (on average) find the binding site more rapidly and escape from secondary minima more easily, but it will also have a larger probability of escaping from the desired minimum. The probability for a particle with mass m to escape from the binding site that produces a potential well with local shape $\omega^2 x^2/2$ surrounded by valleys of height Q equals, in unit time [6.13],

$$P = \left(\frac{\omega}{2\pi}\right) e^{-mQ/(k_B T)}, \tag{6.3}$$

a result widely used in transition-state theory. This rate can be limiting for the assembly process, but it is more frequently the initial adsorption that requires most time. Colloidal particles and solvated molecules gain the mobility required to find binding sites through Brownian motion by collisions with the solvent molecules. Equation (6.1) describes the ideal situation of an infinitely dilute particle suspension, where no interaction between the particles exists. In practice, interactions are very common at higher particle concentrations, and they influence the particle mobility according to [6.14]

$$D(c) = \frac{c}{1-c} \frac{(\partial \mu / \partial c)_{p,T}}{f(c)}, \quad (6.4)$$

which gives the diffusion constant D as a function of the number concentration c, the chemical potential μ, and the friction coefficient f. If the particles interact strongly, the chemical potential will increase with increasing concentration, and so will the diffusion constant. At increased concentrations, the assembly behavior will then change. For gold colloids, the apparent diffusivity can be increased by two orders of magnitude at increased concentrations, but drop radically when the range of stable concentrations is exceeded.

Equation (6.4) is a thermodynamic expression, and the link to the microscopic events at a binding site in templated assembly is not trivial. A reduced diffusion constant can indicate a reduced escape rate from a binding site, but the thermodynamic value obviously does not hold when regarding a single particle, particularly if it is in the proximity of a binding site and encounters additional interactions. Statistical effects (as addressed in Sect. 6.1.4) are also not covered. Two consequences of particle interaction are particularly important: the potential required to increase the concentration of a colloid locally and the limited stability of colloidal suspensions.

Colloid scientists have long studied the case of interacting particles to derive expressions for colloidal stability. Smoluchowski and others derived expressions for the rate of agglomeration as a function of particle mobility and interaction, arriving at a characteristic time for doublet formation of

$$t_p = \frac{\pi \mu a^3 W}{\phi k_B T}, \quad (6.5)$$

which depends on temperature T and viscosity μ [6.14]. The value is inversely proportional to the volume fraction ϕ of the particles and depends strongly on their diameter a and the interaction potentials (expressed via the stability ratio W). Doublets can thus form at time scales ranging from milliseconds to many hours, a very wide range that is reflected in the qualitative statement that a colloid is *stable* or *unstable* towards flocculation. A similar time scale will govern the templated assembly of particles. Different regimes occur, also depending on the hydrodynamic situation. The assembly can be purely Brownian (if there is no flow present), diffusion limited (for a rapid and efficient adsorption process at low concentrations and high surface densities) or reaction limited, if sufficient particles are present. The latter is the most widespread regime in templated assembly that uses chemical patterns on a submerged template with sparse binding sites.

If multiple particles are to be assembled in a single step in close proximity, the same repulsive interactions that prevent agglomeration in the colloid have to be overcome to pack the particles densely. These forces are considerable. In a stable, aqueous colloid, the electrostatic repulsion, characterized by $e\psi_s$, will generally be greater than $10 k_B T$ and often around $100 k_B T$. Overcoming this barrier to reach the energy minimum caused by van der Waals interactions therefore requires a large driving force. Alternatively, the ionic strength can be increased locally to lower the electrostatic interaction and create a funnel through which the particles can reach the densely packed stage. If none of the above is present, sparse packing will result, described by modified random sequential adsorption models, as discussed later.

Many technological applications of self-assembled nanostructures, in particular those in electronics, require high yields of assembly and well-defined arrangements. This is in contrast to biological systems, where defects can be repaired through error-correction mechanisms. In the absence of such mechanisms, however, the yield of assembly has to be very high. This yield depends on the nature of the binding sites, the concentration of particles, and the characteristics of the assembly process. In particular, we can differentiate between *abrupt* assembly processes, where the actual particle deposition and its final immobilization (or quenching) occur almost simultaneously, and *gradual* assembly processes, where the two steps are not coupled. If the assembly takes place in the front of a receding meniscus that moves over a solid template, it will leave the particle dry and immobile, and if a binding site stays empty, there is no second chance for it to be filled. If the template is entirely submerged in the liquid, on the other hand, we can at least theoretically wait until every binding site is filled. For a rough estimate of the assembly rate, we can use Schurr's expression [6.15] for the particle flux J_s to a surface

$$J_s = c_s \sqrt{\frac{k_B T}{2\pi m}}, \quad (6.6)$$

which assumes a Maxwell–Boltzmann-distribution to derive the flux from the particle number concentration at the surface c_s and their mass m. If we know the sticking probability S of the binding site, i.e., the probability for a particle to be adsorbed upon contact, we can directly calculate the half-life of an empty binding site.

If particles readily desorb from binding sites, an equilibrium situation will finally develop. The yield will then never reach unity, and its value will fluctuate over time.

6.1.3 Large Binding Sites

Consider a particle that hits a binding site with area A. If the particle gets sufficiently close to the site and if its interaction with the site is sufficiently large to overcome Brownian motion, the particle will be adsorbed. When we have a large number of such binding sites, particles will be randomly arranged inside the various A, so that the precision of arrangement is limited by the minimum size that (a) the template patterning can produce and (b) allows for sufficiently rapid particle assembly.

If, on the other hand, a funneling effect of the kind discussed above is present, the distribution of the particles might be biased towards a certain part of A. Then, the width of the position distribution is the result of the competition between a stochastic force (Brownian in general) and the directing force.

If the area A is large enough to accommodate multiple particles, particles can either arrange into random submonolayers or into ordered dense layers. The first case, particle adsorption on strongly adsorbing surfaces, is described reasonably well by the random sequential adsorption (RSA) model, which predicts a random particle distribution. Adsorption ceases when there is no space left in the binding area that could accommodate an additional particle. The final packing density is called the *jamming limit*, which can be numerically found to be $\theta_\infty \approx 0.547$ for two-dimensional, circular particles [6.16]. Random sequential adsorption is the subject of numerous reviews, which also discuss its application to anisotropic particle such as rods [6.17, 18].

The RSA model accurately describes many molecular adsorption problems, in particular the adsorption of proteins on surfaces. It does not cover processes that result in dense ordered arrangements, for example, self-assembled monolayers (SAM). In contrast to the RSA model, the molecules that constitute a SAM retain some mobility even after they are adsorbed on the surface. They interact with other molecules even before they adsorb, and they interact with the underlying metal film. Larger particles sometimes behave similarly. The rearrangement of particles in an evaporating liquid film due to capillary forces is a well-known example.

When dense ordered packings are desired, the particle–surface interaction has to be appropriate to avoid RSA-like adsorption. It turns out [6.19] that, by tuning the strength of the interaction, particle arrangements between well-ordered layers and randomly distributed submonolayers can be obtained.

6.1.4 Thermodynamics, Kinetics, and Statistics

Diffusion constants scale inversely with the particle radius. The diffusion constants of nanoparticles are therefore much smaller than those of molecules. A 100 nm-diameter sphere moving in water will exhibit a diffusion constant D of approximately $10^{-12}\,\mathrm{m^2/s}$. Diffusion-limited processes with particles are thus slow, equilibrium situations can often not be reached in observable times, and the kinetics of the assembly process influences the assembly results. From an energy landscape point of view, it is not sufficient to provide a well-defined minimum in an appropriate position; the pathway to this minimum also has to be taken into account.

Most real template–particle systems will have complex energy landscapes with a variety of secondary minima and kinetic traps. A well-known example is a chemically functionalized surface onto parts of which particles should bind specifically. In practice, one finds unspecific deposition and a certain degree of particle accumulation, both caused typically by unspecific van der Waals-type attractions. Countermeasures include stirring, which increases particle mobility and keeps them from settling in secondary minima; rapid processing, which decreases the number of undesired particle collisions and thus the probability of reaching such a minimum; and washing, which removes weakly bound particles.

There is one limitation, however, that cannot be overcome by such mobility-increasing measures. When the number of particles in the volume affected by a binding site is small, the probability of finding at least one particle inside this volume will be small too. In the simple Poisson model of the situation, a volume V would contain a certain number n of particles with probability

$$W(n) = e^{-\nu}\frac{\nu^n}{n!}\,, \tag{6.7}$$

where ν is the average number of particles in the volume, $\nu = Vc$ in the homogenous case. The probability of finding at least a single particle in this volume is therefore smaller than

$$W(n \geq 1) \leq \sum_{n=1}^{\infty} W(n) = 1 - e^{-\nu}\,, \tag{6.8}$$

and the particle concentration has to be above

$$c_c \geq -\frac{\ln(1-\gamma)}{V} \qquad (6.9)$$

to guarantee a certain probability γ for a particle to be present. This limits the yield in assembly methods which only capture particles during a short period of time from the volume V: when there is no particle present, none can be assembled. When we regard a large number of binding sites and require a certain minimum yield, say, 90%, the colloid concentration at the binding sites therefore has to be at least $c = 2.3/V$, independent of any further process details. This concentration can be provided either by an overall larger colloid concentration or (often more practical) by an additional, long-range force that acts on many particles, much like a funnel again. Electrostatic or hydrodynamic forces can increase the particle concentration locally, for example, at a three-phase boundary line, and enable sufficient assembly yields. We will see how this is done experimentally in the next section.

6.2 Classes of Directed Particle Assembly

There are many options and examples of how to assemble particles and small objects into templates. Depending on the synthesis and the material of the particles, and especially on the medium in which the particles are supplied, different strategies can be applied. Furthermore, the material of the target substrate can determine the assembly method to be used.

Nanoparticles can be synthesized and held in the gas phase by a carrier gas as an aerosol. At this point, they can be assembled directly from the gas phase onto a template (Fig. 6.3a). As a dry powder, nanoparticles tend to agglomerate into larger clusters due to strong van der Waals interactions, thus making it almost impossible to arrange patterns of individual particles. Therefore, submicron-sized particles are often delivered as suspensions in a liquid medium, especially when they were synthesized in liquid phase. Usually, nanoparticles are easier to stabilize in liquid, and particle agglomeration is prevented by surface chemicals creating a surface charge or by the addition of surfactants.

For assembly from the liquid phase we differentiate two cases: assembly from the bulk liquid onto the solid template (Fig. 6.3b) or assembly at the solid–liquid–gas boundary, i.e., at the meniscus of a liquid front moving over the substrate (Fig. 6.3c). In the following subsection we will illustrate the different assembly strategies with some instructive examples.

6.2.1 Assembly from the Gas Phase

Particles can be assembled from the gas phase into a pattern by localized surface charges on a substrate, as in xerography. Here however, the fabricated patterns are considerably smaller than in a copier or a laser printer. The *latent image* of charges is produced in a thin-film electret by contacting a nanopatterned electrode with the target substrate [6.20, 21]. The electret material can be a polymer (poly(methylmethacrylate) PMMA or a fluorocarbon layer) or SiO_2. The flexible patterned electrode is made from a patterned silicone elastomer (polydimethylsiloxane PDMS) with a thin conductive gold layer evaporated on top [6.20] or from thin patterned silicon on top of a flat PDMS sheet [6.22]. The flexible electrode is brought into direct contact with the electret and the charge image is produced by an electrical pulse. Charge patterns can also be produced by sequentially writing with a conducting atomic force microscopy (AFM) tip [6.23], although in these exam-

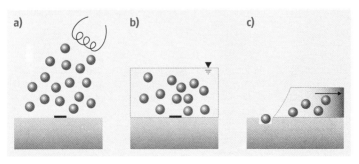

Fig. 6.3a–c Particles can be assembled from different media: they can be synthesized in a vacuum (or a gas) and directly assembled from the gas phase (**a**). Most commonly used are colloidal suspensions, from which particles are assembled at the liquid–solid interface (**b**). Alternatively, the particles can be assembled at the gas–liquid–solid boundary where strong capillary and confinement forces act on them (**c**)

ples the nanoparticles (NPs) were then adsorbed from the liquid phase (see the next section). The charge patterns are reported to be stable for more than 1 week in air [6.23]. Nanoparticle preparation is performed by an evaporative process in a tube furnace, by electrospray or in a plasma system [6.24]. Nanoparticles that have been synthesized in a wet chemical process can be used if they can be aerosolized without agglomerating. An interesting aspect is the combination of gas-phase particle synthesis with particle sorting methods, directly before the particles are assembled [6.25]. Almost monodisperse particle streams with few 10 nm-diameter nanoparticles can be prepared in this way. For the actual assembly, the nanoparticles have to be accelerated towards the target surface by an external field in a particle assembly module. Assembly of nanosized patterns from particles with a narrow size distribution can be achieved in this way.

Templates with an additional material contrast can improve the accuracy of particle assembly from the gas phase [6.24, 25]. The template is prepared from a patterned photoresist on a silicon substrate. In addition to the aerosol of charged nanoparticles, a stream of equally charged ions is introduced into the assembly chamber (Fig. 6.4). The ions are very mobile and fast compared with the nanoparticles and charge the resist structures on the substrate. The electric field of the charged resist pattern guides the nanoparticles into the areas of free silicon substrate. The additional ions improve the contrast between deposition in desired and undesired areas of the template (Fig. 6.5). By controlling the amount of ions it is possible to create and control a funneling effect which focuses the nanoparticles into structures much smaller than the actual template pattern. Among the smallest structures that have been realized by this method are 35 nm features assembled from 10 nm Ag nanoparticles in 200 nm holes (Fig. 6.6) [6.25]. In the majority of these assemblies, multiple nanoparticles are deposited into one assembly site and it is difficult if not impossible to assemble single nanoparticles with high yield.

6.2.2 Assembly in the Liquid Phase

In the majority of examples of templated assembly, particles are deposited from the liquid phase onto a solid template surface. Here, we want to differentiate assembly from the bulk liquid and assembly from the liquid at the liquid–solid–gas boundary (Sect. 6.2.3). For assembly directly from the liquid phase, a great variety of interactions such as electrostatic forces [6.26], capillary forces [6.27], for-

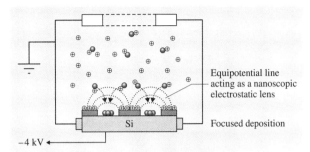

Fig. 6.4 Schematic setup for the assembly of nanoparticles from the gas phase in an electric field and with additional ions in the gas (after [6.25], © Macmillan 2006)

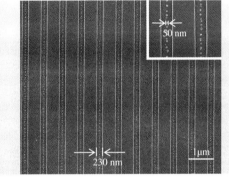

Fig. 6.5 Ag particles (10 nm) assembled in 230 nm-wide lines. The inset shows the funneling effect which reduces the actual width of the assembled particle lines to only 50 nm. The scale bar corresponds to 1 μm (after [6.25], © Macmillan 2006)

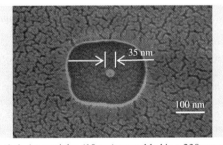

Fig. 6.6 Ag particles (10 nm) assembled in a 230 nm-wide hole. The funneling effect reduces the size of the actual assembly to only 35 nm. The scale bar corresponds to 100 nm (after [6.25], © Macmillan 2006)

mation of covalent bonds [6.28], specific recognition between biomolecules [6.29], supramolecular interac-

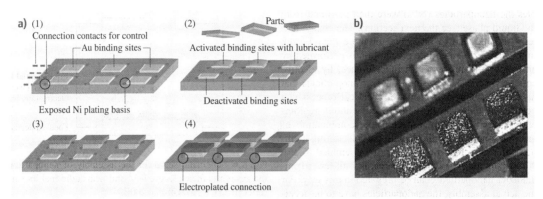

Fig. 6.7a,b Schematic description of the multibatch assembly process (1–4) with SAM-covered binding sites that can be deactivated selectively (**a**). A two-batch assembly result fabricated according to the scheme (**b**) (after [6.32], © IOP 2003)

tions [6.30], and form factor [6.8, 31] have been used. Also, electric fields can be applied to direct the particles or nanoobjects towards the targeted adsorption sites.

Wetting Contrast

For larger particles and objects, ranging from millimeters down to several tens of micrometers, wetting contrast in combination with capillary forces is applied for the assembly [6.27, 32, 33]. Topographic three-dimensional (3-D) features on the template may support the assembly in addition and introduce selectivity in a multicomponent assembly [6.34]. The template has hydrophobic assembly sites which can be selectively covered by a layer of adhesive or solder. The objects to be assembled are agitated in a fluid. In the simplest case – when a low-melting solder or a liquid organic adhesive is used – the fluid is water [6.35]. When higher temperatures are necessary to melt the solder, ethylene glycol [6.33, 34] can be used as a fluid. The suspended objects selectively adhere to the solder or adhesive when they come into contact. Objects to be assembled may also have a combination of hydrophilic and hydrophobic faces, which makes them adsorb with a preferred side or orientation. The strong capillary forces of the solder or adhesive guide the assembled objects into the desired orientation. The geometry of the adsorption sites and of the attached surfaces play a crucial role in this last step because local energy minima might freeze the assembled objects into undesired orientations on the template if the binding sites are not designed carefully.

Böhringer and coworkers devised a method in which hydrophobic assembly sites can be selectively switched off and reactivated later for a second assembly step (Fig. 6.7) [6.32, 36]. In this way, different objects or particles can be assembled onto the same template sequentially. For this purpose, the assembly sites consist of gold electrodes which are covered by a hydrophobic alkanethiol SAM. The alkanethiol SAM can be electrochemically removed from individual electrodes in a selective manner. When dipped into an adhesive, only the hydrophobic SAM-covered sites of the template are wetted and covered with an adhesive layer. In the subsequent assembly step, only the adhesive-covered sites are active and can grab an object from solution. After the first particle assembly, all vacant electrodes can be modified with a SAM, simply by dipping into an alkanethiol solution, and the process can begin again.

Electrostatic Nanoparticle Adsorption

In liquid suspensions, particles are usually stabilized by surface charges. These surface charges prevent the particles from agglomerating and can be exploited to guide the particles by electrostatic interaction to adsorption sites of opposite charge. The template needs to display a contrast in surface charge. This can be achieved by microcontact printing of SAMs with charged end-groups [6.11, 37]. The pattern contrast can be further enhanced through layer-by-layer (LBL) adsorption of polyelectrolyte multilayers onto the printed monolayers [6.26, 38]. Microcontact printing of a polyelectrolyte pattern onto LBL multilayers also results in a pattern of different surface charge on the template [6.39]. Other methods based on nanoimprint lithography (NIL) and subsequent monolayer formation have been described as well (Sect. 6.3) [6.40].

On such charged SAM patterns, oppositely charged 10 μm-diameter gold discs adsorbed selectively onto sites of opposite charge [6.37]. The Au discs were modified by thiol monolayers to control their surface charge. Once the discs have adsorbed onto the surface of the template, there is no more mobility. The discs are fixed to their initial adsorption site. This lack of mobility prevents ordering in the layer of adsorbed discs. For the formation of an ordered monolayer a certain mobility of the discs on the template surface would be required. The same observation is made with smaller particles being adsorbed electrostatically. The adhesion forces are too strong to allow for any mobility of the particles on the surface. Thus, a well-ordered and densely packed layer of particles is inhibited. Well-defined arrays of particles can only be achieved when a single particle or a small number of particles per site are adsorbed. This was demonstrated for particles a few microns in diameter [6.26]. For smaller particles in the nanometer regime this is a very challenging task.

Polar solvents (water, alcohol) are usually necessary to stabilize the colloidal suspension of charged particles. However, additional ions in water have to be avoided since the surface charges of the template are more effectively screened with higher ionic strength in the solvent [6.37].

Sagiv and coworkers fabricated charged adsorption sites by means of writing with a conductive AFM into a self-assembled silane monolayer [6.41]. The otherwise inert monolayer is activated by the charged AFM tip, and functionalized molecules can be coupled onto the patterned areas (Fig. 6.8). The added molecules

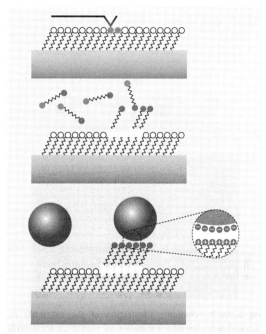

Fig. 6.8 Schematic representation of the modification of SAMs by AFM and subsequent bilayer formation to create assembly sites for selective adsorption of nanoparticles (after [6.41], © American Chemical Society)

can either carry a positive charge to attract negatively charged nanoparticles (Fig. 6.9) or carry a thiol group which binds to gold nanoparticles [6.28]. The latter

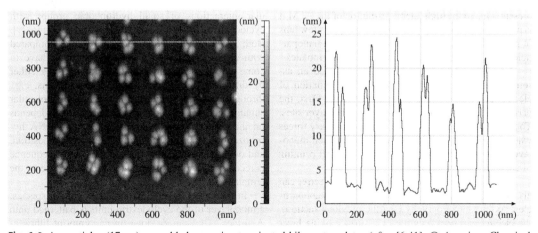

Fig. 6.9 Au particles (17 nm) assembled on amino-terminated bilayer templates (after [6.41], © American Chemical Society 2004)

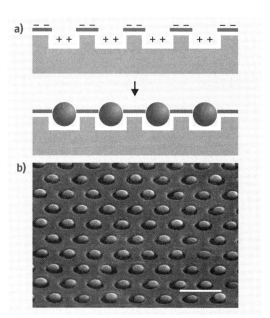

Fig. 6.10a,b Schematic illustration of the assembly of charged particles into a template of charge patterns with additional 3-D features (a). Scanning electron microscopy (SEM) image of 300 nm SiO$_2$ particles assembled into such a template with 450 nm-wide holes (b) (after [6.42], © Wiley-VCH 2007)

the nanometer range, charge is written by a conductive AFM tip [6.23]. Oppositely charged nanoparticles adsorb and adhere to these features. In the case of charges deposited into an electret, only nonpolar solvents such as fluorocarbons can be used. In water, the charge patterns would be neutralized rapidly [6.23].

A typical example [6.42] combines electrostatic patterns with 3-D geometry to define the location and number of adsorbed particles with higher precision. A nanostructured polymer template with 450 nm-wide holes was covered with alternating polyelectrolyte layers by LBL. In the final step, a negatively charged polyelectrolyte layer is printed onto the (positively charged) template with a flat stamp (Fig. 6.10). Thus, only the elevated parts of the template become negatively charged, while the depressions still carry a positively charged polyelectrolyte layer. Nanoparticles with negative surface charge are attracted towards the holes of the template. However, now the number of particles per adsorption site and their exact location is determined by the 3-D geometry of the adsorption site (Fig. 6.10).

case constitutes an example of an adsorption mechanism where the particles are fixed to their adsorption site by a chemical bond and not by Coulomb interaction. In this case the directing force of the surface charge is not present. Due to the high lateral resolution of the AFM it is possible to create very small patterns of just a few tens of nanometers. Still, the adsorption of several nanoparticles per site is routinely observed on these templates.

Aizenberg and coworkers [6.11] published an assembly method that makes use of a combination of electrostatic adsorption and capillary forces. First, the particles are attracted towards charged monolayer sites. Then they are *focused* onto the sites by capillary forces when the solvent dries. In this work, the charged monolayer sites were also prepared by microcontact printing of charged thiol molecules.

Charges deposited into a surface of an electret can also act as a template for the assembly of nanoparticles. As in nanoxerography (see the previous section), the electret can be a polymer or a surface oxide layer and the charge pattern can be formed by a conformal electrode [6.20]. To create very small features in

Specific Interactions

It is sometimes desirable to have particles and binding sites interact more specifically. Certain surface modifications on the binding sites will only interact with appropriately coated particles. Of course, such *specific interactions* are ultimately based on the standard set of interactions such as Coulombic and van der Waals interactions or hydrogen bonds. However, by steric conformation of the interacting entities, selectivity towards a small set of binding partners or even towards a single species arises. As examples, we discuss supramolecular interactions of small hydrophobic groups with cyclodextrins [6.30, 43] and DNA hybridization [6.29]. Both types of interactions are short-range compared with electrostatic forces. Thus, the particles have to come into close vicinity to the binding sites either by diffusion or by other transport mechanisms, e.g., through agitation. While the interaction of molecular subunits with the cyclodextrins is selective and depends on the size and polarity of the guest unit, DNA hybridization is selective towards the exact composition, and only the exact counterpart to the offered sequence is recognized and adsorbed onto the binding site on the template.

In the case of cyclodextrin as recognition species, a monolayer of the cyclodextrin units is patterned onto a surface. This can be done by nanoimprint lithography, which can expose just a fraction of the substrate surface for the formation of the cyclodextrin monolayer [6.30, 43]. Nanoparticles with guest functionality such as ferrocenyl-functionalized silica particles bind selectively to the cyclodextrin-functionalized areas of the substrate [6.43]. Even 3-D structures can be built up sequentially using alternating layers of host- and guest-functionalized particles [6.43].

For DNA recognition, a pattern of single-stranded DNA has to be prepared on the substrate. This can be done by photolithography [6.45] or, for smaller feature sizes, by dip–pen nanolithography (DPN) [6.46]. Gold nanoparticles functionalized with a thiolated DNA strand complementary to the DNA on the template adsorb from solution specifically to the patterned binding sites through DNA hybridization. Usually, the surrounding substrate area has to be functionalized with a second monolayer that prevents nonspecific adsorption of DNA-modified nanoparticles [6.45].

An interesting variant of DNA-mediated assembly is the assembly of nanoparticles onto specific locations of DNA tiles [6.29]. DNA tiles are DNA objects that are formed by assembling smaller subunits into large sheets [6.47] or by folding a long single-stranded DNA with the aid of shorter pieces of DNA (*DNA origami*) [6.48]. DNA tiles can be designed with single-stranded DNA pieces at specific locations. Such a DNA tile with a pattern of single strands can act as a template for the adsorption of nanoparticles functionalized with the complementary strand [6.29,49]. DNA tiles can also carry a pattern of other specific binding sites such as biotin functionalities. In this case templated assembly of streptavidin-functionalized nanoparticles can be carried out [6.50]. The adsorption of nanoparticles onto specific sites of DNA tiles allows for very high resolution in the assembly. However, the problem of assembling the DNA tiles and DNA origami structures themselves onto solid supports at specific locations with designated orientations is not yet fully solved. An interesting approach to this problem using dielectrophoretic assembly is described by *Kuzyk* et al. [6.51].

Dielectrophoretic Assembly

In dielectrophoresis, particles in a solvent are attracted to or repelled from a nonuniform alternating-current (AC) electrical field. The strength and direction of the dielectrophoretic force depends on the dielectric properties of the particles, solvent, electrode configuration, voltage, and frequency. By appropriate design of the electrodes, particles can be forced to desired areas on the template. However, if there is no additional persisting force, the particles leave their positions as soon as the AC field is turned off. *Suzuki* et al. applied a combi-

Fig. 6.11 Schematic illustration of the dielectrophoretic assembly of nanowires onto a template with additional 3-D structures formed in a resist. After assembly, the correctly assembled nanowires are fixed in a plating process and the resist is removed (after [6.44], © Macmillan 2008)

nation of dielectrophoretic assembly and covalent bond formation to overcome this problem [6.52]. The system was designed in such a way that 3 μm polystyrene particles were guided towards areas of weakest electrical field, which was directly underneath the lines of an interdigitated electrode array on the opposite substrate. There, the particles were permanently bonded by a chemical reaction.

Dielectrophoretic assembly lends itself very well to the assembly of nonisotropically shaped objects such as nanorods or nanowires. The electric field can additionally align the wires in a desired orientation. *Mayer* and coworkers have applied this method to align semiconductor and metal nanowires on substrates [6.44]. The electrodes were covered with photoresist, which had openings at the desired binding sites (Fig. 6.11). Nanowires were directed to and adsorbed onto those sites. The topographic structure of the assembly sites helped to maintain the wires in the correct positions upon drying. The assembled wires were fixed in a plating process and lift-off of the resist layer removed those nanowires that adsorbed onto undesired positions. This combination of methods can significantly reduce the error count and increase the yield of the assembly process (Fig. 6.12).

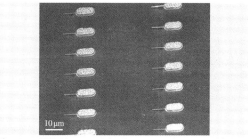

Fig. 6.12 SEM image of rhodium nanowires assembled using the process depicted in Fig. 6.11. The scale bar is 10 μm (after [6.44], © Macmillan 2008)

A very versatile variant of dielectrophoretic assembly was demonstrated by *Chiou* and coworkers [6.53]. They fabricated an assembly setup with rewritable electrode patterns on a photoconductive surface. Simply by projecting an image through a microscope lens, they could define their electrode pattern for dielectrophoretic assembly of 4.5 μm latex beads.

When NPs are assembled from aqueous suspensions, drying is always a critical step where strong capillary forces of the drying droplet may act on the assembled particles and destroy or alter the assembled pattern. On the other hand, the strength and directing capacity of capillary forces may be exploited to control the assembly of nanoparticles very accurately, as shown in the next section.

6.2.3 Assembly at Gas–Liquid Interfaces

At the phase boundary between a colloidal suspension, the template, and the surrounding air, very strong capillary forces may act, depending on the solvent composition used. In many microelectromechanical systems (MEMS) those capillary forces are detrimental to the fabricated microstructures and drying is a very critical step in MEMS fabrication. However, those strong directing forces can be exploited very well for the assembly of particles onto a template.

When the meniscus of an aqueous particle suspension gets pinned on a surface it deposits the particles at the phase boundary in monolayers and multilayers onto the substrate. Convective flow of water transports even more particles towards the edge of the drop, thus forming the well-known coffee-stain-like patterns [6.57]. When the convective flow of water towards the meniscus can be controlled, it is possible to assemble particle monolayers or multilayers in a reproducible manner [6.56, 58]. Particles can even be assembled in spaced arrays when the meniscus only gets pinned at some specific locations on an otherwise smooth and nonwetting substrate. Such pinning locations can be formed by geometric features on the substrate, by a pattern of wetting spots, or by spots of increased particle–substrate interaction. Many researchers have exploited this mechanism for templated particle assembly with different setups (Fig. 6.13).

In most examples of this kind of assembly, the particles are dispersed in an aqueous colloidal suspension. Often, these suspensions contain surfactants to further stabilize the colloids and prevent them from agglomeration and precipitation. When the meniscus of such a particle suspension sweeps over a flat nonwetting surface, no particles are left on the substrate. The meniscus acts like a doctor blade, moving the particles over the surface. At geometrical features on the substrate such as a hole or the step of a raised structure, the water meniscus gets pinned and capillary forces can drive particles into holes or corners.

In the simplest experimental setup, a drop of colloidal suspension is left drying on a topographically

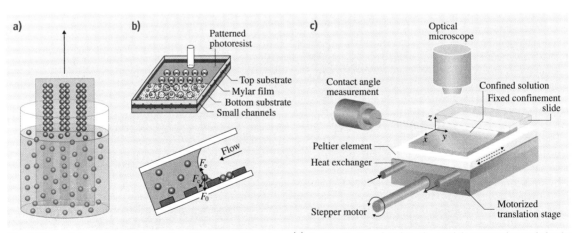

Fig. 6.13a–c Schematic depictions of capillary assembly setups: (**a**) dipping the template into the particle suspension and slowly pulling it out (after [6.54], © Wiley-VCH 2005); (**b**) assembly in a fluidic cell with a constant flow of particle suspension (after [6.55], © American Chemical Society 2001); (**c**) assembly on a motorized stage with controllable assembly speed and temperature (after [6.56], © American Chemical Society 2007)

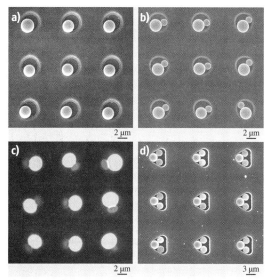

Fig. 6.14a–d Images of particles of different sizes assembled into holes with the device illustrated in Fig. 6.13b (after [6.55], © American Chemical Society 2001)

Xia and coworkers designed a fluidic cell where the colloidal solution is sandwiched between the template and a cover slide (Fig. 6.13b) [6.55, 62]. A thin frame of Mylar film defines the distance between template and cover slide and controls the flow rate at which the dispersion flows through the cell. Depending on the ratio of particle diameter and template geometry, very regular and reproducible arrangements of particles in the assembly sites ranging from pairs to tetrahedral packings can be achieved [6.55]. When the assembly procedure is repeated with a second batch of smaller particles, assemblies of pairs of different particles in the same adsorption site are possible (Fig. 6.14) [6.55].

With a tool that controls colloid temperature and speed of meniscus movement, and allows direct observation of the assembly process through an optical patterned template. As water evaporates the meniscus of the drop sweeps over the template and deposits particles into geometric features on the template. Here, there is only minimal control of the yield and evolution of the deposition process. At the start of the process, a low concentration of particles will be present at the meniscus. Then, with increasing evaporation, more particles are driven to the edge of the drop with the flux of water, and the assembly yield will increase. Finally, as the particle concentration in the drying drop reaches higher values, particles start to agglomerate and deposit in large aggregates. Thus, this simple method only supplies a relative small fraction of the template area with the desired assembly result.

Better assembly yield is achieved by placing the template (almost) vertically into a container of the colloidal suspension (Fig. 6.13a) [6.60]. As the solvent slowly evaporates, the meniscus moves over the template and deposits the particles in a controlled manner. Particles as small as 2 nm in diameter have been successfully assembled into template features of several 10 nm by this method [6.60]. Still, there is no direct control of particle concentration during the assembly and little possibility to react to changing parameters. Better control can be gained by pulling the template in a controlled manner out of the flask of colloidal suspension [6.54, 61].

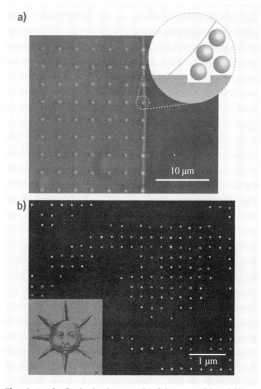

Fig. 6.15a,b Optical micrograph of the assembly of 60 nm Au particles into 3 μm-spaced holes. The bright accumulation zone is clearly visible. (**a**) Optical micrograph (inset) and SEM image of 60 nm Au particles assembled in a setup as illustrated in Fig. 6.13c and transferred to a silicon wafer (after [6.59], © Macmillan 2007)

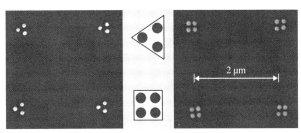

Fig. 6.16 SEM images of isolated 100 nm Au particles after removal of the template. The arrangement of the nanoparticles is determined by the geometry of the template (schematically depicted *in the middle*) (after [6.63], © American Institute of Physics 2006)

microscope, immediate response to changing conditions during assembly is possible (Fig. 6.13c) [6.56]. The template is mounted horizontally on a computer-controlled movable stage with a heatable vacuum chuck, and the colloidal suspension is sandwiched between the template and a glass slide. Observation of the assembly process from above reveals that for good yields a high concentration of particles is required at the meniscus. Particles are transported towards the meniscus by the flux of water in the same direction. As the temperature is increased, the evaporation at the meniscus increases. This causes an even greater flux of water and particles towards the meniscus. Lowering the temperature reduces the water flux and allows the particles to diffuse away from this so-called accumulation zone into the bulk solution. Consequently, assembly yield drops dramatically. Upon renewed increase of temperature and particle flux, the accumulation zone is reestablished and assembly reaches high yields again (Fig. 6.15).

The templates used in this kind of assembly method mostly have topographical (3-D) features that capture the particles. Also, templates which only rely on a chemical contrast have been described [6.54, 61]. The particle trapping relies on electrostatic interactions in this case and many of the adsorption sites also capture particles from the bulk solution [6.19], as described in the previous section. Thus, in the case of templates with chemical patterns, there is a combination of trapping mechanisms. Often, areas prepared for particle adsorption carry a hydrophilic surface functionality which causes lower contact angles in these areas and particle trapping in a mechanism closer to convective assembly.

Combinations of geometrical trapping and wetting contrast are possible too. With a well-designed balance of geometrical features and wetting contrast, it is even possible to control nanoparticle placement within the adsorption site [6.63]. When the adsorption site is large enough, particles are dragged into its corners and thus well-separated particle assemblies in a triangular or quadratic arrangement can be achieved (Fig. 6.16). This can be regarded as a kind of hierarchical assembly, where the assembly mechanism helps to form a substructure with features smaller than those of the actual template.

6.3 Templates

The template carries the positional information on particle arrangement. In most templates used today, there is a simple relation between the position of a binding site and the final particle position. The binding site might be larger than the particle's footprint or differ from its shape; it might accommodate just a single particle or a large number of particles. When working with very small particles, its shape might be irregular due to the limited resolution of the patterning process. The assembly process translates this geometry into a particle arrangement, as discussed in Sect. 6.1. Still, the relation between template and particle position is simple: there has to be a feature on the template exactly where a particle is intended to be placed. Thus, the patterning technique used to make the template has to define the particle locations with an accuracy that is within the range of the particle's dimensions.

Some assembly processes are more complex. A particle might only be deposited on one side of a binding site, only in its center (even if there is enough space for multiple particles around it) or multiple particles might fill a larger binding site with a regular structure. This can be desirable: the fabrication of the template is usually easier if the critical dimensions do not have to be identical to the particle's diameter. In the ideal case, the template pattern would be easy to fabricate but define the desired assembly of very small particles unambiguously, whereupon the assembly process would translate it into a very high-resolution energy landscape for the particles to occupy. If the desired arrangement is very complex, the template will generally have to be rather complex too, but most practical arrangements are highly repetitive and modular, and could be encoded efficiently.

Many assembly templates are fabricated by means of top-down micro- and nanofabrication. The patterns are usually designed in a computer, transferred to a mask (often via electron-beam or laser direct writing), and converted to chemical or topographical patterns on the template surface. The typical resolution limit for templates formed using typical ultraviolet (UV) lithography is around 1 μm, although smaller structures are achievable if artifacts are acceptable. Smaller structures can be written using electron-beam (e-beam) lithography. While photomasks for UV photolithography (produced in large e-beam writers) are readily available commercially, sub-μm e-beam patterns in other materials are less common, and the sequential nature of e-beam writing makes the patterning of large areas time consuming and costly. Electron-beam patterning is very flexible, however, and is widely used for research purposes.

If the primary patterning method is costly, replication techniques are useful both to produce multiple templates from one primary pattern and to cover larger areas with a repeated pattern. For patterns below the optical diffraction limit, molding and printing are popular methods. The primary structure is used either to imprint a polymer layer at increased temperature and under pressure (nanoimprint lithography) or to shape a liquid prepolymer while it is curing (molding, UV imprint lithography, and others). Although some top-down methods directly produce chemical patterns, for example, by oxidizing UV-sensitive monolayers in photolithography or with an e-beam, the most common product is a topographical pattern. Soft lithography is a route both to replicate such a pattern and to convert it to chemical contrast: a silicone prepolymer mixture is cured on the topographical pattern, cured to a solid rubber, and used as a stamp with the inverse pattern. This stamp can then print molecules on various surfaces, a process named microcontact printing.

Even e-beam writing is limited to minimum feature sizes in the range of tens of nanometers. Smaller structures can be formed using probe methods. For example, the tip of an atomic force microscope can mechanically remove (scratch) a monolayer or oxidize its functional groups locally. The resulting template cannot be replicated easily, which makes the process rather uneconomical, but it provides extremely high resolution, for example, for the arrangement of metal clusters in lines to investigate their conductivity.

An efficient alternative to e-beam writing for the patterning of larger areas is the use of interference lithography techniques. They only produce regular interference patterns, but they can do so over large areas in a single exposure step. If small structures are to be created, the radiation wavelength has to be low, its intensity high, and its coherence sufficient. An excellent source is synchrotron radiation. Larger patterns are straightforward to create using simple laser interference. Other efficient routes to certain template geometries include wrinkling patterns and step edges that form when crystals are cleaved along high-index planes. All these templates are limited to specific geometries and, thus, create specific particle arrangements.

6.3.1 Chemical Templates

Chemical templates display a pattern of selective surface chemistries with areas that prevent particle adsorption and others that support it. Additional geometrical features are not necessary per se but can be helpful to increase selectivity. A simple wetting contrast (e.g., hydrophilic patches on a hydrophobic substrates) can be sufficient to assemble colloidal particles at the three-phase boundary [6.61]. In this case, the hydrophilic spots have to be large enough ($> 25\,\mu m$) to cause a significant lowering of the receding contact angle and deposition of particles. Chemical patterns with features of several micrometers can be fabricated by optical lithography. After exposure and development of the photoresist on a Si/SiO_2 surface, the exposed substrate areas can be treated with a silane molecule. Upon removal of the remaining photoresist, a chemical pattern (bare Si/SiO_2 surface versus silane surface) is achieved. For higher-resolution features, nanoimprint lithography or e-beam lithography can be utilized to pattern a polymer resist layer. Again, the accessible substrate areas can then be patterned by a specific surface chemistry, and subsequent removal of the polymer resist provides the template. The areas of bare substrate may also be covered by a surface chemistry orthogonal to the first one (hydrophobic–hydrophilic, anionic–cationic) in order to increase assembly contrast [6.40]. Alternatively, the polymer resist might not be removed at all, thus providing an additional 3-D feature to support assembly onto the template [6.40].

Microcontact printing of organic monolayers is also a viable method for the fabrication of surface-chemical patterns on oxide or noble-metal surfaces. Depending on the quality of the stamp material and architecture, even sub-μm patterns are attainable [6.64].

Many examples of chemical templates do not only rely on a wetting contrast, but provide real adsorption sites for the particles. This can be achieved by patterning a charged molecule or a polyelectrolyte layer, which then attracts the particles by electrostatic forces or with specific supramolecular interactions [6.43, 54].

Self-assembled structures of block copolymers on surfaces can also act as a chemical template when nanoparticles selectively adsorb onto one block of the polymer [6.65].

6.3.2 Charges and Electrodes

Electric charge can be brought onto a surface by means of an electrode. This is the principle of xerography, which has been scaled down using electrically conductive AFM tips to write very small charged areas onto the surface of a dielectric. The charged regions attract particles, which then assemble on the written patterns [6.23]. Small tips (readily available in an AFM) enable high resolution to be attained, although the actual charge pattern can deviate from the intended design.

Actively driven electrodes are the most versatile option for electric-field templates. They require considerable effort in terms of interconnections and electrode design and one has to avoid particle attractions to the wiring, but they can be actively switched. What is more, AC potentials can be applied, so that dielectrophoresis takes place. Nanoparticles [6.66] and nanowires can thus be aligned with two electrodes and can then be connected [6.67]. The electric leads for both assembly electrodes and device interconnects (which can be identical) are fabricated using standard microfabrication techniques. If necessary, they can be combined with additional topographical features, for example, to improve alignment in the assembly of anisotropic particles [6.44].

Dielectrophoresis can also be driven by an external electromagnetic field that is projected onto an appropriate substrate [6.53]. The projected image, microscopically demagnified, causes the assembly forces. Such an image can be modulated and is far more flexible than patterned electrodes; it can even be time dependent to further optimize the assembly process. Its resolution is, however, diffraction limited.

6.3.3 Topographical Templates

Purely topographical templates guide particle assembly by geometrical exclusion and by modulating other forces, for example, capillary interactions. Geometrical confinement can be very precise – the particle cannot enter a template wall – but it is limited by template precision. Compared with electrodes or chemical patterns, topographical templates are simple in structure and fabrication and can be replicated via molding and imprinting techniques.

Topographical templates are used in convective [6.56] and capillary particle assembly [6.62], as an additional guide to the crystal structure in electrophoretic particle assembly [6.68], and as an additional guide in dielectrophoretic assembly [6.44]. In all these cases, the geometries are very simple: holes of uniform depth in an otherwise continuous layer. In most cases, the structures are formed in photoresist by standard UV lithography and used without further processing. If the resist sits on top of a wafer, it is simple to create an additional wetting contrast between the (polymer) top surface and the (silicon oxide) bottom surface of the template holes.

More complex geometries are required to precisely tune the forces in capillary assembly. Step edges, crosses, corner shapes, and other well-defined obstacles trap particles in reproducible arrangements. Such templates are harder to fabricate than holes. They can, however, be replicated in polymers, such as polydimethylsiloxane (PDMS). A single silicon master can then produce many (up to several hundred) single-use assembly templates.

Polymer molding is also the basis of the microfluidic ducts used as templates in the micromolding in capillaries (MIMIC) process. These channels, in which particles are arranged from a microfluidic flow, are first formed as lines in ultrathick resist and then replicated in PDMS. The soft silicone replica adheres to flat surfaces, forming channels into which the particle suspension is sucked by capillary forces.

6.3.4 Advanced Templates

More than one force can be involved in particle assembly, thus assembly templates can guide the assembly in more than one way. Advanced templates combine, for example, a long-range force caused by electrostatic or dielectrophoretic interactions with short-range interactions due to topography that provide high accuracy in the last moments of assembly [6.44]. An electrode array can be created on a flat substrate and a polymer resist patterned on top of the array, so that a hole in the shape of the particle remains. In a similar vein, a hydrophilic substrate can be coated with a hydrophobic resist, which creates a wetting difference that helps to capture a liquid volume in the binding site [6.60].

If one of the particle–template interactions is controllable (as is dielectrophoresis), such templates could be addressable and certain sites *turned off* during assembly [6.32]. In the style of a raster, a general-purpose master with a relatively dense, regular array of binding sites could then be modulated to produce arbitrary particle arrangements.

The ultimate template would not only be controllable, but could also sense whether a given binding site is occupied, and if so, whether the particle alignment is correct. Together with a feedback loop that controls process parameters, the yield of assembly would then be automatically optimized. This might be easier to realize than it seems: if electrodes are present in a template, it seems feasible to measure the dielectric properties of the binding site, which will likely depend on the presence of a particle (and possibly its alignment).

6.4 Processes and Setups

Particle assembly involves bringing the particles into contact with the targeted surface while avoiding non-specific deposition and agglomeration. If the assembly process takes place on a secondary surface (or interface), an additional process transfers the assembled particles onto the target. This target can then be a structured surface or an entirely plain material, which improves the compatibility of self-assembly with other fabrication methods.

There are only a few specialized setups for assembly, and most researchers use standard laboratory equipment to provide the required conditions. Some classical surface-science equipment can be adapted for assembly, however. The purpose of these setups is to bring surface and particles into contact in a controllable way, where convection and other disturbing influences are minimized, and to monitor and control conditions relevant for the assembly process, such as contact angle, ionic strength, temperature, and field strengths. Excess particles are removed without destroying the particles' order and might be reintroduced to increase the assembly yield. Some setups allow inducing a bias, for example, to align anisotropic particles.

If the assembly takes place at the liquid–solid interface, the final removal of the solvent (if so desired) is a critical step. Capillary forces have frequently been found to destroy or change the particle arrangement. In general, a final quenching of the assembled particles is required to permanently retain their order and enable further processing or integration into a device.

The integration of the assembled particles into a functioning device can involve electrical connections, optical coupling, thermal joining, and many other processing steps. The interfaces that are created often govern the device performance. Surface analysis and modification are therefore common subsequent steps in the integration process.

6.4.1 Setups for Particle Assembly

Langmuir–Blodgett troughs are the classic setup to create monolayers at a gas–liquid interface, and they can be used to assemble particles as well, although the interface naturally only provides a uniform surface as a *template*. Depending on the pressure applied, average particle spacings can be adjusted, which influences the overall properties of the film [6.69].

The gas–liquid interface can also be used to assemble particles in geometrical binding sites, where capillary forces and geometrical confinement at the three-phase boundary line guide the particles. This can either be done in simple immersion setups, where the liquid slowly evaporates and the boundary line moves over the vertically immersed patterned surface, or in more involved setups, for example, the *Capillary Assisted Particle Assembly tool* of *Malaquin* et al. [6.56]. The speed of the moving meniscus, the contact angle, and the hydrodynamic situation inside the liquid (all relevant for the assembly process) are more or less well controlled in the different setups.

Hydrodynamic forces are also used for assembly in the absence of capillary bridges but in the presence of gravity, for larger particles well beyond the limits of *Brownian* behavior, at about one micrometer. In so-called *fluidic assembly*, appropriately shaped objects fall into complementary shaped binding sites from a moving liquid [6.8]. This process is not very efficient, and the objects have to be brought to the surface repeatedly. Setups have been devised to move the object *slurry*, agitate it, and recycle it.

Liquid flows also drive the assembly in *micromolding in capillaries* (MIMIC), a term coined by *Whiteside*'s group [6.70]. In this process, microfluidic channels are filled with particles (usually in densely packed structures). *Xia* introduced a similar method to assemble polystyrene spheres first on flat, but later also on patterned surfaces: water is drained slowly from his cell through scratches or small ducts at the side, while ultrasonic agitation provides the mobility needed by the particles to arrange themselves [6.71].

Less widespread are vacuum setups for particle deposition. Here, particle production can be coupled to particle assembly and size selection. Particles are first created (usually directly from a metal with thermal methods), separated by size in an electric field, and assembled in a large vacuum chamber [6.25]. While such setups are considerably more complex than most liquid-phase techniques, they allow particles to be deposited on very clean surfaces.

6.4.2 Particle Printing and Processing

It is often desirable to carry out the particle assembly process away from the target substrate. Assembly usually requires binding sites, which might be undesirable to have on the target; it often involves solution chemistry, which might contaminate the target; or it requires specific surface properties, which the target simply might not have. If the particles are assembled on a secondary surface, these requirements are lifted, but a transfer step becomes necessary to bring the particle arrangement to its final destination.

In the classical case of a Langmuir–Blodgett trough, the formed monolayer is transferred by drawing the target surface through the interface vertically (Langmuir–Blodgett films) or by bringing it into contact with the surface horizontally (the Schaefer approach) [6.72]. Alternatively, a stamp is coated with the layer and then prints it onto a target.

Printing has also been demonstrated for single-particle arrays in a process called *self-assembly, transfer, and integration* (SATI), which uses a multistep adhesion cascade [6.73]. In this approach, particles leave one surface in favor of the other due to differences in adhesion, which has to be tuned. The strategy works with particles covering a wide size range; printing of both $100\,\mu m$ and $60\,nm$ particles has been demonstrated [6.59].

Postprocessing of the (printed or directly assembled) particles is generally required if they are to be electrically connected or need to be embedded, protected, or to act either as a template or building block of a further structure. Parallel electric contacting of many assembled nanowires (e.g., for sensing purposes) has been demonstrated using conventional technology on unconventionally assembled particles [6.44]. Other particles have been used as nucleation sites for nanowire growth [6.59], templates for etches and deposition processes, and transistor and memory elements. In all these cases, the particle surfaces were modified or covered by layers of material. A frequent task is the removal of organic adlayers that remain following liquid-phase assembly, which can be effected using plasma ashing or thermal annealing.

6.5 Conclusions

Building devices and materials from nanoparticles has been proven to be a feasible idea. It resembles industrial production from standardized components. The advantages are similar: building blocks, here nanoparticles, can be produced efficiently in large quantities if they are not too complex, and they can be modified and inspected and then used to build different products. As long as it is impossible to build complex structures directly from atoms and molecules, particle assembly will be one of the most interesting routes to creating nanostructures. Templated particle assembly reduced the process complexity even further and existing methods already provide exquisite control over particle positioning.

There are some niches where templated assembly is used industrially today, be it in the fabrication of RFID tags or for bead-based assays, but sub-μm particle assembly is yet to be introduced into production processes. The main challenges here include the typically very large number of particles that require extremely high yields of assembly, the limited quality of even the best chemically synthesized nanoparticles, and the preparation of suitable templates. Once such obstacles are overcome, templated assembly enables hierarchical, complex structures to be made in large quantities using relatively simple equipment.

All these challenges are currently being addressed. Mechanistic understanding of particle synthesis is increasing as synthesis protocols are being analyzed in detail, scale-up is investigated, and alternative routes become available for many popular nanoparticles. An increasing range of particles is now commercially available in consistent quality.

Templates can already be fabricated with high quality on small areas, and various researchers are working on large-scale nanopatterning and replication methods to produce large areas of identical patterns efficiently. Nanoimprint lithography, for example, can replicate a master many times and is even compatible with *roll-to-roll*-type fabrication, where long plastic sheets are continuously patterned by a rotating drum. As with molding, nanoimprint lithography is not an alternative to templated nanoparticle assembly, since it can only handle a very limited set of materials, but it is ideal for the production of templates.

Finally, the assembly processes themselves are constantly improving. Improved understanding of the interactions during assembly allows researchers to tune the interaction strengths and thereby engineer the energy landscape of the assembly process. Thus, both the stability of the original particles (for example, the colloidal suspension) and their behavior during assembly are optimized towards high yield. In addition, better control of the process parameters during assembly is now possible in modified versions of classical dip-coating setups. When combined with in situ analysis methods, yields and assembly qualities can be optimized by adjusting parameters such as temperature and template velocity.

Today, coatings containing nanoparticles are commonly applied using dip-coating, spin-coating or spray-coating techniques. Such methods are comparatively simple and compatible with a variety of relevant geometries. If templated assembly could be performed using the same deposition techniques, this would render it compatible with established technology and simplify its introduction into other processes. Alternatively, if specialized deposition techniques are required, or if the template cannot be applied to the substrate, assembly can be performed on a specialized template and the particles subsequently transferred onto the target surface. Together, these processes bridge the gap between particle assembly and current standard methods of fabrication. If particles are to be combined with, say, complementary metal–oxide–semiconductor (CMOS)-type components, the assembly and transfer precision has to be adequate to match the underlying structures. In most cases, short-range accuracy is governed by the assembly process and the precision of the template, while long-range order is influenced mainly by the template and the transfer process. All three may have to be optimized to meet the stringent requirements of semiconductor fabrication.

In addition to such improvements, the development of templated assembly processes for increasingly smaller particles with very high accuracy will continue. An important goal here is the assembly of particles well below 10 nm in diameter, which exhibit electronic quantum effects, with a precision that is sufficient to connect them electronically. Ideally, this would be possible on areas far above the square centimeters that have so far been demonstrated, if possible on standard 300 mm wafers. Finally, the assembly (and, if necessary, the transfer) should be compatible with different particle materials and substrates. A truly versatile process would accept any colloidal particle and thus be able to handle a very wide range of materials including oxides, semiconductors, metals, and polymers, amongst many others.

The ideal process would also handle very small particles. How small? We do not know at present. Gold-55 clusters that resemble molecules rather than particles have already been arranged using templated assembly processes, albeit with a precision far worse than the particle diameter [6.28]. Will it be possible at some point to arrange single atoms and molecules on a surface using a reasonably simple template? That such arrangements are stable and lead to interesting effects has already been demonstrated using high-vacuum scanning tunneling microscopy [6.74]. Whether templated assembly can provide a realistic route to such patterning with ultimate precision will remain an active topic of research for years to come.

References

6.1   A.N. Shipway, E. Katz, I. Willner: Nanoparticle arrays on surfaces for electronic, optical, and sensor applications, Chem. Phys. Chem. **1**(1), 18–52 (2000)

6.2   C.B. Murray, C.R. Kagan, M.G. Bawendi: Synthesis and characterization of monodisperse nanocrystals and close-packed nanocrystal assemblies, Annu. Rev. Mater. Sci. **30**, 545–610 (2000)

6.3   F. Caruso: Nanoengineering of particle surfaces, Adv. Mater. **13**(1), 11 (2001)

6.4   K. Molhave, T.M. Hansen, D.N. Madsen, P. Boggild: Towards pick-and-place assembly of nanostructures, J. Nanosci. Nanotechnol. **4**(3), 279–282 (2004)

6.5   P. Alivisatos: The use of nanocrystals in biological detection, Nat. Biotechnol. **22**(1), 47–52 (2004)

6.6   S. Paul, C. Pearson, A. Molloy, M.A. Cousins, M. Green, S. Kolliopoulou, P. Dimitrakis, P. Normand, D. Tsoukalas, M.C. Petty: Langmuir–Blodgett film deposition of metallic nanoparticles and their application to electronic memory structures, Nano Lett. **3**(4), 533–536 (2003)

6.7   J.D. Joannopoulos, S.G. Johnson, J.N. Winn, R.D. Meade: *Photonic Crystals: Molding the Flow of Light* (Princeton Univ. Press, Princeton 2008)

6.8   H.J.J. Yeh, J.S. Smith: Fluidic self-assembly for the integration of GaAs light-emitting-diodes on Si substrates, IEEE Photon. Technol. Lett. **6**(6), 706–708 (1994)

6.9   A. Einstein: The theory of the Brownian motion, Ann. Phys. **19**(2), 371–381 (1906)

6.10   R.M. Mazo: *Brownian Motion. Fluctuations, Dynamics, and Applications* (Clarendon, Oxford 2002)

6.11   J. Aizenberg, P.V. Braun, P. Wiltzius: Patterned colloidal deposition controlled by electrostatic and capillary forces, Phys. Rev. Lett. **84**(13), 2997–3000 (2000)

6.12   N.D. Denkov, O.D. Velev, P.A. Kralchevsky, I.B. Ivanov, H. Yoshimura, K. Nagayama: Mechanism of formation of 2-dimensional crystals from latex-particles on substrates, Langmuir **8**(12), 3183–3190 (1992)

6.13   S. Chandrasekhar: Stochastic problems in physics and astronomy, Rev. Mod. Phys. **15**(1), 1–89 (1943)

6.14   W.R. Russel, D.A. Saville, W.R. Schowalter: *Colloidal Dispersions* (Cambridge Univ. Press, Cambridge 1989)

6.15   J.M. Schurr: Role of diffusion in bimolecular solution kinetics, Biophys. J. **10**(8), 700–716 (1970)

6.16   J. Feder: Random sequential adsorption, J. Theor. Biol. **87**(2), 237–254 (1980)

6.17   J.W. Evans: Random and cooperative sequential adsorption, Rev. Mod. Phys. **65**(4), 1281–1329 (1993)

6.18   J. Talbot, G. Tarjus, P.R. Van Tassel, P. Viot: From car parking to protein adsorption: An overview of sequential adsorption processes, Colloids Surf. A **165**(1-3), 287–324 (2000)

6.19   X.Y. Ling, L. Malaquin, D.N. Reinhoudt, H. Wolf, J. Huskens: An in situ study of the adsorption behavior of functionalized particles on self-assembled monolayers via different chemical interactions, Langmuir **23**(20), 9990–9999 (2007)

6.20   H.O. Jacobs, A.R. Tao, A. Schwartz, D.H. Gracias, G.M. Whitesides: Fabrication of a cylindrical display by patterned assembly, Science **296**(5566), 323–325 (2002)

6.21   C.R. Barry, M.G. Steward, N.Z. Lwin, H.O. Jacobs: Printing nanoparticles from the liquid and gas phases using nanoxerography, Nanotechnology **14**(10), 1057–1063 (2003)

6.22   C.R. Barry, J. Gu, H.O. Jacobs: Charging process and coulomb-force-directed printing of nanoparticles with sub-100-nm lateral resolution, Nano Lett. **5**(10), 2078–2084 (2005)

6.23   P. Mesquida, A. Stemmer: Attaching silica nanoparticles from suspension onto surface charge patterns generated by a conductive atomic force microscope tip, Adv. Mater. **13**(18), 1395–1398 (2001)

6.24   C.R. Barry, H.O. Jacobs: Fringing field directed assembly of nanomaterials, Nano Lett. **6**(12), 2790–2796 (2006)

6.25   H. Kim, J. Kim, H.J. Yang, J. Suh, T. Kim, B.W. Han, S. Kim, D.S. Kim, P.V. Pikhitsa, M. Choi: Parallel patterning of nanoparticles via electrodynamic focusing of charged aerosols, Nat. Nanotechnol. **1**(2), 117–121 (2006)

6.26   I. Lee, H.P. Zheng, M.F. Rubner, P.T. Hammond: Controlled cluster size in patterned particle arrays via directed adsorption on confined surfaces, Adv. Mater. **14**(8), 572–577 (2002)

6.27   U. Srinivasan, D. Liepmann, R.T. Howe: Microstructure to substrate self-assembly using capillary forces, J. Microelectromech. Syst. **10**(1), 17–24 (2001)

6.28   S.T. Liu, R. Maoz, G. Schmid, J. Sagiv: Template guided self-assembly of [Au(55)] clusters on nano-lithographically defined monolayer patterns, Nano Lett. **2**(10), 1055–1060 (2002)

6.29   J.D. Le, Y. Pinto, N.C. Seeman, K. Musier-Forsyth, T.A. Taton, R.A. Kiehl: DNA-templated self-assembly of metallic nanocomponent arrays on a surface, Nano Lett. **4**(12), 2343–2347 (2004)

6.30   P. Maury, M. Peter, O. Crespo-Biel, X.Y. Ling, D.N. Reinhoudt, J. Huskens: Patterning the molecular printboard: patterning cyclodextrin monolayers on silicon oxide using nanoimprint lithography and its application in 3-D multilayer nanostructuring, Nanotechnology **18**(4), 044007 (2007)

6.31   J.J. Talghader, J.K. Tu, J.S. Smith: Integration of fluidically self-assembled optoelectronic devices using a silicon-based process, IEEE Photon. Technol. Lett. **7**(11), 1321–1323 (1995)

6.32   K.F. Böhringer: Surface modification and modulation in microstructures: controlling protein adsorption, monolayer desorption and micro-self-assembly, J. Micromech. Microeng. **13**(4), S1–S10 (2003)

6.33   W. Zheng, H.O. Jacobs: Self-assembly process to integrate and connect semiconductor dies on surfaces with single-angular orientation and contact-pad registration, Adv. Mater. **18**(11), 1387 (2006)

6.34   S.A. Stauth, B.A. Parviz: Self-assembled single-crystal silicon circuits on plastic, Proc. Natl. Acad. Sci. USA **103**(38), 13922–13927 (2006)

6.35   H.O. Jacobs, A.R. Tao, A. Schwartz, D.H. Gracias, G.M. Whitesides: Fabrication of a cylindrical display by patterned assembly, Science **296**(5566), 323–325 (2002)

6.36   X.R. Xiong, Y. Hanein, J.D. Fang, Y.B. Wang, W.H. Wang, D.T. Schwartz, K.F. Böhringer: Controlled multibatch self-assembly of microdevices, J. Microelectromech. Syst. **12**(2), 117–127 (2003)

6.37   J. Tien, A. Terfort, G.M. Whitesides: Microfabrication through electrostatic self-assembly, Langmuir **13**(20), 5349–5355 (1997)

6.38 K.M. Chen, X.P. Jiang, L.C. Kimerling, P.T. Hammond: Selective self-organization of colloids on patterned polyelectrolyte templates, Langmuir **16**(20), 7825–7834 (2000)

6.39 H.P. Zheng, M.F. Rubner, P.T. Hammond: Particle assembly on patterned "plus/minus" polyelectrolyte surfaces via polymer-on-polymer stamping, Langmuir **18**(11), 4505–4510 (2002)

6.40 P. Maury, M. Peter, V. Mahalingam, D.N. Reinhoudt, J. Huskens: Patterned self-assembled monolayers on silicon oxide prepared by nanoimprint lithography and their applications in nanofabrication, Adv. Funct. Mater. **15**(3), 451–457 (2005)

6.41 S.T. Liu, R. Maoz, J. Sagiv: Planned nanostructures of colloidal gold via self-assembly on hierarchically assembled organic bilayer template patterns with in-situ generated terminal amino functionality, Nano Lett. **4**(5), 845–851 (2004)

6.42 Y.H. Kim, J. Park, P.J. Yoo, P.T. Hammond: Selective assembly of colloidal particles on a nanostructured template coated with polyelectrolyte multilayers, Adv. Mater. **19**(24), 4426 (2007)

6.43 X.Y. Ling, I.Y. Phang, D.N. Reinhoudt, G.J. Vancso, J. Huskens: Supramolecular layer-by-layer assembly of 3-D multicomponent nanostructures via multivalent molecular recognition, Int. J. Mol. Sci. **9**, 486–497 (2008)

6.44 M.W. Li, R.B. Bhiladvala, T.J. Morrow, J.A. Sioss, K.K. Lew, J.M. Redwing, C.D. Keating, T.S. Mayer: Bottom-up assembly of large-area nanowire resonator arrays, Nat. Nanotechnol. **3**(2), 88–92 (2008)

6.45 B. Kannan, R.P. Kulkarni, A. Majumdar: DNA-based programmed assembly of gold nanoparticles on lithographic patterns with extraordinary specificity, Nano Lett. **4**, 1521–1524 (2004)

6.46 H. Zhang, Z. Li, C.A. Mirkin: Dip-pen nanolithography-based methodology for preparing arrays of nanostructures functionalized with oligonucleotides, Adv. Mater. **14**, 1472–1474 (2002)

6.47 E. Winfree, F.R. Liu, L.A. Wenzler, N.C. Seeman: Design and self-assembly of two-dimensional DNA crystals, Nature **394**(6693), 539–544 (1998)

6.48 P.W.K. Rothemund: Folding DNA to create nanoscale shapes and patterns, Nature **440**(7082), 297–302 (2006)

6.49 J. Sharma, R. Chhabra, Y. Liu, Y. Ke, H. Yan: DNA-templated self-assembly of two-dimensional and periodical gold nanoparticle arrays, Angew. Chem. Int. Ed. **45**, 730–735 (2006)

6.50 I. Cheng, B. Wei, X. Zhang, Y. Wang, Y. Mi: Patterning of gold nanoparticles on DNA self-assembled scaffolds, Res. Lett. Nanotechnol. **2008**, 827174 (2008)

6.51 A. Kuzyk, B. Yurke, J.J. Toppari, V. Linko, P. Törmä: Dielectrophoretic trapping of DNA origami, Small **4**, 447–450 (2008)

6.52 M. Suzuki, T. Yasukawa, Y. Mase, D. Oyamatsu, H. Shiku, T. Matsue: Dielectrophoretic micropatterning with microparticle monolayers covalently linked to glass surfaces, Langmuir **20**, 11005–11011 (2004)

6.53 P.Y. Chiou, A.T. Ohta, M.C. Wu: Massively parallel manipulation of single cells and microparticles using optical images, Nature **436**(7049), 370–372 (2005)

6.54 P. Maury, M. Escalante, D.N. Reinhoudt, J. Huskens: Directed assembly of nanoparticles onto polymer-imprinted or chemically patterned templates fabricated by nanoimprint lithography, Adv. Mater. **17**(22), 2718–2723 (2005)

6.55 Y.D. Yin, Y. Lu, B. Gates, Y.N. Xia: Template-assisted self-assembly: A practical route to complex aggregates of monodispersed colloids with well-defined sizes, shapes, and structures, J. Am. Chem. Soc. **123**(36), 8718–8729 (2001)

6.56 L. Malaquin, T. Kraus, H. Schmid, E. Delamarche, H. Wolf: Controlled particle placement through convective and capillary assembly, Langmuir **23**, 11513–11521 (2007)

6.57 R.D. Deegan, O. Bakajin, T.F. Dupont, G. Huber, S.R. Nagel, T.A. Witten: Capillary flow as the cause of ring stains from dried liquid drops, Nature **389**(6653), 827–829 (1997)

6.58 B.G. Prevo, O.D. Velev: Controlled, rapid deposition of structured coatings from micro- and nanoparticle suspensions, Langmuir **20**(6), 2099–2107 (2004)

6.59 T. Kraus, L. Malaquin, H. Schmid, W. Riess, N.D. Spencer, H. Wolf: Nanoparticle printing with single-particle resolution, Nat. Nanotechnol. **2**, 570–576 (2007)

6.60 Y. Cui, M.T. Bjork, J.A. Liddle, C. Sonnichsen, B. Boussert, A.P. Alivisatos: Integration of colloidal nanocrystals into lithographically patterned devices, Nano Lett. **4**(6), 1093–1098 (2004)

6.61 C.A. Fustin, G. Glasser, H.W. Spiess, U. Jonas: Parameters influencing the templated growth of colloidal crystals on chemically patterned surfaces, Langmuir **20**, 9114–9123 (2004)

6.62 Y.N. Xia, Y.D. Yin, Y. Lu, J. McLellan: Template-assisted self-assembly of spherical colloids into complex and controllable structures, Adv. Funct. Mater. **13**(12), 907–918 (2003)

6.63 M.J. Gordon, D. Peyrade: Separation of colloidal nanoparticles using capillary immersion forces, Appl. Phys. Lett. **89**(5), 053112 (2006)

6.64 H. Schmid, B. Michel: Siloxane polymers for high-resolution, high-accuracy soft lithography, Macromolecules **33**(8), 3042–3049 (2000)

6.65 C. Minelli, C. Hinderling, H. Heinzelmann, R. Pugin, M. Liley: Micrometer-long gold nanowires fabricated using block copolymer templates, Langmuir **21**(16), 7080–7082 (2005)

6.66 O.D. Velev, E.W. Kaler: In situ assembly of colloidal particles into miniaturized biosensors, Langmuir **15**(11), 3693–3698 (1999)

6.67 P.A. Smith, C.D. Nordquist, T.N. Jackson, T.S. Mayer, B.R. Martin, J. Mbindyo, T.E. Mallouk: Electric-field assisted assembly and alignment of metallic nanowires, Appl. Phys. Lett. **77**(9), 1399–1401 (2000)

6.68 N.V. Dziomkina, G.J. Vancso: Colloidal crystal assembly on topologically patterned templates, Soft Matter **1**(4), 265–279 (2005)

6.69 G. Markovich, C.P. Collier, J.R. Heath: Reversible metal-insulator transition in ordered metal nanocrystal monolayers observed by impedance spectroscopy, Phys. Rev. Lett. **80**(17), 3807–3810 (1998)

6.70 E. Kim, Y.N. Xia, G.M. Whitesides: Micromolding in capillaries: Applications in materials science, J. Am. Chem. Soc. **118**(24), 5722–5731 (1996)

6.71 Y. Lu, Y.D. Yin, B. Gates, Y.N. Xia: Growth of large crystals of monodispersed spherical colloids in fluidic cells fabricated using non-photolithographic methods, Langmuir **17**(20), 6344–6350 (2001)

6.72 I. Langmuir, V.J. Schaefer: Activities of urease and pepsin monolayers, J. Am. Chem. Soc. **60**, 1351–1360 (1938)

6.73 T. Kraus, L. Malaquin, E. Delamarche, H. Schmid, N.D. Spencer, H. Wolf: Closing the gap between self-assembly and microsystems using self-assembly, transfer, and integration of particles, Adv. Mater. **17**(20), 2438–2442 (2005)

6.74 F. Rosei, M. Schunack, Y. Naitoh, P. Jiang, A. Gourdon, E. Laegsgaard, I. Stensgaard, C. Joachim, F. Besenbacher: Properties of large organic molecules on metal surfaces, Prog. Surf. Sci. **71**(5–8), 95–146 (2003)

7. Three-Dimensional Nanostructure Fabrication by Focused Ion Beam Chemical Vapor Deposition

Shinji Matsui

In this chapter, we describe three-dimensional nanostructure fabrication using 30 keV Ga$^+$ focused ion beam chemical vapor deposition (FIB-CVD) and a *phenanthrene* ($C_{14}H_{10}$) source as a precursor. We also consider microstructure plastic art, which is a new field that has been made possible by microbeam technology, and we present examples of such art, including a *micro wine glass* with an external diameter of 2.75 μm and height of 12 μm. The film deposited during such a process is diamond-like amorphous carbon, which has a Young's modulus exceeding 600 GPa, appearing to make it highly desirable for various applications. The production of three-dimensional nanostructure is discussed. The fabrication of microcoils, nanoelectrostatic actuators, and 0.1 μm nanowiring – all potential components of nanomechanical systems – is explained. The chapter ends by describing the realization of nanoinjectors and nanomanipulators, novel nanotools for manipulation and analyzing subcellular organelles.

7.1 Three-Dimensional Nanostructure Fabrication 212
  7.1.1 Fabrication Process 212
  7.1.2 Three-Dimensional Pattern-Generating System 214
7.2 Nanoelectromechanics 215
  7.2.1 Measuring Young's Modulus 215
  7.2.2 Free-Space Nanowiring 217
  7.2.3 Nanomechanical Switch 220
  7.2.4 Nanoelectrostatic Actuator 221
7.3 Nanooptics: Brilliant Blue Observation from a *Morpho* Butterfly Scale Quasistructure 223
7.4 Nanobiology 224
  7.4.1 Nanoinjector 224
  7.4.2 Nanomanipulator 225
7.5 Summary 228
References 228

Electron beams (EBs) and focused ion beams (FIBs) have been used to fabricate various two-dimensional nanostructure devices such as single-electron transistors and metal–oxide–semiconductor (MOS) transistors with nanometer gate lengths. Ten-nanometer structures can be formed by using a commercially available EB or FIB system with 5–10 nm-diameter beams and high-resolution resist [7.1]. Two-dimensional nanostructure fabrication is therefore already an established process. There are various approaches to three-dimensional fabrication using a laser, an EB, or a FIB to perform chemical vapor deposition (CVD). FIB- and EB-CVD are superior to laser-CVD [7.2] in terms of spatial resolution and beam-scan control. *Koops* et al. demonstrated some applications such as an atomic force microscopy (AFM) tip and a field emitter that were realized using EB-CVD [7.3]. *Blauner* et al. demonstrated pillars and walls with high aspect ratios achieved using FIB-CVD [7.4].

The deposition rate of FIB-CVD is much higher than that of EB-CVD due to factors such as the difference in mass between an electron and an ion. Furthermore, the smaller penetration depth of ions compared with electrons makes it easier to create complicated three-dimensional nanostructures. For example, when we attempt to make a coil nanostructure with line width of 100 nm, 10–50 keV electrons pass through the ring of the coil and reach the substrate because of the large range of electrons (at least a few microns), which makes it difficult to create a coil nanostructure using EB-CVD. On the other hand, since the range of ions is a few tens of nanometers or less, the ions are deposited inside the ring. Up to now, the realization of complicated nanostructures using FIB-CVD has not been reported. Therefore, this chapter reports on complicated three-dimensional nanostructure fabrication using FIB-CVD.

7.1 Three-Dimensional Nanostructure Fabrication

We used two commercially available FIB systems (SMI9200, SMI2050, SII Nanotechnology Inc., Tokyo, Japan) with a Ga^+ ion beam operating at 30 keV. The FIB-CVD used a *phenanthrene* ($C_{14}H_{10}$) precursor as the source material. The beam diameter of the SMI9200 system was about 7 nm and that of the SMI2050 system was about 5 nm. The SMI9200 system was equipped with two gas sources in order to increase the gas pressure. The nozzles faced each other and were directed at the beam point. The nozzles were set a distance of 40 μm from each other and positioned about 300 μm above the substrate surface. The inside diameter of a nozzle was 0.3 mm. The *phenanthrene* gas pressure during pillar growth was typically 5×10^{-5} Pa in the specimen chamber, but the local gas pressure at the beam point was expected to be much higher. The crucible of the source was heated to 85 °C. The SMI2050 system, on the other hand, was equipped with a single gas nozzle. The FIB is scanned in order to be able to write the desired pattern via computer control, and the ion dose is adjusted to deposit a film of the desired thickness. The experiments were carried out at room temperature on a silicon substrate.

The deposited film was characterized by observing it with a transmission electron microscope (TEM) and analyzing its Raman spectra. A thin film of carbon (200 nm thick) was deposited on a silicon substrate by 30 keV Ga^+ FIB using *phenanthrene* precursor gas. The cross sections of the structures created and their electron diffraction patterns were observed by using a 300 kV TEM. There were no crystal structures in the TEM images and diffraction patterns. It was therefore concluded that the deposited film was amorphous carbon (a-C).

Raman spectra of the a-C films were measured at room temperature with the 514.5 nm line of an argon-ion laser. The Raman spectra were recorded using a monochromator equipped with a charge-coupled device (CCD) multichannel detector. Raman spectra were measured at 0.1–1.0 mW to avoid thermal decomposition of the samples. A relatively sharp Raman band at 1550 cm^{-1} and a broad-shouldered band at around 1400 cm^{-1} were observed in the spectra excited by the 514.5 nm line. Two Raman bands were plotted after Gaussian line shape analysis. These Raman bands, located at 1550 and 1400 cm^{-1}, originate from the trigonal (sp^2) bonding structure of graphite and tetrahedral (sp^3) bonding structure of diamond. This result suggests that the a-C film deposited by FIB-CVD is diamond-like amorphous carbon (DLC), which has attracted attention due to its hardness, chemical inertness, and optical transparency.

7.1.1 Fabrication Process

Beam-induced chemical vapor deposition (CVD) is widely used in the electrical device industry for repair of chips and masks. This type of deposition is mainly done on two-dimensional (2-D) pattern features, but it can also be used to fabricate a three-dimensional (3-D) object. *Koops* et al. demonstrated nanoscale 3-D structure construction [7.3] by applying electron-beam-induced amorphous carbon deposition onto a micro vacuum tube. However, focused ion beam (FIB)-induced CVD seems to have many advantages for the fabrication of 3-D nanostructures [7.4–6]. The key issue to realizing such 3-D nanostructures is the short penetration depth of the ions (a few tens of nm) into the target material, being much shorter than that of electrons (several hundreds of μm). This short penetration depth reduces the dispersion area of the secondary electrons, and so the deposition area is restricted to roughly several tens of nanometers. A 3-D structure usually contains overhang structures and hollows. Gradual position scanning

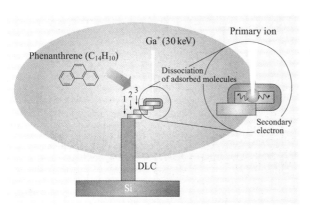

Fig. 7.1 Fabrication process for three-dimensional nanostructure by FIB-CVD

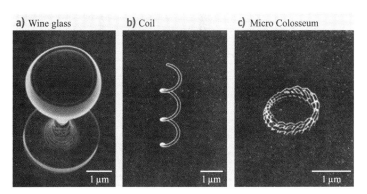

Fig. 7.2 (a) Micro wine glass with an external diameter of 2.75 μm and a height of 12 μm. (b) Microcoil with coil diameter of 0.6 μm, coil pitch of 0.7 μm, and line width of 0.08 μm. (c) *Micro Colosseum*

of the ion beam during the CVD process causes the position of the growth region around the beam point to shift. When the beam point reaches the edge of the wall, secondary electrons appear at the side of the wall and just below the top surface. The DLC then starts to grow laterally; the width of the lateral growth is also about 80 nm. Therefore, by combining the lateral growth mode with rotating beam scanning, it is possible to obtain 3-D structures with rotational symmetry, such as a wine glass.

The process of fabricating three-dimensional structures by FIB-CVD is illustrated in Fig. 7.1 [7.7]. In FIB-CVD processes, the beam is scanned in digital mode. First, a pillar is formed on the substrate by fixing the beam position (position 1). After that, the beam position is moved to within a diameter of the pillar (position 2) and then fixed until the deposited terrace thickness exceeds the range of the ions (a few tens of nm). This process is repeated to make three-dimensional structures. The key point to making three-dimensional structures is to adjust the beam scan speed so that the ion beam remains within the deposited terrace, which means that the terrace thickness always exceeds the range of the ions. Growth in the x- and y-directions is controlled by both beam deflectors. The growth in the z-direction is determined by the deposition rate; that is, the height of the structure is proportional to the irradiation time when the deposition rate is constant.

We intend to open up a new field of microstructure plastic art using FIB-CVD. To demonstrate the possibilities of this field, a *micro wine glass* created on a Si

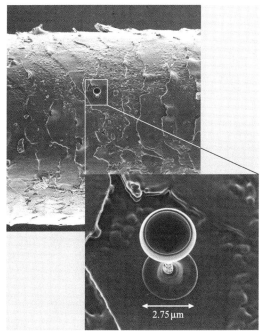

Fig. 7.3 Micro wine glass with an external diameter of 2.75 μm and a height of 12 μm on a human hair

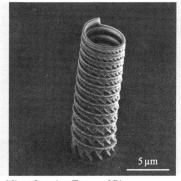

Fig. 7.4 *Micro Leaning Tower of Pisa*

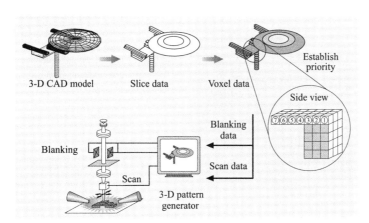

Fig. 7.5 Data flow of 3-D pattern-generating system for FIB-CVD

substrate and on a human hair as works of microstructure plastic art are shown in Figs. 7.2a and 7.3. A micro wine glass with an external diameter of 2.75 μm and a height of 12 μm was formed. The fabrication time was 600 s at a beam current of 16 pA. This beautiful micro wine glass shows the potential of the field of microstructure plastic art. A *micro Colosseum* and a *micro Leaning Tower of Pisa* were also fabricated on a Si substrate, as shown in Figs. 7.2c and 7.4.

Various microsystem parts have been fabricated using FIB-CVD. Figure 7.2b shows a microcoil with a coil diameter of 0.6 μm, a coil pitch of 0.7 μm, and a line width of 0.08 μm. The exposure time was 40 s at a beam current of 0.4 pA. The diameter, pitch, and height of the microcoil were 0.25, 0.20, and 3.8 μm, respectively. The exposure time was 60 s at a beam current of 0.4 pA. The results show that FIB-CVD is a highly promising technique for realizing parts of a microsystem, although their mechanical performance must be measured.

7.1.2 Three-Dimensional Pattern-Generating System

We used ion-beam-assisted deposition of a source gas to fabricate 3-D structures. The 3-D structure is built up as a multilayer structure. In the first step of this 3-D pattern-generating system, a 3-D model of the structure, designed using a 3-D computer-aided design (CAD) system (3-D DXF format), is needed. In this case we realized a structure shaped like a pendulum. The 3-D CAD model, which is a surface model, is cut into several slices, as shown in Fig. 7.5. The thickness of the slices depends upon the resolution in the z-direction (the vertical direction). The x- and y-coordinates of the slices are then used to create the scan data (voxel data). To fabricate the overhanging structure, the ion beam must irradiate the correct positions in the correct order. If the ion beam irradiates a voxel located in mid-air without a support layer, the ions intended for the voxel will be deposited on the substrate. Therefore,

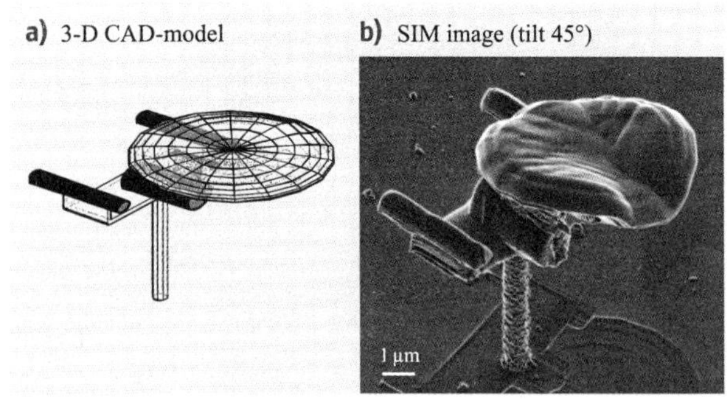

Fig. 7.6a,b *Micro Starship Enterprise* NCC-1701D, 8.8 μm long

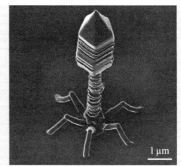

Fig. 7.7 T-4 bacteriophage

the sequence of irradiation is determined, as shown in Fig. 7.5.

The scan data and blanking signal therefore include the scan sequence, the dwell time, the interval time, and the irradiation pitch. These parameters are calculated from the beam diameter, xy-resolution, and z-resolution of fabrication. The z-resolution is proportional to the dwell time and inversely proportional to the square of the irradiation pitch. The scan data are passed to the beam deflector of the FIB-CVD, as are the blanking data. The blanking signal controls the dwell time and interval time of the ion beam.

Figure 7.6 shows a 3-D CAD model and an scanning ion microscope (SIM) image of the starship Enterprise NCC-1701D (from the television series Star Trek), which was fabricated by FIB-CVD at $10 \sim 20$ pA [7.8]. The nanospaceship is 8.8 μm long and was realized at about 1 : 100 000 000 scale on silicon substrate. The dwell time (t_d), interval time (t_i), irradiation pitch (p), and total process time (t_p), were 80 μs, 150 μs, 2.4 nm, and 2.5 h, respectively. The horizontal overhang structure was successfully fabricated.

Figure 7.7 shows a nano T4 bacteriophage, which is an artificial version of the virus fabricated by FIB-CVD on silicon surface. The size of the artificial nano T4 bacteriophage is about ten times that of the real virus.

7.2 Nanoelectromechanics

7.2.1 Measuring Young's Modulus

An evaluation of the mechanical characteristics of such nanostructures is needed for material physics. *Buks* and *Roukes* reported a simple but useful technique [7.9] for measuring the resonant frequencies of nanoscale objects using a scanning electron microscope (SEM). The secondary electron detector in the SEM can detect frequencies up to around 4 MHz, so the sample vibration is measured as the oscillatory output signal of the detector. *Buks* and *Roukes* used this technique to evaluate the Casimir attractive force between two parallel beams fabricated on a nanoscale. We evaluated the mechanical characteristics of DLC pillars in terms of the Young's modulus, determined using resonant vibration and the SEM monitoring technique [7.10, 11].

The system setup for monitoring mechanical vibration is shown in Fig. 7.8b. There were two ways of measuring the pillar vibrations. One is active measurement, where the mechanical vibration is induced by a thin piezoelectric device, 300 μm thick and 3 mm square. The piezo device was bonded to the sidewall of the SEM's sample holder with silver paste. The sample holder was designed to observe cross sections in the SEM (S5000, Hitachi) system. Therefore, the pillar's vibration was observed as a side-view image, as shown in Fig. 7.8a. The range of vibration frequencies involved was 10 kHz up to 2 MHz, which is much faster

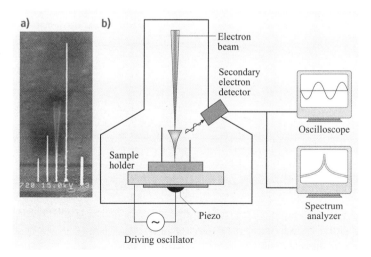

Fig. 7.8 (a) SEM image of the vibration. The resonant frequency was 1.21 MHz. (b) Schematic diagram of the vibration monitoring system

than the SEM raster scanning speed. Thus the resonant vibrations of the pillars can be taken as the trace of the pillar's vibration in the SEM image. The resonant frequency and amplitude were controlled by adjusting the power of the driving oscillator.

The other way to measure pillar vibrations is passive measurement using a spectrum analyzer (Agilent 4395A), where most of the vibration seemed to derive from environmental noise from rotary pumps and air conditioners. Some parts of the vibration result from spontaneous vibration associated with thermal excitations [7.9]. Because of the excitation and residual noise, the pillars on the SEM sample holder always vibrated at a fundamental frequency, even if noise isolation is enforced on the SEM system. The amplitude of these spontaneous vibrations was on the order of a few nanometers at the top of the pillar, and high-resolution SEM can easily detect it at a magnification of 300 000.

We arranged several pillars with varying diameters and lengths. The DLC pillars with the smallest diameter of 80 nm were grown using point irradiation. While we used two FIB systems for pillar fabrication, slight differences in the beam diameters of the two systems did not affect the diameters of the pillars. Larger-diameter pillars were fabricated using an area-limited raster scan mode. Raster scanning of a 160 nm² region produced a pillar with a cross section of about 240 nm², and a 400 nm² scan resulted in a pillar with a cross section of 480 nm². A typical SEM image taken during resonance is shown in Fig. 7.8a. The FIB-CVD pillars seemed very durable against mechanical vibration. This kind of measurement usually requires at least 30 min, including spectrum analysis and photo recording, but the pillars still survived without any change in resonance characteristics. This durability of the DLC pillars should be useful in nanomechanical applications.

The resonant frequency f of the pillar is defined by (7.1) for a pillar with a square cross section, and (7.2) for a circular cross section

$$f_{\text{square}} = \frac{a\beta^2}{2\pi L^2}\sqrt{\frac{E}{12\rho}}, \tag{7.1}$$

$$f_{\text{circular}} = \frac{a\beta^2}{2\pi L^2}\sqrt{\frac{E}{16\rho}}, \tag{7.2}$$

where a is the width of the square pillar or the diameter of the circular-shaped pillar, L is the length of the pillar, ρ is the density, and E is the Young's modulus. The coefficient β defines the resonant mode; $\beta = 1.875$ for the fundamental mode. We used (7.1) for pillars 240 and 480 nm wide, and (7.2) for pillars grown by point-beam irradiation. The relationship of the resonant frequency to the Young's modulus, which depends on the ratio of the pillar diameter to the squared length, is summarized in Fig. 7.9. All of the pillars evaluated in this figure were fabricated using the SMI9200 FIB system under rapid growth conditions. Typical growth rates were about 3–5 μm/min for the 100 nm-diameter and 240 μm-wide pillars, and 0.9 μm/min for the 480 nm-wide pillars. When calculating the data shown in Fig. 7.9, we assumed that the density of the DLC pillars was about 2.3 g/cm³, which is almost identical to that of graphite and quartz. The slope of the line in Fig. 7.9 indicates the Young's modulus for each pillar. The Young's moduli of the pillars were distributed over a range from 65 to 140 GPa, which is almost identical to that of normal metals. Wider pillars tended to have larger Young's moduli.

We found that the stiffness increases significantly as the local gas pressure decreases, as shown in Fig. 7.10. While the absolute value of the local gas pressure at the beam point is very difficult to determine, we found that the growth rate can be a useful parameter for describing the dependence of Young's modulus on pressure. All data points indicated in Fig. 7.10 were obtained from pillars grown using point irradiation. Therefore, the pillar diameters did vary slightly from 100 nm but not by more than 5%. A relatively low gas pressure, with good uniformity, was obtained by using a single gas nozzle and gas reflector. We used a cleaved side-wall of an Si tip as the gas reflector, which was placed 10–50 μm away from the beam point so as to face the gas nozzle. The growth rate was controlled by changing the distance to the wall. While there is a large distribu-

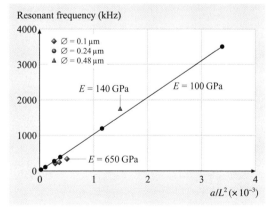

Fig. 7.9 Dependence of resonant frequency on pillar length

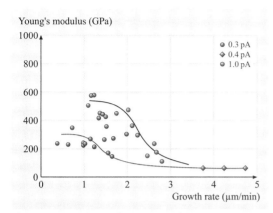

Fig. 7.10 Dependence of Young's modulus on growth rate

tion of data points, the stiffness of the pillar tended to become stiffer as the growth rate decreased. The two curves in Fig. 7.10 represent data points obtained for a beam current of 0.3 and 1 pA, respectively. Both curves show the same tendency; the saturated upper levels of the Young's modulus are different for each ion current at low gas pressure (low growth rate). It should be noted that some of the pillars' Young's moduli exceeded 600 GPa, which is of the same order as that of tungsten carbide. In addition, these estimations assume a pillar density of $2.3\,\mathrm{g/cm^3}$, but a finite amount of Ga was incorporated with the pillar growth. If the calculation takes the increase in pillar density due to the Ga concentration into account, the Young's modulus exceeds 800 GPa. Such a high Young's modulus reaches that of carbon nanotubes and natural diamond crystals. We think that this high Young's modulus is due to surface modification caused by the direct ion impact.

In contrast, when the gas pressure was high enough to achieve a growth rate of more than 3 μm/min, the pillars became soft but the change in the Young's modulus was small. The uniformity of the Young's modulus (as seen in Fig. 7.9) presumably results from the fact that the growth occurred in this insensitive region, where the low levels of source gas limit pillar growth.

7.2.2 Free-Space Nanowiring

All experiments were carried out in a commercially available FIB system (SMI9200: SII NanoTechnology Inc.) using a beam of 30 kV Ga$^+$ ions. The beam was focused to a spot size of 7 nm at a beam current of 0.4 pA, and it was incident perpendicular to the surface. The pattern drawing system (CPG-1000: Crestec Co., Tokyo) was added to the FIB apparatus to draw any patterns. Using the CPG, it is possible to control beam scan parameters such as scanning speed, xy-direction, and blanking of the beam, and so 3-D free space nanowiring can be performed [7.12].

Figure 7.11 illustrates the free-space nanowiring fabrication process using both FIB-CVD and CPG. When *phenanthrene* ($C_{14}H_{10}$) gas or tungsten hexacarbonyl (W(CO)$_6$) gas, which is a reactive organic gas, is evaporated from a heated container and injected into the vacuum chamber by a nozzle located 300 μm above the sample surface at an angle of about 45° with respect to surface, the gas density of the $C_{14}H_{10}$ or W(CO)$_6$ molecules increases on the substrate near the gas nozzle. The nozzle system creates a local high-pressure region over the surface. The base pressure of the sample chamber is 2×10^{-5} Pa and the chamber pressure upon introducing $C_{14}H_{10}$ and W(CO)$_6$ as a source gas was 1×10^{-4} and 1.5×10^{-3} Pa, respectively. If a Ga$^+$

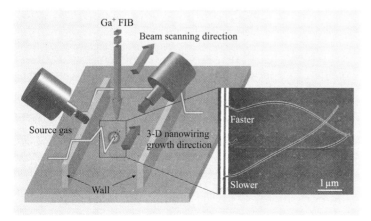

Fig. 7.11 Fabrication of DLC free-space wiring using both FIB-CVD and CPG

ion beam is irradiated onto the substrate, $C_{14}H_{10}$ or $W(CO)_6$ molecules adsorbed on the substrate surface are decomposed, and carbon (C) is mainly deposited onto the surface of the substrate. The direction of deposition growth can be controlling through the scanning direction of the beam. The material deposited using $C_{14}H_{10}$ gas was diamond-like carbon, as confirmed by Raman spectra, and it had a very large Young's modulus of 600 GPa [7.7, 10].

After the two walls shown in Fig. 7.11 were formed, free-space nanowiring was performed by adjusting the beam scanning speed. The ion beam used was a 30 kV Ga^+ FIB, and the irradiation current was 0.8–2.3 pA. The x- and y-scanning directions and the beam scanning speed were controlled by the CPG. The height in the z-direction was proportional to the irradiation time. Deposition is made to occur horizontally by scanning

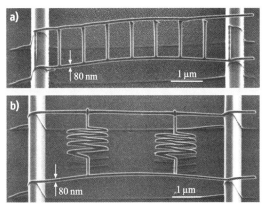

Fig. 7.12 (a) DLC free-space wiring with a bridge shape. (b) DLC free-space wiring with parallel resistances

the beam at a certain fixed speed in a plane. However, if the beam scanning speed is faster than the nanowiring growth speed, it grows downward or drops; conversely, if the scanning speed is too slow, the deposition grows slanting upward. Therefore, it is very important to control the beam scanning speed carefully when growing a nanowire horizontally. It turns out that the optimal beam scanning speed to realize a nanowire growing horizontally, using two $C_{14}H_{10}$ gas guns, was about 190 nm/s. The expected pattern resolution archived using FIB-CVD is around 80 nm, because both the primary Ga^+ ion and secondary-electron scattering occur over distances of around 20 nm [7.10, 13].

Figures 7.12 and 7.13 show examples of free-space nanowiring fabricated by FIB-CVD and CPG. All of the structures shown were fabricated using $C_{14}H_{10}$ as a precursor gas.

Figure 7.12a shows nanobridge free-space wiring. The growth time was 1.8 min and the wiring width was 80 nm. Figure 7.12b shows free-space nanowires with parallel resistances. The growth time was 2.8 min, and the wiring width was also 80 nm.

Figure 7.13a shows free-space nanowiring grown in 16 directions from the center. Figure 7.13b shows a scanning ion microscope (SIM) image of an inductor (L), a resistor (R), and a capacitor (C) in a parallel circuit structure with free-space nanowiring. A coiled structure was fabricated by circle-scanning of the Ga^+ FIB. The growth times of the L, R, and C structures were about 6, 2, and 12 min, and all the nanowiring is about 110 nm wide. From these structures, one can see that it is possible to fabricate nanowiring at an arbitrary position using FIB-CVD and CPG. These results also indicate that various circuit structures can be formed by combining L, C, and R.

The free-space wiring structures were observed using 200 keV TEM. The analyzed area was 20 nm in diameter. Figure 7.14a,b shows TEM images of DLC free-space wiring and a pillar. It became clear from these energy-dispersive x-ray (EDX) measurements that the dark part (A) of Fig. 7.14a corresponds to the Ga core, while the outside part (B) of Fig. 7.14a corresponds to amorphous carbon. This free-space wiring therefore consists of amorphous carbon with a Ga core. The center position of the Ga core is actually located below the center of the wiring. However, in the case of the DLC pillar, the Ga core is located at the center of the pillar. To investigate the difference between these core positions, the Ga core distribution in free-space wiring was observed in detail by TEM. The center position of the Ga core was about 70 nm from the top, which was

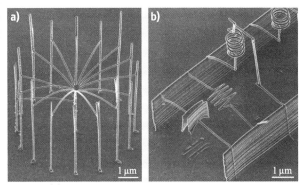

Fig. 7.13 (a) Radial DLC free-space wiring grown in 16 directions from the center. (b) Scanning ion microscope (SIM) micrograph of inductance (L), resistance (R), and capacitor (C) structure

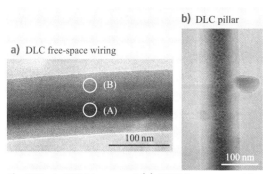

Fig. 7.14a,b TEM images of (**a**) DLC free-space wiring and (**b**) DLC pillar

20 nm below the center of the free-space wiring. We calculated an ion range of 30 kV Ga ions into amorphous carbon, using transport of ions in matter (TRIM), of 20 nm. The calculation indicates that the displacement of the center of the Ga core in the nanowiring corresponds to the ion range.

The electrical properties of free-space nanowiring fabricated by FIB-CVD using a mixture of $C_{14}H_{10}$ and $W(CO)_6$ were measured. Nanowiring was fabricated on an Au electrode. These Au electrodes were formed on a 0.2 μm-thick SiO_2-on-Si substrate by an EB lithography and lift-off process. Two-terminal electrode method was used to measure the electrical resistivity of the nanowiring. Figure 7.15a shows the results for nanowiring fabricated using only $C_{14}H_{10}$ source gas. The growth time here was 65 s and the wiring width was 100 nm. Next, $W(CO)_6$ gas was added to the $C_{14}H_{10}$ gas to create a gas mixture containing a metal in order to obtain lower electrical resistivity. Figure 7.15b–d corresponds to increasing $W(CO)_6$ contents in the gas mixture. The $W(CO)_6$ content rate was controlled by the sublimation temperature of the $C_{14}H_{10}$ gas. As the $W(CO)_6$ content was increased, the nanowiring growth time and width become longer, being 195 s and 120 nm for Fig. 7.15b, 237 s and 130 nm for Fig. 7.15c, and 296 s and 140 nm for Fig. 7.15d. Finally, we tried to fabricate free-space nanowiring using only $W(CO)_6$, but did not obtain continuous wiring, because the deposition rate for a source gas of just $W(CO)_6$ was very slow.

The electrical resistivity (Fig. 7.15a) for nanowiring fabricated using only $C_{14}H_{10}$ source gas was 1×10^2 Ω cm. The elemental contents were 90% C and 10% Ga, as measured using a SEM-EDX spot beam. The I–V curves in Fig. 7.15b–d correspond to increasing $W(CO)_6$ content in the gas mixture. As the $W(CO)_6$ content increases, the electrical resistivity decreases, as shown in the I–V curves (Fig. 7.15b–d). Moreover, the Ga content also increased because the growth of nanowiring slowed; the irradiation time of the Ga^+ FIB became longer. The electrical resistivities calculated from the I–V curves (Fig. 7.15b–d) were 16×10^{-2}, 4×10^{-2}, and 2×10^{-2} Ω cm, respectively. The electrical resistivity in Fig. 7.15e, which was fabricated by using only $W(CO)_6$ source gas, was 4×10^{-4} Ω cm. Increasing the Ga and W metallic content decreases the electrical resistivity, as shown by the SEM-EDX measurements reported in Fig. 7.15. These results indicate that increasing metallic content results in lower resistivity.

Electron holography is a useful technology for direct observation of electrical and magnetic fields at the nanoscale, and also has the property of showing useful information by detecting the phase shift of the electron wave due to the electrical and magnetic field. The technique relies upon an electron biprism, which plays the important role of dividing the electron wave into a reference wave and an objective wave. The biprism is composed of one thin filament and two ground electrodes.

It is important to fabricate as narrow a filament as possible to obtain an interference fringe with high contrast and good fringe quality. However, fabricating the filament with a diameter below 500 nm is very difficult, because a conventional electron biprism is fabricated by pulling a melted glass rod by hand. To overcome this problem, we introduce a new fabrication technique for

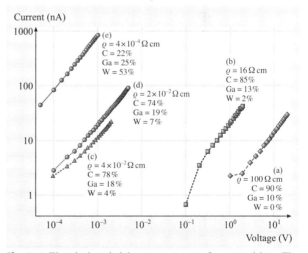

Fig. 7.15 Electrical resistivity measurement for nanowiring. The electrical resistivity ρ was calculated from the I–V curve. Elemental contents of C, Ga, and W were measured by SEM-EDX

Fig. 7.16 Electron biprism fabricated by FIB-CVD

the electron biprism using FIB-CVD, and evaluate the characteristics of the new biprism [7.14].

Figure 7.16 shows an SEM micrograph of the FIB-CVD biprism. We successfully fabricated DLC wiring with a smooth surface in between W rods by free-space wiring fabrication based on FIB-CVD technology. The 80 nm DLC thin wiring works as the filament of the biprism. The diameter and length of the filament are 80 nm and 15 μm, respectively.

Figure 7.17 shows interference fringes obtained using the biprism with a filament of 80 nm diameter (Fig. 7.17a) and 400 nm diameter (Fig. 7.17b), and corresponding fringe profiles. The applied prism voltage was 20 V, respectively. The filament with 400 nm diameter, close to the standard size used in the conventional electron biprism, was fabricated by Pt sputter-coating onto the 80 nm-diameter filament. Interference fringes were successfully obtained. Moreover, the interference region of the fringe obtained using the biprism with the 80 nm-diameter filament is larger than that of the fringe obtained using the biprism with the 400 nm-diameter filament. These results demonstrate the adequacy of the thin filament fabricated by FIB-CVD, and the new biprism will be very useful for accurate observation with high contrast and good fringe quality in electron holography.

7.2.3 Nanomechanical Switch

We have also demonstrated a nanomechanical switch fabricated by FIB-CVD [7.15]. Figure 7.18a shows the principle behind the realization of a nanomechanical switch. First, an Au electrode was formed on

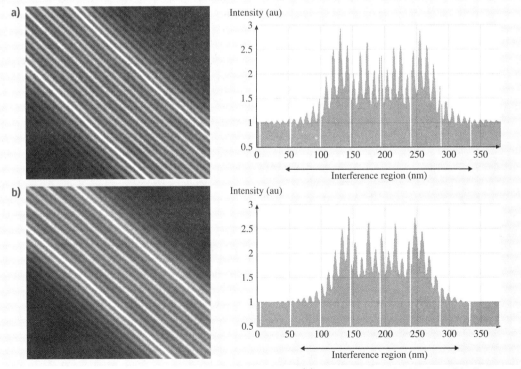

Fig. 7.17a,b Interference fringes and corresponding fringe profiles. (**a**) Obtained using the biprism with diameter of 80 nm, and (**b**) obtained using the biprism with diameter of 400 nm

a 0.2 μm-thick SiO$_2$-on-Si substrate by an electron-beam lithography and lift-off process. After that, a coil and free-space nanowiring were fabricated onto the Au electrode to form a switch function using nanowiring fabrication technology with FIB-CVD and CPG. The coil structure was fabricated by scanning a Ga$^+$ beam in a circle at fixed speed in C$_{14}$H$_{10}$ ambiance gas. An electric charge (positive or negative) was applied to the coil, and the reverse electric charge was applied to the nanowiring. The coil extended upward when a voltage was applied, because these was now an electrical repulsive force between each loop of the coil. At the same time, the coil and the nanowiring gravitated toward one another, because they had opposite charges. This attraction caused the coil to contact with the nanowiring when a certain threshold voltage was reached.

Next, we evaluated the switch function by measuring the current that flowed when the coil and the nanowiring were in contact. Figure 7.18b,c shows SIM micrographs of the nanomechanical switch before and after applying a voltage. These micrographs indicate that the coil and nanowiring make contact when a voltage is applied to the coil. At the same time, $I-V$ measurements of the nanomechanical switch were carried out, as shown in Fig. 7.19a. The current was plotted against the applied voltage at room temperature, and from this graph, it was apparent that the current begins to flow at a threshold voltage of 17.6 V. At this point, the electrical resistance and the resistivity of the nanomechanical switch are about 250 MΩ and 11 Ω cm, respectively. We measured the $I-V$ characteristics for ten nanomechanical switches. The threshold voltage was around 20 V in each case. The switching function was confirmed by performing on/off operations at an ap-

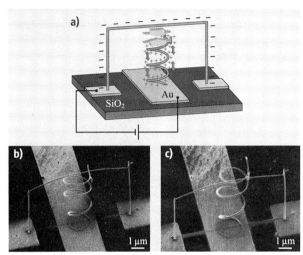

Fig. 7.18 (**a**) Principle of movement of nanomechanical switch. SIM micrographs of nanomechanical switch: (**b**) before applying voltage and (**c**) after applying voltage

plied voltage of 30 V, as shown in Fig. 7.19b. A pulsed current of about 170 nA was detected for this applied voltage.

7.2.4 Nanoelectrostatic Actuator

The fabrication process of 3-D nanoelectrostatic actuators (and manipulators) is very simple [7.16]. Figure 7.20 shows the fabrication process. First, a glass capillary (GD-1: Narishige Co., East meadow, NY) was pulled using a micropipette puller (PC-10: Narishige Co.). The dimensions of the glass capillary were 90 mm in length and 1 mm in diameter. Using this process,

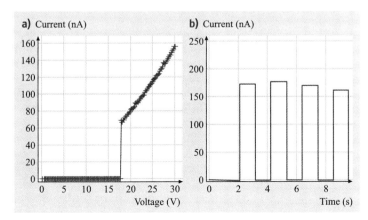

Fig. 7.19 (**a**) $I-V$ curve for the nanomechanical switch. (**b**) Pulsed current to on/off operation for the nanomechanical switch at an applied voltage of 30 V

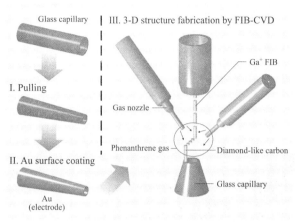

Fig. 7.20 Fabrication process of 3-D nanoelectrostatic actuators

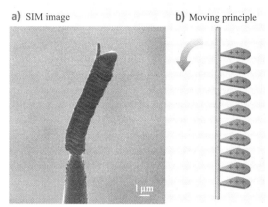

Fig. 7.21a,b Laminated pleats-type electrostatic actuator. (**a**) SIM image of a laminated pleats-type electrostatic actuator fabricated on the tip of a Au-coated glass capillary. (**b**) Illustration of moving principle of the actuator

we obtained a glass capillary tip with a 1 μm diameter. Next, we coated the glass capillary surface with Au by direct-current (DC) sputtering. The Au thickness was ≈ 30 nm. This Au coating serves as the electrode that controls the actuator and manipulator. Then, the 3-D nanoelectrostatic actuators and manipulators were fabricated by FIB-CVD. This process was carried out in a commercially available FIB system (SIM9200: SII NanoTechnology Inc.) with a Ga$^+$ ion beam operating at 30 keV. FIB-CVD was carried out using a *phenanthrene* ($C_{14}H_{10}$) precursor as the source material. The beam diameter was about 7 nm. The inner diameter of each nozzle was 0.3 mm. The *phenanthrene* gas pressure during growth was typically 5×10^{-5} Pa in the specimen chamber. The Ga$^+$ ion beam was controlled by transmitting CAD data on the arbitrary structures to the FIB system.

A laminated pleats-type electrostatic actuator was fabricated by FIB-CVD. Figure 7.21a shows an SIM image of a laminated pleats-type electrostatic actuator fabricated at 7 pA and 60 min exposure time. Figure 7.21b shows the principle behind the movement of this actuator. The driving force is the repulsive force due to the accumulation of electric charge. This electric charge can be stored in the pleats structures of the actuator by applying a voltage across the glass capillary. The pillar structure of this actuator bends due to charge repulsion, as shown in Fig. 7.21b. Figure 7.22 shows the dependence of the bending distance on the

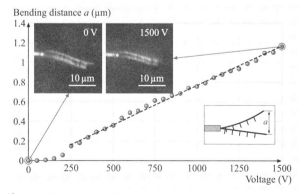

Fig. 7.22 Dependence of bending distance on applied voltage

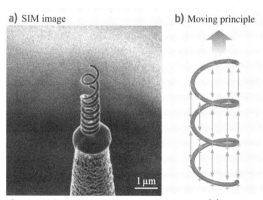

Fig. 7.23a,b Coil-type electrostatic actuator. (**a**) SIM image of a coil-type electrostatic actuator fabricated on the tip of a Au-coated glass capillary. (**b**) Illustration of moving principle for the actuator

applied voltage. The bending distance is defined as the distance a in the inset of Fig. 7.22. The bending rate of this laminated pleats-type electrostatic actuator was about 0.7 nm/V.

A coil-type electrostatic actuator was fabricated by FIB-CVD. Figure 7.23a shows an SIM image of a coil-type electrostatic actuator fabricated at 7 pA and 10 min of exposure time. Figure 7.23b shows the principle behind the movement of this actuator, which is very simple. The driving force is the repulsive force induced by electric charge accumulation; the electric charge can be stored in this coil structure by applying a voltage across the glass capillary. This coil structure expands and contracts due to charge repulsion, as shown in Fig. 7.23b. Figure 7.24 shows the dependence of the coil expansion on the applied voltage. The length of the expansion is the distance a in the inset of Fig. 7.24. The result revealed that the expansion could

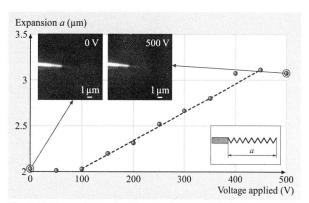

Fig. 7.24 Dependence of coil expansion on applied voltage

be controlled in the applied voltage range from 0 to 500 V.

7.3 Nanooptics: Brilliant Blue Observation from a *Morpho* Butterfly Scale Quasistructure

The *Morpho* butterfly has brilliant blue wings, and the source of this intense color has been an interesting topic of debate for a long time. Due to an intriguing optical phenomenon, the scales reflect interfered brilliant blue color for any angle of incidence of white light. This color is called a structural color, meaning that it is not caused by pigment reflection [7.17]. When we observed the scales with a scanning electron microscope (SEM) (Fig. 7.25a), we found three-dimensional (3-D) nanostructures 2 μm in height, 0.7 μm in width, and with a 0.22 μm grating pitch on the scales. These nanostructures cause a similar optical phenomenon to the iridescence produced by a jewel beetle.

We duplicated the *Morpho* butterfly scale quasistructure with a commercially available FIB system (SMI9200: SII Nanotechnology Inc.) using a Ga$^+$ ion beam operating at 30 kV [7.18]. The beam diameter was about 7 nm at 0.4 pA. The FIB-CVD was performed using *phenanthrene* ($C_{14}H_{10}$) as a precursor.

In this experiment, we used a computer-controlled pattern generator, which converted 3-D computer-aided design (CAD) data into a scanning signal, which was passed to an FIB scanning apparatus in order to fabricate a 3-D mold [7.8]. The scattering range of the Ga primary ions is about 20 nm, and the range of the secondary electrons induced by the Ga ion beam is about

20 nm, so the expected pattern resolution of the FIB-CVD is about 80 nm.

Figure 7.25b shows an SIM image of the *Morpho* butterfly quasistructure fabricated by FIB-CVD using 3-

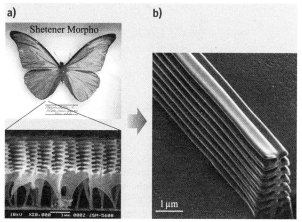

Fig. 7.25a,b *Morpho* butterfly scales. (**a**) Optical microscope image showing top view of *Morpho* butterfly. SEM image showing a cross-sectional view of *Morpho* butterfly scales. (**b**) SIM image showing inclined view of *Morpho* butterfly scale quasistructure fabricated by FIB-CVD

D CAD data. This result demonstrates that FIB-CVD can be used to fabricate the quasistructure.

We measured the reflection intensities from *Morpho* butterfly scales and the *Morpho* butterfly scale quasistructure optically; white light from a halogen lamp was directed onto a sample with angles of incidence ranging from 5° to 45°. The reflection was concentrated by an optical microscope and analyzed using a commercially available photonic multichannel spectral analyzer system (PMA-11: Hamamatsu Photonics K.K., Hamamatsu City, Japan). The intensity of light incident from the halogen lamp had a peak at a wavelength close to 630 nm.

The *Morpho* butterfly scale quasistructure was made of DLC. The reflectivity and transmittance of a 200 nm-thick DLC film deposited by FIB-CVD, measured by the optical measurement system at a wavelength close to 440 nm (the reflection peak wavelength of the *Morpho* butterfly), were 30% and 60%, respectively. Therefore, the measured data indicated that the DLC film had high reflectivity near 440 nm, which is important for the fabrication of an accurate *Morpho* butterfly scale quasistructure.

We measured the reflection intensities of the *Morpho* butterfly scales and the quasistructure with an optical measurement system, and compared their characteristics. Figure 7.26a,b shows the reflection intensities from *Morpho* butterfly scales and the quasistructure, respectively. Both gave a peak intensity near 440 nm and showed very similar reflection intensity spectra for various angles of incidence.

We have thus successfully demonstrated that a *Morpho* butterfly scale quasistructure fabricated using

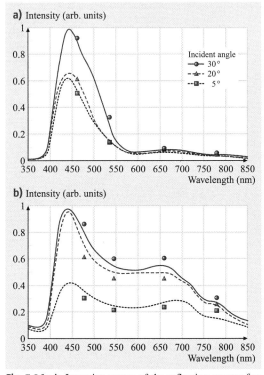

Fig. 7.26a,b Intensity curves of the reflection spectra for: (**a**) *Morpho* butterfly scales, (**b**) *Morpho* butterfly scale quasistructure

FIB-CVD can give almost the same optical characteristics as real *Morpho* butterfly scales.

7.4 Nanobiology

7.4.1 Nanoinjector

Three-dimensional nanostructures on a glass capillary have a number of useful applications, such as manipulators and sensors in various microstructures. We have demonstrated the fabrication of a nozzle nanostructure on a glass capillary for a bioinjector using 30 keV Ga^+ focused ion beam assisted deposition with a precursor of phenanthrene vapor and etching [7.19]. It has been demonstrated that nozzle nanostructures of various shapes and sizes can be successfully fabricated. An inner tip diameter of 30 nm on a glass capillary and a tip shape with an inclined angle have been realized. We reported that diamond-like carbon (DLC) pillars grown by FIB-CVD with a precursor of *phenanthrene* vapor have very large Young's moduli, exceeding 600 GPa, which potentially makes them useful for various applications [7.10]. These characteristics are applicable to the fabrication of various biological devices.

In one experiment, nozzle nanostructure fabrication for biological nanoinjector research was studied. The tip diameters of conventional bioinjectors are greater than 100 nm and the tip shapes cannot be controlled. A bionanoinjector with various nanostructures on the top of a glass capillary has the following potential applications (shown in Fig. 7.27):

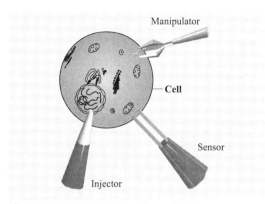

Fig. 7.27 Potential uses for a bionanoinjector

1. Injection of various reagents into a specific organelle in a cell
2. Selective manipulation of a specific organelle outside of a cell by using the nanoinjector as an aspirator
3. Reducing the mechanical stress produced when operating in the cell by controlling the shape and size of the bionanoinjector
4. Measurement of the electric potential of a cell, an organelle, and an ion channel exiting on a membrane, by fabricating an electrode

Thus far, 3-D nanostructure fabrications on a glass capillary have not been reported. We present nozzle nanostructure fabrication on a glass capillary by FIB-CVD and etching in order to confirm the possibility of bionanoinjector fabrication.

The nozzle structures of the nanoinjector were fabricated using a function generator (Wave Factory: NF Electronic Instruments, Yokohama, Japan). Conventional microinjectors are fabricated by pulling a glass capillary (GD-1: Narishige Co.) using a micropipette puller (PC-10: Narishige Co.). The glass capillary was 90 mm in length and 1 mm in diameter.

Conventionally, the tip shape of a microinjector made by pulling a glass capillary, and which is used as an injector into a cell, is controlled by applying mechanical grinding (or not). However, the reliability of this technique for controlling tip shape is very poor and requires experienced workers.

A bionanoinjector tip was fabricated on a glass capillary by FIB-CVD, as shown in Fig. 7.28a–c. First, FIB etching made the tip surface of the glass capillary smooth. Then, a nozzle structure was fabricated at the tip by FIB-CVD. Figure 7.28a shows the surface of a chip smoothed at 120 pA and after 30 s exposure time by FIB etching, with inner hole diameter of 870 nm. A nozzle structure fabricated by FIB-CVD with inner hole diameter of 220 nm is shown in Fig. 7.28b. Figure 7.28c shows a cross section of Fig. 7.28b. These results demonstrate that a bionanoinjector could be successfully fabricated by 3-D nanostructure fabrication using FIB-CVD. The bionanoinjector was used to inject dye into a egg cell (*Ciona intestinalis*) as shown in Fig. 7.29.

7.4.2 Nanomanipulator

An electrostatic 3-D nanomanipulator that can manipulate nanoparts and operate on cells has been developed by FIB-CVD. This 3-D nanomanipulator has four fingers so that it can manipulate a variety of shapes. To move the nanomanipulator, electric charge is accumulated in the structure by applying voltage to the four-fingered structure, and electric charge repulsion causes them to move. Furthermore, we succeeded in catching a microsphere (made from polystyrene latex) with a diameter of 1 μm using this 3-D nanomanipulator with four fingers [7.20].

The glass capillary (GD-1; Narishige Co.) was pulled using a micropipette puller (PC-10; Narishige Co.). A tip diameter of about 1.0 μm could be obtained using this process. Then, the glass capillary surface was coated with Au in order to fabricate an electrode for

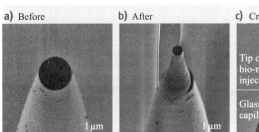
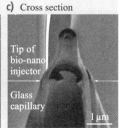

Fig. 7.28a–c SIM images of a bionanoinjector fabricated on a glass capillary by FIB-CVD. (a) Before FIB-CVD, (b) after FIB-CVD, and (c) cross section of (b)

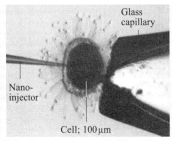

Fig. 7.29 Injection into an egg cell (*Ciona intestinalis*) using a bionanoinjector

nanomanipulator control. The thickness of the Au coating was about 30 nm. Finally, a 3-D nanomanipulator structure with four fingers (Fig. 7.30) was fabricated by FIB-CVD on the tip of the glass capillary with an electrode.

Microsphere (a polystyrene latex ball with a diameter of 1 μm) manipulation was carried out using the 3-D nanomanipulator with four fingers. An illustration of this manipulation experiment is shown in Fig. 7.31. By connecting the manipulator fabricated by FIB-CVD to a commercial manipulator (MHW-3; Narishige Co.), the direction of movement along the *x*-, *y*-, and *z*-axis could be controlled. The microsphere target was fixed to the side of a glass capillary, and the manipulation was observed from the top with an optical microscope.

The optical microscope image of Fig. 7.32 shows the situation during manipulation. First, the 3-D nanomanipulator was made to approach the microsphere; no voltage was applied. Next, the four fingers were opened by applying 600 V in front of the microsphere and the microsphere could be caught by turning off the voltage when the microsphere was in the grasp of the nanomanipulator. The 3-D nanomanipulator was then removed from the side of the glass capillary. Note

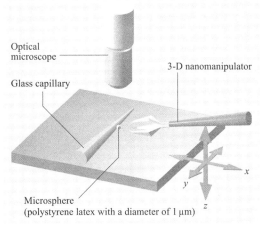

Fig. 7.31 Illustration of 1 μm polystyrene microsphere manipulation by using a 3-D electrostatic nanomanipulator with four fingers

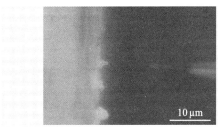

Fig. 7.32 In situ observation of 1 μm polystyrene microsphere manipulation using a 3-D electrostatic nanomanipulator with four fingers

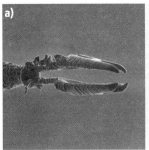 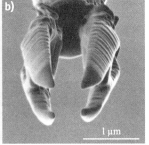

Fig. 7.30a,b SIM image of the 3-D electrostatic nanomanipulator with four fingers before manipulation. (**a**) Side view, (**b**) top view

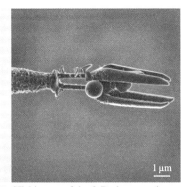

Fig. 7.33 SIM image of the 3-D electrostatic nanomanipulator with four fingers after manipulation

that the action of catching the microsphere occurs due to the elastic force of the manipulator's structure. We succeeded in catching the microsphere, as shown in the SIM image in Fig. 7.33.

Nanonet

Highly functional nanotools are required to perform subcellular operations and analysis in nanospace. For example, nanotweezers have been fabricated on an AFM tip from carbon nanotube [7.21]. We have produced nanotools with arbitrary structures using FIB-CVD. Recently, we have fabricated a nanonet as a novel nanotool for the manipulation and analysis of subcellular organelles; subcellular operations like these are easy to perform using a nanonet [7.22].

To realize the nanonet, a glass capillary (GD-1: Narishige Co.) was pulled with a micropipette puller (PC-10: Narishige Co.). The glass capillary was 90 mm in length and 1 mm in diameter. Using this process, we obtained a 1 μm-diameter tip on the glass capillary. Next, the glass capillary's surface was coated with Au for protection during charging with a Ga$^+$ ion beam. In the final processes, the nanonet structure was fabricated with FIB-CVD. We used a commercially available FIB system (SIM2050MS2: SII NanoTechnology Inc.) that has a Ga$^+$ ion beam operating at 30 kV. The source material for FIB-CVD was a *phenanthrene* ($C_{14}H_{10}$) precursor. Diamond-like carbon (DLC) is deposited by using this source material. The minimum beam diameter was about 5 nm.

The *phenanthrene* gas pressure in the specimen chamber during growth was typically 5×10^{-5} Pa. By transmitting the CAD data for the arbitrary structures to the FIB system, we were able to control the Ga$^+$ ion beam, and therefore to fabricate the nanonet structure on the glass capillary.

The flexibility and practicality of the nanonet is enhanced by fabricating it on a glass capillary, since this is used in many fields, including biology and medicine. The FIB-CVD deposition time was about 40 min at a beam current of 7 pA. The diameter of the ring used to hang the net was about 7 μm, and the width of the net was about 300 nm.

We performed an experiment under an optical microscope in which polystyrene microspheres were captured with the nanonet. Figure 7.34 shows a schematic of the experimental apparatus. Polystyrene microspheres with a diameter of 2 μm were dispersed in distilled water to simulate subcellular organelles. The x-, y-, and z-axis movements of the FIB-CVD nanonet

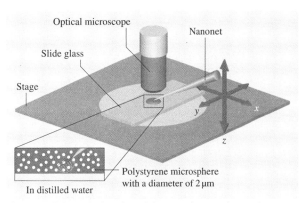

Fig. 7.34 Schematic drawing of the experiment where polystyrene microspheres were captured using a nanonet

were precision-controlled using a commercial manipulator (MHW-3; Narishige Co.).

We performed in situ observations of the capture of 2 μm-diameter polystyrene microspheres using the nanonet. First, the nanonet was brought to the surface of the water. Next, we placed the nanonet into distilled water by controlling its z-axis movement with a commercial manipulator. Then the microspheres were scooped up by moving the nanonet upward. Finally, the nanonet was removed from the surface of the water. At this point, the nanonet had scooped up three microspheres. After the in situ experiments, we observed that the nanonet contained the 2 μm-diameter microspheres. Figure 7.35 shows an SIM image of the nanonet holding the captured microspheres. This proves that we successfully captured the microspheres.

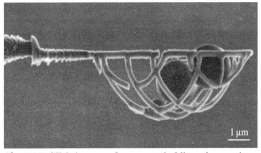

Fig. 7.35 SIM image of nanonet holding three microspheres after capture

7.5 Summary

Three-dimensional nanostructure fabrication using 30 keV Ga$^+$ FIB-CVD and a *phenanthrene* ($C_{14}H_{10}$) source as a precursor has been demonstrated. The film deposited on a silicon substrate was characterized using a transmission microscope and Raman spectra. This characterization indicated that the deposited film is diamond-like amorphous carbon, which has attracted attention due to its hardness, chemical inertness, and optical transparency. Its large Young's modulus, which exceeds 600 GPa, makes it highly desirable for various applications. A nanoelectrostatic actuator and 0.1 μm nanowiring were fabricated and evaluated as parts of nanomechanical system. Furthermore, a nanoinjector and nanomanipulator were fabricated as novel nanotools for manipulation and analysis of subcellular organelles. These results demonstrate that FIB-CVD is one of the key technologies needed to make 3-D nanodevices that can be used in the field of electronics, mechanics, optics, and biology.

References

7.1 S. Matsui: Nanostructure fabrication using electron beam and its application to nanometer devices, Proc. IEEE **85**, 629 (1997)

7.2 O. Lehmann, F. Foulon, M. Stuke: Surface and three-dimensional processing by laser chemical vapor deposition, NATO ASI Ser. Appl. Sci. **265**, 91 (1994)

7.3 H.W. Koops, J. Kretz, M. Rudolph, M. Weber, G. Dahm, K.L. Lee: Characterization and application of materials grown by electron-beam-induced deposition, Jpn. J. Appl. Phys. **33**, 7099 (1994)

7.4 A. Wagner, J.P. Levin, J.L. Mauer, P.G. Blauner, S.J. Kirch, P. Long: X-ray mask repair with focused ion beams, J. Vac. Sci. Technol. B **8**, 1557 (1990)

7.5 I. Utke, P. Hoffmann, B. Dwir, K. Leifer, E. Kapon, P. Doppelt: Focused electron beam induced deposition of gold, J. Vac. Sci. Technol. B **18**, 3168 (2000)

7.6 A.J. DeMarco, J. Melngailis: Lateral growth of focused ion beam deposited platinum for stencil mask repair, J. Vac. Sci. Technol. B **17**, 3154 (1999)

7.7 S. Matsui, T. Kaito, J. Fujita, M. Komuro, K. Kanda, Y. Haruyama: Three-dimensional nanostructure fabrication by focused-ion-beam chemical vapor deposition, J. Vac. Sci. Technol. B **18**, 3181 (2000)

7.8 T. Hoshino, K. Watanabe, R. Kometani, T. Morita, K. Kanda, Y. Haruyama, T. Kaito, J. Fujita, M. Ishida, Y. Ochiai, S. Matsui: Development of three-dimensional pattern-generating system for focused-ion-beam chemical-vapor deposition, J. Vac. Sci. Technol. B **21**, 2732 (2003)

7.9 E. Buks, M.L. Roukes: Stiction, adhesion energy, and the Casimir effect in micromechanical systems, Phys. Rev. B **63**, 033402 (2001)

7.10 J. Fujita, M. Ishida, T. Sakamoto, Y. Ochiai, T. Kaito, S. Matsui: Observation and characteristics of mechanical vibration in three-dimensional nanostructures and pillars grown by focused ion beam chemical vapor deposition, J. Vac. Sci. Technol. B **19**, 2834 (2001)

7.11 M. Ishida, J. Fujita, Y. Ochiai: Density estimation for amorphous carbon nanopillars grown by focused ion beam assisted chemical vapor deposition, J. Vac. Sci. Technol. B **20**, 2784 (2002)

7.12 T. Morita, R. Kometani, K. Watanabe, K. Kanda, Y. Haruyama, T. Hoshino, K. Kondo, T. Kaito, T. Ichihashi, J. Fujita, M. Ishida, Y. Ochiai, T. Tajima, S. Matsui: Free-space-wiring fabrication in nano-space by focused-ion-beam chemical vapor deposition, J. Vac. Sci. Technol. B **21**, 2737 (2003)

7.13 J. Fujita, M. Ishida, Y. Ochiai, T. Ichihashi, T. Kaito, S. Matsui: Graphitization of Fe-doped amorphous carbon pillars grown by focused ion-beam-induced chemical-vapor deposition, J. Vac. Sci. Technol. B **20**, 2686 (2002)

7.14 K. Nakamatsu, K. Yamamoto, T. Hirayama, S. Matsui: Fabrication of fine electron biprism filament by free-space-nanowiring technique of focused-ion-beam + chemical vapor deposition for accurate off-axis electron holography, Appl. Phys. Express **1**, 117004 (2008)

7.15 T. Morita, K. Nakamatsu, K. Kanda, Y. Haruyama, K. Kondo, T. Hoshino, T. Kaito, J. Fujita, T. Ichihashi, M. Ishida, Y. Ochiai, T. Tajima, S. Matsui: Nanomechanical switch fabrication by focused-ion-beam chemical vapor deposition, J. Vac. Sci. Technol. B **22**, 3137 (2004)

7.16 R. Kometani, T. Hoshino, K. Kondo, K. Kanda, Y. Haruyama, T. Kaito, J. Fujita, M. Ishida, Y. Ochiai, S. Matsui: Characteristics of nano-electrostatic actuator fabricated by focused ion beam chemical vapor deposition, Jpn. J. Appl. Phys. **43**, 7187 (2004)

7.17 P. Vukusic, J.R. Sambles: Photonic structures in biology, Nature **424**, 852 (2003)

7.18 K. Watanabe, T. Hoshino, K. Kanda, Y. Haruyama, S. Matsui: Brilliant blue observation from a *Morpho-butterfly*-scale quasi-structure, Jpn. J. Appl. Phys. **44**, L48 (2005)

7.19 R. Kometani, T. Morita, K. Watanabe, K. Kanda, Y. Haruyama, T. Kaito, J. Fujita, M. Ishida, Y. Ochiai, S. Matsui: Nozzle-nanostructure fabrication on glass capillary by focused-ion-beam chemical vapor deposition and etching, Jpn. J. Appl. Phys. **42**, 4107 (2003)

7.20 R. Kometani, T. Hoshino, K. Kondo, K. Kanda, Y. Haruyama, T. Kaito, J. Fujita, M. Ishida, Y. Ochiai, S. Matsui: Performance of nanomanipulator fabricated on glass capillary by focused-ion-beam chemical vapor deposition, J. Vac. Sci. Technol. B **23**, 298 (2005)

7.21 S. Akita, Y. Nakayama, S. Mizooka, Y. Takano, T. Okawa, K.Y. Miyatake, S. Yamanaka, M. Tsuji, T. Nosaka: Nanotweezers consisting of carbon nanotubes operating in an atomic force microscope, Appl. Phys. Lett. **79**, 1691 (2001)

7.22 R. Kometani, T. Hoshino, K. Kanda, Y. Haruyama, T. Kaito, J. Fujita, M. Ishida, Y. Ochiai, S. Matsui: Three-dimensional high-performance nanotools fabricated using focused-ion-beam chemical-vapor-deposition, Nucl. Instrum. Methods Phys. Res. B **232**, 362 (2005)

8. Introduction to Micro-/Nanofabrication

Babak Ziaie, Antonio Baldi, Massood Z. Atashbar

This chapter outlines and discusses important micro- and nanofabrication techniques. We start with the most basic methods borrowed from the integrated circuit (IC) industry, such as thin-film deposition, lithography and etching, and then move on to look at microelectromechanical systems (MEMS) and nanofabrication technologies. We cover a broad range of dimensions, from the micron to the nanometer scale. Although most of the current research is geared towards the nanodomain, a good understanding of top-down methods for fabricating micron-sized objects can aid our understanding of this research. Due to space constraints, we focus here on the most important technologies; in the microdomain these include surface, bulk, and high-aspect-ratio micromachining; in the nanodomain we concentrate on e-beam lithography, epitaxial growth, template manufacturing, and self-assembly. MEMS technology is maturing rapidly, with some new technologies displacing older ones that have proven to be unsuited to manufacture on a commercial scale. However, the jury is still out on methods used in the nanodomain, although it appears that bottom-up methods are the most feasible, and these will have a major impact in a variety of application areas such as biology, medicine, environmental monitoring, and nanoelectronics.

8.1	Basic Microfabrication Techniques	232
	8.1.1 Lithography	232
	8.1.2 Thin-Film Deposition and Doping	233
	8.1.3 Etching and Substrate Removal	238
	8.1.4 Substrate Bonding	243
8.2	MEMS Fabrication Techniques	244
	8.2.1 Bulk Micromachining	244
	8.2.2 Surface Micromachining	248
	8.2.3 High-Aspect-Ratio Micromachining	252
8.3	Nanofabrication Techniques	256
	8.3.1 E-Beam Nanofabrication	257
	8.3.2 Epitaxy and Strain Engineering	257
	8.3.3 Scanning Probe Techniques	258
	8.3.4 Self-Assembly and Template Manufacturing	261
8.4	Summary and Conclusions	265
	References	265

Recent innovations in the area of micro/nanofabrication have created a unique opportunity for manufacturing structures in the nm–mm range. The available six orders of magnitude dimensional span can be used to fabricate novel electronic, optical, magnetic, mechanical, and chemical/biological devices with applications ranging from sensors to computation and control. In this chapter, we will introduce major micro/nanofabrication techniques currently used to fabricate structures from the nm to several hundred μm range. We will mainly focus on the most important and widely used techniques and will not discuss specialized methods. After a brief introduction to basic microfabrication, we will discuss MEMS fabrication techniques used to build microstructures down to about 1 μm in dimensions. Following this, we will discuss several top-down and bottom-up nanofabrication methods not discussed in other chapters of this Handbook.

8.1 Basic Microfabrication Techniques

Most micro/nanofabrication techniques have their roots in the standard fabrication methods developed for the semiconductor industry [8.1–3]. Therefore, a clear understanding of these techniques is necessary for anyone starting to embark on a research and development path in the micro/nano area. In this section, we will discuss the major microfabrication methods used most frequently in the manufacturing of micro/nanostructures. Some of these techniques such as thin-film deposition and etching are common between the micro/nano and very large-scale integration (VLSI) microchip fabrication disciplines. However, several other techniques which are more specific to the micro/nanofabrication area will also be discussed in this section.

8.1.1 Lithography

Lithography is the technique used to transfer a computer-generated pattern onto a substrate (silicon, glass, GaAs, etc.). This pattern is subsequently used to etch an underlying thin film (oxide, nitride, etc.) for various purposes (doping, etching, etc.). Although photolithography, i.e., lithography using an ultraviolet (UV) light source, is by far the most widely used lithography technique in the microelectronic fabrication, electron-beam (e-beam) and x-ray lithography are two other alternatives which have attracted considerable attention in the MEMS and nanofabrication areas. We will discuss photolithography in this section and postpone discussion of e-beam and x-ray techniques to subsequent sections dealing with MEMS and nanofabrication.

The starting point subsequent to the creation of the computer layout for a specific fabrication sequence is the generation of a photomask. This involves a sequence of photographic processes (using optical or e-beam pattern generators), which results in a glass plate having the desired pattern in the form of a thin (≈ 100 nm) chromium layer. Following the generation of the photomask, the lithography process can proceed as shown in Fig. 8.1. This sequence demonstrates the pattern transfer onto a substrate coated with silicon dioxide; however, the same technique is applicable to other materials. After depositing the desired material on the substrate, the photolithography process starts with spin-coating the substrate with a photoresist. This is a polymeric photosensitive material which can be spun onto the wafer in liquid from (usually an adhesion promoter such as hexamethyldisilazane HMDS is used prior to the application of the resist). The spinning speed and photoresist viscosity will determine the final resist thickness, which is typically in the range 0.5–2.5 μm. Two different kinds of photoresist are available: positive and negative. With positive resist, UV-exposed areas will be dissolved in the subsequent development stage, whereas with negative photoresist, the exposed areas will remain intact after UV development. Due to its better performance with regard to process control in small geometries, positive resist is the most extensively used in the VLSI processes. After spinning the photoresist onto the wafer, the substrate is soft-baked (5–30 min at 60–100 °C) in order to remove the solvents from the resist and improve adhesion. Subsequently, the mask is aligned to the wafer and the photoresist is exposed to a UV source.

Depending on the separation between the mask and the wafer three different exposure systems are available:

Fig. 8.1 Lithography process flow

1. Contact
2. Proximity
3. Projection

Although contact printing gives better resolution compared with the proximity technique, the constant contact of the mask with the photoresist reduces the process yield and can damage the mask. Projection printing uses a dual-lens optical system to project the mask image onto the wafer. Since only one die at a time can be exposed, this requires a step-and-repeat system to cover the whole wafer area. Projection printing is by far the most widely used system in microfabrication and can yield superior resolutions compared with the contact and proximity methods. The exposure source for photolithography depends on the resolution. Above $0.25\,\mu m$ minimum line width, a high-pressure mercury lamp is adequate (436 nm g-line and 365 nm i-line). However, between 0.25 and $0.13\,\mu m$, deep-UV sources such as excimer lasers (248 nm KrF and 193 nm ArF) are required. Although there has been extensive competition for the below-$0.13\,\mu m$ regime (including e-beam and x-ray), extreme UV (EUV) with wavelength of 10–14 nm seems to be the preferred technique, although major technical challenges still remain [8.4]. Immersion lithography (i.e., using a liquid in the space between the lens and substrate in order to increase the numerical aperture), a recent innovation, has allowed the minimum feature size to be reduced to 32 nm without the requirement for EUV sources [8.5].

After exposure, the photoresist is developed in a process similar to the development of photographic films. The resist is subsequently hard-baked (20–30 min at 120–180 °C) in order to further improve adhesion. The hard-bake step concludes the photolithography sequence by creating the desired pattern on the wafer. Next, the underlying thin film is etched and the photoresist is stripped using acetone or other organic removal solvents. Figure 8.2 shows a schematic of the photolithography steps with a positive photoresist.

8.1.2 Thin-Film Deposition and Doping

Thin-film deposition and doping are extensively used in micro/nanofabrication technologies. Most fabricated micro/nanostructures contain materials other than that of the substrate, which are obtained by various deposition techniques or by modification of the substrate. Following is a list of a few typical applications for the deposited and/or doped materials used in micro/nanofabrication, which gives an idea of the required properties:

- Mechanical structure
- Electrical isolation
- Electrical connection
- Sensing or actuating
- Mask for etching and doping
- Support or mold during deposition of other materials (sacrificial materials)
- Passivation

Most of the deposited thin films have properties different from those of their corresponding *bulk* forms (for example, metals shows higher resistivities in thin-film

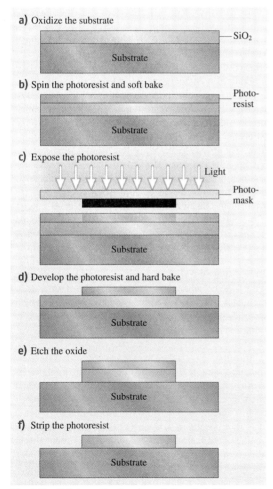

Fig. 8.2a–f Schematic drawing of the photolithographic steps with a positive PR

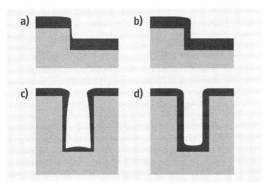

Fig. 8.3a–d Step coverage and conformality: (**a**) poor step coverage, (**b**) good step coverage, (**c**) nonconformal layer, and (**d**) conformal layer

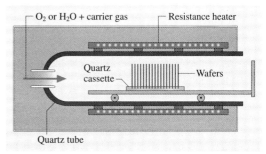

Fig. 8.4 Schematic representation of a typical oxidation furnace

form). In addition, the techniques utilized to deposit these materials have a great impact on their final properties. For instance, internal stress (compressive or tensile) in a film is strongly process dependent. Excessive stress may crack or detach the film from the substrate and therefore must be minimized, although it may also be useful for certain applications. Adhesion is another important issue that needs to be taken into account when depositing thin films. In some cases such as the deposition of noble metals (e.g., gold) an intermediate layer (chromium or titanium) may be needed to improve adhesion. Finally, step coverage and conformality are two properties that can also influence the choice of one or another deposition technique. Figure 8.3 illustrates these concepts.

Oxidation

Oxidation of silicon is a process used to obtain a thin film of SiO_2 with excellent quality (very low density of defects) and thickness homogeneity. Although it is not properly a deposition, the result is the same; i.e., a thin layer of a new material covering the surface is produced. The oxidation process is typically carried out at temperatures in the range of 900–1200 °C in the presence of O_2 (dry oxidation) or H_2O (wet oxidation). The reactions for oxide formation are

$$Si_{(solid)} + O_{2\,(gas)} \Rightarrow SiO_{2\,(solid)}$$

and

$$Si_{(solid)} + 2H_2O_{(steam)} \Rightarrow SiO_{2\,(solid)} + 2H_{2\,(gas)}\,.$$

Although the rate of oxide growth is higher for wet oxidation, this is achieved at the expense of lower oxide quality (density). Since silicon atoms from the substrate participate in the reaction, the substrate is consumed as the oxide grows ($\approx 44\%$ of the total thickness lies above the line of the original silicon surface). The oxidation of silicon also occurs at room temperature, however a layer of about 20 Å (native oxide) is enough to passivate the surface and prevent further oxidation. To grow thicker oxides, wafers are introduced into an electric resistance furnace such as that represented in Fig. 8.4. Tens of wafers can be processed in a single batch in such equipment. By strictly controlling the timing, temperature, and gas flow entering the quartz tube the desired thickness can be achieved with high accuracy. Thicknesses ranging from a few tens of Angstroms to 2 µm can be obtained in reasonable times. Despite the good quality of the SiO_2 obtained by silicon oxidation (also called *thermal oxide*), the use of this process is often limited to the early stages of the fabrication, since some of the materials added during the formation of structures may not withstand the high temperatures. The contamination of the furnace, when the substrates have been previously in contact with certain etchants such as KOH or when materials such as metals have been deposited, also poses limitations in most cases.

Doping

The introduction of certain impurities in a semiconductor can change its electrical, chemical, and even mechanical properties. Typical impurities or *dopants* used in silicon include boron (to form p-type regions) and phosphorous or arsenic (to form n-type regions). Doping is the main process used in the microelectronic industry to fabricate major components such as diodes and transistors. In micro/nanofabrication technologies doping has additional applications such as the formation of piezoresistors for mechanical transducers or the creation of etch stop-layers. Two different techniques

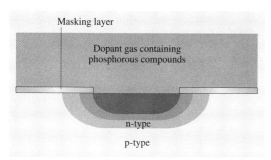

Fig. 8.5 Formation of an n-type region on a p-type silicon substrate by diffusion of phosphorous

are used to introduce impurities into a semiconductor substrate: diffusion and ion implantation.

Diffusion is the process which became dominant in the initial years following the invention of the integrated circuit to form n- and p-type regions in the silicon. The diffusion of impurities into silicon occurs only at high temperature (above 800 °C). Furnaces used to carry out this process are similar to those used for oxidation. Dopants are introduced in the furnace gaseous atmosphere from liquid or solid sources. Figure 8.5 illustrates the process of creating an n-type region by diffusion of phosphor from the surface into a p-type substrate. A masking material is previously deposited and patterned on the surface to define the areas to be doped. However, because diffusion is an isotropic process, the doped area will also extend underneath the mask. In microfabrication, diffusion is mainly used for the formation of very highly boron-doped regions (p^{++}), which are usually used as an etch stop in bulk micromachining.

Ion implantation allows more precise control of the dose (the total amount of impurities introduced per area unit) and the impurity profile (the concentration versus depth). In ion implantation the impurities are ionized and accelerated towards the semiconductor surface. The penetration of impurities into the material follows a Gaussian distribution. After implantation, an annealing process is needed to activate the impurities and repair the damage in the crystal structure produced by ion collisions. A *drive-in* process to redistribute the impurities, done in a standard furnace such as those used for oxidation or diffusion, may be required as well.

Chemical Vapor Deposition and Epitaxy

As its name suggests, chemical vapor deposition (CVD) includes all deposition techniques using the reaction of chemicals in gas phase to form the deposited thin film. The energy needed for the chemical reaction to occur is usually supplied by maintaining the substrate at elevated temperature. Alternative energy sources such as plasma or optical excitation are also used, with the advantage of requiring a lower temperature at the substrate. The most common CVD processes in microfabrication are low-pressure CVD (LPCVD) and plasma-enhanced (PECVD).

The LPCVD process is typically carried out in electrically heated tubes, similar to oxidation tubes, equipped with pumping capabilities to achieve the low pressures required (0.1–1.0 Torr). Large numbers of wafers can be processed simultaneously and the material is deposited on both sides of the wafers. The process temperature depends on the material to be deposited, but generally is in the range 550–900 °C. As in oxidation, high temperatures and contamination issues can restrict the type of processes used previous to the LPCVD. Typical materials deposited by LPCVD include silicon oxide (e.g., $SiCl_2H_2 + 2N_2O \Rightarrow SiO_2 + 2N_2 + 2HCL$ at 900 °C), silicon nitride (e.g., $3SiH_4 + 4NH_3 \Rightarrow Si_3N_4 + 12H_2$ at 700–900 °C), and polysilicon (e.g., $SiH_4 \Rightarrow Si + 2H_2$ at 600 °C). Due to its faster etch rate in HF, in situ phosphorous-doped LPCVD oxide (phosphosilicate glass, PSG) is extensively used in surface micromachining as the sacrificial layer. Conformality in this process is excellent, even for very high-aspect-ratio structures. Mechanical properties of LPCVD materials are good compared with others such as PECVD, and are often used as structural materials in microfabricated devices. Stress in the deposited layers depends on the material, deposition conditions, and subsequent thermal history (e.g., postdeposition annealing). Typical values are: 100–300 MPa (compressive) for oxide, ≈ 1 GPa (tensile) for stoichiometric nitride, and ≈ 200–300 MPa (tensile) for polysilicon. The stress in nitride layers can be reduced to nearly zero by using a silicon-rich composition. Since the stress values can vary over a wide range, one has to measure and characterize the internal stress of deposited thin films for any specific equipment and deposition conditions.

The PECVD process is performed in plasma systems such as that represented in Fig. 8.6. The use of radiofrequency (RF) energy to create highly reactive species in the plasma allows for the use of lower temperature at the substrate (150–350 °C). Parallel-plate plasma reactors normally used in microfabrication can only process a limited number of wafers per batch. The wafers are positioned horizontally on top of the lower electrode so only one side gets deposited. Typical materials deposited with PECVD include silicon oxide,

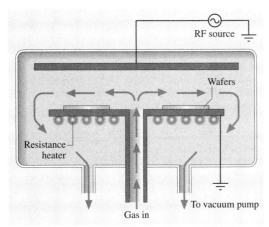

Fig. 8.6 Schematic representation of a typical PECVD system

nitride, and amorphous silicon. Conformality is good for low-aspect-ratio structures, but becomes very poor for deep trenches (20% of the surface thickness inside through-wafer holes with an aspect ratio of 10). Stress depends on deposition parameters and can be either compressive or tensile. PECVD nitrides are typically nonstoichiometric (Si_xN_y) and are much less resistant to etchants in masking applications.

Another interesting type of CVD is epitaxial growth. In this process, a single-crystalline material is grown as an extension of the crystal structure of the substrate. It is possible to grow dissimilar materials if the crystal structures are somehow similar (lattice matched). Silicon-on-sapphire (SoS) substrates and some heterostructures are fabricated in this way. However, most common in microfabrication is the growth of silicon on another silicon substrate. Of particular interest for the formation of microstructures is selective epitaxial growth. In this process the silicon crystal is allowed to grow only in windows patterned on a masking material. Many CVD techniques have been used to produce epitaxial growth. The most common for silicon is thermal chemical vapor deposition or vapor-phase epitaxy (VPE). Metalorganic chemical vapor deposition (MOCVD) and molecular-beam epitaxy (MBE) are the most common for growing high-quality III–V compound layers with nearly atomic abrupt interfaces. The former uses vapors of organic compounds with group III atoms such as trimethylgallium ($Ga(CH_3)_3$) and group V hydrides such as AsH_3 in a CVD chamber with fast gas switching capabilities. The latter typically uses molecular beams from thermally evaporated elemental sources aiming at the substrate in an ultrahigh-vacuum chamber. In this case, rapid on/off control of the beams is achieved by using shutters in front of the sources. Finally, it should be mentioned that many metals (molybdenum, tantalum, titanium, and tungsten) can also be deposited using LPCVD. These are attractive for their low resistivities and their ability to form silicides with silicon. Due to its application in new interconnect technologies, copper CVD is an active area of research.

Physical Vapor Deposition (Evaporation and Sputtering)

In physical deposition systems the material to be deposited is transported from a source to the wafers, both being in the same chamber. Two physical principles are used to do this: evaporation and sputtering.

In evaporation, the source is placed in a small container with tapered walls, called the crucible, and is heated up to a temperature where evaporation occurs. Various techniques are utilized to reach the high temperatures needed, including the induction of high currents with coils wound around the crucible and the bombardment of the material surface with an electron beam (e-beam evaporators). This process is mainly used to deposit metals, although dielectrics can also be evaporated. In a typical system the crucible is located at the bottom of a vacuum chamber whereas the wafers are placed lining the dome-shaped ceiling of the chamber (Fig. 8.7). The main characteristic of this process is very poor step coverage, including shadow effects as illustrated in Fig. 8.8. As will be explained in subsequent

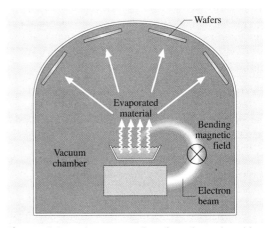

Fig. 8.7 Schematic representation of an e-beam deposition system

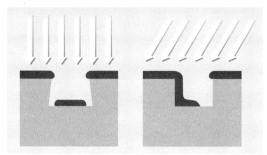

Fig. 8.8 Shadow effects observed in evaporated films. *Arrows* show the trajectory of the material atoms being deposited

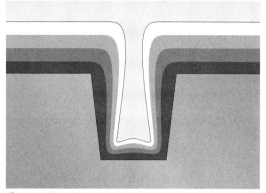

Fig. 8.9 Typical cross section evolution of a trench while being filled with sputter deposition

sections, some microfabrication techniques utilize these effects to pattern the deposited layer. One way to improve the step coverage is by rotating and/or heating the wafers during deposition.

In sputtering, a target of the material to be deposited is bombarded with high-energy inert ions (usually argon). The outcome of the bombardment is that individual atoms or clusters are removed from the surface and ejected towards the wafer. The physical nature of this process allows its use with virtually any existing material. Examples of interesting materials for microfabrication that are frequently sputtered include metals, dielectrics, alloys (such as shape memory alloys), and all kinds of compounds (for example, piezoelectric lead zirconate titanate (PZT)). The inert ions bombarding the target are produced in direct-current (DC) or RF plasma. In a simple parallel-plate system the top electrode is the target and the wafers are placed horizontally on top of the bottom electrode. In spite of its lower deposition rate, step coverage in sputtering is much better than in evaporation. However, the films obtained with this deposition process are nonconformal. Figure 8.9 illustrates successive sputtering profiles in a trench.

Both evaporation and sputtering systems are often able to deposit more than one material simultaneously or sequentially. This capability is very useful to obtain alloys and multilayer films (e.g., multilayer magnetic recording heads are sputtered). For certain low-reactivity metals such as Au and Pt the previous deposition of a thin layer of another metal is needed to improve adhesion. Ti and Cr are two frequently used adhesion promoters. Stress in evaporated or sputtered layers is typically tensile. The deposition rates are much higher than for most CVD techniques. However, due to stress accumulation and cracking, thickness beyond $2\,\mu m$ is rarely deposited with these processes. For thicker deposition a technique described in the next section is sometimes used.

Electroplating

Electroplating (or electrodeposition) is a process typically used to obtain thick (tens of μm) metal structures. The sample to be electroplated is introduced into a solution containing a reducible form of the ion of the desired metal and is maintained at a negative potential (cathode) relative to a counterelectrode (anode). The ions are reduced at the sample surface and the insoluble metal atoms are incorporated into the surface. As an example, copper electrodeposition is frequently done in copper sulfide-based solutions. The reaction taking place at the surface is $Cu^{2+} + 2e^- \rightarrow Cu_{(s)}$. Recommended current densities for electrodeposition processes are on the order of $5-100\,mA/cm^2$.

As can be deduced from the process mechanism, the surface to be electroplated has to be electrically conductive, and preferably of the same material as the deposited one if good adhesion is desired. In order to electrodeposit metals on top of an insulator (the most frequent case) a thin film of the same metal, called the seed layer, is previously deposited on the surface. Masking of the seed layer with a resist permits selective electroplating of the patterned areas. Figure 8.10 illustrates a typical sequence of the steps required to obtain isolated metal structures.

Pulsed Laser and Atomic Layer Deposition

Pulsed laser and atomic layer deposition techniques have attracted a considerable amount of attention recently. These two techniques offer several unique advantages compared with other thin-film deposition

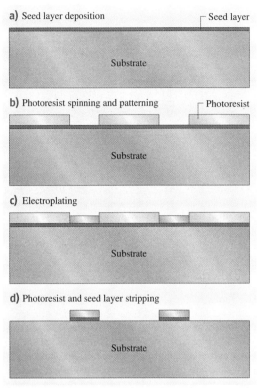

Fig. 8.10a–d Formation of isolated metal structures by electroplating through a mask: (**a**) seed layer deposition, (**b**) photoresist spinning and patterning, (**c**) electroplating, and (**d**) photoresist and seed layer stripping

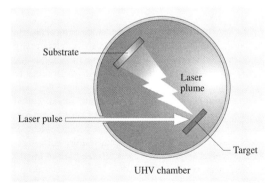

Fig. 8.11 A typical PLD deposition setup

methods that are particularly useful for next-generation nanoscale device fabrication. Pulsed laser deposition (PLD) is a simple technique that uses an intense (1 GW within 25 ns) UV laser (e.g., a KrF excimer) to ablate a target material [8.6]. Plasma is subsequently formed from the target and is deposited on the substrate. Multitarget systems with Auger and reflection high-energy electron diffraction (RHEED) spectroscopes are commercially available. Figure 8.11 shows a typical PLD deposition setup. The main advantages of the PLD are its simplicity and ability to deposit complex materials with preserved stoichiometry (so-called stoichiometry transfer). In addition, fine control over film thickness is also possible by controlling the number of pulses. The stoichiometry-transfer property of the PLD allows many complex targets such as ferroelectrics, superconductors, and magnetostrictives to be deposited. Other deposited materials include oxides, carbides, polymers, and metallic systems (e.g., FeNdB).

Atomic layer deposition (ALD) is a gas-phase self-limiting deposition method capable of depositing atomic layer thin films with excellent large-area uniformity and conformality [8.7]. It enables simple and accurate control over film composition and thickness at the atomic layer level (typical growth rates of a few Å/cycle). Although most of the attention recently has been directed towards depositing high-k dielectric materials (Al_2O_3, and HfO_2) for next-generation complementary metal–oxide–semiconductor (CMOS) electronics, other materials can also be deposited. These include transition metals (Cu, Co, Fe, and Ni), metal oxides, sulfides, nitrides, and fluorides. Atomic-level control over film thickness and composition are also attractive features for applications in MEMS such as conformal three-dimensional (3-D) packaging and air-gap structures. ALD is a modification of the CVD process and is based on two or more vapor-phase reactants that are introduced into the deposition chamber in a sequential manner. One growth cycle consists of four steps. First, a precursor vapor is introduced into the chamber, resulting in the deposition of a self-limiting monolayer on the surface of the substrate. Then, the extra unreacted vapor is pumped out and a vapor dose of a second reactant is introduced. This reacts with the precursor on the surface in a self-limiting fashion. Finally, the extra unreacted vapor is pumped out and the cycle is repeated.

8.1.3 Etching and Substrate Removal

Thin-film and bulk substrate etching is another fabrication step that is of fundamental importance to both VLSI processes and micro/nanofabrication. In the VLSI area, various conducting and dielectric thin films deposited for passivation or masking purposes need to be removed

a) Profile for isotropic etch through a photoresist mask

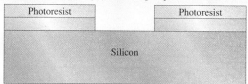

b) Profile for anisotropic etch through a photoresist mask

Fig. 8.12a,b Profile for (**a**) isotropic and (**b**) anisotropic etching through a photoresist mask

at some point or another. In micro/nanofabrication, in addition to thin-film etching, very often the substrate (silicon, glass, GaAs, etc.) also needs to be removed in order to create various mechanical micro/nanostructures (beams, plates, etc.). Two important figures of merit for any etching process are selectivity and directionality. Selectivity is the degree to which the etchant can differentiate between the masking layer and the layer to be etched. Directionality has to do with the etch profile under the mask. In an isotropic etch, the etchant attacks the material in all directions at the same rate, hence creating a semicircular profile under the mask (Fig. 8.12a). In contrast, in an anisotropic etch, the dissolution rate depends on specific directions and one can obtain vertical side-walls or other noncircular profiles (Fig. 8.12b). One can also divide the various etching techniques into wet and dry categories. In this chapter, we will use this classification and discuss different wet etchants first followed by dry etching techniques used most often in the micro/nanofabrication.

Wet Etching

Historically, wet etching techniques preceded the dry ones. These still constitute an important group of etchants for micro/nanofabrication in spite of their less frequent application in the VLSI processes. Wet etchants are by and large isotropic and show superior selectivity to the masking layer as compared with various dry techniques. In addition, due to the lateral undercut, the minimum feature size achievable with wet etchants is limited to $> 3\,\mu\text{m}$. Silicon dioxide is commonly etched in dilute (6:1, 10:1, or 20:1 by volume) or buffered HF (BHF: $HF + NH_4F$) solutions (etch rate of $\approx 1000\,\text{Å/min}$ in BHF). Photoresist and silicon nitride are the two most common masking materials for the wet oxide etch. The wet etchant for silicon nitride is hot (140–200 °C) phosphoric acid with silicon oxide as the masking material. Nitride wet etch is not very common (except for blanket etch) due to the masking difficulty and nonrepeatable etch rates. Metals can be etched using various combinations of acid and base solutions. There are also many commercially available etchant formulations for aluminum, chromium, and gold which can easily be used. A comprehensive table of various metal etchants can be found in [8.8].

Anisotropic and isotropic wet etching of crystalline (silicon and gallium arsenide) and noncrystalline (glass) substrates is an important topic in micro/nanofabrication [8.9–12]. In particular, the realization of the possibility of anisotropic wet etching of silicon is considered to mark the beginning of the micromachining and MEMS discipline. Isotropic etching of silicon using $HF/HNO_3/CH_3COOH$ (various different formulations have been used) dates back to the 1950s and is still frequently used to thin down the silicon wafer. The etch mechanism for this combination has been elucidated and is as follows: HNO_3 is used to oxidize the silicon, which is subsequently dissolved away in the HF. The acetic acid is used to prevent the dissociation of HNO_3 (the etch works as well without the acetic acid). For short etch times, silicon dioxide can be used as the masking material; however, one needs to use silicon nitride if a longer etch time is desired. This etch also shows dopant selectivity, with the etch rate dropping at lower doping concentrations ($< 10^{17}\,\text{cm}^{-3}$ n- or p-type). Although this effect can potentially be used as an etch-stop mechanism in order to fabricate microstructures, the difficulty in masking has prevented widespread application of this approach. Glass can also be isotropically etched using the HF/HNO_3 combination with the etch surfaces showing considerable roughness. This has been extensively used in fabricating microfluidic components (mainly channels). Although Cr/Au is usually used as the masking layer, long etch times require a more robust mask (bonded silicon has been used for this purpose).

Silicon anisotropic wet etch constitutes an important technique in bulk micromachining. The three most important silicon etchants in this category are potassium hydroxide (KOH), ethylenediamine pyrochatechol (EDP), and tetramethyl ammonium hydroxide (TMAH). These are all anisotropic etchants which attack silicon along preferred crystallographic directions. In addition, they all show marked reduction of the etch rate in heavily ($> 5 \times 10^{19}$ cm^{-3}) boron-doped (p^{++}) regions. The chemistry behind the action of these etchants is not yet very clear but it seems that silicon atom oxidation at the surface and its reaction with hydroxyl ions (OH$^-$) is responsible for the formation of a soluble silicon complex (SiO$_2$(OH)$^{2-}$). The etch rate depends on the concentration and temperature and is usually around 1 μm/min at temperatures of 85–115 °C. Common masking materials for anisotropic wet etchants are silicon dioxide and nitride, with the latter being superior for longer etch times. The crystallographic plane which shows the slowest etch rate is the (111) plane. Although the lower atomic concentration along these planes has been speculated to be the reason for this phenomena, the evidence is inconclusive and other factors must be included to account for this remarkable etch-stop property. The anisotropic behavior of these etchants with respect to the (111) planes has been extensively used to create beams, membranes, and other mechanical and structural components. Figure 8.13 shows the typical cross sections of (100) and (110) silicon wafers etched with an anisotropic wet etchant. As can be seen, the (111) slow planes are exposed in both situations, one creating 54.7° sloped side-walls in the (100) wafer and the other creating vertical side-walls in the (110) wafer. Depending on the dimensions of the mask opening, a V-groove or a trapezoidal trench is formed in the (100) wafer. A large enough opening will allow the silicon to be etched all

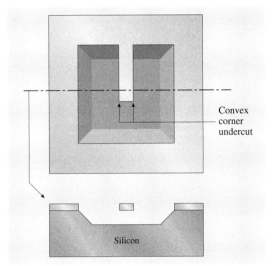

Fig. 8.14 Top view and cross section of a dielectric cantilever beam fabricated using convex corner undercut

the way through the wafer, thus creating a thin dielectric membrane on the other side. It should be mentioned that exposed convex corners have a higher etch rate than concave ones, resulting in an undercut which can be used to create dielectric (e.g., nitride) cantilever beams. Figure 8.14 shows a cantilever beam fabricated using the convex corner undercut on a (100) wafer.

The three above-mentioned etchants show different directional and dopant selectivities. KOH has the best (111) selectivity (400/1), followed by TMAH and EDP. However, EDP has the highest selectivity with respect to deep boron diffusion regions. Safety and CMOS compatibility are other important criteria for choosing a particular anisotropic etchant. Among the three mentioned etchants TMAH is the most benign, whereas EDP is extremely corrosive and carcinogenic. Silicon can be dissolved in TMAH in order to improve its selectivity with respect to aluminum. This property has made TMAH very appealing for post-CMOS micromachining where aluminum lines have to be protected. Finally, it should be mentioned that one can modulate the etch rate using a reversed-biased p–n junction (electrochemical etch stop). Figure 8.15 shows the setup commonly used to perform electrochemical etching. The silicon wafer under etch consists of an n-epi region on a p-type substrate. Upon the application of a reverse-bias voltage to the structure (p-substrate is in contact with the solution and n-epi is protected using a watertight fixture), the p-substrate is etched away. When the n-epi

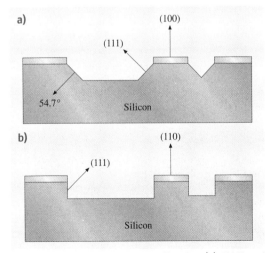

Fig. 8.13a,b Anisotropic etch profiles for: **(a)** (100) and **(b)** (110) silicon wafers

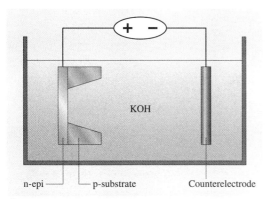

Fig. 8.15 Electrochemical etch setup

regions are exposed to the solution an oxide passivation layer is formed and etching is stopped. This technique can be used to fabricate single-crystalline silicon membranes for pressure sensors and other mechanical transducers.

Dry Etching

Most of the dry etching techniques are plasma based. They have several advantages when compared with wet etching. These include smaller undercut (allowing smaller lines to be patterned) and higher anisotropy (allowing high-aspect-ratio vertical structures). However, the selectivity of dry etching techniques is lower than that of wet etchants, and one must take into account the finite etch rate of the masking materials. The three basic dry etching techniques, namely high-pressure plasma etching, reactive-ion etching (RIE), and ion milling, utilize different mechanisms to obtain directionality.

Ion milling is a purely physical process which utilizes accelerated inert ions (e.g., Ar^+) striking perpendicular to the surface to remove the material ($p \approx 10^{-4}$–10^{-3} Torr) (Fig. 8.16a). The main characteristics of this technique are very low etch rates (in the order of a few nm/min) and poor selectivity (close to 1 : 1 for most materials); hence it is generally used to etch very thin layers. In high-pressure (10^{-1}–5 Torr) plasma etchers highly reactive species are created that

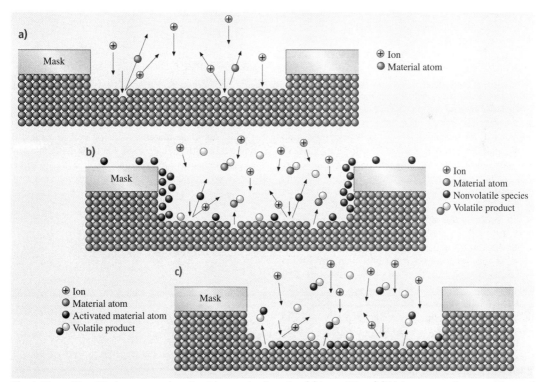

Fig. 8.16a–c Simplified representation of etching mechanisms for: (**a**) ion milling, (**b**) high-pressure plasma etching, and (**c**) RIE

react with the material to be etched. The products of the reaction are volatile so that they diffuse away and new material is exposed to the reactive species. Directionality can be achieved, if desired, with the side-wall passivation technique (Fig. 8.16b). In this technique nonvolatile species produced in the chamber deposit on and passivate the surfaces. The deposit can only be removed by physical collision with incident ions. Because the movement of the ions has a vertical directionality the deposit is removed mainly on horizontal surfaces, while vertical walls remain passivated. In this fashion, the vertical etch rate becomes much higher than the lateral one.

RIE etching, also called ion-assisted etching, is a combination of physical and chemical processes. In this technique the reactive species react with the material only when the surfaces are *activated* by the collision of incident ions from the plasma (e.g., by breaking bonds at the surface). As in the previous technique, the directionality of the ion's velocity produces much more collisions on the horizontal surfaces than on the walls, thus generating faster etching rates in the vertical direction (Fig. 8.16c). To increase the etch anisotropy further, in some cases side-wall passivation methods are also used. An interesting case is the deep reactive-ion etching (DRIE) technique, capable of achieving aspect ratios of 30 : 1 and silicon etching rates of $2-3\,\mu\text{m/min}$ (through wafer etch is possible). In this technique, the passivation deposition and etching steps are performed sequentially in a two-step cycle, as shown in Fig. 8.17. In commercial silicon DRIE etchers SF_6/Ar is typically used for the etching step and a combination of Ar and a fluoropolymer (nCF_2) for the passivation step. A Teflon-like polymer about 50 nm thick is deposited during the latter step, covering only the side-walls (Ar^+ ion bombardment removes the Teflon on the horizontal surfaces). Due to the cyclic nature of this process, the side-walls of the etched features show a periodic *wave-shaped* roughness in the range of 50–400 nm. More recently, *Aimi* et al., reported on a similar method for deep etching of titanium. In this case titanium oxide was used as a side-wall passivation layer [8.13].

Dry etching can also be performed in nonplasma equipment if the etching gases are reactive enough. The so-called vapor-phase etching (VPE) processes can be carried out in a simple chamber with gas feeding and pumping capabilities. Two examples of VPE are xenon difluoride (XeF_2) etching of silicon and HF vapor etching of silicon dioxide. Due to its isotropic nature, these processes are typically used for etching sacrificial layers

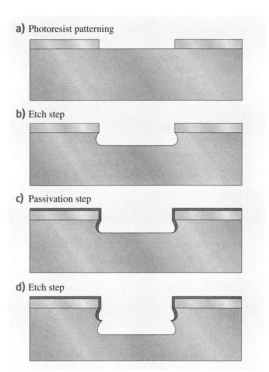

Fig. 8.17a–d DRIE cyclic process: (**a**) photoresist patterning, (**b**) etch step, (**c**) passivation step, and (**d**) etch step

and releasing structures while avoiding stiction problems (Sects. 8.2.1 and 8.2.2).

Most important materials can be etched with the aforementioned techniques, and for each material a variety of chemistries are available. Table 8.1 lists some of the most common materials along with selected etch

Table 8.1 Typical dry etch chemistries

Si	CF_4/O_2, CF_2Cl_2, CF_3Cl, $SF_6/O_2/Cl_2$, $Cl_2/H_2/C_2F_6/CCl_4$, C_2ClF_5/O_2, Br_2, SiF_4/O_2, NF_3, ClF_3, CCl_4, $C_3Cl_3F_5$, C_2ClF_5/SF_6, C_2F_6/CF_3Cl, CF_3Cl/Br_2
SiO_2	CF_4/H_2, C_2F_6, C_3F_8, CHF_3/O_2
Si_3N_4	$CF_4/O_2/H_2$, C_2F_6, C_3F_8, CHF_3
Organics	O_2, CF_4/O_2, SF_6/O_2
Al	BCl_3, BCl_3/Cl_2, $CCl_4/Cl_2/BCl_3$, $SiCl_4/Cl_2$
Silicides	CF_4/O_2, NF_3, SF_6/Cl_2, CF_4/Cl_2
Refractories	CF_4/O_2, NF_3/H_2, SF_6/O_2
GaAs	BCl_3/Ar, $Cl_2/O_2/H_2$, $CCl_2F_2/O_2/Ar/He$, H_2, CH_4/H_2, $CClH_3/H_2$
InP	CH_4/H_2, C_2H_6/H_2, Cl_2/Ar
Au	$C_2Cl_2F_4$, Cl_2, $CClF_3$

recipes [8.14]. For each chemistry the etch rate, directionality, and selectivity with respect to the mask materials depend on parameters such as the flow rates of the gases entering the chamber, the working pressure, and the RF power applied to the plasma.

8.1.4 Substrate Bonding

Substrate (wafer) bonding (silicon–silicon, silicon–glass, and glass–glass) is among the most important fabrication techniques in microsystem technology [8.15, 16]. It is frequently used to fabricate complex 3-D structures both as a functional unit and as a part of the final microsystem package and encapsulation. The two most important bonding techniques are silicon–silicon fusion (or silicon direct bonding) and silicon–glass electrostatic (or anodic) bonding. In addition to these techniques, several other alternative methods which utilize an intermediate layer (eutectic, adhesive, and glass frit) have also been investigated. All these techniques can be used to bond the substrates at the wafer level. In this section we will only discuss wafer-level techniques and will not treat device-level bonding methods (e.g., e-beam and laser welding).

Silicon Direct Bonding

Direct silicon or fusion bonding is used in the fabrication of micromechanical devices and silicon-on-insulator (SOI) substrates. Although it is mostly used to bond two silicon wafers with or without an oxide layer, it has also been used to bond different semiconductors such as GaAs and InP [8.16]. One main requirement for a successful bond is sufficient planarity ($< 10\,\text{\AA}$ surface roughness and $< 5\,\mu\text{m}$ bow across a 4 inch wafer) and cleanliness of the surfaces. In addition, thermal expansion mismatch also needs to be considered if bonding of two dissimilar materials is contemplated. The bonding procedure is as follows: the silicon or oxide-coated silicon wafers are first thoroughly cleaned. Subsequently the surfaces are hydrated (activated) in HF or boiling nitric acid (Radio Corporation of America (RCA) clean also works). This renders the surfaces hydrophilic by creating an abundance of hydroxyl ions. Then the substrates are brought together in close proximity (starting from the center to avoid void formation). The close approximation of the bonding surfaces allows the short-range attractive van der Waals forces to bring the surfaces into intimate contact on the atomic scale. Following this step, hydrogen bonds between the two hydroxyl-coated silicon wafers bond the substrates together. These steps can be performed at room temperature; however, in order to increase the bond strength, a high-temperature (800–1200 °C) anneal is usually required. A major advantage of silicon fusion bonding is the thermal matching of the substrates.

Anodic Bonding

Silicon–glass anodic (electrostatic) bonding is another major substrate joining technique which has been extensively used for microsensor packaging and device fabrication. The main advantage of this technique is its lower bonding temperature, which is around 300–400 °C. Figure 8.18 shows the bonding setup. A glass wafer (usually Pyrex 7740 because of thermal expansion match to silicon) is placed on top of a silicon wafer and the sandwich is heated to 300–400 °C. Subsequently, a voltage of $\approx 1000\,\text{V}$ is applied to the glass–silicon sandwich with the glass connected to the cathode. The bond starts immediately after the application of the voltage and spreads outward from the cathode contact point. The bond can be observed visually as a dark-grayish front which expands across the wafer.

The bonding mechanism is as follows. During the heating period, glass sodium ions move toward the cathode and create a depletion layer at the silicon–glass interface. A strong electrostatic force is therefore created at the interface, which pulls the substrates into intimate contact. The exact chemical reaction responsible for anodic bonding is not yet clear, but covalent silicon–oxygen bonds at the interface seem to be responsible for the bond. Silicon–silicon anodic bonding using sputtered or evaporated glass interlayer is also possible.

Bonding with Intermediate Layers

Various other wafer bonding techniques utilizing an intermediate layer have also been investigated [8.16]. Among the most important ones are adhesive, eutectic, and glass frit bonds. Adhesive bonding using a poly-

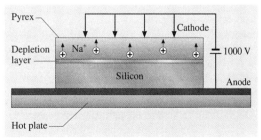

Fig. 8.18 Glass–silicon anodic bonding setup

mer (e.g., polyimides, epoxies, thermoplastic adhesives, and photoresists) in between the wafers has been used to join different wafer substrates [8.17]. Complete curing (in an oven or using dielectric heating) of the polymer before or during the bonding process prevents subsequent solvent outgassing and void formation. Although reasonably high bonding strengths can be obtained, these bonds are nonhermetic and unstable over a period of time.

In eutectic bonding process, gold-coated silicon wafers are bonded together at temperatures greater than the silicon–gold eutectic point (363 °C, 2.85% silicon and 97.1% Au) [8.18]. This process can achieve high bonding strength and good stability at relatively low temperatures. For good bond uniformity silicon dioxide must be removed from the silicon surface prior to the deposition of the gold. In addition, all organic contaminants on the gold surface must be removed (using UV light) prior to the bond. Pressure must also be applied in order to achieve a better contact. Although eutectic bonding can be accomplished at low temperatures, achieving uniformity over large areas has proven to be challenging.

Glass frit can also be used as an interlayer in substrate bonding. In this technique, first a thin layer of glass is deposited and preglazed. The glass-coated substrates are then brought into contact and the sandwich is heated to above the glass melting temperature (typically < 600 °C). As for the eutectic process, pressure must be applied for adequate contact [8.19].

8.2 MEMS Fabrication Techniques

In this section, we will discuss various important MEMS fabrication techniques commonly used to build various microdevices (microsensors and microactuators) [8.9–12]. The dimensional spectrum of the microstructures that can be fabricated using these techniques spans from 1 mm to 1 µm. As mentioned in the introduction, we will mostly emphasize the more important techniques and will not discuss specialized methods.

8.2.1 Bulk Micromachining

Bulk micromachining is the oldest MEMS technology and hence probably one of the more mature ones [8.20]. It is currently by far the most commercially successful one, helping to manufacture devices such as pressure sensors and inkjet printheads. Although there are many different variations, the basic concept behind bulk micromachining is selective removal of the substrate (silicon, glass, GaAs, etc.). This allows the creation of various micromechanical components such as beams, plates, and membranes which can be used to fabricate a variety of sensors and actuators. The most important microfabrication techniques used in bulk micromachining are wet and dry etching and substrate bonding. Although one can use various criteria to categorize bulk micromachining techniques, we will use a historical timeline for this purpose. Starting with the more traditional wet etching techniques, we will proceed to discuss the more recent ones using deep RIE and wafer bonding.

Bulk Micromachining
Using Wet Etch and Wafer Bonding

The use of anisotropic wet etchants to remove silicon can be marked as the beginning of the micromachining era. Back-side etch was used to create movable structures such as beams, membranes, and plates (Fig. 8.19). Initially, etching was timed in order to create a specified thickness. However, this technique proved to be inadequate for the creation of thin structures (< 20 µm). Subsequent use of various etch-stop techniques allowed the creation of thinner membranes in a more controlled fashion. As was mentioned in *High-Aspect Ratio Micromachining*, heavily boron-doped regions and electrochemical bias can be used to slow down the etch process drastically and hence create controllable thickness microstructures. Figure 8.20a,b shows the cross section of two piezoresistive pressure sensors fabricated using electrochemical and p^{++} etch-stop techniques. The use of the p^{++} method requires epitaxial growth

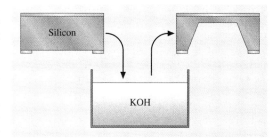

Fig. 8.19 Wet anisotropic silicon back-side etch

of a lightly doped region on top of a p^{++} etch-stop layer. This layer is subsequently used for the placement of piezoresistors. However, if no active component is required one can simply use the p^{++} region to create a thin membrane (Fig. 8.20c).

The p^{++} etch-stop technique can also be used to create isolated thin silicon structures through the dissolution of the entire lightly doped region [8.21]. This technique was successfully used to fabricate silicon recording and stimulating electrodes for biomedical applications. Figure 8.21 shows the cross section of such a process which relies on deep (15–20 μm) and shallow boron (2–5 μm) diffusion steps to create microelectrodes with flexible connecting ribbon cables. An extension of this process which uses a combination of p^{++} etch-stop layers and silicon–glass anodic bonding has also been developed. This process is commonly known as the dissolved wafer process and has been used to fabricate a variety of microsensors and microactuators [8.22]. Figure 8.22 shows

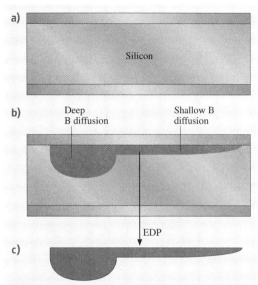

Fig. 8.21a–c Free-standing microstructure fabrication using deep and shallow boron diffusion and EDP release (**a**) silicon wafer, (**b**) deep and shallow boron diffusion, and (**c**) EDP etch

the cross section of this process. Figure 8.23 shows a scanning electron microscopy (SEM) photograph of a microaccelerometer fabricated using the dissolved wafer process.

It is also possible to merge wet bulk micromachining and microelectronics fabrication processes to build micromechanical components on the same substrate as the integrated circuits (CMOS, bipolar, or biopolar complementary metal oxide semiconductor (BiCMOS)) [8.23]. This is very appealing since it allows the integration of interface and signal-processing circuitry with MEMS structures on a single chip. However, important fabrication issues such as process compatibility and yield have to be carefully considered. Among the most popular techniques in this category is postprocessing of CMOS integrated circuits by front-side etching in TMAH solutions. As was mentioned previously, silicon-rich TMAH does not attack aluminum and therefore can be used to undercut microstructures in an already processed CMOS chip. Figure 8.24 shows a schematic of such a process in which a front-side wet etch and electrochemical etch stop are used to produce suspended beams. This technique has been extensively used to fabricate a variety of microsensors (e.g., humidity, gas, chemical, and

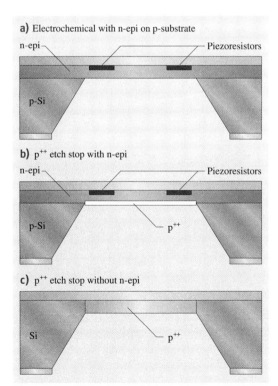

Fig. 8.20a–c Wet micromachining etch-stop techniques: (**a**) electrochemical with n-epi on p-substrate, (**b**) p^{++} etch stop with n-epi, and (**c**) p^{++} etch stop without n-epi

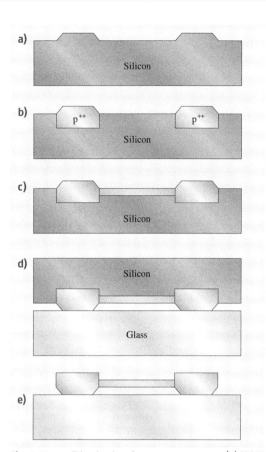

Fig. 8.23 SEM image of a microaccelerometer fabricated using the dissolved wafer process (after [8.22])

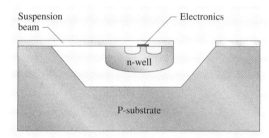

Fig. 8.24 Suspended island created on a prefabricated CMOS chip using front-side wet etch and electrochemical etch stop

Fig. 8.22a–e Dissolved wafer process sequence: (**a**) KOH etch, (**b**) deep B diffusion, (**c**) shallow B diffusion, (**d**) silicon–glass anodic bond, and (**e**) release in EDP

pressure). Figure 8.25 shows a photograph of a post-CMOS-processed chemical sensor.

Bulk Micromachining Using Dry Etch

Bulk silicon micromachining using dry etching is a very attractive alternative to the wet techniques described in the previous section. These techniques were developed during the mid 1990s subsequent to successful efforts geared towards the development of processes for anisotropic dry silicon etch. More recent advances in deep silicon RIE and the availability of SOI wafers with a thick top silicon layer have increased the applicability of these techniques. These techniques allow the fabrication of high-aspect-ratio vertical structures in isolation or along with on-chip electronics. Process compatibility with active microelectronics is less of a concern in dry

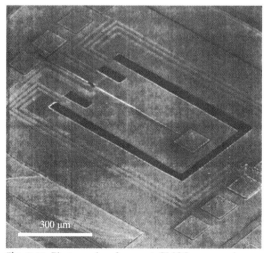

Fig. 8.25 Photograph of a post-CMOS-processed cantilever beam resonator for chemical sensing (after [8.23])

methods since many of them do not damage the circuit or its interconnect.

The simplest dry bulk micromachining technique relies on front-side undercut of microstructures using a XeF$_2$ vapor-phase etch [8.25]. As was mentioned before, this however, is an isotropic etch and therefore has limited applications. A combination of isotropic/anisotropic dry etch is more useful and can be used to create a variety of interesting structures. Two successful techniques using this combination are single-crystal reactive etching and metallization (SCREAM) [8.26] and post-CMOS dry release using aluminum/silicon dioxide laminate [8.27]. The first technique relies on the combination of isotropic/anisotropic dry etch to create single-crystalline suspended structures. Figure 8.26

Fig. 8.27 SEM image of a structure fabricated using the SCREAM process: *A* comb-drive actuator, *B* suspended spring, *C* spring support, *D* moving suspended capacitor plate, and *E* fixed capacitor plate (after [8.24])

shows the cross section of this process. It starts with an anisotropic (Cl$_2$/BCl$_3$) silicon etch using an oxide mask (Fig. 8.26b). This is followed by a conformal PECVD oxide deposition (Fig. 8.26c). Subsequently an anisotropic oxide etch is used to remove the oxide at the bottom of the trenches leaving the side-wall oxide intact (Fig. 8.26d). At this stage an isotropic silicon etch (SF$_6$) is performed, which results in undercut and release of the silicon structures (Fig. 8.26e). Finally, if electrostatic actuation is desired, a metal can be sputtered to cover the top and side-wall of the microstructure and bottom of the cavity formed below it (Fig. 8.26f). Figure 8.27 shows an SEM photograph of a comb-drive actuator fabricated using SCREAM technology.

The second dry release technique relies on the masking capability of aluminum interconnect lines in a CMOS integrated circuit to create suspended microstructures. Figure 8.28 shows a cross section of this process. As can be seen the third level Al of a prefabricated CMOS chip is used as a mask to etch the underlying oxide layers anisotropically all the way to the silicon (CHF$_3$/O$_2$) (Fig. 8.28b). This is followed by an anisotropic silicon etch to create a recess in the silicon, which will be used in the final step to facilitate the undercut and release (Fig. 8.28c). Finally, an isotropic silicon etch is used to undercut and release the structures (Fig. 8.28d). Figure 8.29 shows an SEM photograph of a comb-drive actuator fabricated using this technology.

In addition to the methods described above, recent advancements in the development of deep reactive-ion etching of silicon (DRIE) have created new op-

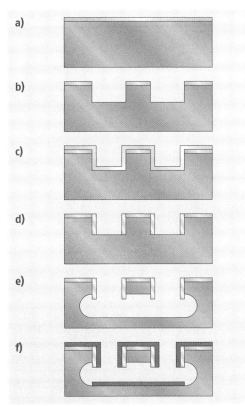

Fig. 8.26a–f Cross section of the SCREAM process (**a**) silicon wafer, (**b**) anisotropic silicon etch, (**c**) conformal passivation, (**d**) anisotropic etching of the passivation (hence protecting the sidewall), (**e**) isotropic silicon etch, and (**f**) metal deposition

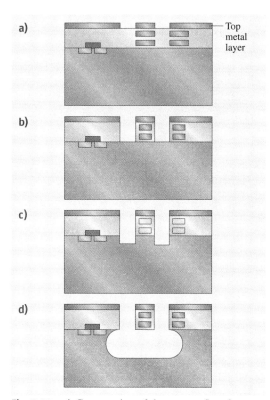

Fig. 8.28a–d Cross section of the process flow for post-CMOS dry microstructure fabrication

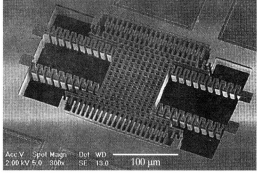

Fig. 8.29 SEM image of a comb-drive actuator fabricated using aluminum-mask post-CMOS dry release (after [8.29])

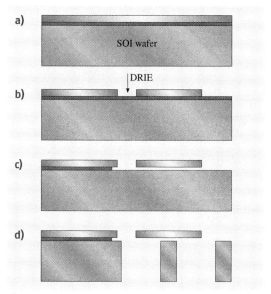

Fig. 8.30a–d DRIE processes using SOI wafers

in various top silicon thicknesses [8.28]. Figure 8.30 shows a cross section of a typical process using DRIE and SOI wafers. The top silicon layer is patterned and etched all the way to the buried oxide (Fig. 8.30b). The oxide is subsequently removed in HF, hence releasing suspended single-crystalline microstructures (Fig. 8.30c). In a modification of this process, the substrate can also be removed from the back-side, allowing easy access from both sides (which allows easier release and prevents stiction) (Fig. 8.30d).

8.2.2 Surface Micromachining

Surface micromachining is another important MEMS microfabrication technique which can be used to create movable microstructures on top of a silicon substrate [8.30]. This technique relies on the deposition of structural thin films on a sacrificial layer which is subsequently etched away, resulting in movable micromechanical structures (beams, membranes, plate, etc.). The main advantage of surface micromachining is that extremely small sizes can be obtained. In addition, it is relatively easy to integrate the micromachined structures with on-chip electronics for increased functionality. However, due to the increased surface nonplanarity with any additional layer, there is a limit to the number of layers that can be deposited. Although one of the earliest reported MEMS structures

portunities for dry bulk micromachining techniques (Sect. 8.2.3). One of the most important ones uses thick silicon SOI wafers which are commercially available

was a surface-micromachined resonant gate transistor [8.31], material-related difficulties resulted in the termination of efforts in this area. In the mid 1980s, improvements in the field of thin-film deposition rekindled interest in surface micromachining [8.32]. Later in the same decade polysilicon surface micromachining was introduced which opened the door to the fabrication of a variety of microsensors (accelerometers, gyroscopes, etc.) and microactuators (micromirrors, RF switches, etc.). In this section, we will concentrate on the key process steps involved in surface-micromachining fabrication and the various materials used. In addition, monolithic integration of CMOS with MEMS structures and 3-D surface micromachining are also discussed.

Basic Surface-Micromachining Processes

The basic surface-micromachining process is illustrated in Fig. 8.31. The process begins with a silicon substrate, on top of which a sacrificial layer is grown and patterned (Fig. 8.31a). Subsequently, the structural material is deposited and patterned (Fig. 8.31b). As can be seen the structural material is anchored to the substrate through the openings created in the sacrificial layer during the previous step. Finally, the sacrificial layer is removed, resulting in the release of the microstructures (Fig. 8.32c). In wide structures, it is usually necessary to provide access holes in the structural layer for fast sacrificial layer removal. It is also possible to seal microcavities created by the surface-micromachining technique [8.11]. This can be done at the wafer level and is a big advantage in applications such as pressure sensors which require a sealed cavity. Figure 8.32 shows two different techniques that can be used for this purpose. In the first technique, following the etching of the sacrificial layer, a LPCVD dielectric layer (oxide or nitride) is deposited to cover and seal the etch holes in the structural material (Fig. 8.33a). Since the LPCVD deposition is performed at reduce pressures, a subatmospheric pillbox microcavity can be created. In the second technique, also called reactive sealing, the polysilicon structural material is oxidized following the sacrificial layer removal (Fig. 8.33b). If access holes are small enough the grown oxide can seal the cavity. Due to the consumption of oxygen during the growth process, in this case also the cavity is subatmospheric.

The most common sacrificial and structural materials are phosphosilicate glass (PSG) and polysilicon, respectively (low-temperature oxide, LTO, is also frequently used as the sacrificial layer). However, there are several other sacrificial/structural combinations that have been used to create a variety of surface-micromachined structures. Important design issues related to the choice of the sacrificial layer are:

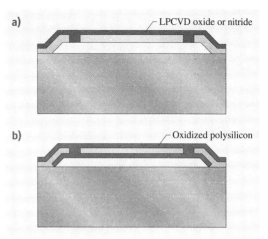

Fig. 8.32a,b Two sealing techniques for cavities created by surface micromachining

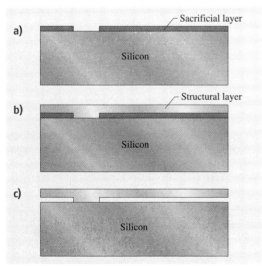

Fig. 8.31a–c Basic surface-micromachining fabrication process (a) silicon wafer with patterned sacrificial layer, (b) deposition and patterning of the structural layer, and (c) removal of the sacrificial layer

1. Quality (pinholes, etc.)
2. Ease of deposition
3. Deposition rate
4. Deposition temperature

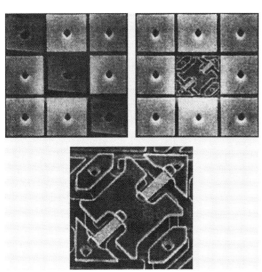

Fig. 8.33 SEM images of the Texas Instrument micromirror array (after [8.30])

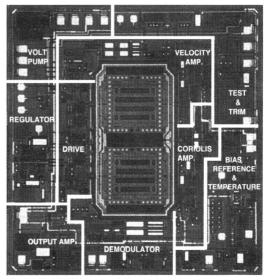

Fig. 8.34 SEM image of the Analog Devices gyroscope (after [8.33])

5. Etch difficulty and selectivity (sacrificial layer etchant should not attack the structural layer)

The particular choice of material for the structural layer depends on the desired properties and specific application. Several important requirements are:

1. Ease of deposition
2. Deposition rate
3. Step coverage
4. Mechanical properties (internal stress, stress gradient, Young's moduli, fracture strength, and internal damping)
5. Etch selectivity
6. Thermal budget and history
7. Electrical conductivity
8. Optical reflectivity

Two examples of commercially available surface-micromachined devices illustrate various successful sacrificial/structural combinations. The Texas Instruments (TI) deformable mirror display (DMD) spatial light modulator uses aluminum as the structural material (good optical reflectivity) and photoresist as the sacrificial layer (easy dry etch and low processing temperatures, allowing easy post-IC integration with CMOS) [8.34] (Fig. 8.33), whereas the Analog Devices microgyroscope uses polysilicon structural material and a PSG sacrificial layer (Fig. 8.34). Two recent additions to the collection of available structural layers are polysilicon–germanium and polygermanium [8.35, 36]. These are intended as substitutes for polysilicon in applications where the high polysilicon deposition temperature (around 600 °C) is prohibitive (e.g., CMOS integration). Unlike in LPCVD of polysilicon, polygermanium (poly-Ge) and polysilicon–germanium (poly-$Si_{1-x}Ge_x$) can be deposited at temperatures as low as 350 °C (poly-Ge deposition temperature is

Table 8.2 Several important surface-micromachined sacrificial–structural combinations

System	Sacrificial layer	Structural layer	Structural layer etchant	Sacrificial layer etchant
1	PSG or LTO	Poly-Si	RIE	Wet or vapor HF
2	Photoresist, polyimide	Metals (Al, Ni, Co, Ni-Fe)	Various metal etchants	Organic solvents, plasma O_2
3	Poly-Si	Nitride	RIE	KOH
4	PSG or LTO	Poly-Ge	H_2O_2 or RCA1	Wet or vapor HF
5	PSG or LTO	Poly-Si-Ge	H_2O_2 or RCA1	Wet or vapor HF

usually lower than that for poly-SiGe). Table 8.2 summarizes important surface-micromachined sacrificial/structural combinations.

An important consideration in the design and processing of surface-micromachined structures is the issue of stiction [8.11, 37, 38]. This can happen during the release step if a wet etchant is used to remove the sacrificial layer or during the device lifetime. The reason for stiction during release is the surface tension of the liquid etchant, which can hold the microstructure down and cause stiction. This usually happens when the structure is compliant and does not possess enough spring constant to overcome the surface tension force of the rinsing liquid (i. e., water). There are several ways one can alleviate the release-related stiction problem. These include:

1. The use of dry or vapor phase etchant
2. The use of solvents with lower surface tension
3. Geometrical modifications
4. CO_2 critical drying
5. Freeze-drying
6. Self-assembled monolayer (SAM) or organic thin-film surface modification

The first technique prevents stiction by not using a wet etchant, although in the case of vapor-phase release, condensation is still possible and can cause some stiction. The second method uses rinsing solvents such as methanol with a lower surface tension than water. This is usually followed by rapid evaporation of the solvent on a hot-plate. However, this technique is not optimum and many structures still stick. The third technique is geometrical, providing dimples in the structural layer in order to reduce the contact surface area and hence reduce the attractive force. The fourth and fifth techniques rely on phase change (in one case CO_2 and the other butyl alcohol) which avoids the liquid phase altogether by directly going to the gas phase. The last technique uses self-assembled monolayers or organic thin films to coat the surfaces with a hydrophobic layer. The stiction that occurs during the operating lifetime of the device (in-use stiction) is due to condensation of moisture on the surfaces, electrostatic charge accumulation, or direct chemical bonding. Surface passivation using self-assembled monolayers or organic thin films can be used to reduce the surface energy and reduce or eliminate capillary forces and direct chemical bonding. These organic coatings also reduce electrostatic forces if a thin layer is applied directly to the semiconductor (without an intervening oxide layer). Commonly used organic coatings include fluorinated fatty acids (TI aluminum micromirrors), silicone polymeric layers (Analog Devices accelerometers), and siloxane self-assembled monolayers.

Surface-Micromachining Integration with Active Electronics

Integration of surface-micromachined structures with on-chip circuitry can increase performance and simplify packaging. However, issues related to process compatibility and yield have to be carefully considered. The two most common techniques are MEMS-first and MEMS-last techniques. In the MEMS-last technique, the integrated circuit is first fabricated and surface-micromachined structures are subsequently built on top of the silicon wafer. An aluminum structural layer with a sacrificial photoresist layer is an attractive combination due to the low thermal budget of the process (TI micromirror array). However, in applications where mechanical properties of Al are not adequate, polysilicon structural material with an LTO or PSG sacrificial layer must be used. Due to the rather high deposition temperature of polysilicon, this combination requires special attention with regard to the thermal budget. For example, aluminum metallization must be avoided and substituted with refractory metals such as tungsten. This can only be achieved at the cost of greater process complexity and lower transistor performance.

The MEMS-first technique alleviates these difficulties by fabricating the microstructures at the very beginning of the process. However, if the microstructures are processed first, they have to be buried in a sealed trench to eliminate the interference of microstructures with subsequent CMOS processes. Figure 8.35 shows a cross section of a MEMS-first fabrication process developed at the Sandia National Laboratory [8.39]. The process starts with shallow anisotropic etching

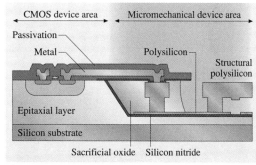

Fig. 8.35 Cross section of the Sandia MEMS-first integrated fabrication process

of trenches in a silicon substrate to accommodate the height of the polysilicon structures fabricated later on. A silicon nitride layer is then deposited to provide isolation at the bottom of the trenches. Next, several layers of polysilicon and sacrificial oxide are deposited and patterned in a standard surface-micromachining process. Subsequently, the trenches are completely filled with sacrificial oxide and the wafers are planarized with chemical–mechanical polishing (this avoids complication in the following lithographic steps). After an annealing step, the trenches are sealed with a nitride cap. At this point, a standard CMOS fabrication process is performed. At the end of the CMOS process the nitride cap is etched and the buried structures released by etching the sacrificial oxide.

Three-Dimensional Microstructures Using Surface Micromachining

Three-dimensional surface microstructures can be fabricated using surface micromachining. The fabrication of hinges for the vertical assembly of MEMS was a major advance towards achieving 3-D microstructures [8.41]. Optical microsystems have greatly benefited from surface-micromachined 3-D structures. These microstructures are used as passive or active components (micromirror, Fresnel lens, optical cavity, etc.) on a silicon optical bench (silicon microphotonics). An example is a Fresnel lens that has been surface micromachined in polysilicon and then erected using hinge structures and locked in place using micromachined tabs, thus liberating the structure from the horizontal plane of the wafer [8.40, 42]. Various microactuators (e.g., comb drive, and vibromotors) have been used to move these structures out of the silicon plane and into position. Figure 8.36 shows an SEM photograph of a bar-code microscanner using a silicon optical microbench with 3-D surface-micromachined structures.

8.2.3 High-Aspect-Ratio Micromachining

The bulk and surface micromachining technologies presented in the previous sections fulfill the requirements of a large group of applications. Certain applications, however, require the fabrication of high-aspect-ratio structures that is not possible with the aforementioned technologies. In this section, we describe three technologies, LIGA, HEXSIL, and HARPSS, capable of producing structures with vertical dimensions much larger than their lateral dimensions by means of x-ray lithography (LIGA) and DRIE etching (HEXSIL and HARPSS).

LIGA

LIGA is a high-aspect-ratio micromachining process which relies on x-ray lithography and electroplating (in German: *Lithographie, Galvanoformung, Abformung*) [8.43, 44]. We already introduced the concept of the plating-through-mask technique in *Surface Micromachining* (Fig. 8.10). With standard UV photolithography and photoresists, the maximum thickness achievable is on the order of a few tens of microns and the resulting metal structures show tapered walls. LIGA is a technology based on the same plating-through-mask idea but can be used to fabricate metal structures of thickness ranging from a few microns to a few millimeters with almost vertical side-walls. This is achieved using x-ray lithography and special photoresists. Due to their short wavelength, x-rays are able to penetrate through a thick photoresist layer with no scattering and define features with lateral dimensions down to 0.2 μm (aspect ratio > 100 : 1).

The photoresists used in LIGA should comply with certain requirements, including sensitivity to x-rays, resistance to electroplating chemicals, and good adhesion to the substrate. Based on such requirements poly-(methyl methacrylate) (PMMA) is considered to be an optimal choice for the LIGA process. Application of the thick photoresist on top of the substrate can be performed by various techniques such as multiple spin-coating, precast PMMA sheets, and plasma polymerization coating. The mask structure and materials for x-ray lithography must also comply with certain requirements. The traditional masks based on glass plates with a patterned chrome thin layer are not suitable be-

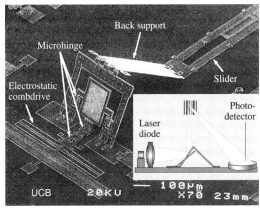

Fig. 8.36 Silicon pin-and-sample hinge scanner with 3-D surface-micromachined structures (after [8.40])

Fig. 8.37 SEM of assembled LIGA-fabricated nickel structures (after [8.44])

membrane is supported by a silicon frame which can be fabricated using bulk micromachining techniques. Once the photoresist is exposed to the x-rays and developed, the process proceeds with electroplating of the desired metal. Ni is the most commonly used, although other metals and metallic compounds such us Cu, Au, NiFe, and NiW are also electroplated in LIGA processes. Good agitation of the plating solution is the key to obtaining a uniform and repeatable result during this step. A paddle plating cell, based on a windshield-wiper-like device moving only a millimeter away from the substrate surface, provides extremely reproducible agitation. Figure 8.37 shows an SEM of a LIGA microstructure fabricated with electroplating nickel.

Due to the high cost of the x-ray sources (synchrotron radiation), LIGA technology was initially intended for the fabrication of molds that could be used many times in hot-embossing or injection-molding processes. However, it has been also used in many applications to directly form high-aspect-ratio metal structures on top of a substrate. A cheaper alternative to the LIGA process (with somehow poorer qualities) called UV-LIGA or *poor man's LIGA* has been proposed [8.45, 46]. This process uses SU-8 negative photoresists (available for spin-coating at various thickness ranging from 1 to 500 μm) and standard contact lithography equipment. Using this technique, aspect ra-

cause x-rays are not absorbed by the chromium layer and the glass plate is not transparent enough. Instead, x-ray lithography uses a silicon nitride mask with gold as the absorber material (typically formed by electroplating gold to a thickness of 10–20 μm). The nitride

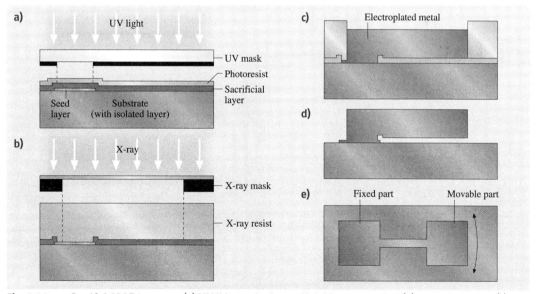

Fig. 8.38a–e Sacrificial LIGA process: (**a**) UV lithography for sacrificial layer patterning, (**b**) x-ray lithography, (**c**) electroplating, (**d**) structure releasing, and (**e**) top view of the movable structure

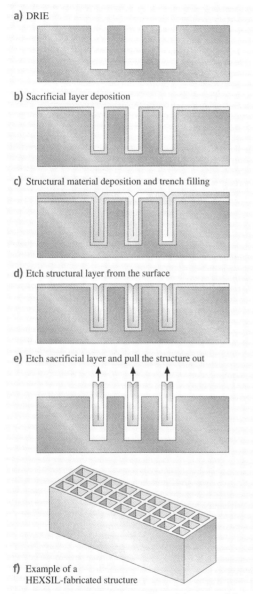

f) Example of a HEXSIL-fabricated structure

Fig. 8.39a–f HEXSIL process flow: (**a**) DRIE, (**b**) sacrificial layer deposition, (**c**) structural material deposition and trench filling, (**d**) etch structural layer from the surface, (**e**) etch sacrificial layer and pull the structure out, and (**f**) example of a HEXSIL-fabricated structure

photoresist after plating. Various methods, showing different degrees of success, have been proposed. These include: wet etching with special solvents, burning at high temperatures (600 °C), dry etching, use of a release layer, and high-pressure water-jet etching.

A variation of the basic LIGA process, shown in Fig. 8.38, permits the fabrication of electrically isolated movable structures, and thus opens more possibilities for sensor and actuator design using this technology [8.48]. This so-called sacrificial LIGA (SLIGA) starts with the patterning of the seed layer. Subsequently a sacrificial layer (e.g., titanium) is deposited and patterned. The process then proceeds as in standard LIGA until the last step, when the sacrificial layer is removed. The electroplated structures that overlap with the sacrificial layer are released in this step.

HEXSIL

The second method for fabricating high-aspect-ratio structures, which is based on a template replication technology, is hexagonal honeycomb polysilicon (HEXSIL) [8.49]. Figure 8.39 shows a simplified process flow. A high-aspect-ratio template is first formed in a silicon substrate using DRIE. Next, a sacrificial multilayer is deposited to allow the final release of the structures. The multilayer is composed of one or more PSG nonconformal layers to provide fast etch release ($\approx 20\,\mu\text{m/min}$ in 49% HF) alternated with conformal layers of either oxide or nitride to provide enough thickness for proper release of the structures. The total thickness of the sacrificial layer has to be larger than the shrinkage or elongation of the structures caused by

Fig. 8.40 SEM micrograph of an angular microactuator fabricated using the HEXSIL process (after [8.47])

tios larger than 20 : 1 have been demonstrated. A major problem of this alternative is the removal of the SU-8

the relaxation of the internal (compressive or tensile) stress during the release step. Otherwise the structures will clamp themselves to the walls of the template and their retrieval will not be possible. Any material that can be conformally deposited and yet not damaged during the HF release step is suitable for the structural layer. Structures made of polysilicon, nitride, and electroless nickel [8.50] have been reported. Nickel can only be deposited in combination with polysilicon since a con-

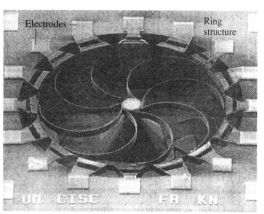

Fig. 8.42 SEM photograph of a microgyroscope fabricated using the HARPSS process (after [8.51])

ductive surface is needed for the deposition to occur. After deposition of structural materials a blanket etch (poly-Si or nitride) or a mechanical lapping (nickel) is performed to remove the excess materials from the surface. Finally, a 49% HF with surfactant is used to dissolve the sacrificial layers. The process can be repeated many times using the same template, thus considerably lowering fabrication costs. Figure 8.40 shows an SEM photograph of a microactuator fabricated using the HEXSIL process.

HARPSS

The high-aspect-ratio combined poly- and single-crystal silicon (HARPSS) technology is another technique capable of producing high-aspect-ratio electrically isolated polycrystalline and single-crystal silicon microstructures with capacitive air gaps ranging from submicrometer to tens of micrometers [8.52]. The structures, tens to hundreds of micrometers thick, are defined by trenches etched with DRIE and filled with oxide and poly layers. The release of the microstructures is done at the end by means of a directional silicon etch followed by an isotropic etch. The small vertical gaps and thick structures possible with this technology find application in the fabrication of a variety of MEMS devices, particularly inertial sensors [8.53] and RF beam resonators [8.54]. Figure 8.41 shows the process flow in a cross section of a single-crystal silicon beam resonator. The HARPSS process starts with deposition and patterning of a silicon nitride layer that will be used to isolate the poly structure's connection pads from the substrate. High-aspect-ratio trenches ($\approx 5\,\mu\mathrm{m}$ wide) are then etched into the substrate using a DRIE etch. Next,

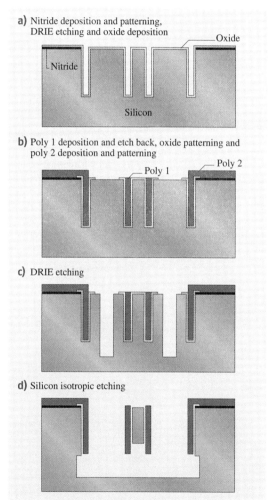

Fig. 8.41a–d HARPSS process flow: (a) nitride deposition and patterning, DRIE etching and oxide deposition, (b) poly 1 deposition and etch back, oxide patterning, and poly 2 deposition and patterning, (c) DRIE etching, and (d) silicon isotropic etching

a conformal oxide layer (LPCVD) is deposited. This layer has two functions, to:

1. Protect the structures during the dry etch release
2. Define the submicrometer gap between silicon and polysilicon structures

Following the oxide deposition, the trenches are completely filled with LPCVD polysilicon. The polysilicon is etched back and the underlying oxide is patterned to provide anchor points for the structures. A second layer of polysilicon is then deposited and patterned.

Finally, the structures are released using a DRIE step followed by an isotropic silicon etch through a photoresist mask that exposes only the areas of silicon substrate surrounding the structures. It should be noted that single-crystal silicon structures are not protected at the bottom during the isotropic etch. This causes the single-crystal silicon structures to be etched vertically from the bottom, and thus be shorter than the polysilicon structures. Figure 8.42 shows an SEM photograph of a microgyroscope fabricated using the HARPSS process.

8.3 Nanofabrication Techniques

The microfabrication techniques discussed so far were mostly geared towards fabricating devices in the 1 mm to 1 μm dimensional range (submicrometer dimensions being possible in certain techniques such as HARPSS using a dielectric sacrificial layer). Recent years have witnessed a tremendous surge of interest in fabricating submicro- (1 μm–100 nm) and nanostructures (100–1 nm range) [8.55]. This interest arises from both practical and fundamental viewpoints. At the more scientific and fundamental level, nanostructures provide an interesting tool for studying the electrical, magnetic, optical, thermal, and mechanical properties of matter at the nanometer scale. These include important quantum-mechanical phenomena (e.g., conductance quantization, bandgap modification, coulomb blockade, etc.) arising from confinement of charged carriers in structures such as quantum wells, wires, and dots (Fig. 8.43). On the practical side, nanostructures can provide significant improvements in the performance of electronic/optical devices and sensors. In the device area investigators have been mostly interested in fabricating nm-sized transistors in anticipation of technical difficulties forecasted in extending Moore's law beyond 32 nm resolution. In addition, optical sources and detectors having nm-size dimensions exhibit improved characteristics not achievable in larger devices (e.g., lower threshold current, improved dynamic behavior, and improved emission line width in quantum dot lasers). These improvements create novel possibilities for next-generation computation and communication devices. In the sensors area, shrinking dimensions beyond conventional optical lithography can provide major improvements in sensitivity and selectivity.

One can broadly divide various nanofabrication techniques into top-down and bottom-up categories. The first approach starts with a bulk or thin-film material and removes selective regions to fabricate nanostructures (similar to micromachining techniques). The second method relies on molecular recognition and self-assembly to fabricate nanostructures out of smaller building blocks (molecules, colloids, and clusters). As can be anticipated, the top-down approach is an off-shoot of standard lithography and micromachining techniques. On the other hand, the bottom-up approach has a more chemical engineering and material science flavor and relies on fundamentally different principles. In this chapter, we will discuss several nanofabrication

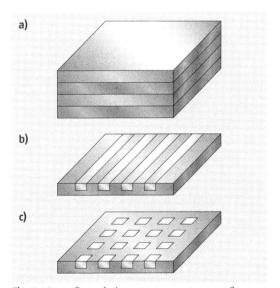

Fig. 8.43a–c Several important quantum confinement structures: (**a**) quantum well, (**b**) quantum wire, and (**c**) quantum dot

techniques that are not covered in other chapters of this Handbook. These include:

1. E-beam nanofabrication
2. Epitaxy and strain engineering
3. Scanning-probe techniques
4. Self-assembly and template manufacturing

8.3.1 E-Beam Nanofabrication

In previous sections, we discussed several important lithography techniques used commonly in MEMS and microfabrication. These included various forms of UV (regular, deep, and extreme) and x-ray lithography. However, due to the lack of resolution (in case of the UV) or difficulty in manufacturing mask and radiation sources (x-ray), these techniques are not suitable for nm-scale fabrication. E-beam lithography is an alternative and attractive technique for fabricating nanostructures [8.56]. It uses an electron beam to expose an electron-sensitive resist such as poly(methyl methacrylate) (PMMA) dissolved in trichlorobenzene (positive) or polychloromethylstyrene (negative) [8.57]. The e-beam gun is usually part of a scanning electron microscope (SEM), although transmission electron microscopes (TEM) can also be used. Although electron wavelengths of the order of 1 Å can be easily achieved, electron scattering in the resist limits the attainable resolutions to > 10 nm. Beam control and pattern generation is achieved through a computer interface. E-beam lithography is serial and hence has low throughput. Although this is not a major concern in fabricating devices used in studying fundamental microphysics, it severely limits large-scale nanofabrication. E-beam lithography in conjunction with processes such as lift-off, etching, and electrodeposition can be used to fabricate various nanostructures.

8.3.2 Epitaxy and Strain Engineering

Atomic-precision deposition techniques such as molecular-beam epitaxy (MBE) and metalorganic chemical vapor deposition (MOCVD) have proven to be effective tools in fabricating a variety of quantum confinement structures and devices (quantum well lasers, photodetector, resonant tunneling diodes, etc.) [8.58–60]. Although quantum wells and superlattices are the structures that lend themselves most easily to these techniques (Fig. 8.43a), quantum wires and dots have also been fabricated by adding subsequent steps such as etching and selective growth. Fabrication of quantum well and superlattice structures using epitaxial growth is a mature and well-developed field and therefore will not be discussed in this chapter. Instead, we will concentrate on quantum wire and dot nanostructure fabrication using basic epitaxial techniques [8.61, 62].

Quantum Structure Nanofabrication Using Epitaxy on Patterned Substrates

There have been several different approaches to the fabrication of quantum wires and dots using epitaxial layers. The most straightforward technique involves e-beam lithography and etching of an epitaxial grown layer (e.g., InGaAs on GaAs substrate) [8.63]. However, due to damage and/or contamination during lithography, this method is not very suitable for active device fabrication (e.g., quantum dot lasers). Several other methods involving regrowth of epitaxial layers over nonplanar surfaces such as step-edge, cleaved-edge, and patterned substrate have been used to fabricate quantum wires and dots without the need for lithography and etching of the quantum confinement structure [8.62, 64]. These nonplanar surface templates can be fabricated in a variety of ways such as etching through a mask or cleavage along crystallographic planes. Subsequent epitaxial growth on top of these structures results in a set of planes with different growth rates depending on the geometry or surface diffusion and adsorption effects. These effects can significantly enhance or limit the growth rate on certain planes, resulting in lateral patterning and confinement of deposited epitaxial layers and formation of quantum wires (in V-grooves) and dots

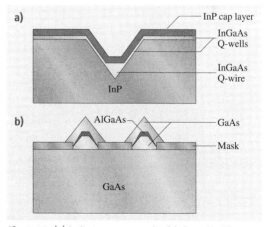

Fig. 8.44 (a) InGaAs quantum wire fabricated in V-groove InP, and (b) AlGaAs quantum wire fabricated by epitaxial growth on a masked GaAs substrate

(in inverted pyramids). Figure 8.44a shows a schematic cross section of an InGaAs quantum wire fabricated in a V-groove in InP. As can be seen the growth rate on the side-walls is lower than that of the top and bottom surfaces. Therefore the thicker InGaAs layer at the bottom of the V-groove forms a quantum wire confined from the sides by a thinner layer having a wider bandgap. Figure 8.44b shows a quantum wire formed using epitaxial growth over a dielectric patterned planar substrate. In both of these techniques it is relatively easy to create quantum wells; however, in order to create quantum wires and dots one still needs e-beam lithography to pattern the grooves and window templates.

Quantum Structure Nanofabrication Using Strain-Induced Self-Assembly

A more recent technique for fabricating quantum wires and dots involves strain-induced self-assembly [8.62, 65]. The term *self-assembly* represents a process where a strained two-dimensional (2-D) system reduces its energy by a transition into a 3-D morphology. The most commonly used material combination for this technique is the $In_xGa_{1-x}As/GaAs$ system, which offers a large lattice mismatch (7.2% between InAs and GaAs) [8.66, 67], although recently Ge dots on Si substrate have also attracted considerable attention [8.68]. This method relies on lattice mismatch between an epitaxially grown layer and its substrate to form an array of

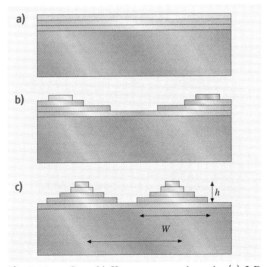

Fig. 8.45a–c Stranski–Krastanow growth mode, (**a**) 2-D wetting layer, (**b**) growth front roughening and breakup, and (**c**) coherent 3-D self-assembly

quantum dots or wires. Figure 8.45 shows a schematic of the strain-induced self-assembly process. When the lattice constant of the substrate and the epitaxial layer differ considerably, only the first few deposited monolayers crystallize, in the form of an epitaxial strained layer in which the lattice constants are equal. When a critical thickness is exceeded, a significant strain that occurs in the layer leads to the breakdown of this ordered structure and to the spontaneous formation of randomly distributed islets of regular shape and similar size (usually < 30 nm in diameter). This mode of growth is usually referred to as the Stranski–Krastanow mode. The quantum dot size, separation, and height depend on the deposition parameters (i. e., total deposited material, growth rate, and temperature) and material combinations. As can be seen, this is a very convenient method to grow perfect crystalline nanostructures over a large area without any lithography and etching. One major drawback of this technique is the randomness of the quantum dot distribution. It should be mentioned that this technique can also be used to fabricate quantum wires by strain relaxation bunching at step edges.

8.3.3 Scanning Probe Techniques

The invention of scanning probe microscopy in the 1980s revolutionized atomic-scale imaging and spectroscopy. In particular scanning tunneling and atomic force microscopes (STM and AFM) have found widespread applications in physics, chemistry, material science, and biology. The possibility of atomic-scale manipulation, lithography, and nanomachining using such probes was considered from the beginning and has matured considerably over the past decade. In this section after a brief introduction to scanning probe microscopes, we will discuss several important nanolithography and machining techniques which have been used to create nm-sized structures.

Scanning probe microscopy (SPM) systems are capable of controlling the movement of an atomically sharp tip in close proximity to or in contact with a surface with subnanometer accuracy. Piezoelectric positioners are typically used in order to achieve such accuracy. High-resolution images can be acquired by raster scanning the tips over a surface while simultaneously monitoring the interaction of the tip with the surface. In scanning tunneling microscope systems a bias voltage is applied to the sample and the tip is positioned close enough to the surface that a tunneling current develops through the gap (Fig. 8.46a). Because this current is extremely sensitive to the distance between the

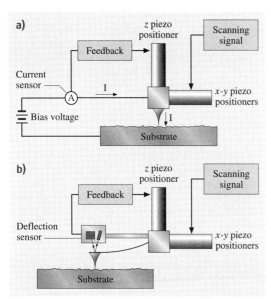

Fig. 8.46a,b Scanning probe systems: (**a**) STM and (**b**) AFM

tip and the surface, scanning the tip in the x–y-plane while recording the tunnel current permits the mapping of the surface topography with resolution on the atomic scale. In a more common mode of operation the amplified current signal is connected to the z-axis piezoelectric positioner through a feedback loop so that the current and therefore the distance are kept constant throughout the scanning. In this configuration the picture of the surface topography is obtained by recording the vertical position of the tip at each x–y-position.

The STM system only works for conductive surfaces because of the need to establish a tunneling current. The atomic force microscopy was developed as an alternative for imaging either conducting or nonconducting surfaces. In AFM the tip is attached to a flexible cantilever and is brought into contact with the surface (Fig. 8.46b). The force between the tip and the surface is detected by sensing the cantilever deflection. A topographic image of the surface is obtained by plotting the deflection as a function of the x–y-position. In a more common mode of operation a feedback loop is used to maintain a constant deflection while the topographic information is obtained from the cantilever vertical displacement. Some scanning probe systems use a combination of the AFM and the STM modes, i.e., the tip is mounted on a cantilever with electrical connection so that both the surface forces and tunneling

currents can be controlled or monitored. STM systems can be operated in ultrahigh vacuum (UHV STM) or in air, whereas AFM systems are typically operated in air. When a scanning probe system is operated in air, water adsorbed onto the sample surface accumulates underneath the tip, forming a meniscus between the tip and the surface. This water meniscus plays an important role in some of the scanning probe techniques described below.

Scanning-Probe-Induced Oxidation

Nanometer-scale local oxidation of various materials can be achieved using scanning probes operating in air and biased at a sufficiently high voltage (Fig. 8.47). A tip bias of -2 to -10 V is normally used, with a writing speed of 0.1–$100\,\mu$m/s in ambient humidity of 20–40%. It is believed that the water meniscus formed at the contact point serves as an electrolyte such that the biased tip anodically oxidizes a small region of the surface [8.70]. The most common application of this principle is the oxidation of hydrogen-passivated silicon. A dip in HF solution is typically used to passivate the silicon surfaces with hydrogen atoms. Patterns of oxide *written* on a silicon surface can be used as a mask for wet or dry etching. Patterns with 10 nm line width have been successfully transferred to a silicon substrate in this fashion [8.71]. Various metals have also been locally anodized using this approach, such us aluminum or titanium [8.72]. An interesting variation of this process is anodization of deposited amorphous silicon [8.73]. Amorphous silicon can be deposited at low

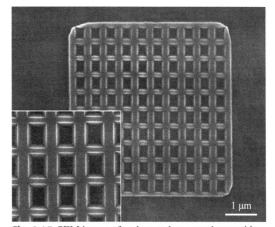

Fig. 8.47 SEM image of an inverted truncated pyramid array fabricated on a silicon SOI wafer by SPM oxidation and subsequent etch in TMAH (pitch is 500 nm) (after [8.69])

temperature on top of many materials. The deposited silicon layer can be patterned and used as, for example, the gate of a 0.1 μm CMOS transistor [8.74], or it can be used as a mask to pattern an underlying film. The major drawback of this technique is poor reproducibility due to tip wear during the anodization. However, using AFM in noncontact mode has proved to overcome this problem [8.70].

Probe Resist Exposure and Lithography

Electrons emitted from a biased SPM tip can be used to expose a resist in the same way e-beam lithography that does (Fig. 8.48) [8.74]. Various systems have been used for this lithographic technique; these include constant-current STM, noncontact AFM, and AFM with constant tip–resist force and constant current. The systems using AFM cantilevers have the advantage that they can perform imaging and alignment tasks without exposing the resist. Resists well characterized for e-beam lithography (e.g., PMMA or SAL601) have been used with scanning probe lithography to achieve reliable sub-100 nm lithography. The procedure for this process is as follows. The wafers are cleaned and the native oxide (in the case of silicon or poly) is removed with a HF dip. Subsequently 35–100 nm-thick resist is spin-coated on top of the surface. The exposure is done by moving the SPM tip over the surface while applying a bias voltage sufficiently high to produce emission of electrons from the tip (a few tens of volts). Development of the resist is performed in standard solutions following the exposure. Features below 50 nm in width have been achieved with this procedure.

Dip-Pen Nanolithography

In dip-pen nanolithography (DPN) the tip of an AFM operated in air is *inked* with a chemical of interest and brought into contact with a surface. The ink molecules flow from the tip onto the surface as with a fountain pen. The water meniscus that naturally forms between the tip and the surface enables the diffusion and transport of the molecules, as shown in Fig. 8.49. Inking can be done by dipping the tip into a solution containing a low concentration of the molecules followed by a drying step (e.g., blow-drying with compressed difluoroethane). Line widths down to 12 nm with spatial resolution of 5 nm have been demonstrated with this technique [8.75]. Species patterned with DPN include conducting polymers, gold, dendrimers, DNA, organic dyes, antibodies, and alkanethiols. Alkanethiols have been also used as an organic monolayer mask to etch a gold layer and subsequently etch the exposed silicon substrate. One can also use a heated AFM cantilever to control the deposition of a solid organic ink. This technique was recently reported by *Sheehan* et al. in which 100 nm lines of octadecylphosphonic acid (melting point 100 °C) were written using a heated AFM probe [8.76].

Other Scanning Probe Nanofabrication Techniques

A great variety of nanofabrication techniques using scanning probe systems have been demonstrated. Some of these are proof-of-concept demonstrations and their utility as viable and repeatable fabrication processes has yet to be evaluated. For example, a substrate can be mechanically machined using a STM/AFM tip acting as a plow or engraving tool [8.77]. This can be used to create structures directly in the substrate, although it is more commonly used to pattern resist for a subsequently etch, lift-off or electrodeposition step. Mechanical nanomachining with SPM probes can be facilitated by heating the tip above the glass-transition temperature of a polymeric substrate material. This approach has been applied to SPM-based high-density data storage in polycarbonate substrates [8.78].

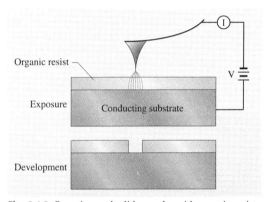

Fig. 8.48 Scanning probe lithography with organic resist

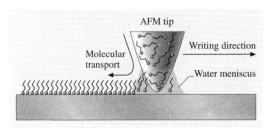

Fig. 8.49 Schematic representation of the working principle of dip-pen nanolithography

Electric fields strong enough to induce the emission of atoms from the tip can be easily generated by applying voltage pulses above 3 V. This phenomenon has been used to transfer material from the tip to the surface and vice versa. Mounds (10–20 nm) of metals such as Au, Ag, and Pt have been deposited or removed from a surface in this fashion [8.79]. The same approach has been used to extract single atoms from a semiconductor surface and redeposit them elsewhere [8.80]. Manipulation of nanoparticles, molecules, and single atoms on top of a surface has also been achieved by simply pushing or sliding them with the SPM tip [8.81]. Metals can also be locally deposited by the STM chemical vapor deposition technique [8.82]. In this technique a precursor organometallic gas is introduced into the STM chamber. A voltage pulse applied between the tip and the surface dissociates the precursor gas into a thin layer of metal. Local electrochemical etching [8.83] and electrodeposition [8.84] is also possible using SPM systems. A droplet of suitable solution is first placed on the substrate. Then the STM tip is immersed into the droplet and a voltage is applied. In order to reduce Faradaic currents the tip is coated with wax such that only the very end is exposed to the solution. Sub-100 nm feature size has been achieved using this technique.

Using a single tip to serially produce the desired modification in a surface leads to very slow fabrication processes that are impractical for mass production. Many of the scanning probe techniques developed so far, however, could also be performed by an array of tips, which would increase throughput and make them more competitive compared with other parallel nanofabrication processes. This approach has been demonstrated for imaging, lithography [8.85], and data storage [8.86] using both one- and two-dimensional arrays of scanning probes. With the development of larger arrays with advances in individual control of force, vertical position, and current, we might see these techniques being incorporated as standard fabrication processes in the industry.

8.3.4 Self-Assembly and Template Manufacturing

Self-assembly is a nanofabrication technique that involves aggregation of colloidal nanoparticles into the final desired structure [8.87]. This aggregation can be either spontaneous (entropic) and due to the thermodynamic minima (energy minimization) constraints or chemical and due to the complementary binding of organic molecules and supramolecules (molecular self-assembly) [8.88]. Molecular self-assembly is one of the most important techniques used in biology for the development of complex functional structures. Since these techniques require that the target structures be thermodynamically stable, it tends to produce structures that are relatively defect-free and self-healing. Self-assembly is by no means limited to molecules or the nanodomain and can be carried out on just about any scale, making it a powerful bottom-up assembly and manufacturing method (multiscale ordering). Another attractive feature of this technique relates to the possibility of combining self-assembly properties of organic molecules with the electronic, magnetic, and photonic properties of inorganic components. Template manufacturing is another bottom-up technique which utilizes material deposition (electroplating, CVD, etc.) into nanotemplates in order to fabricate nanostructures. Due to the simplicity and flexibility of electrochemistry for plating and surface finishing of a broad range of materials, its principle has recently been widely used for electrochemical fabrication of various metallic nanostructures based on various templates. For example, electrochemical deposition has been used to deposit large arrays of nanostructures in nanoporous templates, such as porous alumina. This template-based deposition typically provides metal nanowires as small as 25 nm in diameter and a few micrometers in length [8.89].

The nanotemplates used to fabricate nanostructures are usually prepared using self-assembly techniques. In the following sections, we will discuss various important self-assembly and template manufacturing techniques currently under heavy investigation.

Physical and Chemical Self-Assembly

The central theme behind the self-assembly process is spontaneous (physical) or chemical aggregation of colloidal nanoparticles [8.90]. Spontaneous self-assembly exploits the tendency of monodispersed nano or submicro colloidal spheres to organize into a face-centered cubic (fcc) lattice. The force driving this process is the desire of the system to achieve a thermodynamically stable state (minimum free energy). In addition to spontaneous thermal self-assembly, gravitational, convective, and electrohydrodynamic forces can also be used to induce aggregation into complex 3-D structures. Chemical self-assembly requires the attachment of a single-molecular organic layer (self-assembled monolayer or SAM) to the colloidal particles (organic or inorganic) and subsequent self-assembly of these components into a complex structures using molecular recognition and binding.

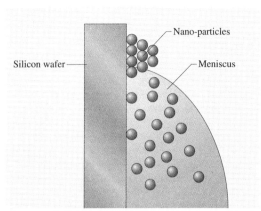

Fig. 8.50 Colloidal particle self-assembly onto solid substrates upon drying in vertical position

Fig. 8.51 Cross-sectional SEM image of a thin planar opal silica template (spheres 855 nm in diameter) assembled directly on a Si wafer (after [8.91])

Physical Self-Assembly

This is an entropy-driven method that relies on spontaneous organization of colloidal particles into a relatively stable structure through noncovalent interactions; for example, colloidal polystyrene spheres can be assembled into a 3-D structure on a substrate which is held vertically in the colloidal solution (Fig. 8.50) [8.91, 92]. Upon the evaporation of the solvent, the spheres aggregate into a hexagonal close-packed (hcp) structure. The interstitial pore size and density are determined by the polymer sphere size. The polymer spheres can be etched into smaller sizes after forming the hcp arrays, thereby altering the template pore separations [8.93]. This technique can fabricate large patterned areas in a quick, simple, and cost-effective way. A classic example is the natural assembly of on-chip silicon photonic-bandgap crystals [8.91] which are capable of reflecting the light arriving in any direction in a certain wavelength range [8.94]. In this method, a thin layer of silica colloidal spheres is assembled on a silicon substrate. This is achieved by placing a silicon wafer vertically in a vial containing an ethanolic suspension of silica spheres. A temperature gradient across the vial aids the flow of silica spheres. Figure 8.51 shows a cross-sectional SEM image of a thin planar opal template assembled directly on a Si wafer from 855 nm spheres. Once such a template is prepared, LPCVD can be used to fill the interstitial spaces with Si, so that the high refractive index of silicon provides the necessary bandgap.

One can also deposit colloidal particles onto a patterned substrate (template-assisted self-assembly, TASA) [8.95, 96]. This method is based on the principle that, when an aqueous dispersion of colloidal particles is allowed to dewet from a solid surface which is already patterned, the colloidal particles are trapped by the recessed regions and assemble into aggregates with shapes and sizes determined by the geometric confinement provided by the template. The patterned arrays of templates can be fabricated using conventional contact-mode photolithography which gives control over the shape and dimensions of the templates, thereby allowing the assembly of complex structures from colloidal particles. The cross-sectional view of a fluidic cell used in TASA is shown in Fig. 8.52. The fluidic cell has two parallel glass substrates to confine the aqueous dispersion of the colloidal particles. The surface of the bottom substrate is patterned with a 2-D array of templates. When the aqueous dispersion is allowed to dewet slowly across the cell, the capillary force exerted on the liquid pushes the colloidal spheres across the surface of the bottom substrate until they are physically trapped by the templates. If the concentration of the colloidal dispersion is high enough, the template will be filled by

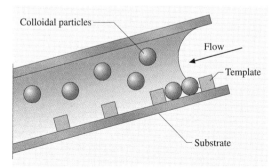

Fig. 8.52 A cross-sectional view of the fluidic cell used for template-assisted self-assembly

the maximum number of colloidal particles determined by the geometrical confinement. This method can be used to fabricate a variety of polygonal and polyhedral aggregates which are difficult to generate [8.97].

Chemical Self-Assembly

Organic and supramolecular SAMs play a critical role in colloidal particle self-assembly. SAMs are robust organic molecules which are chemically adsorbed onto solid substrates [8.98]. Most often they have a hydrophilic (polar) head which can be bonded to various solid surfaces and a long hydrophobic (nonpolar) tail which extends outward. SAMs are formed by the immersion of a substrate in a dilute solution of the molecule in an organic solvent or water (liquid phase) or by exposure to an atmosphere containing such a molecule (gas phase). The resulting film is a dense organization of molecules arranged to expose the end group. The durability of the SAM is highly dependent on the effectiveness of the anchoring to the surface of the substrate. SAMs have been widely studied because the end group can be functionalized to form precisely arranged molecular arrays for various applications ranging from simple, ultrathin insulators and lubricants to complex biological sensors. Chemical self-assembly uses organic or supramolecular SAMs as the binding and recognition sites for fabricating complex 3-D structures from colloidal nanoparticles. The most commonly used organic monolayers include:

1. Organosilicon compounds on glass and oxidized silicon
2. Alkanethiols, dialkyl disulfides, and dialkyl sulfides on gold
3. Fatty acids on alumina and other metal oxides
4. DNA

Octadecyltrichlorosilane (OTS) is the most common organosilane used in the formation of SAMs, mainly because of the fact that it is simple, readily available, and forms good, dense layers [8.99, 100]. Alkyltrichlorosilane monolayers can be prepared on clean silicon wafers whose surface is SiO_2 (with almost 5×10^{14} SiOH groups/cm^2). Figure 8.53 shows a schematic representation of the formation of alkylsiloxane monolayers by adsorption of alkyltrichlorosilane from solution onto Si/SiO$_2$ substrates. Since the silicon–chloride bond is susceptible to hydrolysis, a limited amount of water has to be present in the system in order to obtain good-quality monolayers. Monolayers made of methyl- and vinyl-terminated alkylsilanes are autophobic to the hydrocarbon solution and hence emerge uniformly dry from the solution, whereas monolayers made of ester-terminated alkylsilanes emerge wet from the solution used in their formation. The disadvantage of this method is that, if the alkyltrichlorosilane in the solvent adhering to the substrate is exposed to water, a cloudy film is deposited on the surface due to the formation of a gel of polymeric siloxane. One solution to this problem is the use of alkyldimethylchlorosilanes, which have a single anchoring point, and thus cannot form polymers. Chlorosilanes are sometimes preferred over alkoxysilanes because of their higher reactivity. However, the reactivity of chlorosilanes severely limits the range of functional groups that can be introduced at the end of the hydrocarbon tail. On the contrary, methoxysilanes and ethoxysilanes are commonly available with many functional groups including amino, mercapto, epoxy, and thiocyanate groups, which are often necessary for subsequent binding of colloidal particles and biomolecules. Gas-phase deposition of these molecules yields more uniform layers compared with liquid-phase procedures [8.101]. Soft-lithography-like molds have been used to obtain reactive silane patterns from the gas phase by taking advantage of the characteristic permeability of PDMS to volatile molecules [8.102].

Another important organic SAM system is the alkanethiols $(X(CH_2)_nSH$, where X is the end group) on gold [8.98, 103–105]. A major advantage of using gold as the substrate material is that it does not have a stable oxide and thus can be handled in ambient conditions. When a fresh, clean, hydrophilic gold substrate is im-

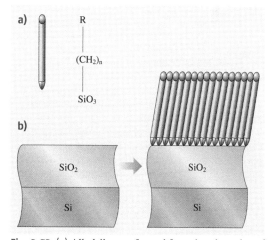

Fig. 8.53 (a) Alkylsiloxane formed from the adsorption of alkyltrichlorosilane on Si/SiO$_2$ substrates. (b) Schematic representation of the process

mersed (several min to several h) into a dilute solution (10^{-3} M) of the organic sulfur compound (alkanethiols) in an inorganic solvent a close-packed, oriented monolayers can be obtained. Sulfur is used as the head group because of its strong interaction with the gold substrate (44 kcal/mol), resulting in the formation of a close-packed, ordered monolayer. The end group of alkanethiol can be modified to render the adsorbed layer hydrophobic or hydrophilic. Another method for depositing alkanethiol SAM is soft lithography. This technique is based on inking a PDMS stamp with alkanethiol and its subsequent transfer to planar or nonplanar substrates. Alkanethiol-functionalized surfaces (planar, nonplanar, spherical) can also be used to self-assemble a variety of intricate 3-D structures [8.106].

Carboxylic acid derivatives self-assemble on surfaces (e.g., glass, Al_2O_3, and Ag_2O) through an acid–base reaction, giving rise to monolayers of fatty acids [8.107]. The time required for the formation of a complete monolayer increases with decreasing concentration. Higher concentration of the carboxylic acid is required to form a monolayer on gold as compared with on Al_2O_3. This is due to differences in the affinity of the COOH groups (more affinity to Al_2O_3 and glass than gold) and also the surface concentration of the salt-forming oxides of the two substrates. In the case of amorphous metal oxide surfaces, the chemisorption of alkanoic acids is not unique. For example, on Ag_2O, the two oxygen atoms of the carboxylate bind to the substrate in a nearly symmetrical manner, thus resulting in ordered monolayers with a chain tilt angle of $15-25°$ from the surface normal. However, on CuO and Al_2O_3, the oxygen atoms bind themselves symmetrically and the chain tilt angle is close to $0°$. The structure of the monolayers is thus a balance of the various interactions taking place in the polymer chains.

Deoxyribonucleic acid (DNA), the framework on which all life is built, can be used to self-assemble nanomaterials into useful macroscopic aggregates that display a number of desirable physical properties [8.108]. DNA consists of two strands, which are coiled around each other to form a double helix. When the two strands are uncoiled singular strands of nucleotides are left. These nucleotides consist of a sugar (pentose ring) a phosphate (PO_4), and a nitrogenous base. The order and architecture of these components is essential for the proper structure of a nucleotide. There are typically four nucleotides found in DNA: adenine (A), guanine (G), cytosine (C), and thymine (T). A key property of the DNA structure is that the nucleotides described bind specifically to another nucleotide when arranged in the two-strand double helix (A to T, and C to G). This specific bonding capability can be used to assemble nanophase material and nanostructures [8.109]. For example, nucleotide-functionalized nanogold particles have been assembled into complex 3-D structures by attaching DNA strands to the gold via an enabler or linker [8.110]. In a separate work DNA was used to assemble nanoparticles into macroscopic materials. This method uses alkane dithiol as the linker molecule to connect the DNA template to the nanoparticle. The thiol groups at each end of the linker molecule covalently attach themselves to the colloidal particles to form aggregate structures [8.111].

Template Manufacturing

Template manufacturing refers to a set of techniques that can be used to fabricate 3-D organic or inorganic structures from a nanotemplate. These templates differ in material, pattern, feature size, overall template size, and periodicity. Although nanotemplates can be fabricated using e-beam lithography, the serial nature of this technique prohibits its widespread application. Self-assembly is the preferred technique to produce large-area nanotemplates in a massively parallel fashion. Several nanotemplates have been investigated for use in template manufacturing. These include polymer colloidal spheres, alumina membranes, and nuclear-track etched membranes. Colloidal spheres can be deposited in a regular 3-D array using the techniques described in the previous section (Figs. 8.50–8.52). Porous aluminum oxide membranes can be fabricated by the anodic oxidation of aluminum [8.112]. The oxidized film consists of columnar arrays of hexagonal close-packed pores with separation comparable to the pore size. By controlling the electrolyte species, temperature, anodizing voltage, and time, different pore sizes, densities, and heights can be obtained. The pore size and depth can further be adjusted by etching the oxide in an appropriate acid. Templates of porous polycarbonate or mica membranes can be fabricated by nuclear-track etched membranes [8.113]. This technique is based on the passage of high-energy decay fragments from a radioactive source through a dielectric material. The particles leave behind chemically active damaged tracks which can subsequently be etched to create pores through the thickness of the membrane [8.114, 115]. Unlike other methods, the pore separation and hence the pore density is independent of the pore size. The pore density is only determined by the irradiation process. More recently, electrochemical deposition of metallic nanowires on the step edges of highly oriented pyrolytic graphite (HOPG) templates have also been demonstrated, which produces metallic nanowires with diameters as small as 15 nm. This method has been successfully used for metals such as Cu, Ni, Au, and Pd [8.116, 117].

Subsequent to template fabrication, the interstitial spaces (in the case of colloidal spheres) or pores (in the case of alumina and polycarbonate membranes) in the template are filled with the desired material [8.93, 118]. This can be done by using a variety of deposition techniques such as electroplating and CVD. The final structure can be a composite of nanotemplate and deposited material or the template can be selectively etched, resulting in an air-filled 3-D complex structure. For example, nickel [8.119], iron [8.120], and cobalt [8.121] nanowires have been electrochemically grown into porous template matrices. Three-dimensional photonic crystals have been fabricated by electrochemical deposition of CdSe and silicon into polystyrene and silica colloidal assembly templates [8.91, 122]. An interesting example of template-assisted manufacturing is the synthesis of nanometer metallic barcodes [8.123]. These nanobarcodes are prepared by electrochemical reduction of metallic ions into the pores of an aluminum oxide membrane, followed by their release through etching of the template [8.93].

8.4 Summary and Conclusions

In this chapter, we have discussed various micro/nanofabrication techniques used to manufacture structures of a wide range of dimensions (mm–nm). Starting with some of the most common microfabrication techniques (lithography, deposition, and etching), we have presented an array of micromachining and MEMS technologies which can be used to fabricate microstructures down to $\approx 1\,\mu\text{m}$. These techniques have attained an adequate level of maturity to enable a variety of MEMS-based commercial products (pressure sensors, accelerometers, gyroscopes, etc.). More recently, nm-size structures have attracted an enormous amount of interest. This is mainly due to their unique electrical, magnetic, optical, thermal, and mechanical properties. These could lead to a variety of electronic, photonic, and sensing devices with a superior performance compared with their macro counterparts. Subsequent to our discussion on MEMS and micromachining, we presented several important nanofabrication techniques currently under intense investigation. Although e-beam and other high-resolution lithography techniques can be used to fabricate nm-size structures, their serial nature and/or cost preclude their widespread application. This has forced investigators to explore alternative and potentially superior techniques such as strain engineering, self-assembly, and nanoimprint lithography. Among these self-assembly is the most promising method due to its low cost and the ability to produce nanostructures at different length scales.

References

8.1 S.A. Campbell: *The Science and Engineering of Microelectronic Fabrication* (Oxford Univ. Press, New York 2001)
8.2 C.J. Jaeger: *Introduction to Microelectronic Fabrication* (Prentice Hall, New Jersey 2002)
8.3 J.D. Plummer, M.D. Deal, P.B. Griffin: *Silicon VLSI Technology* (Prentice-Hall, New Jersey 2000)
8.4 J.E. Bjorkholm: EUV lithography: the successor to optical lithography, Intel Technol. J. **2**, 1–8 (1998)
8.5 S. Owa, H. Nagasaka: Advantage and feasibility of immersion lithography, J. Microlithogr. Microfabr. Microsyst. **3**, 97–103 (2004)
8.6 H.U. Krebs, M. Störmer, J. Faupel, E. Süske, T. Scharf, C. Fuhse, N. Seibt, H. Kijewski, D. Nelke, E. Panchenko, M. Buback: Pulsed laser deposition (PLD) – A versatile thin film technique, Adv. Solid State Phys. **43**, 505–517 (2003)
8.7 M. Leskela, M. Ritala: Atomic layer deposition chemistry: recent developments and future challenges, Angew. Chem. Int. Ed. **42**, 5548–5554 (2003)
8.8 J.L. Vossen: *Thin Film Processes* (Academic, New York 1976)
8.9 M. Gad-el-Hak (Ed.): *The MEMS Handbook* (CRC, Boca Raton 2002)
8.10 T.-R. Hsu: *MEMS and Microsystems Design and Manufacture* (McGraw Hill, New York 2002)
8.11 G.T.A. Kovacs: *Micromachined Transducers Sourcebook* (McGraw Hill, New York 1998)

8.12   P. Rai-Choudhury (Ed.): *Handbook of Microlithography, Micromachining and Microfabrication. Vol. 2: Micromachining and Microfabrication* (SPIE/IEE, Bellingham, Washington/London 1997)

8.13   M.F. Aimi, M.P. Rao, N.C. MacDonald, A.S. Zuruzi, D.P. Bothman: High-aspect-ratio bulk micromachining of titanium, Nat. Mater. **3**, 103–105 (2004)

8.14   T.J. Cotler, M.E. Elta: Plasma-etch technology, IEEE Circuits Devices Mag. **6**, 38–43 (1990)

8.15   U. Gosele, Q.Y. Tong: Semiconductor wafer bonding, Annu. Rev. Mater. Sci. **28**, 215–241 (1998)

8.16   Q.Y. Tong, U. Gosele: *Semiconductor Wafer Bonding: Science and Technology* (Wiley, New York 1999)

8.17   F. Niklaus, P. Enoksson, E. Kalveston, G. Stemme: Void-free full-wafer adhesive bonding, J. Micromech. Microeng. **11**, 100–107 (2000)

8.18   C.A. Harper: *Electronic Packaging and Interconnection Handbook* (McGraw-Hill, New York 2000)

8.19   W.H. Ko, J.T. Suminto, G.J. Yeh: Bonding techniques for microsensors. In: *Micromachining and Micropackaging for Transducers*, ed. by W.H. Ko (Elsevier, Amsterdam 1985)

8.20   G.T.A. Kovacs, N.I. Maluf, K.A. Petersen: Bulk micromachining of silicon, Proc. IEEE **86**(8), 1536–1551 (1998)

8.21   K. Najafi, K.D. Wise, T. Mochizuki: A high-yield IC-compatible multichannel recording array, IEEE Trans. Electron Devices **32**, 1206–1211 (1985)

8.22   A. Selvakumar, K. Najafi: A high-sensitivity z-axis capacitive silicon microaccelerometer with a torsional suspension, J. Microelectromech. Syst. **7**, 192–200 (1998)

8.23   H. Baltes, O. Paul, O. Brand: Micromachined thermally based CMOS microsensors, Proc. IEEE **86**(8), 1660–1678 (1998)

8.24   N.C. MacDonald: SCREAM MicroElectroMechanical Systems, Microelectron. Eng. **32**, 51–55 (1996)

8.25   B. Eyre, K.S.J. Pister, W. Gekelman: Multi-axis microcoil sensors in standard CMOS, Proc. SPIE Conf. Micromach. Devices Compon., Austin (1995) pp.183–191

8.26   K.A. Shaw, Z.L. Zhang, N.C. MacDonnald: SCREAM: A single mask, single-crystal silicon process for microelectromechanical structures, Proc. IEEE Workshop Micro Electro Mech. Syst., Fort Lauderdale (1993) pp.155–160

8.27   G.K. Fedder, S. Santhanam, M.L. Reed, S.C. Eagle, D.F. Guillo, M.S.C. Lu, L.R. Carley: Laminated high-aspect-ratio microstructures in a conventional CMOS process, Proc. IEEE Workshop Micro Electro Mech. Syst., San Diego (1996) pp.13–18

8.28   B.P. Van Drieenhuizen, N.I. Maluf, I.E. Opris, G.T.A. Kovacs: Force-balanced accelerometer with mG resolution fabricated using silicon fusion bonding and deep reactive ion etching, Proc. Int. Conf. Solid-State Sens. Actuators, Chicago (1997) pp.1229–1230

8.29   X. Huikai, L. Erdmann, Z. Xu, K.J. Gabriel, G.K. Fedder: Post-CMOS processing for high-aspect-ratio integrated silicon microstructures, J. Microelectromech. Syst. **11**, 93–101 (2002)

8.30   J.M. Bustillo, R.S. Muller: Surface micromachining for microelectromechanical systems, Proc. IEEE **86**(8), 1552–1574 (1998)

8.31   H.C. Nathanson, W.E. Newell, R.A. Wickstrom, J.R. Davis: The resonant gate transistor, IEEE Trans. Electron Devices **14**, 117–133 (1967)

8.32   R.T. Howe, R.S. Muller: Polycrystalline silicon micromechanical beams, Proc. Electrochem. Soc. Spring Meet., Montreal (1982) pp.184–185

8.33   J.A. Geen, S.J. Sherman, J.F. Chang, S.R. Lewis: Single-chip surface-micromachined integrated gyroscope with 50/h root Allan deviation, IEEE J. Solid-State Circuits **37**, 1860–1866 (2002)

8.34   P.F. Van Kessel, L.J. Hornbeck, R.E. Meier, M.R. Douglass: A MEMS-based projection display, Proc. IEEE **86**(8), 1687–1704 (1998)

8.35   A.E. Franke, D. Bilic, D.T. Chang, P.T. Jones, R.T. Howe, G.C. Johnson: Post-CMOS integration of germanium microstructures, Proc. Micro Electro Mech. Syst., Orlando (1999) pp. 630–637

8.36   S. Sedky, P. Fiorini, M. Caymax, S. Loreti, K. Baert, L. Hermans, R. Mertens: Structural and mechanical properties of polycrystalline silicon germanium for micromachining applications, J. Microelectromech. Syst. **7**, 365–372 (1998)

8.37   N. Tas, T. Sonnenberg, H. Jansen, R. Legtenberg, M. Elwenspoek: Stiction in surface micromachining, J. Micromech. Microeng. **6**, 385–397 (1996)

8.38   R. Maboudian, R.T. Howe: Critical review: Adhesion in surface micromechanical structures, J. Vac. Sci. Technol. B **15**(1), 1–20 (1997)

8.39   J.H. Smith, S. Montague, J.J. Sniegowski, J.R. Murray, P.J. McWhorter: Embedded micromechanical devices for the monolithic integration of MEMS with CMOS, Proc. Int. Electron Devices Meet., Washington (1995) pp. 609–612

8.40   R.S. Muller, K.Y. Lau: Surface-micromachined microoptical elements and systems, Proc. IEEE **86**(8), 1705–1720 (1998)

8.41   K.S.J. Pister, M.W. Judy, S.R. Burgett, R.S. Fearing: Microfabricated hinges: 1mm vertical features with surface micromachining, Proc. 6th Int. Conf. Solid-State Sens. Actuators, San Francisco (1991) pp. 647–650

8.42   L.Y. Lin, S.S. Lee, M.C. Wu, K.S.J. Pister: Micromachined Integrated optics for free space interconnection, Proc. IEEE Micro Electro Mech. Syst. Workshop, Amsterdam (1995) pp.77–82

8.43   E.W. Becker, W. Ehrfeld, P. Hagmann, A. Maner, D. Munchmeyer: Fabrication of microstructures with high aspect ratios and great structural heights by synchrotron radiation lithography, galvanoforming, and plastic molding (LIGA process), Microelectron. Eng. **4**, 35–56 (1986)

8.44 H. Guckel: High-aspect-ratio micromachining via deep x-ray lithography, Proc. IEEE **86**(8), 1586–1593 (1998)

8.45 K.Y. Lee, N. LaBianca, S.A. Rishton, S. Zolgharnain, J.D. Gelorme, J. Shaw, T.H.P. Chang: Micromachining applications of a high resolution ultra-thick photoresist, J. Vac. Sci. Technol. B **13**, 3012–3016 (1995)

8.46 K. Roberts, F. Williamson, G. Cibuzar, L. Thomas: The fabrication of an array of microcavities utilizing SU-8 photoresist as an alternative 'LIGA' technology, Proc. 13th Bienn. Univ./Gov./Ind. Microelectron. Symp. (IEEE), Minneapolis (1999) pp. 139–141

8.47 D.A. Horsley, M.B. Cohn, A. Singh, R. Horowitz, A.P. Pisano: Design and fabrication of an angular microactuator for magnetic disk drives, J. Microelectromech. Syst. **7**, 141–148 (1998)

8.48 C. Burbaum, J. Mohr, P. Bley, W. Ehrfeld: Fabrication of capacitive acceleration sensors by the LIGA technique, Sens. Actuators A Phys. **A27**, 559–563 (1991)

8.49 C.G. Keller, R.T. Howe: Hexsil bimorphs for vertical actuation, Dig. Tech. Pap. 8th Int. Conf. Solid-State Sens. Actuators Eurosens. IX, Stockholm (1995) pp. 99–102

8.50 C.G. Keller, R.T. Howe: Nickel-filled hexsil thermally actuated tweezers, Dig. Tech. Pap. 8th Int. Conf. Solid-State Sens. Actuators Eurosens. IX, Stockholm (1995) pp. 376–379

8.51 N. Yazdi, F. Ayazi, K. Najafi: Micromachined inertial sensors, Proc. IEEE **86**, 1640–1659 (1998)

8.52 F. Ayazi, K. Najafi: High aspect-ratio combined poly and single-crystal silicon (HARPSS) MEMS technology, J. Microelectromech. Syst. **9**, 288–294 (1998)

8.53 F. Ayazi, K. Najafi: A HARPSS polysilicon vibrating ring gyroscope, J. Microelectromech. Syst. **10**, 169–179 (2001)

8.54 Y.S. No, F. Ayazi: The HARPSS process for fabrication of nano-precision silicon electromechanical resonators, Proc. 2001 1st IEEE Conf. Nanotechnol., Maui (2001) pp. 489–494

8.55 G. Timp: *Nanotechnology* (Springer, New York 1998)

8.56 P. Rai-Choudhury (Ed.): *Handbook of Microlithography, Micromachining and Microfabrication. Vol. 1: Microlithography* (SPIE/IEE, Bellingham, Washington/London 1997)

8.57 L. Ming, C. Bao-qin, Y. Tian-Chun, Q. He, X. Qiuxia: The sub-micron fabrication technology, Proc. 6th Int. Conf. Solid-State Integr.-Circuit Technol., San Francisco, Vol. 1 (2001) pp. 452–455

8.58 M.A. Herman: *Molecular Beam Epitaxy: Fundamentals and Current Status* (Springer, New York 1996)

8.59 J.S. Frood, G.J. Davis, W.T. Tsang: *Chemical Beam Epitaxy and Related Techniques* (Wiley, New York 1997)

8.60 S. Mahajan, K.S. Sree Harsha: *Principles of Growth and Processing of Semiconductors* (McGraw Hill, New York 1999)

8.61 S. Kim, M. Razegi: Advances in quantum dot structures. In: *Processing and Properties of Compound Semiconductors*, ed. by K. Willardson, E.R. Weber (Academic, New York 2001)

8.62 D. Bimberg, M. Grundmann, N.N. Ledentsov: *Quantum Dot Heterostructures* (Wiley, New York 1999)

8.63 G. Seebohm, H.G. Craighead: Lithography and patterning for nanostructure fabrication. In: *Quantum Semiconductor Devices and Technologies*, ed. by T.P. Pearsall (Kluwer, Boston 2000)

8.64 E. Kapon: Lateral patterning of quantum well heterostructures by growth on nonplanar substrates. In: *Epitaxial Microstructures*, ed. by A.C. Gossard (Academic, New York 1994)

8.65 F. Guffarth, R. Heitz, A. Schliwa, O. Stier, N.N. Ledentsov, A.R. Kovsh, V.M. Ustinov, D. Bimberg: Strain engineering of self-organized InAs quantum dots, Phys. Rev. B **64**, 085305-1–085305-7 (2001)

8.66 M. Sugawara: *Self-Assembled InGaAs/GaAs Quantum Dots* (Academic, New York 1999)

8.67 B.C. Lee, S.D. Lin, C.P. Lee, H.M. Lee, J.C. Wu, K.W. Sun: Selective growth of single InAs quantum dots using strain engineering, Appl. Phys. Lett. **80**, 326–328 (2002)

8.68 K. Brunner: Si/Ge nanostructures, Rep. Prog. Phys. **65**, 27–72 (2002)

8.69 F.S.S. Chien, W.F. Hsieh, S. Gwo, A.E. Vladar, J.A. Dagata: Silicon nanostructures fabricated by scanning-probe oxidation and tetra-methyl ammonium hydroxide etching, J. Appl. Phys. **91**, 10044–10050 (2002)

8.70 M. Calleja, J. Anguita, R. Garcia, K. Birkelund, F. Perez-Murano, J.A. Dagata: Nanometer-scale oxidation of silicon surfaces by dynamic force microscopy: reproducibility, kinetics and nanofabrication, Nanotechnology **10**, 34–38 (1999)

8.71 E.S. Snow, P.M. Campbell, F.K. Perkins: Nanofabrication with proximal probes, Proc. IEEE **85**, 601–611 (1997)

8.72 H. Sugimura, T. Uchida, N. Kitamura, H. Masuhara: Tip-induced anodization of titanium surfaces by scanning tunneling microscopy: A humidity effect on nanolithography, Appl. Phys. Lett. **63**, 1288–1290 (1993)

8.73 N. Kramer, J. Jorritsma, H. Birk, C. Schonenberger: Nanometer lithography on silicon and hydrogenated amorphous silicon with low energy electrons, J. Vac. Sci. Technol. B **13**, 805–811 (1995)

8.74 H.T. Soh, K.W. Guarini, C.F. Quate: *Scanning Probe Lithography* (Kluwer Academic, Boston 2001)

8.75 C.A. Mirkin: Dip-pen nanolithography: automated fabrication of custom multicomponent, sub-100-nanometer surface architectures, MRS Bulletin **26**, 535–538 (2001)

8.76 P.E. Sheehan, L.J. Whitman, W.P. King, B.A. Nelson: Nanoscale deposition of solid inks via thermal dip

8.76 pen nanolithography, Appl. Phys. Lett. **85**, 1589–1591 (2004)

8.77 L.L. Sohn, R.L. Willett: Fabrication of nanostructures using atomic-force-microscope-based lithography, Appl. Phys. Lett. **67**, 1552–1554 (1995)

8.78 H.J. Mamin, B.D. Terris, L.S. Fan, S. Hoen, R.C. Barrett, D. Rugar: High-density data storage using proximal probe techniques, IBM J. Res. Dev. **39**, 681–699 (1995)

8.79 K. Bessho, S. Hashimoto: Fabricating nanoscale structures on Au surface with scanning tunneling microscope, Appl. Phys. Lett. **65**, 2142–2144 (1994)

8.80 I.W. Lyo, P. Avouris: Field-induced nanometer- to atomic-scale manipulation of silicon surfaces with the STM, Science **253**, 173–176 (1991)

8.81 M.F. Crommie, C.P. Lutz, D.M. Eigler: Confinement of electrons to quantum corrals on a metal surface, Science **262**, 218–220 (1993)

8.82 A. de Lozanne: Pattern generation below 0.1 micron by localized chemical vapor deposition with the scanning tunneling microscope, Jpn. J. Appl. Phys. **33**, 7090–7093 (1994)

8.83 L.A. Nagahara, T. Thundat, S.M. Lindsay: Nanolithography on semiconductor surfaces under an etching solution, Appl. Phys. Lett. **57**, 270–272 (1990)

8.84 T. Thundat, L.A. Nagahara, S.M. Lindsay: Scanning tunneling microscopy studies of semiconductor electrochemistry, J. Vac. Sci. Technol. A **8**, 539–543 (1990)

8.85 S.C. Minne, S.R. Manalis, A. Atalar, C.F. Quate: Independent parallel lithography using the atomic force microscope, J. Vac. Sci. Technol. B **14**, 2456–2461 (1996)

8.86 M. Lutwyche, C. Andreoli, G. Binnig, J. Brugger, U. Drechsler, W. Häberle, H. Rohrer, H. Rothuizen, P. Vettiger: Microfabrication and parallel operation of 5×5 2-D AFM cantilever arrays for data storage and imaging, Proc. MEMS **98**, 8–11 (1998)

8.87 G.M. Whitesides, B. Grzybowski: Self-assembly at all scales, Science **295**, 2418–2421 (2002)

8.88 P. Kazmaier, N. Chopra: Bridging size scales with self-assembling supramolecular materials, MRS Bulletin **25**, 30–35 (2000)

8.89 G.A. Gelves, Z.T.M. Murakami, M.J. Krantz, J.A. Haber: Multigram synthesis of copper nanowires using ac electrodeposition into porous aluminium oxide templates, J. Mater. Chem. **16**, 3075–3083 (2006)

8.90 R. Plass, J.A. Last, N.C. Bartelt, G.L. Kellogg: Self-assembled domain patterns, Nature **412**, 875 (2001)

8.91 Y.A. Vlasov, X.-Z. Bo, J.G. Sturm, D.J. Norris: On-chip natural self-assembly of silicon photonic bandgap crystals, Nature **414**, 289–293 (2001)

8.92 C. Gigault, K. Dalnoki-Veress, J.R. Dutcher: Changes in the morphology of self-assembled polystyrene microsphere monolayers produced by annealing, J. Colloid Interface Sci. **243**, 143–155 (2001)

8.93 J.C. Hulteen, C.R. Martin: A general template-based method for the preparation of nanomaterials, J. Mater. Chem. **7**, 1075–1087 (1997)

8.94 J.D. Joannopoulos, P.R. Villeneuve, S. Fan: Photonic crystals: putting a new twist on light, Nature **386**, 143–149 (1997)

8.95 T.D. Clark, R. Ferrigno, J. Tien, K.E. Paul, G.M. Whitesides: Template-directed self-assembly of 10-μm-sized hexagonal plates, J. Am. Chem. Soc. **124**, 5419–5426 (2002)

8.96 S.A. Sapp, D.T. Mitchell, C.R. Martin: Using template-synthesized micro- and nanowires as building blocks for self-assembly of supramolecular architectures, Chem. Mater. **11**, 1183–1185 (1999)

8.97 Y. Yin, Y. Lu, B. Gates, Y. Xia: Template assisted self-assembly: A practical route to complex aggregates of monodispersed colloids with well-defined sizes, shapes and structures, J. Am. Chem. Soc. **123**, 8718–8729 (2001)

8.98 J.L. Wilbur, G.M. Whitesides: Self-assembly and self-assembles monolayers in micro- and nanofabrication. In: *Nanotechnology*, ed. by G. Timp (Springer, New York 1999)

8.99 S.R. Wasserman, Y.T. Tao, G.M. Whitesides: Structure and reactivity of alkylsiloxane monolayers formed by reaction of alkyltrichlorosilanes on silicon substrates, Langmuir **5**, 1074–1087 (1989)

8.100 C.P. Tripp, M.L. Hair: An infrared study of the reaction of octadecyltrichlorosilane with silica, Langmuir **8**, 1120–1126 (1992)

8.101 S. Fiorilli, P. Rivolo, E. Descrovi, C. Ricciardi, L. Pasquardini, L. Lunelli, L. Vanzetti, C. Pederzolli, B. Onida, E. Garrone: Vapor-phase self-assembled monolayers of aminosilane on plasma-activated silicon substrates, J. Colloid Interface Sci. **321**, 235–241 (2008)

8.102 R. de la Rica, A. Baldi, E. Mendoza, A. San Paulo, A. Llobera, C. Fernandez-Sanchez: Silane nanopatterns via gas-phase soft lithography, Small **4**(8), 1076–1079 (2008)

8.103 D.R. Walt: Nanomaterials: Top-to-bottom functional design, Nature **1**, 17–18 (2002)

8.104 J. Noh, T. Murase, K. Nakajima, H. Lee, M. Hara: Nanoscopic investigation of the self-assembly processes of dialkyl disulfides and dialkyl sulfides on Au(111), J. Phys. Chem. B **104**, 7411–7416 (2000)

8.105 M. Himmelhaus, F. Eisert, M. Buck, M. Grunze: Self-assembly of n-alkanethiol monolayers: A study by IR-visible sum frequency spectroscopy (SFG), J. Phys. Chem. **104**, 576–584 (1999)

8.106 A.K. Boal, F. Ilhan, J.E. DeRouchey, T. Thurn-Albrecht, T.P. Russell, V.M. Rotello: Self-assembly of nanoparticles into structures spherical and network aggregates, Nature **404**, 746–748 (2000)

8.107 A. Ulman: *An Introduction to Ultrathin Organic Films: From Langmuir-Blodgett to Self-Assembly* (Academic, New York 1991)

8.108   E. Winfree, F. Liu, L.A. Wenzler, N.C. Seeman: Design and self-assembly of two-dimensional DNA crystals, Nature **394**, 539–544 (1998)

8.109   J.H. Reif, T.H. LaBean, N.C. Seeman: Programmable assembly at the molecular scale: self-assembly of DNA lattices, Proc. 2001 IEEE Int. Conf. Robot. Autom., Seoul (2001) pp. 966–971

8.110   A.P. Alivisatos, K.P. Johnsson, X. Peng, T.E. Wilson, C.J. Loweth, M.P. Bruchez Jr, P.G. Schultz: Organization of 'nanocrystal molecules' using DNA, Nature **382**, 609–611 (1996)

8.111   H. Cao, Z. Yu, J. Wang, J.O. Tegenfeldt, R.H. Austin, E. Chen, W. Wu, S.Y. Chou: Fabrication of 10 nm enclosed nanofluidic channels, Appl. Phys. Lett. **81**, 174–176 (2002)

8.112   H. Masuda, H. Yamada, M. Satoh, H. Asoh: Highly ordered nanochannel-array architecture in anodic alumina, Appl. Phys. Lett. **71**, 2770–2772 (1997)

8.113   R.L. Fleischer: *Nuclear Tracks in Solids: Principles and Applications* (Univ. of California Press, Berkeley 1976)

8.114   R.E. Packard, J.P. Pekola, P.B. Price, R.N.R. Spohr, K.H. Westmacott, Y.Q. Zhu: Manufacture observation and test of membranes with locatable single pores, Rev. Sci. Instrum. **57**, 1654–1660 (1986)

8.115   L. Sun, P.C. Searson, C.L. Chien: Electrochemical deposition of nickel nanowire arrays in single-crystal mica films, Appl. Phys. Lett. **74**, 2803–2805 (1999)

8.116   E.C. Walter, M.P. Zach, F. Favier, B.J. Murray, K. Inazu, J.C. Hemminger, R.M. Penner: Metal nanowire arrays by electrodeposition, ChemPhysChem **4**(2), 131–138 (2003)

8.117   M.Z. Atashbar, D. Banerji, S. Singamaneni, V. Bliznyuk: Deposition of parallel arrays of palladium nanowires and electrical characterization using microelectrode contacts, Nanotechnology **15**(3), 374–378 (2004)

8.118   Y. Du, W.L. Cai, C.M. Mo, J. Chen, L.D. Zhang, X.G. Zhu: Preparation and photoluminescence of alumina membranes with ordered pore arrays, Appl. Phys. Lett. **74**, 2951–2953 (1999)

8.119   M. Guowen, C. Anyuan, C. Ju-Yin, A. Vijayaraghavan, J.J. Yung, M. Shima, P.M. Ajayan: Ordered Ni nanowire tip arrays sticking out of the anodic aluminum oxide template, J. Appl. Phys. **97**, 64303-1–64303-5 (2005)

8.120   S. Yang, H. Zhu, D. Yu, Z. Jin, S. Tang, Y. Du: Preparation and magnetic property of Fe nanowire array, J. Magn. Magn. Mater. **222**, 97–100 (2000)

8.121   M. Sun, G. Zangari, R.M. Metzger: Cobalt island arrays with in-plane anisotropy electrodeposited in highly ordered alumina, IEEE Trans. Magn. **36**, 3005–3008 (2000)

8.122   P.V. Braun, P. Wiltzius: Electrochemically grown photonic crystals, Nature **402**, 603–604 (1999)

8.123   S.R. Nicewarner-Pena, R.G. Freeman, B.D. Reiss, L. He, D.J. Pena, I.D. Walton, R. Cromer, C.D. Keating, M.J. Natan: Submicrometer metallic barcodes, Science **294**, 137–141 (2001)

9. Nanoimprint Lithography – Patterning of Resists Using Molding

Helmut Schift, Anders Kristensen

Nanoimprint lithography (NIL) is an emerging high-resolution parallel patterning method, mainly aimed towards fields in which electron-beam and high-end photolithography are costly and do not provide sufficient resolution at reasonable throughput. In a top-down approach, a surface pattern of a stamp is replicated into a material by mechanical contact and three-dimensional material displacement. This can be done by shaping a liquid followed by a curing process for hardening, by variation of the thermomechanical properties of a film by heating and cooling, or by any other kind of shaping process using the difference in hardness of a mold and a moldable material. The local thickness contrast of the resulting thin molded film can be used as a means to pattern an underlying substrate at the wafer level by standard pattern transfer methods, but also directly in applications where a bulk modified functional layer is needed. This makes NIL a promising technique for volume manufacture of nanostructured components. At present, structures with feature sizes down to 5 nm have been realized, and the resolution is limited by the ability to manufacture the stamp relief. For historical reasons, the term nanoimprint lithography refers to a hot embossing process (thermal NIL). In ultraviolet (UV)-NIL, a photopolymerizable resin is used together with a UV-transparent stamp. In both processes thin-film squeeze flow and capillary action play a central role in understanding the NIL process. In this chapter we will give an overview of NIL, with emphasis on general principles and concepts rather than specific process issues and state-of-the-art tools and processes. Material aspects of stamps and resists are discussed. We discuss specific applications where imprint methods have significant advantages over other structuring methods. We conclude by discussing areas where further development in this field is required.

9.1	Emerging Nanopatterning Methods	273
	9.1.1 Next-Generation Lithography	274
	9.1.2 Variants of Nanoimprint Lithography	275
9.2	Nanoimprint Process	277
	9.2.1 Limits of Molding	277
	9.2.2 Squeeze Flow of Thin Films	279
	9.2.3 Residual Layer Thickness Homogeneity	281
	9.2.4 Demolding	282
	9.2.5 Curing of Resists	283
	9.2.6 Pattern Transfer	283
	9.2.7 Mix-and-Match Methods	285
	9.2.8 Multilayer and Multilevel Systems	286
	9.2.9 Reversal NIL	287
9.3	Tools and Materials for Nanoimprinting	288
	9.3.1 Resist Materials for Nanoimprinting	288
	9.3.2 Stamp Materials	290
	9.3.3 Stamp Fabrication	290
	9.3.4 Antiadhesive Coatings	291
	9.3.5 Imprinting Machines	292
9.4	Nanoimprinting Applications	294
	9.4.1 Types of Nanoimprinting Applications	294
	9.4.2 Patterned Magnetic Media for Hard-Disk Drives	295
	9.4.3 Subwavelength Metal-Strip Gratings	297
	9.4.4 High-Brightness Light-Emitting Diodes	298
	9.4.5 Polymer Optics	299
	9.4.6 Bio Applications	300
9.5	Conclusions and Outlook	302
References		304

Take a piece of wax between your fingers and imprint your fingerprints into it from both sides. The pressure produced is sufficiently high to replicate the soft surface pattern of your skin into the wax by mechanical deformation. The process is facilitated by the heat resulting from our blood circulation, which softens the wax in order to make it deform until it conforms to the three-dimensional (3-D) pattern of our skin. Of course, the fidelity of the original pattern is distorted during molding, but even an incomplete molding allows the identification of the person according to the purely two-dimensional (2-D) code of their fingerprint. The pattern resolution of below 1 mm is similar to that of the first records fabricated over 100 years ago in celluloid. In 1887 *Berliner* applied for a patent on a so-called gramophone, which resembles Edison's phonograph with its wax-coated roll [9.1, 2]. The information is inscribed into wax coated onto a zinc disk. The tracks are cut through the wax down to the solid zinc and are etched before using the zinc disk as a mold to press thermoplastic foils. With a playing time of a little more than 1 min, those disks had track widths below 1 mm and resolution in the sub-100 μm range. Over the years the track size was reduced to below 200 μm. The materials changed from shellac to vinyl filled with carbon black [9.3]. Today's compact discs (CD) have pit sizes of below 400 nm [in a digital versatile disc (DVD)] and are fabricated in polycarbonate (PC) in a few seconds by injection molding. Disc formats such as blu-ray (BD) with further reduced pit sizes are currently commercialized [9.4, 5].

In this Introduction some basic concepts of molding polymers are illustrated, ranging from shaping by mechanical pressure, stamps, materials, to pattern transfer. A softened hard material can be deformed by pressure, and even if a soft, flexible stamp is used, the difference in mechanical properties makes it possible to replicate its surface pattern in a parallel, reproducible way. The squeezing of a thin film of wax leads to a lateral flow of material, but because of the high viscosity, the process will slow down quickly and a residual layer which cannot be thinned down to zero will always remain. Furthermore the softness of the stamp and the viscosity of the material will determine the completeness of molding and thus the replication fidelity. Similar concepts of molding processes can be observed in daily life, such as imprinting a footprint into snow or clay, making waffles in a pressure process with subsequent thermocuring, or replicating a seal into wax (Fig. 9.1). Even these examples show the variety of molding processes. One common important prerequisite of these

Fig. 9.1 Printing a seal into viscous wax is a way of replication using hot embossing. The image shows a seal (stamp), wax tube (candle), and embossed pattern

molding processes is that the mechanical properties of the molded material can be changed by pressure, temperature or chemical processing. The material must be shaped in a viscous state but should keep its form during demolding. The imprint in snow is a hard molding by local densification, while clay hardens by squeezing out and evaporation of water. The waffle is cured due to thermochemical changes in the dough, and the seal can be demolded with high fidelity because the heat of the wax dissipates into the seal and the wax hardens during cooling. The processes described here are very similar to the molding of viscous thermoplastic materials in the nanoimprint lithography (NIL) process [9.6], also referred to as hot embossing lithography (HEL) [9.7], where a thickness profile in a thin polymer film is generated by pressure, however, with the surprising difference that features below 10 nm can be replicated with unprecedented precision (Fig. 9.2). In contrast to conventional methods based on exposure and development, limitations imposed by the wavelength of exposure or by chemical reactions can be overcome simply by inducing local displacement of material by mechanical force.

The example of the fingerprint may even serve to illustrate (soft lithography) microcontact printing (SL or μCP). While for NIL a hard stamp would assure more complete molding, here the softness of the stamp is essential to assure conformal contact with any protrusion, but at the expense of a possible reduction of feature resolution due to deformation of the stamp. These issues are treated in more detail in [9.8, 9].

In this chapter we provide an overview of the different processes currently called nanoimprinting, from hot embossing of thermoplastic materials to imprinting and

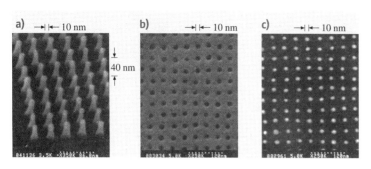

Fig. 9.2a–c Micrographs showing the basic steps of NIL, demonstrated by *Chou* and *Krauss* [9.6]. (a) NIL stamp in silicon with a 40 nm-period array of pillars with 40 nm height, (b) imprinted 10 nm-diameter holes in a thin polymer film (PMMA), (c) 10 nm metal dots after pattern transfer (lift-off), using the thin polymer layer as a mask

curing of liquid resins. After this Introduction into the basics of molding, Sect. 9.1 places the two main NIL techniques into the context of the emerging nanopatterning methods for lithography. Section 9.2 is the main section, where the NIL process is described in detail, beginning with a discussion about polymer properties, giving an insight into squeeze flow of thin films. As a first step towards applications major pattern transfer techniques used in NIL are presented. Section 9.3 presents materials and tools for NIL, ranging from materials for stamps and resists, to imprint machines. Section 9.4 presents typical applications which are currently envisaged both at an industrial and at laboratory scale. Although for many people the main driving force behind NIL is its use as next-generation lithography (NGL) for complementary metal–oxide–semiconductor (CMOS) chip fabrication, the reader will be introduced to different other applications which do not have the demanding overlay requirements imposed by multilevel processes. We conclude with an outlook in Sect. 9.5, in which we discuss the prospects of NIL and aspects of its commercialization. Further information can be found in the references, in publications dealing with so-called lithography, electroforming, and molding (LIGA, from its German abbreviation) technology [9.10] and optical storage fabrication, but not least within this Handbook in Chaps. 8 and 10 about silicon micromachining and soft lithography. In this chapter we restrict ourselves to lithographic patterning of thin films on hard substrates. We present basic concepts rather than state-of-the-art tools and hot scientific issues. As a complement to this chapter, the reader is advised to refer to two publications: A recent review on NIL [9.11] deals with a range of process issues relevant for research and industry, and a deeper insight into advanced concepts of printing. Specific NIL processes and process flows for a variety of applications are presented in the NaPa Library of processes (NaPa LoP) [9.12].

9.1 Emerging Nanopatterning Methods

Nanoimprint lithography (NIL) is a replication technique which has proven to provide a resolution unmatched by many other techniques, while at the same time offering parallel and fast fabrication of micro- and nanostructures [9.13]. On the one hand, this enables its application to fields where large areas covered by nanostructures or a number of identical structures for statistical evaluation are needed. This was often impossible due to the low throughput of lithographic research tools. On the other hand the resolution achieved so far by molding is much higher than that used in industrial fabrication of processors and memory chips with high-end photolithography (PL). This makes NIL a promising technology for NGL [9.14]. Apart from these advantages molding offers more: By creating a three-dimensional (3-D) resist pattern by mechanical displacement of material, the patterning of a range of specific functional materials and polymers becomes possible, without loss of their chemical properties during molding. Furthermore this ability can be used to fabricate complex structures, e.g., by building up devices with embedded channels. These processes are presented in more detail in Sect. 9.2.

In this section we present the basic concepts of NIL and how it can conform to the requirements of state-of-the-art nanofabrication techniques. NIL uses, as do other lithographic techniques, the concept of resist patterning (which can also be found in different chapters in this Handbook). The resist patterns are generated by molding of a viscous material and fixed by cooling and curing, while in PL the resist is patterned by selective local chemical modification of a positive or negative resist

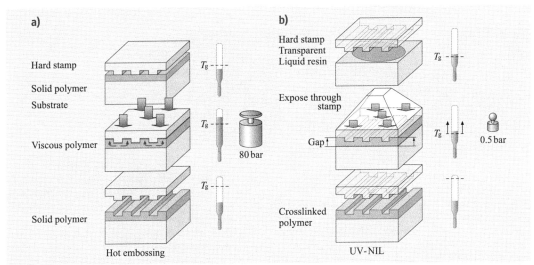

Fig. 9.3a,b Schematic of NIL process: (**a**) thermal NIL (hot embossing) and (**b**) UV-NIL. In both cases a thickness profile is generated in the thin polymer layer. After removing the residual layer, the remaining polymer can serve as a masking layer which can be used as a resist for pattern transfer

by exposure and wet development. The two main NIL methods are outlined in Fig. 9.3. For lithographic applications, as needed in microelectronics and hard disks, NIL is in competition with other emerging patterning techniques. Its success will mainly depend on the ability to solve processing issues such as resolution and throughput. It is also important to develop reliable tools with a long lifetime, which are available and can be used in combination with other cleanroom process technologies, and to establish standard processes which can be scaled up to common wafer sizes.

9.1.1 Next-Generation Lithography

With its integration into the International Technology Roadmap for Semiconductors (ITRS) on NGL in 2003 for the 32 nm node and beyond, NIL has become more than a simple high-resolution method [9.14] (Table 9.1).

Table 9.1 ITRS roadmap showing the resolution of different lithographic patterning techniques, with focus on large area, parallel techniques, and practical and actual resolution limits for different lithography methods (after [9.15], revised and updated (state of the art 2009))

Lithography type	Practical resolution limit (nm)	Ultimate resolution limit (nm)
UV-proximity photolithography (365 nm)	2500	125 (hard contact)
Deep-UV projection (DUV, 193 nm)	45	20–30 (immersion)
EUV projection (soft X-rays, 13.6 nm, with reflective mask)	45	20–30
EUV interference lithography (with diffraction grating)	20	10
X-ray proximity (0.8 nm, 1 : 1 mask)	70	10
Electron beam (low-energy beam arrays)	40–50	Resist: 7–20
Ion beam projection (mask-less patterning)	25	20
Thermal nanoimprint (hot embossing)	20–40	5
UV nanoimprint (hard stamp)	20–40	2–5
UV nanoimprint (soft stamp)	100	50
Soft lithography (contact printing)	50–100	10–50
Scanning probe methods (e.g. millipede)	15	0.5 (atomic resolution)

It is now considered a candidate for replacing or complementing advanced optical lithographic methods for the fabrication of processors and solid-state memory chips, which over the years have been developed and pushed to higher resolution with a vast investment of resources. Over more than 40 years, Moore's law has described with amazing accuracy the reduction of feature size (and cost per transistor), and therefore serves as a roadmap for the developments needed for future microchips [9.15–17]. This development is driven by economic considerations, and leads to competition between different candidate fabrication methods. These do not only have to provide the resolution of the smallest feature size (node), but also satisfy issues such as alignment (overlay of several masking levels), critical dimensions (CD), simple mask fabrication, high throughput (mass fabrication), and low cost of ownership (CoO, e.g. no dependence of expensive machines such as synchrotrons, back-ups and tool and mask redundancy), which become increasingly difficult to meet if smaller exposure wavelengths have to be used (Fig. 9.4).

The financial and physical barriers to these techniques are now so great that alternatives such as NIL are considered as a way out of this spiral of rising investments for next-generation chips with even smaller feature sizes. This means that all technical issues connected with NIL for integration into chip manufacturing must satisfy the requirements for full compatibility, similar specifications, yield, and throughput. The investments are expected to be lower than for the current frontrunners: extreme-ultraviolet (EUV) lithography or parallel electron-beam exposure.

9.1.2 Variants of Nanoimprint Lithography

Molding of Thermoplastic Resists by Thermal NIL

NIL was first reported as thermoplastic molding [9.20–22], and is therefore often referred to as thermal NIL (here also named NIL or T-NIL) or hot embossing lithography (HEL) [9.7, 22, 23]. The unique advantage of a thermoplastic material is that the viscosity can be changed to a large extent by simply varying the temperature. Figure 9.5 shows viscosity plotted against temperature for various thermoplastic polymers, i.e., poly(methyl methacrylate) (PMMA) and polystyrene (PS) with different molecular weights, and some com-

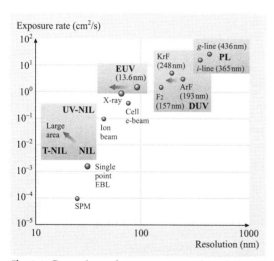

Fig. 9.4 Comparison of exposure rate and resolution of different lithographic techniques. To date, NIL provides a high resolution of below 10 nm, and achieves wafer-scale patterning within some minutes down to seconds (after [9.15])

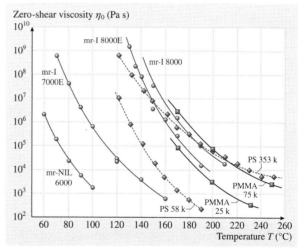

Fig. 9.5 Graph showing zero-shear viscosities for some standard resists for thermal NIL for different polymers, taken from different sources: PMMA with M_w of 25 and 75 k [9.18], PS of 58 and 353 k [9.19], and the commercial resists mr-I 7000E, 8000, 8000E, and mr-NIL6000 [9.11], showing the potential of rheology and of the large variation of viscosity of thermoplastic polymers with temperature. These curves present the temperature range which characterizes the viscous state above T_g. The process window for imprinting is limited by high viscosity, where unwanted viscoelastic effects become dominant and molding becomes slow. Viscosities below 10^3 Pa s are often not useful, often being achieved with too low a M_w or too high a $T_{imprint}$ (after [9.11])

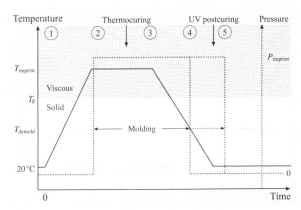

Fig. 9.6 Typical process sequence: schematics of process sequence used for hot embossing (temperature/pressure diagram with time dependence), (1) begin heating, (2) begin embossing, (3) begin cooling, (4) demolding at elevated T, and (5) demolding at ambient

mercial resists [9.11, 18, 19]. Switching between a solid and a highly viscous state is possible within a range of some tens of degrees Celsius, and can be reversed [9.18]. The first stage of the NIL process is the molding of a thin thermoplastic film using a hard master. During a process cycle the resist material is made viscous by heating, and shaped by applying pressure (Fig. 9.6). Here the thermoplastic film is compressed between the stamp and substrate, and the viscous polymer is forced to flow into the cavities of the mold, conforming exactly to the surface relief of the stamp. When the cavities of the stamps are filled, the polymer is cooled down, while the pressure is maintained. Thus the molten structure is frozen. After relieving the pressure, the stamp can be retrieved (demolded) without damage and reused for the next molding cycle.

Molding of UV-Curable Resists by UV-NIL

With the integration of light sources into imprint machines, UV-NIL was developed for curable re-

Table 9.2 Comparison of hot embossing (NIL) and UV imprinting (UV-NIL), with typical parameters of current processes

Type of NIL (properties)	Thermal NIL (hot embossing)	UV-NIL (hard stamp)
Basic process sequence (Fig. 9.6)	1) Spin-coat thermoplastic film 2) Place stamp on film 3) Heat until viscous 4) Emboss at high pressure 5) Cool until solid 6) Demold stamp	1) Dispense liquid resin 2) Parallel alignment of stamp with defined gap 3) Imprint at low pressure 4) Expose with UV light through stamp and crosslink 5) Demold stamp
Pressure p	20–100 bar	0–5 bar
Temperature T_{mold}	100–200 °C	20 °C (ambient)
Temperature T_{demold}	20–80 °C	20 °C (ambient)
Resist	Solid, thermoplastic $T_g \approx 60$–$100\,°C$	Liquid, UV-curable
Viscosity η	10^3–10^7 Pa s	10^{-2}–10^{-3} Pa s
Stamp material	Si, SiO$_2$ opaque	Glass, SiO$_2$ transparent
Stamp area	Full wafer, > 200 mm diameter	25×25 mm^2, limited by control of gap
Stamp contact	Facilitated by bending	Planarization layer
Embossing time	From s to min	< 1 min (per exposure)
Advantage	Low-cost, large-area equipment and stamps	Low viscosity, low pressure, alignment through stamp
Challenge	Process time, thermal expansion due to thermal cycle	Step and repeat needed for large areas
Development needed	Alignment, residual layer homogeneity	Material variety
Step and repeat	Step and stamp with 4×4 mm^2 stamps	Step and flash (SFIL) with 30×45 mm^2
Hybrid approaches	Thermoset resists: Embossing and curing before demolding	Thermoplastic resists: Hot molding and UV-curing before demolding
Advantage	Low-temperature-variation cycle: Demolding at high temperature possible	Solid resist: Single-step wafer-scale imprint possible

sists [9.24–27]. The basic difference between UV-NIL and NIL is that a resin that is liquid at room temperature is shaped by a moderate pressure, and by exposing light through the transparent stamp the resin is cross-linked and hardened. The stamp either sinks down to the substrate or must be kept at a constant distance from the substrate during both filling and exposure, due to the low resist viscosities. The mechanical setup has to be able to compensate for wedge errors in a low-imprint-pressure process. Patterning on nonflat substrates or over topography therefore requires a planarization strategy and often small stamps. Because of the small pressures used, both hard stamps or stamps with protrusions made by soft material on a rigid backbone can be used.

Resist Window Opening for Pattern Transfer
A basic characteristic of NIL is the patterning of a thin layer of material, in which the dimensions (lateral structure size and height) become similar to the film thickness used. In a second step, the thickness profile of the polymer film can now be used as a resist for pattern transfer. For this, the residual layer remaining in the thin areas of the resist has to be removed, which is done by homogeneously thinning down the entire resist using an (ideally) anisotropic etching process. In this way, process windows are opened to the substrate and the polymer can be used as a masking layer for further processing steps. There are an increasing number of process variations, which are mostly variants of the established thermal NIL and UV-NIL processes, particularly those using special methods of pattern transfer (e.g., reversal imprint) and hybrid processes (combinations of different processes). All the processes have their specific advantages, e.g., while UV-NIL can be performed at room temperature, hot embossing is low cost since nontransparent stamps can be used. The major characteristics of typical processes, along with those of hybrid approaches, are summarized in Table 9.2 and presented in more detail in [9.11].

9.2 Nanoimprint Process

Molding techniques based on imprint processes make use of the differences between the mechanical properties of a structured stamp and a molding material. The viscous molding material is shaped by pressing the hard stamp into it until the polymer conforms to the stamp surface. In hot embossing processes we mostly deal with thermoplastic materials whose mechanical properties can be repeatedly and reversibly changed from a solid into a viscous state by simply varying the temperature. In order to achieve a reasonable process time and yield, this is normally carried out under high pressure. Thermal NIL deals with a viscosity range which is considered as sufficiently low to enable significant squeeze flow over large distances, but high enough that bending of wafers can be used to equilibrate surface undulations of common substrates and pattern density variations in stamps. The rheological processes described here for thermoplastic materials can be considered to be similar for thermoset or UV-curable materials as long as the thermomechanical properties can be changed without affecting the chemical ones. While squeeze flow governs high-viscosity molding (where pressure is the driving force to displace the viscous material), in UV-NIL low pressure or even mold filling by simple capillary action (where surface energy controls the wetting and spreading of the viscous material) is possible if very low-viscosity resins are used. In this section we want to take a closer look at the squeeze flow of thin polymer films as used for thermal NIL, a concept which is quite general and enables an insight into possible parameter variations for process optimization. We will give a brief introduction to the theory of polymers [9.28, 29] and discuss the implications for NIL. This will enable the reader to understand rheology in NIL from a practical point of view. More fundamental questions of squeeze flow are discussed in [9.30, 31]. We conclude this section by presenting pattern transfer processes used in combination with NIL and show examples of the fabrication of simple devices. In the section on curable resists, we will introduce concepts mainly used in UV-NIL such as soft UV-NIL and droplet dispensing.

9.2.1 Limits of Molding

Resists used in NIL are polymers, which are defined by their chemical composition and physical properties. In the case of molding these are often long-chain molecules with molecular weight M_w. The polymer M_w is important because it determines many physical properties. Some examples include the temperatures for transitions from liquid to viscoelastic rubber to solid, and mechanical properties such as stiffness, strength, viscoelasticity, toughness, and viscosity. However, if the M_w is too low, the transition temperatures will be too low and the mechanical properties of the polymer material will be insufficient to be useful as a hard resist for pattern transfer. The examples given in this section are simple and meant to illustrate the specific terms needed to understand polymer behavior in molding.

It has been known for a long time that polymers can replicate topographies with high fidelity. Up to now 5 nm resolution of polymer ridges with a pitch of 14 nm has been demonstrated [9.32]. In contrast with methods such as electron-beam lithography (EBL), where nanoscale chemical contrast can be produced by local irradiation-induced chain scission, polymer chains are only moved and deformed during molding, thus retaining their chemical properties such as M_w. Molding topographic details down to a few nanometers means that single polymer chains have to deform or flow. This deformation can be illustrated by comparing the polymer with a pot full of cooked spaghetti, and instead of the viscosity change with temperature we simply take the different mobility of the filaments when wet or dry. When a water glass, representing the 10 nm pillar stamp shown in Fig. 9.2a, is pressed into this pot, single spaghetti filaments have to be moved before the glass can sink into the entangled network. If the polymers can slide along each other, the deformation can be permanent after drying and demolding. If stress is frozen, the matrix around the cylindrical hole will relax after demolding. Note that this simple example can also be used to illustrate the difference between totally amorphous and semicrystalline polymers.

A polymer is a large molecule made up of many small, simple chemical units, joined together by chemical reaction. For example, polyethylene [$CH_3-(CH_2)_n-CH_3$] is a long chain-like molecule composed of ethylene molecules [$CH_2=CH_2$]. Most artificially produced polymers are a repetitive sequence of particular atomic groups, and take the form [$-A-A-A-$]. The basic unit A of this sequence is called the monomer unit, and the number of units n in the sequence is called the degree of polymerization. The molecular weight M_w of a polymer is defined by the weight of a molecule expressed in atomic mass units (amu). It may be calculated from the molecular formula of the substance; it is the sum of the atomic weights of the atoms making up the molecule. For example, poly(methyl methacrylate) (PMMA), a classic resist material, exhibits very good resolution for both EBL and NIL. A high-M_w PMMA, typically above 500 kg/mol (also denoted 500 k), is normally used for EBL, since the development contrast between exposed and unexposed areas increases with M_w [9.33, 34]. A lower M_w, of some tens of kg/mol, is patterned in NIL, due to the strong increase in temperature-dependent viscosity with M_w [9.35]. Apart from their mobility it is expected that shorter chains, which in the case of amorphous polymers are normally present as entangled coils, can move more easily into small mold cavities. A convenient way of expressing the size of a macromolecule present as a statistical coil aggregate is by its radius of gyration R_g, which is calculated from the statistical mean path of the chain in a random-walk model using a self-avoiding walk. R_g can be measured directly in experiments by small-angle neutron scattering [9.36]. It can also be defined not only for a linear chain but also for polymers with branched structure, etc. It also equals the square of the average distance between the segments and the center of mass of the polymer [9.28], which means it can be used to give a rough estimate of the mean distance between different coils. Since entire coils are both moved and deformed, R_g will only give a rough estimate of the achievable minimum resolution of a pattern in an amorphous polymer film. As an example we take a PMMA macromolecule with M_w of 25 kg/mol. Here the chain contains $N = 250$ MMA monomer elements [$C_5H_8O_2$] with a weight of 100 g/mol each and has a total length of about $L = 80$ nm. Both with simple considerations based on the volume of a single molecule in the bulk PMMA and formulas for the random walk [$R_g = (N/6)^{1/2} \cdot (L/N)$], a R_g value of 2 nm can be calculated.

A polymeric liquid, whilst retaining the properties of a liquid, follows a rubber-like elasticity. An example is the melted cheese on a pizza. If melted cheese is dripped vertically, it flows slowly, just like a liquid. However, if it is pulled and then the tension removed, melted cheese will contract just like rubber. In other words, although melted cheese is a liquid, it also has elasticity. Substances like this, which have both viscous and elastic properties, are called viscoelastic substances. In order to calculate the flow of a fluid when an external force is applied, we need an equation relating the stress in the fluid to its deformation. This type of equation is called a constitutive equation. For example, if a polymeric liquid undergoing a steady flow is stopped, the stress does not immediately become 0, but decays with a relaxation time τ. Here τ depends strongly on the M_w of the polymer and the temperature, and can be on the order of several minutes to hours in some cases.

In the case of NIL, this relaxation has the effect that structures can still deform after molding. Considering the fact that molding is achieved by deformation of a polymer network at the molecular level, the question is how the polymer can be permanently shaped and whether the replicated structure will deform back due to internal reordering and relaxation of polymer chains.

The reduced viscosity of polymers at higher temperatures is a result of the increasing ability of the chains to move freely, while entanglements and van der Waals interactions of the chains are reduced. The glass transition of a thermoplastic polymer is related to the thermal energy required to allow changes in the conformation of the molecules at a microscopic level, and above T_g there is sufficient thermal energy for these changes to occur. However, the transition is not sharp, nor is it thermodynamically defined. It is therefore different from melting (defined by T_m), which is an equilibrium transition mostly present in polymers with crystalline entities. The glass transition is a thermodynamic transition in the sense that it is marked by discontinuities in thermodynamic quantities (Fig. 9.7) [9.37]. A distinct change from rubbery (above T_g) to glassy (below T_g) behavior is readily observable in a wide range of polymers over a relatively narrow temperature range. For thin films, however, T_g can be different from bulk values [9.38, 39].

Most of our considerations here are valid for a range of practical process parameters, as used in current hot embossing processes, where linear behavior can be assumed (Newtonian flow regime). This is in particular the case at molding temperatures well above the T_g. For thermoplastic molding, however, the T_g is only a rough indication of a temperature for fast molding. More suitable than T_g is the flow temperature T_f, which characterizes the point at which the viscosity drops to practical values needed for fast NIL (i.e., 10^3–10^7 Pa s, about 50 °C above T_g for 25 k PMMA; Fig. 9.7) [9.11].

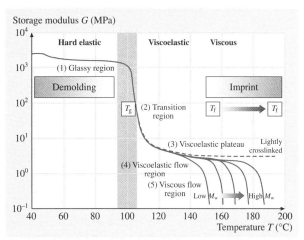

Fig. 9.7 Mechanical properties of polymers dependent on temperature, molecular weight, and cross-linking (after [9.37]). Schematic for a polymer with a T_g around 100 °C for normal process conditions. Particularly important for thermal NIL are the large drops of G at two temperatures: T_g and T_f. At T_g the thermomechanical properties between stamp and polymer become sufficiently different for repeated molding. T_f characterizes the point at which the viscosity drops to practical values needed for molding in fast imprinting

9.2.2 Squeeze Flow of Thin Films

During embossing linear movement of the stamp is transformed into complex squeeze flow of the viscous material. In the thin polymer films used in NIL, a small vertical displacement of the stamp results in a large lateral flow. The two surfaces of the stamp and the substrate have to come entirely into contact with each other and keep this contact until the desired residual layer thickness is reached. Furthermore new concepts are possible such as roll embossing and soft embossing using flexible stamps. In Fig. 9.8, the embossing of a stamp with line cavities is schematically shown.

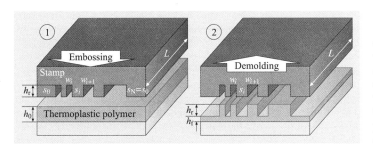

Fig. 9.8 Geometrical definitions used for the description of the flow process for a stamp with line cavities and protrusions: (1) before molding, and (2) after demolding. In the case of viscous molding, where volume conservation can be anticipated, the residual layer thickness can be calculated from geometrical parameters such as the initial film thickness and the size and density of cavities

Before embossing, the polymer film has an initial thickness h_0 and the depth of the microrelief is h_r. For a fully inserted stamp, the film thickness under the single stamp protrusions (elevated structures) with width s_i is h_f. We can calculate this specific residual layer height h_f by applying the continuity equation with the assumption that the polymer melt is incompressible (conservation of polymer volume). It can be directly deduced from the fill factor v, i.e., the ratio of the area covered by cavities to the total stamp area

$$h_f = h_0 - v h_r \quad \text{with} \quad v = \frac{\sum_i w_i}{\sum_i (s_i + w_i)}. \tag{9.1}$$

This formula only applies for rigid stamps with constant fill factor.

A simple model for the squeezed polymer flow underneath the stamp protrusion is obtained by treating the polymer as an incompressible liquid of constant viscosity, and solving the Navier–Stokes equation with nonslip boundary conditions at the stamp and substrate surfaces. According to this model, given for line-shaped stamp protrusions and cavities in [9.13, 18, 40, 41], we find the following expression, known as the Stefan equation [9.42], for the film thickness $h(t)$ underneath the stamp protrusion when a constant imprint force F is applied to the single stamp protrusion

$$\frac{1}{h^2(t)} = \frac{1}{h_0^2} + \frac{2F}{\eta_0 L s^3} t. \tag{9.2}$$

Inserting the final thickness $h_f \equiv h(t_f)$ into (9.2) gives the embossing time

$$t_f = \frac{\eta_0 L s^3}{2F} \left(\frac{1}{h_f^2} - \frac{1}{h_0^2} \right). \tag{9.3}$$

For many practical cases, where a constant pressure under each stamp protrusion $p = F/(sL)$ is assumed, this formula gives

$$t_f = \frac{\eta_0 s^2}{2p} \left(\frac{1}{h_f^2} - \frac{1}{h_0^2} \right). \tag{9.4}$$

As a direct consequence of the Stefan equation it can be seen that, at identical pressure, small (narrow) stamp protrusions will sink faster than large (wide) ones. The stamp geometry can therefore be optimized by reducing the dimensions of the protrusions. While stamps with nanopillar arrays, as shown in Fig. 9.2, would allow fast embossing of some microseconds, using standard NIL process parameters, already protrusions of some hundreds of microns would increase embossing times to

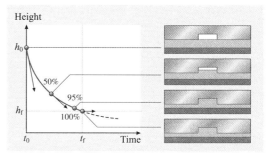

Fig. 9.9 Schematic (*right*) of the squeeze flow of a compressed polymer film into one cavity. Once the cavity is filled the stamp continues to sink but at a much slower rate (*left*), as a direct consequence of the Stefan equation

some hours. The strong dependence of the embossing time on the pressing area has the consequence that, for a fully inserted stamp relief (full contact over the total stamp area), the flow practically stops (as shown schematically in Fig. 9.9). For this case, s becomes large and flow continues only towards the stamp borders. It is also evident that there is only a weak influence of the embossing force ($t_f \propto 1/F$). At first sight there is a similar weak influence for η_0. However, the viscosity can be changed significantly by varying the temperature. For practical use, it is quite important that tradeoffs are possible between structure height, resist thickness, pressure, and temperature. For example, within certain limits, a low imprint pressure can be compensated by a longer time or a higher temperature.

For completeness we now give the expression similar to (9.3), but derived for a cylindrical stamp protrusion with radius R, i.e., with a stamp protrusion area of πR^2

$$\frac{1}{h^2(t)} = \frac{1}{h_0^2} + \frac{4F}{3\pi \eta_0 R^4} t. \tag{9.5}$$

We present an example illustrating the consequences of these equations. In Fig. 9.10a we show a stamp which contains an array of small structures in the center while the large single stamp protrusions surrounding the array dominate the sinking velocity (large s_i). The array in Fig. 9.10a is equivalent to the microcavity in Fig. 9.10b, which has the same volume as the total volume of the cavity array. This simplification can be used for the calculation of embossing times. The fill factor should be kept constant, both locally (at length scales corresponding to the cavity dimensions) and also across the wafer, i.e., for large stamp protrusions, to ensure better flow of the polymer and shorter embossing

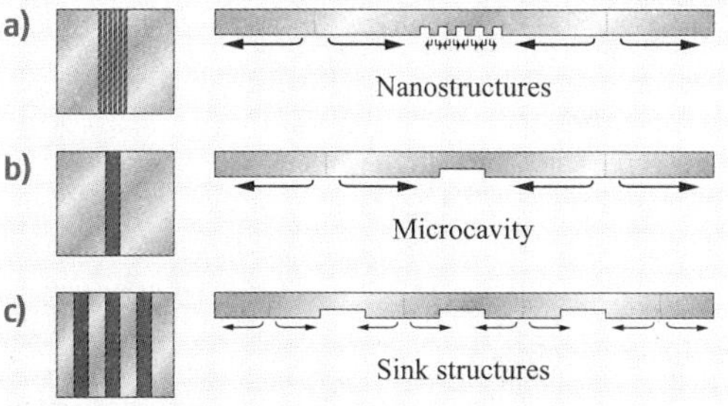

Fig. 9.10a–c Comparison of the squeeze flow for a nano- and microcavities (schematics). In the case of an array of nanocavities and a single microcavity, surrounded by large unstructured stamp areas, the polymer has to flow over large distances, thus leading to long molding time. By the introduction of additional sink microstructures, or a denser arrangement of cavities, faster and more homogeneous molding can be achieved (*left*: top view; *right*: side view)

times. For this purpose, additional protrusions or cavities can be placed in intermediate areas not needed for the device function (Fig. 9.10c), or structures can be repeated several times. We would also like to draw the reader's attention to the fact that the different sinking rates of protrusions of different sizes means that the stamp, which is normally backed by an elastic silicone mattress, can bend locally. This will result in a residual layer height that is not uniform over the entire embossing area.

The implications of squeeze flow are discussed in more detail in [9.43–46], including rheological issues [9.47–59], bending of stamps in large-area imprinting [9.44, 60–69], and the influence of vacuum and self-assembly [9.70–79]. More information can also be found in Sect. 9.2.6 about pattern transfer and Sect. 9.3.1 for NIL materials.

9.2.3 Residual Layer Thickness Homogeneity

The main difference between NIL and lithography based on exposure and development is that a residual layer below the stamp protrusions is left after demolding. As seen before, this is a result of the molding process slowing down due to the squeeze flow. For many applications, when pattern transfer has to be achieved after the embossing, it is important to determine the final residual thickness h_f of this polymer layer (Fig. 9.8) before the next process step. Furthermore it is important to know the thickness variation over the embossed area; otherwise, parts of the structure will be lost during pattern transfer. As will be shown in the following, bending of stamps has to be taken into account, as well as effects such as air inclusions, dewetting, and self-assembly of resist [9.18, 72]. In most cases a homogeneous residual layer can be achieved by optimizing the pattern design, but also by using adapted processes which create thin residual layers independent of the design. In contrast to this, pattern transfer processes which are insensitive to thickness variations have to be used, e.g., by using a resist with high etch resistance or an intermediate layer as a hard mask. The following examples demonstrate how soft and hard elements for equilibration are used to achieve homogeneous molding.

Bending of Stamps in High-Pressure Imprinting

In NIL, the stamp is often considered as a hard tool which is inflexible over millimeter distances. However, this is only true for special cases, e.g., when density and size of stamp protrusions are homogeneous over the whole stamp surface. Furthermore it is strongly dependent on the pressure used, and therefore only plays a significant role in current hot embossing processes. Local bending of some nanometers occurring due to small local variations of the stamp geometry has to be considered as the general case during hot embossing of thin films [9.18, 53, 80, 81]. Both the global movement of up to a few hundred nanometers, and the compensation of local height variations of a few tens of nanometers, are easy to implement with a compliance-type mechanism. In presses with a stiff mechanism based on hydraulic, air, and screw-driven hard stampers, the build-up of the whole stack includes the use of an elastic compliance layer (e.g., flexible graphite, rubber or teflon), which is needed for surface equilibration due to the lack of flatness of common substrates. Other concepts use an air-pressurized membrane as a soft cushion, which equilibrates local pressure variations during the sinking of the stamp in a more controlled way.

For a typical case where the grating is surrounded by a large unstructured area, stamp bending results in an inhomogeneous residual layer at the border of the grating. Figure 9.10a shows such a case, in which a grating area, typically of some square millimeters, is surrounded by a large nonstructured area. In the ideal case of a totally rigid stamp, the final thickness would be determined by the fill factor of the grating averaged over the whole stamp area, which could be calculated by the simple rule of conservation of polymer volume. This can only be achieved if the polymer can flow easily over large distances; otherwise, parts of the grating will not be filled. In the other extreme case of a totally flexible stamp and low lateral transport of polymer, both stamp areas could be calculated independently. While the center of the grating would sink to half of the depth of the cavities (assuming a fill factor of 50%), in the nonstructured area almost no sinking would occur. In between, at the border of the grating, the stamp tries to accommodate this mismatch by bending. Depending on the thickness and elastic behavior of the stamp, as well as the design of the stamp, characteristic distances can be calculated.

In many cases rules of thumb for design and process optimizations are sufficient for achieving homogenous molding. However, for more complex cases, simulations are needed to predict the filling of both small and large structures in the vicinity of one another. Furthermore the dynamic behavior of filling has to be taken into account. The task becomes even more challenging if embossing over topography has to be considered. In this case, a planarization layer can be used below the NIL resist.

Resist Density Adaptation in Low-Pressure Imprinting

UV-NIL processes are performed at room temperature, at which resist precursors are present as liquid films or droplets. When using hard stamps as in step and flash imprint lithography (SFIL, a step and repeat UV-NIL process [9.26, 27]), or jet-and-flash imprint lithography (JFIL, for single step wafer scale imprint), a homogeneous residual layer thickness can be achieved by locally varying the amount of liquid resin necessary to fill the cavities of the stamp. Particularly suitable for this is an array of droplets formed by dispensing low-viscosity UV curable monomer (with $\eta_0 \leq 5$ mPa s) onto the substrate surface prior to imprinting. By contact of the stamp with the dense droplet array, a continuous film is formed by capillary action. To handle pattern density variations, the drop-on-demand UV-NIL process at atmospheric environmental pressure has been developed [9.82].

Soft Lithography with Conformable Stamps in Low-Pressure Imprinting

The forces on a stamp protrusion with liquid resists are induced by capillary action rather than by squeeze flow and are therefore low. Therefore in UV-NIL, compliant stamps made from elastomeric materials, e.g., polydimethylsiloxane (PDMS), a UV-transparent rubber, can also be applied. The concept of layered stamps – a thin PDMS relief coated on a harder substrate – is particularly useful in full wafer concepts. It combines the complementary mechanical properties of a soft surface relief for the achievement of local conformal contact and a rigid but bendable backbone, which can be used for mounting and alignment. A process working with moderate resist viscosities (with $\eta_0 = 50$ mPa s and below) for providing liquid films by spin-coating has been developed and can be applied at reduced environmental pressure [9.69, 83].

9.2.4 Demolding

During demolding the rigid stamp is detached from the molded structure, which can be done in a parallel way when using small rigid stamps, or by delamination if thin wafer-like substrates are used. If fully molded, the thickness profile in the resist exhibits the inverse polarity of the relief of the stamp surface. The demolding process, also called de-embossing, is normally performed in the frozen state, i.e., when both the mold and molded material are considered solid. For thermoplastic materials this happens at temperature well below T_g, but high enough that frozen stress due to thermal contraction does not lead to damage during demolding. In cases in which the resist is cured before demolding, i.e., cross-linked by exposure or heat, demolding can take place at temperatures similar to the molding temperature. A successful demolding process relies on a controlled balance of forces at the interfaces between the stamp, substrate, and molded polymer film. Therefore mechanical, physical, and chemical mechanisms responsible for adhesion have to be overcome. The following effects have to be avoided or reduced [9.11, 18, 53, 80, 81] (Fig. 9.11):

- Undercuts or negative slopes in the stamp may lead to mechanical interlocking of structures, which in the frozen state are elastically elongated and deformed before ripping. Sidewalls with positive or at best vertical inclination are a prerequisite for demolding without distortion.
- Friction due to surface roughness may occur during the sliding of molded structures along vertical cavity walls. The effect of this can only be overcome if the surface of the molded material is elastic and enables gliding of the wall without sticking.
- The enlarged surface area of the patterned stamp leads to an increase of hydrogen bridges and van der Waals forces, or other chemical bonding effects due to ionic, atomic, and metallic binding. This effect

can only be overcome if the stamp surface can be provided with sufficient antiadhesive properties.

The most critical point is that demolding forces largely depend on the geometry of the mold, and the overall design of a stamp structure has to be taken into account. Therefore structures with high aspect ratio may be more prone to ripping, and if many neighboring structures exert high forces on the underlying substrate, whole areas of resist may be detached from the substrate surface. Antiadhesion layers on the mold can reduce friction forces, but have to be thin and durable. In thermal NIL the expansion coefficient of the substrate $\alpha_{substrate}$ and of the stamp α_{stamp} should be similar, to avoid distortion due to mechanical stress induced by cooling. In the case of very thin polymer layers, the lateral thermal expansion of the resist is determined by the substrate. For structures with higher aspect ratio the demolding temperature T_{demold} should be well below T_g, to enable the demolding of a hardened resist without distortion, but as near as possible to T_g, because the stress induced by thermal shrinkage should not exceed a maximum value in critical areas where structures tend to break.

9.2.5 Curing of Resists

Curing by UV exposure, by thermal treatment or by chemical initiation is a way to cross-link polymers and make them durable for demolding [9.24, 25, 27, 82–96]. A high reaction speed, as caused by a high exposure dose, high initiator content or curing at high temperatures, leads to fast but weak cross-linking, whereas a slow reaction leads to highly polymerized, tougher materials because the slow polymerization enables a more complete process. As shown in Sect. 9.1.2, various process strategies have been developed. In most of them the curing step is independent of the molding step, and can be initiated by light or a specific temperature after molding is complete. Because curing involves a change in the physical conformation of the polymer, it always goes along with volumetric shrinkage of the polymer; e.g., acrylate polymerization is known to be accompanied by volumetric shrinkage that is the result of chemical bond formation. Consequently, the size, shape, and placement of the replicated features may be affected. In the following the main processes which involve curing are presented in more detail:

- In the UV-NIL process, as used in SFIL [9.26, 27], the resist is cured after molding but before demolding of the stamp. The process relies on the photopolymerization of a low-viscosity, acrylate-based solution. Shrinkage was found to be less than 10% of total volume in most cases. The current liquid is a multicomponent solution. The silylated monomer provides etch resistance in the O_2 transfer etch, and is therefore called the etch barrier. Cross-linker monomers provide thermal stability to the cured etch barrier and also improve its cohesive strength. Organic monomers serve as mass-persistent components and lower the viscosity of the etch barrier formulation. The photoinitiators dissociate to form radicals upon UV irradiation, and these radicals initiate polymerization.

- If a solid curable resist exhibits thermoplastic behavior, it can be molded at an elevated temperature and then cross-linked, either before or after demolding. The advantage of the process is that low-M_w resists with low T_g can be provided, which can be processed at moderate temperatures. However, before pattern transfer, hardening is often necessary. They can also be used for mix-and-match with PL or for polymeric stamp copies.

- Thermoset resists can be cross-linked by heat. Here it is of advantage that the temperature for molding is lower than the curing temperature. Then the structure is first molded and then heated to its cross-linking temperature to induce cross-linking, before the stamp is demolded from the hardened surface relief.

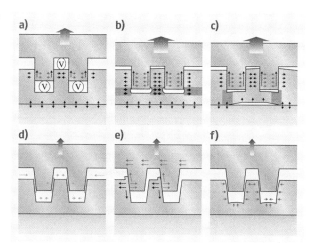

Fig. 9.11a–f Demolding issues: (**a**) generation of vacuum voids (V), (**b**) elongation and ripping of single structures, (**c**) ripping of resist from substrate, (**d**) penetration of air into voids (inclined sidewalls), (**e**) shrinkage and generation of rims, and (**f**) relaxation of frozen-in strain (after [9.11])

More information about curing and multilayer resists can be found in Sect. 9.3.

9.2.6 Pattern Transfer

In many cases the lithographic process is only complete when the resist pattern is transferred to another material. This process, in which the resist is transformed into a patterned masking layer, allows the substrate to be attacked by plasma, etching solutions, electroplating, deposition of materials, and other substrate-altering processes. A unique advantage of molding instead of exposure is that complex stamp profiles, such as stair cases, V-grooves, and pyramids, both convex and concave, can be replicated. They can be used for the generation of 3-D structures such as for T-gate transistors or contact holes, or serve for stepwise etching of underlying layers with variation of the opening width. As long as undercuts and 3-D patterning are not necessary, in most cases this pattern transfer is therefore similar to in EBL. However, in this section we emphasize methods where NIL has some specific process advantages over conventional lithographic methods, or where the use of NIL implies some major changes in the fabrication process or properties of the devices:

- In NIL, etching is used for both the removal of the residual layer and the pattern transfer of the resist pattern to the underlying substrate [9.7, 97–106]. In the first case the polymer layer has to be homogeneously thinned down until openings to the underlying substrate are generated. This is also called a window-opening or breakthrough etch. In the second case the thickness contrast of the remaining polymer is used to mask the substrate against the etching medium. Both processes have to be highly anisotropic, i. e., during the transfer step the lateral size of the structure has to be preserved, including the slope of the original pattern. Apart from opening windows using reactive-ion etching (RIE), other pattern transfer strategies have been found which circumvent the residual layer problem.
- Lift-off is a patterning technique adding thin layers of a solid material (e.g., metal) locally to the window openings in the resist [9.107–114]. Undercuts, as can be generated in PL and EBL, are a prerequisite for good lift-off. However, in NIL, where sidewalls are at best vertical, a high thickness contrast (aspect ratio) of the structures is needed. Lift-off resists are a means to generate defined undercuts using a bilayer resist system, by selectively dissolving a sacrificial bottom layer through the structured openings of a top layer.
- Electroforming and electroplating, like lift-off, are processes that add material to the areas not covered by the resist [9.97, 115, 116]. Electroforming provides a good alternative to the lift-off process because metal structures can be generated with considerable height and good surface quality. If a conductive seed layer is deposited below the resist, during electroplating the metal layer starts to grow from within the window regions and conforms to the

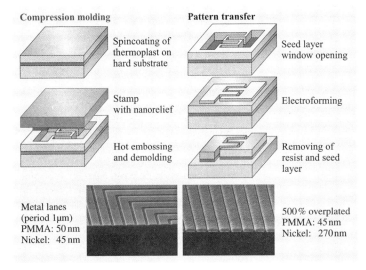

Fig. 9.12 NIL and electroforming: Electrode structures have been fabricated in Ni by using a plating base of Cr and Ge. After plating on top of the Ge, both layers of the plating base can be etched using RIE (Cr: chlorine chemistry; Ge: SF_6). Even with 500% overplating, the thick electrodes stay separated (after [9.97])

outlines of the cavities in the resist. Depending on the extent of electroplating, the structure height can be either preserved or increased.

Some of the examples presented here for pattern transfer already give insight into simple demonstrators, particularly when the application is based on a simple pattern transfer or NIL is used as the first patterning step of a nonstructured surface. Examples of applications are:

- Large-area metal gratings, as needed for polarizers or interdigitated electrode structures, can be fabricated by etching of a metal layer, lift-off or electroplating; in Fig. 9.12, e.g., electrode structures have been fabricated in Ni by using a plating base of Cr and Ge. After plating on top of the Ge, both layers of the plating base can be etched using RIE [9.97].
- Surface patterns with chemical contrast can be generated by locally depositing silanes onto a SiO_2 surface by lift-off (Fig. 9.13). By patterning molecules with biofunctionality, integrated biodevices such as biosensors and biochips can be fabricated. In [9.113] the combination of NIL and molecular assembly patterning by lift-off (MAPL) is demonstrated.
- By etching, the NIL process can be used to draw copies from a stamp original [9.44, 99, 115]. Often the deposition of a metal layer for subsequent etching is used as a hard mask to generate copies with an enhanced aspect ratio (Figs. 9.14 and 9.18).

More specific applications, where one or several of these pattern transfer processes were used, are presented in more detail in Sect. 9.4.

9.2.7 Mix-and-Match Methods

Mix-and-match approaches are used to combine the advantages of two or more lithographic processes or simply to avoid their mutual disadvantages [9.117–126]. It is also a way to improve throughput and reliability; e.g., since the fabrication of large-area nanostructures is often costly, the definition of microstructures can be done with PL, while the nanopatterning of critical structures in small areas is done by NIL. In many cases NIL would be used as the first process step and, by adding alignment structures along with the nanopatterns, the less critical structures can be added after the NIL step using PL with an accuracy given by the mask aligner (in the range of $1\,\mu m$). NIL allows different variants of mix-and-match:

- In a sequential approach the second resist process is added to the first structured pattern after pattern transfer. Specific problems such as overlay or nonflat surfaces have to be solved. An example of mix-and-match can be seen in Fig. 9.14, where a nanoporous membrane was fabricated by NIL (pore definition) and PL (windows for silicon etching and membrane release) [9.106].
- By using a UV-sensitive thermoplastic resist, the nanopattern can be created by NIL and the micropattern added into the molded resist by PL in subsequent patterning steps. Thus, using this bilithographic step, the pattern transfer can be done for the whole structure after the resist structuring is complete. The resists used for this purpose are cross-linked during exposure, which makes it possible to dissolve the unexposed areas [9.117].

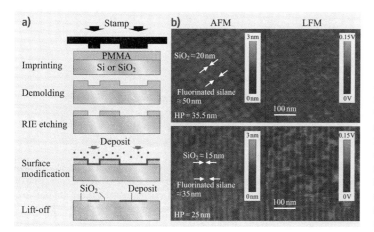

Fig. 9.13a,b NIL and lift-off for the generation of nanopatterns with chemical contrast. (**a**) Process scheme for local silane deposition from gas phase and (**b**) AFM/LFM (atomic/lateral force microscope) images for chemically patterned surfaces modified with a fluorinated silane, showing sub-50 nm areas with hydrophobic (silane) and hydrophilic (SiO_2) properties (after [9.110–112])

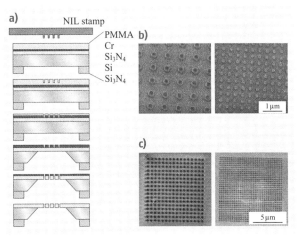

Fig. 9.14a–c Mix- and match of NIL and silicon micromachining: (**a**) process scheme for the fabrication of nanopores in a Si_3N_4 membrane. SEM images (**b**) of the NIL stamps (pillars) and (**c**) the corresponding nanopores (after [9.106])

- A specific mix-and-match approach is possible if UV exposure is done before the stamp is detached from the molded resist. This is possible when parts of the stamp are transparent (e.g., the recessed areas), while the protrusions are coated with an opaque layer (e.g., a metal masking layer such as that used for etching the stamp structures) [9.119]. This makes it possible to cross-link the thick resist areas while the residual layer can be dissolved.

9.2.8 Multilayer and Multilevel Systems

Multilayer resist systems are used if the etching selectivity of a masking layer has to be enhanced, e.g., for the fabrication of high-aspect-ratio structures, undercuts have to be generated, e.g., for lift-off, or a planarization layer has to be employed for printing over topography [9.127–134].

The most important application of double resists is for low-pressure processes such as UV molding (Fig. 9.15). For prestructured substrates with topography, a planarization layer is needed, because the low pressure of below 1 bar is often not sufficient to achieve conformal contact of the transparent mask with the non-flat substrate surface; otherwise parts of the resist stay unmolded. Multilayer resist approaches with a thick polymer planarization layer on top of the substrate require complex processes with multiple steps but also entail deep etching steps to etch through the thick planarization layer, which often degrades the resolution and fidelity of the pattern. Bilayer resists are also used for better lift-off. For this purpose lift-off resists (LOR) have been developed [9.135]; they are coated below the top layer and can be selectively removed by wet development through the patterned top layer. The developers used are adapted to generate undercuts in LOR layers of some tens of nanometers up to some microns. Then even a curable resist which is cross-linked (equivalent to a negative resist) can be used as a top layer, while the sacrificial bottom layer makes it possible to release the top layer as well as the metal layer used for lift-off.

Often top layers with high etching resistance, e.g., silicon-containing resists (similar to hardening by silanization), are chosen for UV-NIL. After molding the top layer, the pattern is transferred to the underlying planarization (transfer) layer. The top layer can be kept thin, while the etching depth can be further increased by choosing a thick bottom layer. Normally the tone of a stamp pattern is inverted when etching

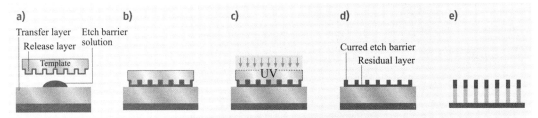

Fig. 9.15a–e Process scheme of UV imprinting and pattern transfer, using a double layer (also called direct SFIL). The molded top layer, also called the etch barrier, is coated on a transfer layer, which serves as a planarization layer. It has also antireflective properties for the UV exposure through the stamp. (**a**) dispensing of viscous resist droplet, (**b**) imprint, (**c**) UV-exposure and curing, (**d**) demolding of hardened resist, (**e**) residual layer etch and transfer into bottom layer (breakthrough etch/window opening)

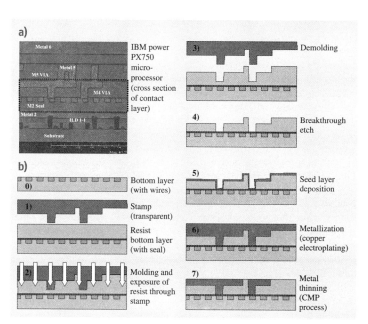

Fig. 9.16a,b Modified SFIL process proposed by Sematech to replace a dual top hard damascene process for copper contact plating by a two-tiered stamp [9.27, 137]. (**a**) *left* (*top*): SEM of a contact layer of a microchip (cross section) with interconnecting copper layers, (**b**) process scheme (Source: *Trybulla*, Sematech, [9.137])

is used for pattern transfer. The tone can be preserved if another tone reversal process is used. This can be achieved by imprinting a pattern into the thick transfer layer, and by spin-coating a silicon-containing resin on top of it. If the top residual layer of the planarized film is etched away, the high etch resistance of the silicon remaining in the trenches of the bottom layer will enable the patterning of the transfer layer with reversed tone. This strategy has the advantage that stamp contamination containing silicon residues is avoided [9.136].

The 3-D patterning capability of NIL makes it possible to reduce the number of process steps in contact layer fabrication of microchips by using innovative pattern transfer. The connection of the transistors is done using several levels of lateral wires, each contacted vertically by through-holes. This contact layer of a chip is fabricated using lithography and copper electroplating. For the wiring scheme of a chip, as shown in Fig. 9.16, eight levels of wiring layers are needed, each of which is done in a so-called dual hard damascene process. A process has been proposed which reduces the number of process steps necessary for one level from 16 to 7 [9.137]. A two-tiered stamp with three height levels makes it possible to pattern the through-holes as well as the wires in one step [9.138]. In this way, several exposure steps can be replaced by a single imprint with patterns of different residual polymer layer thickness. In total, the reduction from 128 process steps down to 56 results in a cost reduction that justifies the introduction of a new technology and serves as an example that the 3-D pattern capability can be a decisive argument over resolution for the introduction of NIL into chip manufacturing. Figure 9.16 shows the pattering scheme for one level of the contact layer of an IBM power PC microprocessor. Obducat has used a similar process for the generation of micrometer-sized contact holes in printed circuit boards (PCB).

9.2.9 Reversal NIL

In contrast to NIL, in reversal NIL the resist is patterned either directly onto the stamp or onto an auxiliary substrate, e.g., by spin-coating, casting or imprint, and then transferred from the mold to a different substrate by bonding. Thus patterned resist structures are obtained as in direct NIL, and even embedded channels can be created. The concept is well presented in [9.139–143]. In reversal NIL it is possible to transfer patterns onto substrates that are not suitable for spin-coating or have surface topographies. However, complete transfer does not only depend on a good balance of the surface energies, but also on the pattern density and roughness of the structures. As an example, embedded channels generated by reversal NIL are shown in Fig. 9.17 [9.140].

9.3 Tools and Materials for Nanoimprinting

Mechanical nanofabrication techniques based on molding need tools and materials with matched mechanical properties. The mold has to be made from a material which is sufficiently hard to sustain at least one processing cycle. From the viewpoint of mass fabrication, a mold is considered as a tool which survives the molding process unaltered and uncontaminated, and thus can be reused many times after each molding step. In this way many identical replicas can be drawn (copied) from one mold. Due to the conformal molding, the surface of these copies is the negative structure of the original (inverted polarity). Therefore a true replica of the mold is generated, when a negative is again molded into a positive structure. Here, we use the terms replica and copy in the more general sense that also negatives are considered as true copies of an original.

As the terms imprinting, embossing, molding, and replication are often used for the same process, different names for the replication tools exist depending on their origins: mold or mold insert for those coming from polymer processing; master or stamp (stamper) from CD fabrication; and template, mask, and die from the lithography community.

In this section we will have a closer look at concepts for tools, machines, and processes used for NIL. We will start with a discussion of resist materials for NIL, and then proceed with materials used for stamps. We describe fabrication methods, both for original stamps and for stamp copies, and the use and application of antiadhesive coatings. We will then present concepts for NIL machines, and how a homogeneous pressure distribution is achieved for nanoreplication. For thermal imprinting as well as UV imprinting single-step wafer-scale processing and step-and-repeat approaches have been developed. The aim is to make the reader familiar with concepts rather than presenting machines and materials sold on the market.

9.3.1 Resist Materials for Nanoimprinting

Resists used for NIL are either used as an intermediate masking layer for the substrate or as a functional layer

Table 9.3 Properties of thermoplastic polymers for thermal NIL

Material (other names)	Solvent	Glass transition temperature T_g and imprint temperature $T_{imprint}$ (°C)	Viscosity at typical imprint temperature (Pa s) (Fig. 9.5)	Comments
Poly(methyl methacrylate) (PMMA) [9.135, 144–150]	Chlorobenzene, safe solvents	100 (at 160–190)	3×10^4 (25 k at 180 °C)	The *classic* NIL resist, refractive index $n = 1.49$ [9.6, 18]
Polystyrene (PS) [9.145]	Toluene	104 (150–170) [9.21]	1.8×10^3 (58 k at 170 °C)	Integrated optics, biology, $n = 1.59$ [9.147]
Polycarbonate (PC) [9.148, 149]	Cyclohexanone [9.23, 41, 150], 1,1,2,2-tetra-chloroethane	148 (160–190)		Integrated optics, $n = 1.59$ high etching resistance [9.147]
mr-I T85 [9.144]	Toluene [9.151]	85 (140–170) [9.152]	2×10^4 (at 170 °C) [9.151]	Chemically resistant, low water absorption, highly transparent, $n = 1.497$ [9.153–155]
mr-NIL 6000 [9.144]	Safe solvent	40 (100–110)	2×10^3 (at 100 °C)	UV-curable, low-T_g NIL resist for mix-and-match, multilevel patterning [9.156, 157]
mr-I 7000 [9.144] and E	Safe solvent	60 (125–150)	3×10^3 (E grade at 140 °C)	Low T_g NIL resist, $n = 1.415$
mr-I 8000 [9.144] and E	Safe solvent	115 (170–190)	7×10^4 (E grade at 180 °C)	$n = 1.415$, NIL resist with thermal properties similar to PMMA, but higher etch resistance
mr-I 9000 [9.144] and E	Safe solvent	65 (140–160)		$n = 1.417$, thermocurable NIL resist [9.90]
NEB22 [9.158]	PGMEA [9.159]	80 (95–130) [9.160]		Negative EBL resist based on poly(hydroxystyren), high etch resistance in fluoro- and chloro-based plasmas [9.150], low M_w (3k)

Table 9.4 Comparison of different materials for stamps

Material	Young's modulus (GPa)	Poisson's ratio	Thermal expansion (10^{-6} K^{-1})	Knoop microhardness (kg mm^{-2})	Thermal conductivity (W m^{-1} K^{-1})	Specific heat (J kg^{-1} K^{-1})
Silicon (Si)	131	0.28	2.6	1150	170	705
Fused silica (bulk) (SiO$_2$)	73	0.17	0.6	500	1–6	700
Quartz (fused silica)	70–75	0.17	0.6	> 600 (8 GPa)	1.4	670
Silicon nitride (Si$_3$N$_4$)	170–290	0.27	3	1450	15	710
Diamond	1050	0.104	1.5	8000–8500	630	502
Nickel (Ni)	200	0.31	13.4	700–1000	90	444
TiN	600	0.25	9.4	2000	19	600
PDMS	0.00036–0.00087	0.5	310	22	0.15	1460

for a specific application. Both the processing properties as well as those for the final application purpose have to be considered. Many of the resists, as used for PL and EBL [9.161, 162], exhibit thermoplastic behavior. A typical example is PMMA, a regular linear homopolymer, with a short side-chain. It is used as a high-resolution standard material for EBL and also as a bulk material for hot embossing and injection molding. For a long time it has been known that sub-10 nm resolution can be achieved [9.35]. PMMA is a low-cost material, and available with different M_w values. It is compatible with other cleanroom processes, exhibits good coating properties using safer solvents, and can be coated from solution to a thickness ranging from 20 nm to several µm. It has well-characterized optical, mechanical, and chemical properties, and proved reliability in many different applications. When used as an etching mask, e.g., for Si, it exhibits a sufficient, but not high etching resistance. The glass-transition temperature T_g of PMMA (105 °C) is low enough to enable molding at temperatures below 200 °C, but high enough to ensure sufficient thermal stability in etching processes. Acrylate-based polymers can also be used with cross-linking agents. A further advantage of PMMA is that the process window, defined as the temperature range between the lower temperature for viscoelastic molding where relaxation due to frozen-in strain has to be expected and the higher temperature where the viscosity is so low that the onset of capillary bridges (viscous fingering) will affect the residual layer homogeneity [9.19], is quite large. This enables imprinting to be optimized by using tradeoffs between different parameters according to Stefan's equation. Apart from PMMA, during the first 10 years of NIL, a number of resists have been developed and characterized; they exhibit different T_g values, and have been optimized for greater etching resistance or better flow at lower temperatures. In Table 9.3 we give an overview of NIL resists with references to further information on these materials. Further information can be found in [9.90, 91, 163–170].

UV-curable NIL materials are composed of a mixture of monomers (or prepolymers) and a suitable photoinitiator, and often chemicals are added which decrease the effect of radical scavengers on photopolymerization [9.11, 48, 171–178]. Immediately during contact of the stamp with the liquid mixture, filling of the mold starts by capillary forces, which pulls the stamp towards the substrate. Therefore, the general strategy is that low viscosities are needed for both rapid dispensing and filling of mold cavities. Thin resin layers on top of a thicker transfer layer are used to achieve

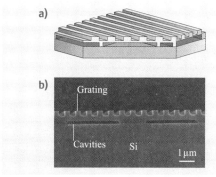

Fig. 9.17a,b Reverse microfluidic channels fabricated by double-sided imprinting: (a) 3-D schematic of a resist with a top grating and embedded channels. SEM micrographs of cross-sections of imprinted nanofluidic channels: (b) 3000 nm (width) × 200 nm (height) channels, with a 700 nm-pitch grating on top (after [9.140])

a homogeneous film thickness. Cross-linking and photopolymer conversion is adapted to achieve high curing speed and high etch resistance in the following breakthrough plasma etching process.

In UV-NIL a chemical reaction between the stamp and resist cannot be excluded. Small feature sizes along with high silicon content and a large degree of crosslinking make any residual imprint polymer left on the mold almost impossible to remove from the template without damaging the expensive quartz template. It has been shown that a fluorosilane release layer applied to a UV-NIL stamp undergoes attack by acrylate, methacrylate, and vinyl ether UV-curable resist systems, indicating that its degradation is intrinsic to the chemistries involved. Future resist chemistries have to satisfy the criterion of low reactivity toward antiadhesive coatings and stamp materials [9.179, 180].

9.3.2 Stamp Materials

Not only the mechanical but also the optical and chemical properties are important when choosing a stamp material for NIL. Critical mechanical parameters and their implications for NIL are hardness and thermal stability (lifetime and wear), thermal expansion coefficient and Poisson's ratio (dimension mismatch leading to distortions during demolding), roughness (higher demolding force and greater damage), Young's modulus (bending), and notch resistance (lifetime and handling). Issues related to fabrication are processibility (etching processes, selectivity, cleanroom environment) and surface quality (resolution). Use of a stamp material in a NIL process is determined by additional properties such as transparency, conductivity, antisticking properties (with or without an antiadhesive coating, e.g., a covalent coating), availability and cost (standard materials and sizes, tolerances, processing equipment and time), and how easy it is to employ in NIL (e.g., fixing by clamping, thermobonding, gluing). In Table 9.4 we give a brief overview of the mechanical and thermal properties of materials used for stamps. Further information can be found in [9.27, 100, 156, 181–194].

9.3.3 Stamp Fabrication

Any kind of process generating a surface profile in a hard material can be used to fabricate stamps for NIL. The most common lithographic processes are based on resist patterning with subsequent pattern transfer. Therefore the requirements for these processes such as resolution, aspect ratio, depth homogeneity, sidewall roughness, and sidewall inclination are similar to the processes presented before in this chapter. For highest resolution, both serial and parallel fabrication methods are available, however, with different area, throughput, and freedom of design. The processes are standard processes for nanolithography, which also can be used directly for patterning. When using them for the fabrication of stamps, apart from higher throughput, greater flexibility and reproducibility can be achieved. Using stamp copies instead of the original is a way to enhance the lifetime of a stamp, simply because the original is reserved for the copying process. There are different methods to generate copies from hard masters with proved resolutions below 100 nm:

- Electroplating is a commercially successful method to copy an original into a metal replica. The nickel shims used in CD manufacturing support tens of thousands of molding cycles without significant wear. The original, a patterned resist or etched relief on a glass master, is often lost during the transfer to nickel, therefore only after a first-generation nickel copy is drawn can further generations be repeatedly copied from it.

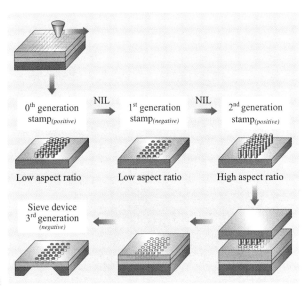

Fig. 9.18 Process chain from stamp origination to application: the example of a porous membrane chip as shown in Fig. 9.14. The low-aspect-ratio original stamp fabricated by EBL and RIE is transferred into a high-aspect-ratio stamp by two consecutive NIL copying steps, providing increased lifetime of the original and greater flexibility

- Using the hard master with an etched surface relief directly as a mold is a straightforward approach if the mechanical setup allows or favors the use of silicon (Si) or fused silica (SiO_2). Stamp copies can be fabricated using NIL and subsequent pattern transfer (Fig. 9.18). Molds made from silicon wafers are well suited to use as stamps in NIL, and have even shown their mass-fabrication capability in CD injection molding. For UV-NIL such as SFIL, molds were successfully made using standard mask blanks (fused silica).
- As a third solution a polymer or sol–gel layer with an imprinted surface relief can be directly used as a replication tool. This is possible if the thermomechanical replication process does not exert high forces on the relief structure. Resist hardened by light, heat or by chemical initiation may support high temperatures and can be used repeatedly in NIL. However, the lifetime of polymeric molds is still low, and good solutions for antiadhesive coatings have to be found.
- Hybrid molds use different materials for the surface relief and the support. They consist of a substrate plate as a mechanical support covered with a thin polymer layer with nanostructured relief. In the case of NIL they have the advantage that a substrate material can be chosen with thermomechanical properties adapted to the substrate to be patterned. Furthermore this approach is useful if thin flexible substrates are needed.

The methods differ mostly in the properties of the materials used for the stamps (mechanical robustness, thermal expansion coefficient, transparency, fabrication tolerance) and the surface properties of the patterned relief (antiadhesive coating possibility). Although for many applications electroplating of metal molds is favored because of their great flexibility and robustness compared with silicon, the effort to fabricate high-quality mold inserts with defined outlines is often only justified for production tools.

9.3.4 Antiadhesive Coatings

One of the most important tasks for NIL is to provide stamps with good antisticking surface properties [9.195–202]. The stamp surface should allow the molded surfaces to detach easily from the mold, and once released, provide a low friction resulting in a continuous vertical sliding movement without sticking. Nanoscopic interlocking of structures caused by sidewall roughness should be elastically absorbed by the molded material, while the surface maintains its antisticking properties. Because the molded polymer film is squeezed between the two surfaces of stamp and substrate, they need to exhibit opposite surface properties. The adhesion at both interfaces must be different to an extent that, while the polymer film adheres perfectly to the substrate surface, the stamp can be separated from the structures without any damage at any location of the stamp. If the stamp material does not exhibit good antisticking properties to the molded material, the stamp has to be coated with a thin antiadhesive layer. A low-surface-energy release layer on stamp surfaces not only helps to improve imprinting quality, but it also significantly increases stamp lifetime by preventing surface contamination. An antiadhesive coating has to be chemically inert and hydrophobic but at the same time allow filling of the mold cavities when the polymer is in its viscous state.

One of the major advantages of using Si or SiO_2 stamps for NIL is that they can be coated with antisticking films using silane chemistry. Damage to the molded structure during demolding is highly dependent on the quality of the antiadhesive layer. Fluorinated trichlorosilanes with different carbon chain lengths are commonly used due to their low surface energy, high surface reactivity, and high resistance to temperature and pressure. They support multiple long embossing sequences with repeated temperature cycles higher than 200 °C. Currently it seems that, as long as mechanical abrasion can be avoided, the silanes match the *normal use lifetime* of a Si stamp, which is some tens of cycles for NIL in a laboratory environment or thousands if automated step-and-repeat imprinting processes or injection molding processes are used. However, the low-energy surface that a fluorosilane layer presents is not unreactive, and it is rapidly and easily degraded during use, particularly at high temperatures (above 200 °C) and by chemical attack by the abundant free radicals present in curable resists. Therefore not only the chemistry of resists has to be taken into account to improve the lifetime of stamps, but strategies such as recoating have to be considered. Apart from silicon wafers, which have the advantage that they are suitable for standard cleanroom processing, other materials to be used as NIL stamps, e.g., nickel (Ni) shim or duroplastic polymers, can also be coated with silanes if an intermediate SiO_2 layer is deposited onto the materials. The silane coating can be performed by immersion in a solution of iso-octane, or by chemical vapor deposition (CVD), either at ambient pressure by heating the silane on a hot plate

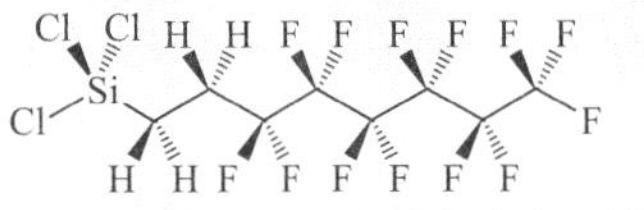

F$_{13}$-OTCS

(Tridecafluoro-1,1,2,2-tetrahydroOctyl)TriChloroSilane

Fig. 9.19 Molecular structures of a fluorinated silane with a reactive trichlorosilane head group and a long alkyl chain with fluorine substituents (length about 2 nm). The silane binds covalently to the silicon oxide of the stamp surface and is used as the standard coating of silicon stamps in NIL

or by applying a moderate vacuum of some mbar. One of the most prominent advantages of the vapor deposition method is that it is not affected by the wetting ability of a surface, and that it is suitable for stamp surfaces with extremely small nanostructures.

A commercially available silane that is used is shown in Fig. 9.19. F$_{13}$-OTCS = (tridecafluoro-1,1,2,2-tetrahydrooctyl)-trichlorosilane is the standard material for antiadhesive coatings on silicon (ABCR SIT 8174) [9.203].

9.3.5 Imprinting Machines

NIL can be carried out using three different types of machines: single step (Fig. 9.20a,b), step-and-repeat (Fig. 9.20c), and roller imprinting. An imprinting machine needs a precise pressing mechanism with high requirements on mechanical stiffness, uniformity, and homogeneity over large areas [9.25, 68, 204–209]. At the same time it should adapt to local variations of pressure and temperature, due to imperfections and tolerances in stamps and substrates, and simply because the stamp protrusions are inhomogeneously distributed. In molding of microstructures, where deep channels with lateral and vertical sizes in the range of 50 μm have to be molded, the stamps are made stiff, and precise reproducible vertical piston movements within some tens of μm have to be realized with good fidelity. NIL would need precision of a few tens of nm, which does not correspond to the tolerances of some μm usual for substrates and tools. Therefore NIL stamps have to be flexible, and must be made to adapt to small vertical deviations from an ideally flat surface over a long lateral range. These deviations are the dimensions of the fabrication tolerances of common templates for stamps and substrates, and the density variations of the stamp surface relief.

Embossing machines generate a desired pressure pattern over the total area of the stamp. High throughput for manufacturing devices at the full wafer scale can be achieved either by parallel patterning of large areas or by fast repeated patterning using a semiserial stepping process. The pressure field can also be applied sequentially by using a rigid but stepped embossing mechanism, as used in millipede stamps (Chap. 45), or a continuously scanned pressure field, as used in roll embossing (Fig. 9.21). In all cases a defined area of the molding material is sandwiched between the solid stamp and substrate, which are backed by a pressing mechanism. The major differences lie in the fact that single-step imprinting processes might not be easily transferable to continuously repeated imprints, where

Fig. 9.20a–c Three examples of NIL presses. (**a**) Simple hydraulic press, with temperature-controlled pressing plates. (**b**) Semiautomated, hydraulic full-wafer NIL press, based on an anodic bonder. (**c**) Automated step-and-flash UV-NIL production tool

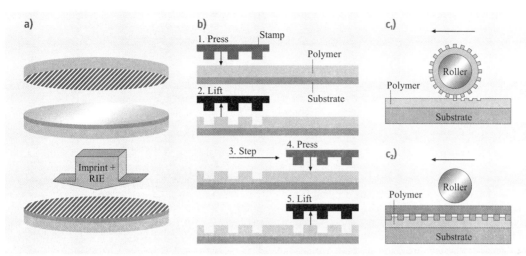

Fig. 9.21a–c Outline of the three most common types of NIL machines: (**a**) full-wafer parallel press, (**b**) step-and-repeat press, and (**c**) two roll-embossing setups

previously structured areas should not be affected by imprints in the close vicinity (e.g., reheating of already molded resist over T_g in thermal NIL and cross-linking of resist outside the stamp area in UV-NIL). In PL, stepping was needed because of the limitation of the maximum field size to be exposed, and because the continuous reduction of structure sizes and diffraction effects was only possible by optical reduction of the masking structures into the resist by high-resolution optics. Furthermore this enabled a noncontact process to be established, while 1 : 1 imaging of a mask structure would have lead to unwanted reduction of the proximity gap.

Single-Step Wafer-Scale NIL

Single-step NIL machines pattern the surface on an entire wafer in one step. Thus the stamp must have the same size as the wafer to be patterned. The simplest mechanism for full wafer imprinting is a parallel-plate embossing system. A linear movement of the piston behind the stamp leads to local thinning of the polymer under the stamp protrusions, which is possible because the polymer is moved from squeezed areas into voids in the stamp. This movement can be generated using pneumatic, hydraulic, or motor-driven pistons. The pressure must be maintained during the whole molding process, until the voids are filled, and the molded structures are *fixed* during the cooling or curing step, depending on the method used. However, under normal process conditions, embossing with a hard master does not work without a cushioning mechanism. This cushion balances thickness variations due to both tolerances of the setup and the nature of the molding process. The latter is caused by the fact that the size and shape of the stamp surface relief leads to local pressure variations during the squeeze flow and, if the stamp can bend, to local differences in the sinking velocity. When using thick polymer plates, for which molding leads to surface modulation of a bulk material, the cushion is formed by the viscous material itself. However, in NIL, a thickness profile has to be generated in a resist whose thickness is often lower than the thickness tolerances of the substrates and mechanical setup used. Furthermore height defects in the range up to some µm, such as dust particles, have to be equilibrated. Therefore the cushioning has to be achieved by the pressing mechanism, and its ability to compensate has to be larger than the defects and tolerances of the stamps and substrates. Lateral spreading and dispersion of the applied pressure can be achieved by using a spring mechanism, which can consist of an additional plastic or elastic layer; e.g., a mattress made of rubber (silicone, polydimethylsiloxane (PDMS), Viton), polytetrafluoroethylene (PTFE, Teflon) or elastic graphite can be used. The thickness has to be chosen in order to achieve equilibration of a few micrometers, for which some 100 µm are sufficient. Due to the high pressure used in NIL, compensation of small wedges, i.e., nonparallel alignment, is not needed. The applied pressure of the large backing plate is then spread into infinitesimal small area elements behind the stamp, and is able to compensate for pressure variations occurring during the lateral flow of the molding material. By using this method the height requirements on the substrate surface and material can be minimized and continuous imprinting in all areas is enabled. Even better pressure homogeneity can be obtained when the cushion effect is generated by compressed air or liquid. This can be realized by forming one stamp by a pressure chamber sealed against the backside of the stamp. In practice this is realized by placing a metallic or polymeric membrane between the pressure chamber and the stamp, which deforms around the stamp and substrate, and which is sealed with the counterforce of the stamper [9.11, 13]. The advantage of this soft stamping method is that a very gentle contact between stamp and substrate can be achieved by adjusting the air pressure, so that the surface can assume parallel alignment before the molding starts. During molding the pressure is equilibrated without delay, which assures a constant pressure in all areas of the stamp, only limited by the bending of the stamp.

All press concepts can be realized with heating elements for NIL, or with a UV exposure tool that enables exposure of the resist during molding. Furthermore, combinations of thermoplastic molding and UV exposure are possible. The main difference between thermoplastic molding and UV imprinting is the pressure needed for embossing. Pressures from 1 to 100 bar are used in NIL, while < 1 bar is sufficient in UV-NIL.

Step-and-Repeat NIL

Step-and-repeat NIL machines are physically identical to single-step NIL machines. They pattern a smaller area of a wafer at a time, and then move to an unpatterned area, where the process is repeated (Fig. 9.22). The process is continued until the whole wafer is patterned. This enables the imprinted area to be enlarged by repeated printing with a smaller stamp, as long as subsequent imprints do not affect adjacent patterned areas.

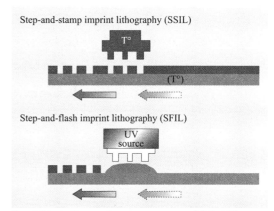

Fig. 9.22a,b Step-and-repeat processes. (**a**) In NIL: step-and-stamp imprinting lithography (SSIL), and (**b**) in UV-NIL: step-and-flash imprint lithography (SFIL). While in SFIL the liquid resin is cured locally by exposure through the stamp, in SSIL the resist is locally heated above its glass-transition temperature by the hot stamp ($T°$ denotes a temperature often set above room temperature)

While this setup enables the use of smaller and more cost-effective molds, with which higher alignment accuracy can be achieved, higher process times and stitching errors at the borders of the patterned fields have to be taken into account. In the case of NIL heating and cooling times can be reduced because of the lower thermal mass, and or in the case of UV-NIL smaller exposure fields may be an advantage.

In thermal NIL the thermal mass of the parts being thermally cycled should be minimized, in order to reduce the obtainable process time. This problem is readily addressed in step-and-stamp (SSIL) and in roll-embossing (roll-to-roll) approaches, but has also found a solution in the concept of heatable stamps [9.12, 210] or by surface heating by means of pulsed laser light [9.211].

9.4 Nanoimprinting Applications

9.4.1 Types of Nanoimprinting Applications

NIL applications can be as manifold as those of other lithographic patterning methods. The applications can be divided into two main categories: pattern-transfer applications and polymer devices. In the first category, pattern-transfer applications, the nanoimprinted resist structure is used as a temporary masking layer for a subsequent pattern-transfer step. In the second category, polymer devices, the imprinted pattern adds functionality to the polymer film, which is the end product.

In many pattern-transfer applications, the main issue is high throughput at nanoscale resolution. Disregarding this issue, it is of minor importance whether the resist film is patterned by means of electromagnetic radiation, electrons or by mechanical deformation. Only a few steps in the process flow are different, for example, the dry etch step to remove the 10–100 nm-thick residual polymer layer after the imprint. Both additive and subtractive processes have been demonstrated, as discussed in Sect. 9.2. Sometimes even the resist is the same, for example, PMMA, which is a widely used resist for both EBL and NIL. The advantages of NIL come into play if high resolution is needed over a large area. For such applications, NIL is a cost-effective alternative to current cutting-edge lithography techniques such as deep-ultraviolet (DUV) lithography [9.212], dedicated to CMOS chip manufacturing. The cost of ownership for NGL technologies, such as extreme-ultraviolet (EUV) lithography [9.213], is reaching a level that requires extremely high production volumes to be economically viable. This development has already forced several branches of the electronics industry to explore NIL as an alternative fabrication method. Examples of such products are patterned media for hard-disk drives (HDD) [9.214, 215], surface acoustic wave (SAW) filters for cell phones [9.27, 216], and subwavelength wire grid polarizers for high-definition TV (HDTV) projectors [9.217]. Even the semiconductor industry was considering NIL as possible NGL to deliver the 32 nm node and beyond [9.14]. For chip manufacture the ability to print smaller features sizes is the most important issue, because NIL simply does not have the restrictions encountered by optical methods and already now offers a resolution higher than the next technical nodes. Among other the major technological challenges to be solved are: overlay accuracy, low defect density, error detection in high resolution stamps and imprints, fast imprint cycles, and critical dimension (CD) control. In addition to the high resolution, the NIL technique also offers capability for 3-D or multilevel imprinting, when the stamp is patterned with structures of different heights (Sect. 9.2.8).

The NIL process offers new possibilities to form polymer devices with microscale to nanoscale features. Nanoscale-patterned polymer films find a wide range of applications within optics, electronics, and nanobiotechnology. The capability to form 3-D polymer structures, with curved surfaces and high aspect ratio, paves the way for new classes of polymer-based passive optical devices, such as lenses and zone plates [9.126], photonic crystals (PhC) [9.100, 218, 219], and integrated polymer optics [9.147]. The NIL technique allows for choosing a wide range of polymers with optimized optical properties [9.153, 219], and allows for patterning thin films of organic light-emitting materials and polymers doped with laser dyes to create organic light-emitting devices (OLED) [9.220, 221] and lasers [9.155, 222, 223]. NIL is also suitable for nanoscale patterning of conducting organic films for cost-effective organic electronics [9.224].

Within the rapidly growing field of lab-on-a-chip applications [9.225], NIL offers an attractive, cost-effective method for molding of complex structures, integrating micro- and nanofluidics, optics, mechanics, and electronics on a single chip [9.226]; for example, the micro- to nanoscale fabrication capabilities are used to create single-use polymer devices containing nanopillar arrays [9.227] and nanofluidic channels [9.228] for DNA separation and sequencing.

In this section we will give an overview of different fields of applications. We start with two examples of pattern-transfer applications that are close to production: patterned media for HDD, and subwavelength metal wire gratings for HDTV projectors. We then discuss a few examples of laboratory-scale potential high-impact applications of NIL. These examples were selected from a large number of NIL applications. The number of laboratory-scale NIL applications is rapidly growing, reflecting a wealth of new possible device architectures becoming feasible by NIL. Some of the applications are directly relevant for industrial production, and others are directed towards research. Even in research the nanostructuring capability of replication processes are needed. Further insight into this field is given in Sect. 9.2.6 about pattern transfer and in Sect. 9.5 about commercialization aspects of NIL.

9.4.2 Patterned Magnetic Media for Hard-Disk Drives

Since the first demonstration of NIL, patterned magnetic media for HDD has been a key application for NIL technology [9.229]. After the invention of the HDD in 1957, the storage capacity, quantified in areal density of bits, has been increased to the current (2008) level of 178 Gb/inch^2 in data storage applications. The size of the individual bits, defined by local magnetization of a homogeneous (unpatterned) thin magnetic film, was reduced, and the bit density increased, by the application of multilayer magnetic films as recording media; the sensitivity of the read head was increased by exploiting the giant-magnetoresistance effect in multilayer thin-film conductors [9.230]; and the magnetization was applied perpendicular to the surface of the recording media, while microelectromechanical systems (MEMS) technology for the mechanical parts has been developed to a rather extreme level: In current HDDs the read–write head flies at a height of 2–3 nm above the surface of the disk plate. An overview of HDD technology is given in [9.231]. This current level of storage density is projected to increase by three orders of magnitude over the next 10 years, in order to meet market requirements.

The possibilities to increase the bit density with current technology, where bits are written by local magnetization of an unpatterned thin magnetic film, are mainly limited by the read–write width, the positioning of the magnetic head, and by thermal instability induced by superparamagnetism in the grains of the magnetic film. These challenges are addressed by patterning the magnetic film.

Discrete track recording (DTR) media [9.232], where the magnetic film is patterned with a spiral land and groove track, have been developed to overcome the problems associated with the read–write width and positioning of the magnetic head (Fig. 9.23a). The idea of DTR media is more than 40 years old [9.232], but has not been implemented in production due to the lack of a nanolithography process that meets the demanding requirements for the surface smoothness of the disk surface [9.233] and that is suitable for large-scale low-cost fabrication. Researchers at WD Media (formerly Komag Inc.) have demonstrated a cost-effective process for volume manufacturing of DTR media, based on double-sided thermal NIL with a commercially available resist and wet etching on a 95 nm-diameter nickel phosphorous (NiP)-plated Al:Mg disk [9.214]. The process steps are outlined in Fig. 9.23b. The nickel stamps with track pitches down to 127 nm, corresponding to an areal density of 200 Gb/inch^2, were electroformed from a silicon master, which was patterned either by laser-beam or electron-beam writing, equipped with a rotating stage with radial beam positioning. After etching, the polymer was removed by oxygen plasma, and the disk was then sputter-coated with a CrX/Co-alloy double-layer magnetic thin film. These devices were designed for in-plane, i.e., longitudinal magnetic polarization, but DTR media for perpendicular polarization have also been realized by EBL and RIE etching of the magnetic film [9.234].

The DTR media technology offers the possibility to regain the loss in electrical signal-to-noise ratio, as the magnetic bit size is reduced. However, with decreasing bit size that is necessary to follow the roadmap, the technology will be limited by thermal instabilities, or superparamagnetism. The magnetic film consists of small, weakly coupled magnetic grains, which behave as single-domain magnetic particles. Each bit consists of the order of 100 grains (domains with single crystalline orientation) to obtain a reasonable signal-to-noise ratio. In order to keep this ratio of grains per bit, the grain size must be reduced with the bit size. The magnetic energy of a single grain scales with the volume of the grain. This implies that the bit can be erased thermally, when the grain size becomes sufficiently small and weakly coupled to neighboring grains. This is referred to as the superparamagnetic limit.

The superparamagnetic limit can be overcome by lithographically defining each bit, as a magnetic nanoparticle, or *nanomagnet* [9.230, 235, 236]. In such a *quantized magnetic disk* [9.235] each magnetic nanoparticle is a single magnetic domain with a well-defined shape and uniaxial magnetic anisotropy, so the magnetization only has two possible stable states, equal in magnitude but opposite in direction, as illustrated in Fig. 9.24. Such defined bits can be thermally stable for sizes down below 10 nm [9.215].

The feasibility of NIL for fabrication of patterns of magnetic nanostructures for quantized magnetic disks has been investigated by several research groups, as re-

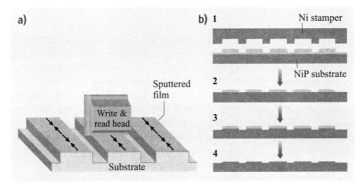

Fig. 9.23 (a) Outline of a DTR medium showing the land and groove structure, patterned into a NiP-plated Al:Mg substrate. The magnetic thin film is sputtered onto the patterned substrate. An improved signal-to-noise ratio can be obtained by making the magnetic read and write heads wider than the land width. (b) Outline of the NIL-based fabrication process (after [9.214])

cently reviewed in [9.237–239]. The imprinted pattern has been transformed to magnetic nanoparticles by electroplating into etched holes [9.99], by lift-off [9.240], and by deposition onto etched pillars [9.215, 241].

In Fig. 9.25 we show the outline of the process flow for large-area fabrication of 55 nm-diameter, 11 nm-high CoPt magnetic islands [9.215], by means of UV-NIL. A SiO$_2$ master containing three $50 \times 50\,\mu m^2$ areas of hexagonal 100 nm-pitch array of 30 nm-high, 55 nm-diameter pillars was fabricated by defining the dot pattern by means of EBL in a 160 nm-thick film of M_w 950 k PMMA. The patterned PMMA film was used in a lift-off process, to define a Cr etch mask. The pillars were etched by tetrafluoromethane (CF$_4$) RIE, and the metal mask was removed. The master was used to form a stamp in a photopolymer material. This stamp is used to UV-imprint the dot pattern in a photopolymer film on a SiO$_2$ substrate, leaving a replica in the photocured polymer, with 28 nm-high pillars on top of a 10 nm-thick residual layer. The pattern was transferred to the SiO$_2$ substrate by CF$_4$ RIE to remove the residual layer, followed by a $(7:1)/(CF_4:CH_4)$ RIE. Finally a CoPt magnetic multilayer structure (Pt$_{1\,nm}$(Co$_{0.3\,nm}$Pt$_{1\,nm}$)$_7$Pt$_{1\,nm}$) was deposited by electron-beam evaporation. The devices were characterized by magnetic force microscopy (MFM), revealing that the film on each pillar is a magnetically isolated single domain that switches independently.

9.4.3 Subwavelength Metal-Strip Gratings

Metallic wire gratings with a period below 200 nm can be used to create polarizers, polarization beam splitters, and optical isolators in the visible range. Such devices have many applications in compact and integrated optics. One example is the use of subwavelength wire-grid

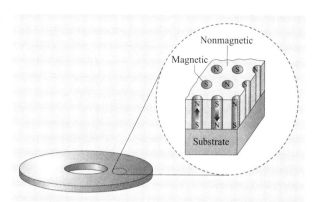

Fig. 9.24 Outline of a patterned magnetic disk for high-density data storage. Each bit is a lithographically defined, single-domain magnetic nanostructure, embedded in a nonmagnetic matrix (after [9.229])

polarization beam splitters in liquid crystal on silicon (LCoS) projection displays for HDTV, yielding higher contrast, uniformity, and brightness of the displayed image (Fig. 9.26).

The polarizing functionality of subwavelength wire gratings is based on form birefringence, an optical anisotropy which appears when isotropic material is structured on a length scale much smaller than the wavelength of light λ. In this limit, the description of light propagation based on the laws of diffraction, refraction, and reflection is not valid, and a rigorous solution of Maxwell's equations with the relevant boundary conditions must be applied. For a review of subwavelength optics see [9.242]. The subwavelength linear grating of period $d < \lambda/2$, line width a, and height h, as illustrated in Fig. 9.27, will behave as a film of birefringent material with refractive indices n_s and n_p for the s-polarized (E-field parallel to the grating) and p-polarized (E-field

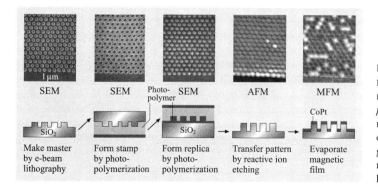

Fig. 9.25 Outline of the process flow for fabrication of 55 nm diameter magnetic islands by UV-NIL. The *top panel* shows SEM, AFM, and MFM micrographs at the different stages of the process. The MFM micrograph shows quantized up and down magnetization of isolated domains. Reproduced from [9.215]

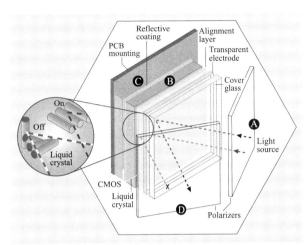

Fig. 9.26 LCoS display for HDTV projection. A light source shines through an external polarizing layer (A) that blocks all light except waves oriented in one plane. The liquid crystal layer (B) twists some waves and lets others proceed unchanged to the reflective layer (C), depending on each pixel's charge; from there they bounce back to another external polarizing layer (D). Here the untwisted light passes through, and the twisted light is blocked (after [9.243])

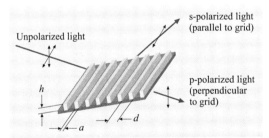

Fig. 9.27 Subwavelength wire grid polarizer. By application of subwavelength gratings, with a pitch below 100 nm for visible light, first-order diffraction with a high acceptance angle and low dispersion birefringence is obtained

perpendicular to the grating) light

$$n_p^2 = \frac{d}{a} n_1^2 + \left(1 - \frac{d}{a}\right) n_2^2,$$

$$n_s^2 = \frac{n_1^2 n_2^2}{\frac{d}{a} n_2^2 + \left(1 - \frac{d}{a}\right) n_1^2}, \quad (9.6)$$

where n_1 and n_2 are the refractive indices of the isotropic grating and fill materials, respectively.

Subwavelength wire gratings have several advantages in terms of large acceptance angle, large extinction (ratio T_p/T_s between the transmittance of the s- and p-polarized components), and long-term stability at high light-flux levels, temperature, and humidity. They can be manufactured in large volume by semiconductor fabrication processes. For applications in liquid-crystal display (LCD) and LCoS projection devices, it is a key challenge to obtain a sufficiently high extinction ratio, larger than 2000, at the shorter wavelengths, i.e., for blue light ($\lambda \approx 450\,\text{nm}$). This requires a pitch d of 100 nm or smaller, which is not practical for producing with conventional optical lithography. *Yu* et al. [9.244], demonstrated a large-area ($100 \times 100\,\text{mm}^2$) $d = 100\,\text{nm}$ by NIL. The stamp gratings were formed by interference lithography using an Ar-ion laser ($\lambda = 351.1\,\text{nm}$) to achieve a pitch around 200 nm, which was transferred to a SiO_2 film by RIE. The pitch was subsequently halved by *spatial frequency doubling* [9.244]: conformal CVD deposition of Si_3N_4, and anisotropic trifluoromethane/oxygen (CHF_3/O_2) RIE (Fig. 9.28). Researchers have realized $d = 100\,\text{nm}$ Al wire grating polarizers by thermal NIL. The process is outlined in Fig. 9.28. The large-area grating stamp is fabricated by laser interference lithography in photoresist, and transferred into the underlying 200 nm-thick SiO_2 film using CF_4 and O_2 RIE [9.217]. The $50 \times 50\,\text{mm}^2$ devices (Figs. 9.29 and 9.30) have an extinction ratio over 2000 and a transmittance above 85% in the blue, at $\lambda = 450\,\text{nm}$. In comparison, commercially available $d = 140\,\text{nm}$ wire grid polarization beam splitters [9.245], fabricated by optical interference lithography, have an extinction ratio around 1000 in the blue. Nanoimprinted subwavelength polarizers for the infrared ($1.0\,\mu\text{m} < \lambda < 1.8\,\mu\text{m}$) are also commercially available [9.246], with a transmittance above 97% and transmission extinction better than 40 dB.

9.4.4 High-Brightness Light-Emitting Diodes

GaN-based light-emitting diodes (LEDs) have large potential as energy-efficient, long-lifetime, environmental-friendly, and stable light sources, and are currently entering a range of applications, such as full-color displays and projectors, traffic lights, and automotive and architectural lighting. Due to the high refractive index of the semiconductor material, the emitted light is easily trapped in waveguide modes inside the device, which strongly reduces the external efficiency of the light source. The light extraction from the device can be significantly enhanced by a patterning the surface

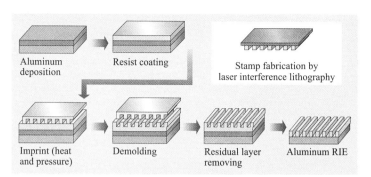

Fig. 9.28 Outline of the NIL process to fabricate $d = 100$ nm-pitch aluminum wire grating polarizers

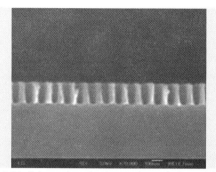

Fig. 9.29 Subwavelength wire grating polarizer with $d = 100$ nm pitch. The aluminum ribs are 100 nm high

with a 2-D photonic crystal [9.247, 248] – an array of holes – with a photonic bandgap that prohibits propagation of photons of frequencies within the bandgap, leading to enhanced extraction of photons through the surface of the device. *Kim* et al. [9.248] demonstrated a ninefold enhancement of photoluminescence intensity of GaN-based green LEDs by means of a 2-D PhC structures of 180 nm-diameter, 100 nm-deep holes arranged in a square lattice with a period of 295 nm. The PhC pattern was defined by thermal NIL and RIE etching through a Cr mask. The NIL stamp was patterned by laser interference lithography.

9.4.5 Polymer Optics

NIL is ideally suited for the fabrication of polymer nanophotonics and waveguide devices with submicron critical dimensions, defined over large areas. It is also compatible with many polymer materials, giving great freedom to choose a material with specific optical properties [9.222, 223, 249].

In Fig. 9.31 we show a polymer microring resonator fabricated by NIL [9.147]. This type of device has been realized in PMMA, PC, and PS on SiO_2 substrates. The resonator consists of a planar waveguide and an adjacent microring waveguide. The waveguide and microring are coupled though the evanescent field in the coupling region. Resonant dips in the transmission through the waveguide occur when the phase pick-up in a trip round the microring is equal to $2\pi m$, where m is an integer. The device works as a narrow-bandwidth filter, and finds applications within integrated optics and for biosensing [9.47].

The evanescent coupling coefficient between the waveguide and microring depends exponentially on the size of the gap. The devices are realized with 1.5 μm-high waveguides, and a coupling air-gap of 100–200 nm. The process flow is outlined in Fig. 9.31c. A thin initial polymer layer is spin-cast onto a SiO_2 substrate layer. The stamp has a very large fill factor and

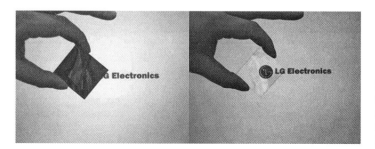

Fig. 9.30 Large-area 100 nm-pitch wire grid polarizer with 85% transmission and extinction ratio larger than 2000 at wavelength $\lambda = 450$ nm (blue light)

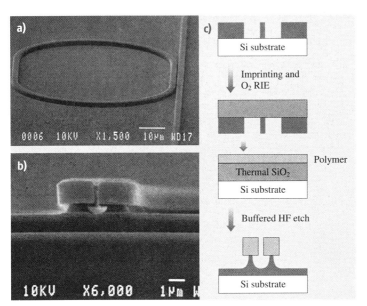

Fig. 9.31a–c Nanoimprinted polymer microring resonator. (a) SEM picture of the imprinted device (b) Cross sectional SEM picture of the polymer waveguides in the coupling region of the microring device (c) Outline of the process flow (after [9.147])

large protrusion areas, implying that a large polymer flow is needed to fill the stamp cavities. A thin residual polymer layer is obtained by combining a high imprint pressure, a high process temperature, and a long imprint time. The mode confinement in the PS waveguides is enhanced by etching the substrate oxide layer isotropically in hydrofluoric acid (HF), to create a pedestal structure. The Q-factor of the resonator device depends critically on the surface scattering losses in the waveguides. The surface roughness of the polymer waveguides can be reduced by a controlled thermal reflow. The device is heated to 10–20 °C below the glass transition, and the surface reflows under the action of surface tension. A loss reduction of more than 70 dB/cm was achieved by this approach [9.250].

9.4.6 Bio Applications

Micro- and nanofabrication technology has enabled methods to manipulate and probe individual molecules and cells on a chip [9.251–255]. This type of application often requires a large area covered with nanostructures. Sometimes a large number of identical devices are needed for statistical evaluation, or to give redundancy, e.g., against clogging of nanofluidic channels. With these requirements, NIL is advantageous, or sometimes the only viable lithography method, even for laboratory-scale experiments and prototyping. Another example is devices for investigation of cell response to nanostructured surface topography, which require nanometer-scale patterned surface areas in the mm^2 to cm^2 range.

Nanofluidic channels can be used to stretch DNA [9.255, 256] for high-throughput linear analysis, measuring the length L of individual DNA molecules, or possibly sequencing by detection of fluorescent labels attached to specific DNA sequences [9.228]. The linear analysis relies on uniform stretching of DNA molecules without coiling as they are driven through a narrow channel. This implies that the nanofluidic channel should have cross-sectional dimensions D close to or smaller than the persistence length of DNA, $L_p \approx 50$ nm [9.257]. The assumption of uniform stretching of the molecule also puts strong requirements on channel sidewall smoothness.

Tegenfeldt et al. [9.255] investigated the dynamics of genomic-length DNA molecules in 100 nm-wide nanochannels, defined by NIL. The device layout is shown in Fig. 9.32. Two microfluidic channels, A–B and D–E, are connected by a 5×1 mm^2 array of 100 nm-wide nanofluidic channels. The nanofluidic channel array is defined by thermal NIL, and the pattern is transferred into the silica substrate by metallization, lift-off, and CF$_4$: H$_2$ RIE. The microfluidic channels are defined on a second silica substrate by UV photolithography (PL) and RIE, and fluidic access ports are sandblasted. The two silica substrates are bonded by cleaning the surfaces, using the so-called RCA protocol

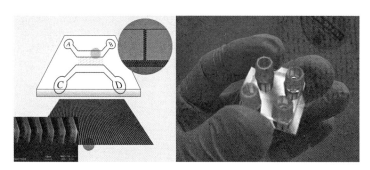

Fig. 9.32 Nanofluidic device for high-throughput linear DNA analysis. Microfluidic channels A–B and C–D are connected via an array of 100 nm-wide nanofluidic channels. The 5×1 mm^2 nanofluidic channel array is defined by NIL. The picture to the right shows the finished device package (after [9.255])

(standard wet chemical process for removal of particles and organic surface contamination [9.258]) before bonding at room temperature, and annealing at 100 °C. The microfluidic channels allow for fast transport of the DNA from the input port to the nanofluidic channels. External electrodes are fitted in the access ports A–E, in order to apply a driving electric field, pulling the DNA through the nanochannels. The DNA is marked with fluorescent dye molecules, which makes it possible to detect individual DNA molecules optically in the nanochannels, by means of an optical microscope.

Similar nanofluidic devices for DNA stretching can be fabricated in polymer at low cost and high throughput in a single NIL process, as demonstrated

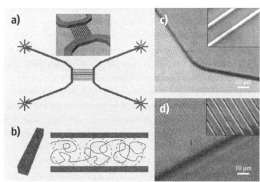

Fig. 9.33a–d Nanofluidic channels fabricated in PMMA by a single thermal NIL step using a two-level stamp (after [9.259]). (a) V-shaped, microfluidic channels (50 μm wide and 1 μm deep) are connected by an array of nanofluidic channels, 250 nm wide and 250 nm deep. (b) Schematics showing the conformation of linear DNA when confined inside the poly(methyl methacrylate) (PMMA) nanochannels (de Gennes regime). (c) SEM picture of two nanoimprint stamp; (d) SEM picture of the imprinted device before the channels are sealed by thermal polymer bonding of lid (after [9.259])

by *Thamdrup* et al. [9.259] (Fig. 9.33). The devices were fabricated by thermal NIL in low-M_w (50 k) PMMA using a 100 mm-diameter two-level hybrid stamp. The fluidic structures were sealed using thermal fusion bonding. The line array of stamp protrusions to imprint the (250×250 nm^2) nanochannels was defined by EBL in SU-8 [9.135] and RIE etching in a thermally grown oxide layer on a silicon wafer. The 1 μm-high, 50 μm-wide stamp protrusions for the microfluidic load channels were subsequently formed by UV-PL in a sol–gel process, using an organic–inorganic hybrid polymer commercialized under the name Ormocomp [9.144]. The stamp is compatible with molecular vapor deposition (MVD), used for applying a durable chlorosilane-based antistiction coating, and allows for imprinting up to a temperature of 270 °C. To benchmark the device performance to conventional fused-silica devices the extension of YOYO-1-stained T4 GT7 bacteriophage DNA inside the PMMA nanochannels was experimentally investigated using epifluorescence microscopy. The measured average extension length amounts to 20% of the full contour length, with a standard deviation of 4%. These results are in good agreement with results obtained by stretching DNA in conventional fused-silica nanochannels.

Cell growth and adhesion can be strongly influenced by surface topography on the micrometer to nanometer length scale [9.260]. This has been exploited by *Gadegaard* et al. [9.261] to create a three-dimensional tubular scaffold for tissue engineering of blood vessels that reproduce the basic structure of natural blood vessels: a layer of smooth muscle cells (fibroblasts) coaxially embedded between an outer collagen mesh and an inner linen of endothelial cells (Fig. 9.34). Such artificially grown blood vessels with tight control of cellular attachment, migration, and growth are expected to reduce problems with cellular debris and inflammation. This would be a major improvement compared with current medical procedures, where polymer tubes are used

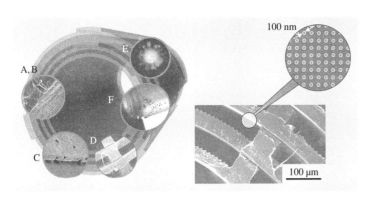

Fig. 9.34 *Swiss roll* tubular construct for vascular tissue engineering

for vascular grafting. To facilitate cell growth which mimics the structure of blood vessels, the 3-D tubular scaffold consists of a coaxial polymer layers with different surface topography, which selectively stimulates growth of a particular cell type: muscle and endothelial cells. A surface topography with nanometer-scale features on the inner surface favors adhesion and growth of endothelial cells. The bulk of the tube wall consists of microchannels with embedded micrometer-sized grooves which stimulate growth and adhesion of muscle cells (fibroblasts). The scaffold is fabricated by multilevel thermal NIL in an approximately 30 μm-thick film polycaprolactone (PCL), which is a US Food and Drug Administration (FDA)-approved biodegradable polymer, and a thermoplastic with glass-transition temperature $T_g \approx 60\,°C$, which was subsequently rolled up to form the required 3-D tubular structure. The desired surface structure was realized on flat silicon stamps by UV-PL and EBL, and negative stamp replicas were formed in PDMS by casting and peel-off. The PDMS stamp replicas were used for double-side embossing on the PCL sheet. After embossing, the PCL sheet was rolled up to form the tubular scaffold structure.

9.5 Conclusions and Outlook

Technological development is heavily based on so-called enabling techniques. For example, Gutenberg's book printing with movable metal letters was based on a combination of different existing techniques (large wine presses and metallurgy for letter casting), solving throughput and flexibility problems, and was developed at a time of globalization when information needed to be spread (around 1450 AD, only years before Columbus discovered the sea route to America) [9.262]. In a similar way, a new lithographic technique with micro- and nanopatterning capability, such as NIL, is not entirely new, but is based on patterning techniques coming from silicon micromachining and compact-disc molding. In a time of technological dynamics it will lead to advances in different fields:

- In research, as long as machines are affordable and reliable enough that they can replace or complement standard lithographic techniques. Many research institutes and universities now have access to silicon processing technology, which often comprises tools such as resist process technology, pattern generators, mask aligners, and etching and deposition facilities in a cleanroom environment. In the case of thermal NIL it is advantageous that nanostructures can be replicated with simple molding tools, e.g., hot presses without alignment, thus making it possible to integrate NIL into a simple device manufacturing. More sophisticated NIL machines are now available, typically for laboratory-type small-scale production. Standard mask aligners can be upgraded to perform UV-NIL with moderate pressures. In combination with anodic bonders or microembossing tools they can be used for thermal NIL. These setups allow alignment and provide increased reproducibility. The enterprises offering equipment, stamps and materials for NIL are listed in [9.144, 263–271].
- In industry, if they help to cross technological barriers, reduce cost, and enable to step into fields reserved for high-throughput applications. Success will also depend on whether they fit into the pro-

cess chain already established in a silicon cleanroom environment. Furthermore substrate sizes, throughput, and yield have to correspond to the production needs. As in research, many of the machines already available can be used for moderate-scale production. They can be scaled up to substrate sizes of 200 mm and higher in combination with batch-mode operation. More sophisticated are machines based on step-and-repeat NIL, which can help to solve equilibration and overlay issues. Further improvements can be expected if new resists and process schemes are developed. In order to achieve a critical mass of technological expertise, the integration of NIL into a consortium of technology providers is of advantage, making it possible for manufacturers to buy standard equipment and materials, along with process knowhow.

Until now NIL was considered as a very promising patterning method, because it combines resolution with large area and throughput. As long as it is seen as an alternative to establish high-end photolithographies, the strategy will most likely be to replace single lithographic steps by imprinting. The only consequence in the multimasking process sequence needed in microchip fabrication would then be modifying the pattern transfer process, e.g., by adding the residual layer etch (Sect. 9.2.6). The requirements of the ITRS roadmap are so high that other more established NGLs might make faster advances towards the next node, and the introduction of NIL into large-scale fabrication would be further postponed (NIL was first added to the 2003 ITRS roadmap for the 32 nm node [9.14, 272]). However, NIL has other capabilities, as demonstrated in Sect. 9.4, even if not all requirements of the ITRS roadmap are met at once:

- Enterprises with applications ranging from templates for hard-disk production to SAW filters for mobile phones, polarizers for flat-panel screens, and templates for biodevices are now heading into replication techniques based on NIL processes. Most of these processes are based on single layers covered with nanostructures, mostly regular high-resolution gratings and dot arrays, and need single-step wafer-scale replication tools for large areas.

- The 3-D patterning capability (Sect. 9.2.8) makes it possible to develop innovative pattern-transfer processes, thus leading to significant cost reduction. Similar advances could be achieved if materials with new properties are patterned. This is mainly due to the fact that NIL has the concept of displacing material at the nanoscale rather than removing material selectively and locally. Often this goes along with some tradeoffs on resolution and alignment, which is justified depending on the application.

NIL has now passed a barrier from the laboratory scale to industrial preproduction. Although it seems that room-temperature processes based on UV exposure have an advantage over processes based on thermocycles, to date it is difficult to say which process will become a standard process and make it to the production line. For example, isothermal processes at elevated temperatures using hybrid processes that use both thermal NIL and hardening by UV curing have been established [9.273]. With state-of-the-art UV-NIL equipment [9.268], more than six wafers per hour with a diameter of 200 mm can now be achieved in a step-and-repeat modus (using a stamp area of $45 \times 60\,\text{mm}^2$). Single-step wafer-scale hot embossing has similar capabilities, and can even push throughput further if heatable stamps with low thermal mass are used [9.210, 274]. However, NIL is currently such a fast-moving field that prejudgment about the final success of one technique is not possible or advisable. Innovative solutions are still needed to solve process and stamp lifetime issues for many different applications. Probably, not only a single NIL process will be successfully implemented, but many variants of NIL. This includes hybrid approaches, e.g., NIL in combination with other lithographic processes, or the fabrication and copying of stamps using NIL.

The aim of this chapter was to give an insight into the concepts used in NIL, along with presenting the advantages and limitations of processes ranging from tool fabrication to pattern transfer. Although more referring to the older thermoplastic molding process, which is the authors' original field of expertise, it was intended to be general enough that future developments can be judged. The interested reader, however, will find more detailed information at technological conferences and in scientific publications, and also in the patent literature.

References

9.1 E. Berliner: Gramophone, US Patent 372786 (1887), http://www.audioannals.com/berlinere.htm

9.2 E. Berliner: Process for producing records of sound, US Patent 382790 (1888), http://www.audioannals.com/berlinere.htm

9.3 J.C. Ruda: Record manufacturing: making sound for everyone, J. Audio Eng. Soc. **25**(10/11), 702–711 (1977)

9.4 K.C. Pohlmann: *The Compact Disc Handbook*, Comput. Music Dig. Audio Ser., Vol. 5, 2nd edn. (A-R Editions, Madison 1992)

9.5 H. Schift, C. David, M. Gabriel, J. Gobrecht, L.J. Heyderman, W. Kaiser, S. Köppel, L. Scandella: Nanoreplication in polymers using hot embossing and injection molding, Microelectron. Eng. **53**, 171–174 (2000)

9.6 S.Y. Chou, P.R. Krauss: Imprint lithography with sub-10 nm feature size and high throughput, Microelectron. Eng. **35**, 237–240 (1997)

9.7 R.W. Jaszewski, H. Schift, J. Gobrecht, P. Smith: Hot embossing in polymers as a direct way to pattern resist, Microelectron. Eng. **41/42**, 575–578 (1998)

9.8 Y. Xia, G.M. Whitesides: Soft lithography, Angew. Chem. Int. Ed. **37**, 550–575 (1998)

9.9 B. Michel, A. Bernard, A. Bietsch, E. Delamarche, M. Geissler, D. Juncker, H. Kind, J.-P. Renault, H. Rothuizen, H. Schmid, P. Schmidt-Winkel, R. Stutz, H. Wolf: Printing meets lithography: Soft approaches to high-resolution, IBM J. Res. Dev. **45**(5), 697–719 (2001)

9.10 W. Menz, J. Mohr, O. Paul: *Microsystem Technology* (Wiley-VCH, Weinheim 2001)

9.11 H. Schift: Nanoimprint lithography: An old story in modern times? A review, J. Vac. Sci. Technol. B **26**(2), 458–480 (2008)

9.12 H. Schift (Ed.): *NaPa Library of Processes* (NaPa-consortium, 2008), http://www.napanil.org (last access December 2009)

9.13 C. Sotomayor Torres: Alternative lithography – Unleashing the potential of nanotechnology. In: *Nanostructure Science and Technology*, ed. by D.J. Lockwood (Kluwer, New York 2003)

9.14 International Technology Roadmap for Semiconductors, http://public.itrs.net/ (last accessed May 8, 2008)

9.15 R. Compañó (Ed.): *Technology Roadmap for Nanoelectronics*, European Commission IST Programme, Future and Emerging Technologies, 2nd edn. (European Commission, Brussels 2000)

9.16 H. Moore: Cramming more components onto integrated circuits, Electronics **38**(8), 114–117 (1965)

9.17 S. Okazaki: Resolution limits of optical lithography, J. Vac. Sci. Technol. B **9**(6), 2829–2833 (1991)

9.18 L.J. Heyderman, H. Schift, C. David, J. Gobrecht, T. Schweizer: Flow behaviour of thin polymer films used for hot embossing lithography, Microelectron. Eng. **54**, 229–245 (2000)

9.19 H. Schulz, M. Wissen, N. Bogdanski, H.-C. Scheer, K. Mattes, C. Friedrich: Impact of molecular weight of polymers and shear rate effects for nanoimprint lithography, Microelectron. Eng. **83**, 259–280 (2006)

9.20 S.Y. Chou, P.R. Krauss, P.J. Renstrom: Imprint of sub-25 nm vias and trenches in polymers, Appl. Phys. Lett. **67**(21), 3114–3116 (1995)

9.21 S.Y. Chou, P.R. Krauss, P.J. Renstrom: Nanoimprint lithography, J. Vac. Sci. Technol. B **14**(6), 4129–4133 (1996)

9.22 S.Y. Chou: Nanoimprint lithography, US Patent 5772905 (1995)

9.23 L. Baraldi, R. Kunz, J. Meissner: High-precision molding of integrated optical structures, Proc. SPIE **1992**, 21–29 (1993)

9.24 J. Haisma, M. Verheijen, K. van den Heuvel, J. van den Berg: Mold-assisted lithography: A process for reliable pattern replication, J. Vac. Sci. Technol. B **14**, 4124–4128 (1996)

9.25 M. Colburn, S. Johnson, M. Stewart, S. Damle, T. Bailey, B. Choi, M. Wedlake, T. Michealson, S.V. Sreenivasan, J. Ekerdt, C.G. Willson: Step and flash imprint lithography: A new approach to high-resolution patterning, Proc. SPIE **3676**, 379–389 (1999)

9.26 D.J. Resnick, W.J. Dauksher, D. Mancini, K.J. Nordquist, T.C. Bailey, S. Johnson, N. Stacey, J.G. Ekerdt, C.G. Willson, S.V. Sreenivasan, N. Schumaker: Imprint lithography: Lab curiosity or the real NGL?, Proc. SPIE **5037**, 12–23 (2003)

9.27 D.J. Resnick, S.V. Sreenivasan, C.G. Willson: Step and flash imprint lithography, Mater. Today **8**, 34–42 (2005)

9.28 M. Doi: *Introduction to Polymer Physics* (Clarendon, Oxford 1996)

9.29 D.W. van Krevelen: *Properties of Polymers* (Elsevier, Amsterdam 1990)

9.30 H. Schift, L.J. Heyderman: Nanorheology – squeezed flow in hot embossing of thin films. In: *Alternative Lithography*, Nanostruct. Sci. Technol., ed. by C. Sotomayor Torres (Kluwer, New York 2003) pp. 46–76

9.31 H.-C. Scheer, H. Schulz, T. Hoffmann, C.M. Sotomayor Torres: Nanoimprint techniques. In: *Handbook of Thin Film Materials*, Vol. 5, ed. by H.S. Nalwa (Academic, New York 2002) pp. 1–60, Chap. 1

9.32 M.D. Austin, H. Ge, W. Wu, M. Li, Z. Yu, D. Wasserman, S.A. Lyon, S.Y. Chou: Fabrication of 5 nm linewidth and 14 nm pitch features by nanoimprint lithography, Appl. Phys. Lett. **84**(26), 5299–5301 (2004)

9.33 E.A. Dobisz, S.L. Brandow, R. Bass, J. Mitterender: Effects of molecular properties on nanolithography

9.33 in polymethyl methacrylate, J. Vac. Sci. Technol. B **18**, 107–111 (2000)

9.34 A. Olzierski, I. Raptis: Development and molecular-weight issues on the lithographic performance of poly (methyl methacrylate), Microelectron. Eng. **73/74**, 244–251 (2004)

9.35 M. Khoury, D.K. Ferry: Effect of molecular weight on poly(methyl methacrylate) resolution, J. Vac. Sci. Technol. B **14**, 75–79 (1996)

9.36 L.J. Fetters, D.J. Lohse, D. Richter, T.A. Witten, A. Zirkel: Connection between polymer molecular weight, density, chain dimensions, and melt viscoelastic properties, Macromolecules **27**, 4639–4647 (1994)

9.37 A. Franck: *Kunststoff-Kompendium*, 4th edn. (Vogel, Würzburg 1996) p. 255, in German

9.38 C.B. Roth, J.R. Dutcher: Mobility on different length scales in thin polymer films. In: *Soft Materials: Structure and Dynamics*, ed. by J.R. Dutcher, A.G. Marangoni (Dekker, New York 2004)

9.39 J.N. D'Amour, U. Okoroanyanwu, C.W. Frank: Influence of substrate chemistry on the properties of ultrathin polymer films, Microelectron. Eng. **73/74**, 209–217 (2004)

9.40 R.B. Bird, C.F. Curtis, R.C. Armstrong, O. Hassager: *Dynamics of Polymeric Liquids* (Wiley, New York 1987)

9.41 L.G. Baraldi: Heißprägen in Polymeren für die Herstellung integriert-optischer Systemkomponenten. Ph.D. Thesis (ETH Zurich, Zurich 1994), Vol. 10762, in German

9.42 M.J. Stefan: Parallel Platten Rheometer, Akad. Wiss. Math.-Naturwiss. Vienna **2**(69), 713–735 (1874), in German

9.43 J.-H. Jeong, Y.-S. Choi, Y.-J. Shin, J.-J. Lee, K.-T. Park, E.-S. Lee, S.-R. Lee: Flow behavior at the embossing stage of nanoimprint lithography, Fibers Polym. **3**(3), 113–119 (2002)

9.44 H. Schift, S. Park, J. Gobrecht: Nano-imprint – Molding resists for lithography, J. Photopolym. Sci. Technol. **16**(3), 435–438 (2003)

9.45 H.-C. Scheer, H. Schulz, T. Hoffmann, C.M. Sotomayor Torres: Problems of the nanoimprinting technique for nanometer scale pattern definition, J. Vac. Sci. Technol. B **16**, 3917–3921 (1998)

9.46 H.-C. Scheer, H. Schulz: A contribution to the flow behaviour of thin polymer films during hot embossing lithography, Microelectron. Eng. **56**, 311–332 (2001)

9.47 L.J. Guo: Recent progress in nanoimprint technology and its applications, J. Phys. D **37**, R123–R141 (2004)

9.48 L.J. Guo: Nanoimprint lithography: Methods and material requirements, Adv. Mater. **19**, 495–513 (2007)

9.49 C. Gourgon, C. Perret, G. Micouin, F. Lazzarino, J.H. Tortai, O. Joubert, J.-P.E. Grolier: Influence of pattern density in nanoimprint lithography, J. Vac. Sci. Technol. B **21**(1), 98–105 (2003)

9.50 A. Lebib, Y. Chen, J. Bourneix, F. Carcenac, E. Cambril, L. Couraud, H. Launois: Nanoimprint lithography for a large area pattern replication, Microelectron. Eng. **46**, 319–322 (1999)

9.51 C. Gourgon, J.H. Tortai, F. Lazzarino, C. Perret, G. Micouin, O. Joubert, S. Landis: Influence of residual solvent in polymers patterned by nanoimprint lithography, J. Vac. Sci. Technol. B **22**(6), 602–606 (2004)

9.52 Y. Hirai, M. Fujiwara, T. Okuno, Y. Tanaka, M. Endo, S. Irie, K. Nakagawa, M. Sasago: Study of the resist deformation in nanoimprint lithography, J. Vac. Sci. Technol. B **19**(6), 2811–2815 (2001)

9.53 Y. Hirai, T. Konishi, T. Yoshikawa, S. Yoshida: Simulation and experimental study of polymer deformation in nanoimprint lithography, J. Vac. Sci. Technol. B **22**(6), 3288–3293 (2002)

9.54 H.D. Rowland, W.P. King: Polymer deformation and filling modes during microembossing, J. Micromech. Microeng. **14**, 1625–1632 (2004)

9.55 S. Zankovych, T. Hoffmann, J. Seekamp, J.-U. Bruch, C.M. Sotomayor Torres: Nanoimprint lithography: challenges and prospects, Nanotechnology **12**(2), 91–95 (2001)

9.56 M. Beck, M. Graczyk, I. Maximov, E.-L. Sarwe, T.G.I. Ling, M. Keil, L. Montelius: Improving stamps for 10 nm level wafer scale nanoimprint, lithography, Microelectron. Eng. **61/62**, 441–448 (2002)

9.57 D.-Y. Khang, H.H. Lee: Room-temperature imprint lithography by solvent vapor treatment, Appl. Phys. Lett. **76**(7), 870–872 (2000)

9.58 D.-Y. Khang, H. Yoon, H.H. Lee: Room-temperature imprint lithography, Adv. Mater. **13**(10), 749–751 (2001)

9.59 D.-Y. Khang, H. Kang, T.-I. Kim, H.H. Lee: Low-pressure nanoimprint lithography, Nano Lett. **4**(4), 633–637 (2004)

9.60 H. Lee, G.Y. Jung: Full wafer scale near zero residual nano-imprinting lithography using UV curable monomer solution, Microelectron. Eng. **77**(1), 42–47 (2005)

9.61 L. Tan, Y.P. Kong, S.W. Pang, A.F. Yee: Imprinting of polymer at low temperature and pressure, J. Vac. Sci. Technol. B **22**(5), 2486–2492 (2004)

9.62 C. Finder, C. Mayer, H. Schulz, H.-C. Scheer, M. Fink, K. Pfeiffer: Non-contact fluorescence measurements for inspection and imprint depth control in nanoimprint lithography, Proc. SPIE **4764**, 218–223 (2002)

9.63 D. Jucius, V. Grigaliunas, A. Guobiene: Rapid evaluation of imprint quality using optical scatterometry, Microelectron. Eng. **71**, 190–196 (2004)

9.64 A. Fuchs, B. Vratzov, T. Wahlbrink, Y. Georgiev, H. Kurz: Interferometric in situ alignment for UV-based nanoimprint, J. Vac. Sci. Technol. B **22**(6), 3242–3245 (2002)

9.65 Z. Yu, H. Gao, S.Y. Chou: In situ real time process characterisation in nanoimprint lithography using time-resolved diffractive scatterometry, Appl. Phys. Lett. **85**(18), 4166–4168 (2004)

9.66 F. Lazzarino, C. Gourgon, P. Schiavone, C. Perret: Mold deformation in nanoimprint lithography, J. Vac. Sci. Technol. B **22**(6), 3318–3322 (2002)

9.67 C. Perret, C. Gourgon, F. Lazzarino, J. Tallal, S. Landis, R. Pelzer: Characterization of 8 in wafers printed by nanoimprint lithography, Microelectron. Eng. **73/74**, 172–177 (2004)

9.68 C. Gourgon, C. Perret, J. Tallal, F. Lazzarino, S. Landis, O. Joubert, R. Pelzer: Uniformity across 200 mm silicon wafers printed by nanoimprint lithography, J. Phys. D **38**, 70–73 (2005)

9.69 U. Plachetka, M. Bender, A. Fuchs, B. Vratzov, T. Glinsner, F. Lindner, H. Kurz: Wafer scale patterning by soft UV-nanoimprint lithography, Microelectron. Eng. **73/74**, 167–171 (2004)

9.70 N. Roos, M. Wissen, T. Glinsner, H.-C. Scheer: Impact of vacuum environment on the hot embossing process, Proc. SPIE **5037**, 211–218 (2003)

9.71 D. Pisignano, A. Melcarne, D. Mangiullo, R. Cingolani, G. Gigli: Nanoimprint lithography of chromophore molecules under high-vacuum conditions, J. Vac. Sci. Technol. B **22**(1), 185–188 (2004)

9.72 H. Schift, L.J. Heyderman, M. Auf der Maur, J. Gobrecht: Pattern formation in hot embossing of thin polymer films, Nanotechnology **12**, 173–177 (2001)

9.73 S.Y. Chou, L. Zhuang: Lithographically induced self-assembly of periodic polymer micropillar arrays, J. Vac. Sci. Technol. B **17**, 3197–3202 (1999)

9.74 S.Y. Chou, L. Zhuang, L.J. Guo: Lithographically induced self-construction of polymer microstructures for resistless patterning, Appl. Phys. Lett. **75**, 1004–1006 (1999)

9.75 L. Wu, S.Y. Chou: Electrohydrodynamic instability of a thin film of viscoelastic polymer underneath a lithographically manufactured mask, J. Non-Newton. Fluid Mech. **125**, 91–99 (2005)

9.76 E. Schäffer, T. Thurn-Albrecht, T.P. Russell, U. Steiner: Electrically induced structure formation and pattern transfer, Nature **403**, 874–877 (2000)

9.77 E. Schäffer, T. Thurn-Albrecht, T.P. Russell, U. Steiner: Method and apparatus for forming submicron patterns on films, US Patent 07880075001 (1999)

9.78 E. Schäffer, U. Steiner: Methods and apparatus for the formation of patterns in films using temperature gradients, European Patent PCT 124205.6 (2000)

9.79 K.Y. Suh, H.H. Lee: Capillary force lithography: large-area patterning, self-organization, and anisotropic dewetting, Adv. Funct. Mater. **6/7**, 405–413 (2002)

9.80 Y. Hirai, S. Yoshida, N. Takagi: Defect analysis in thermal nanoimprint lithography, J. Vac. Sci. Technol. B **21**(6), 2765–2770 (2003)

9.81 Y. Hirai, T. Yoshikawa, N. Takagi, S. Yoshida: Mechanical properties of poly-methyl methacrylate (PMMA) for nanoimprint lithography, J. Photopolym. Sci. Technol. **16**(4), 615–620 (2003)

9.82 M. Colburn, B.J. Choi, S.V. Sreenivasan, R.T. Bonnecaze, C.G. Willson: Ramifications of lubrication theory on imprint lithography, Microelectron. Eng. **75**, 321–329 (2004)

9.83 A. Fuchs, M. Bender, U. Plachetka, U. Hermanns, H. Kurz: Ultraviolet-based nanoimprint at reduced environmental pressure, J. Vac. Sci. Technol. B **23**(6), 2925–2928 (2005)

9.84 M. Colburn, I. Suez, B.J. Choi, M. Meissl, T. Bailey, S.V. Sreenivasan, J.G. Ekerdt, C.G. Willson: Characterization and modelling of volumetric and mechanical properties for step and flash imprint lithography photopolymers, J. Vac. Sci. Technol. B **19**(6), 2685–2689 (2001)

9.85 D.J. Resnick, W.J. Dauksher, D. Mancini, K.J. Nordquist, T.C. Bailey, S. Johnson, N. Stacey, J.G. Ekerdt, C.G. Willson, S.V. Sreenivasan, N. Schumaker: Imprint lithography for integrated circuit fabrication, J. Vac. Sci. Technol. B **21**(6), 2624–2631 (2003)

9.86 M. Otto, M. Bender, B. Hadam, B. Spangenberg, H. Kurz: Characterization and application of a UV-based imprint technique, Microelectron. Eng. **57/58**, 361–366 (2001)

9.87 B. Vratzov, A. Fuchs, M. Lemme, W. Henschel, H. Kurz: Large scale ultraviolet-based nanoimprint lithography, J. Vac. Sci. Technol. B **21**(6), 2760–2764 (2003)

9.88 M. Komuro, J. Taniguchi, S. Inoue, N. Kimura, Y. Tokano, H. Hiroshima, S. Matsui: Imprint characteristics by photo-induced solidification of liquid polymer, Jpn. J. Appl. Phys. **39**, 7075–7079 (2000)

9.89 H. Schulz, H.-C. Scheer, T. Hoffmann, C.M. Sotomayor Torres, K. Pfeiffer, G. Bleidießel, G. Grützner, C. Cardinaud, F. Gaboriau, M.-C. Peignon, J. Ahopelto, B. Heidari: New polymer materials for nanoimprinting, J. Vac. Sci. Technol. B **18**(4), 1861–1865 (2000)

9.90 H. Schulz, D. Lyebyedyev, H.-C. Scheer, K. Pfeiffer, G. Bleidießel, G. Grützner, J. Ahopelto: Master replication into thermosetting polymers for nanoimprinting, J. Vac. Sci. Technol. B **18**(6), 3582–3585 (2000)

9.91 K. Pfeiffer, M. Fink, G. Bleidießel, G. Grützner, H. Schulz, H.-C. Scheer, T. Hoffmann, C.M. Sotomayor Torres, F. Gaboriau, C. Cardinaud: Novel linear and crosslinking polymers for nanoimprinting with high etch resistance, Microelectron. Eng. **53**, 411–414 (2000)

9.92 S. Rudschuck, D. Hirsch, K. Zimmer, K. Otte, A. Braun, R. Mehnert, P. Bigl: Replication of 3-D-micro- and nanostrucutres using different UV-curable polymers, Microelectron. Eng. **53**, 557–560 (2000)

9.93 M. Sagnes, L. Malaquin, F. Carcenac, C. Vieu, C. Fournier: Imprint lithography using thermo-polymerisation of MMA, Microelectron. Eng. **61/62**, 429–433 (2002)

9.94 A. Abdo, S. Schuetter, G. Nellis, A. Wei, R. Engelstad, V. Truskett: Predicting the fluid behavior during the dispensing process for step-and-flash imprint lithography, J. Vac. Sci. Technol. B **22**(6), 3279–3282 (2002)

9.95 Y. Hirai, H. Kikuta, T. Sanou: Study on optical intensity distribution in photocuring nanoimprint lithography, J. Vac. Sci. Technol. B **21**(6), 2777–2782 (2003)

9.96 C.-H. Chang, R.K. Heilmann, R.C. Fleming, J. Carter, E. Murphy, M.L. Schattenburg, T.C. Bailey, J.G. Ekerdt, R.D. Frankel, R. Voisin: Fabrication of sawtooth diffraction gratings using nanoimprint lithography, J. Vac. Sci. Technol. B **21**(6), 2755–2759 (2003)

9.97 L.J. Heyderman, H. Schift, C. David, B. Ketterer, M. Auf der Maur, J. Gobrecht: Nanofabrication using hot embossing lithography and electroforming, Microelectron. Eng. **57/58**, 375–380 (2001)

9.98 P.R. Krauss, S.Y. Chou: Nano-compact disks with 400 Gbit/in^2 storage density fabricated using nanoimprint lithography and read with proximal probe, Appl. Phys. Lett. **71**(21), 3174–3176 (1997)

9.99 W. Wu, B. Cui, X. Sun, W. Zhang, L. Zhuang, L. Kong, S.Y. Chou: Large area high density quantized magnetic disks fabricated using nanoimprint lithography, J. Vac. Sci. Technol. B **16**(6), 3825–3829 (1998)

9.100 H. Schift, S. Park, C.-G. Choi, C.-S. Kee, S.-P. Han, K.-B. Yoon, J. Gobrecht: Fabrication process for polymer photonic crystals using nanoimprint lithography, Nanotechnology **16**, S261–S265 (2005)

9.101 M. Hartney, D. Hess, D. Soane: Oxygen plasma etching for resist stripping and multilayer lithography, J. Vac. Sci. Technol. B **7**, 1–13 (1989)

9.102 W. Pilz, J. Janes, K.P.M. Müller, J. Pelka: Oxygen reactive ion etching of polymers – Profile evolution and process mechanisms, Proc. SPIE **1392**, 84–94 (1990)

9.103 B. Heidari, I. Maximov, E.-L. Sarwe, L. Montelius: Large scale nanolithography using imprint lithography, J. Vac. Sci. Technol. B **17**, 2961–2964 (1999)

9.104 D. Lyebyedyev, H.-C. Scheer: Mask definition by nanoimprint lithography, Proc. SPIE **4349**, 82–85 (2001)

9.105 X.-M. Yan, S. Kwon, A.M. Contreras, J. Bokor, G.A. Somorjai: Fabrication of large number density platinum nanowire arrays by size reduction lithography and nanoimprint lithography, Nano Lett. **5**(4), 745–748 (2005)

9.106 L.J. Heyderman, B. Ketterer, D. Bächle, F. Glaus, B. Haas, H. Schift, K. Vogelsang, J. Gobrecht, L. Tiefenauer, O. Dubochet, P. Surbled, T. Hessler: High volume fabrication of customised nanopore membrane chips, Microelectron. Eng. **67/68**, 208–213 (2003)

9.107 H. Schift, R.W. Jaszewski, C. David, J. Gobrecht: Nanostructuring of polymers and fabrication of interdigitated electrodes by hot embossing lithography, Microelectron. Eng. **46**, 121–124 (1999)

9.108 L. Montelius, B. Heidari, M. Graczyk, E.-L. Sarwe, T.G.I. Ling: Nanoimprint- and UV-lithography: mix&match process for fabrication of interdigitated nanobiosensors, Microelectron. Eng. **53**, 521–524 (2000)

9.109 M. Beck, F. Persson, P. Carlberg, M. Graczyk, I. Maximov, T.G.I. Ling, L. Montelius: Nanoelectrochemical transducers for (bio-) chemical sensor applications fabricated by nanoimprint lithography, Microelectron. Eng. **73/74**, 837–842 (2004)

9.110 H. Schift, L.J. Heyderman, C. Padeste, J. Gobrecht: Chemical nano-patterning using hot embossing lithography, Microelectron. Eng. **61/62**, 423–428 (2002)

9.111 S. Park, H. Schift, C. Padeste, J. Gobrecht: Nanostructuring of anti-adhesive layer by hot embossing lithography, Microelectron. Eng. **67/68**, 252–258 (2003)

9.112 S. Park, S. Saxer, C. Padeste, H.H. Solak, J. Gobrecht, H. Schift: Chemical patterning of sub 50 nm half pitches via nanoimprint lithography, Microelectron. Eng. **78/79**, 682–688 (2005)

9.113 D. Falconnet, D. Pasqui, S. Park, R. Eckert, H. Schift, J. Gobrecht, R. Barbucci, M. Textor: A novel approach to produce protein nanopatterns by combining nanoimprint, lithography and molecular self-assembly, Nano Lett. **4**(10), 1909–1914 (2004)

9.114 J.D. Hoff, L.-J. Cheng, E. Meyhofer, L.J. Guo, A.J. Hunt: Nanoscale protein patterning by imprint lithography, Nano Lett. **4**(5), 853–857 (2004)

9.115 T. Schliebe, G. Schneider, H. Aschoff: Nanostructuring high resolution phase zone plates in nickel and germanium using cross-linked polymers, Microelectron. Eng. **30**, 513–516 (1996)

9.116 G. Simon, A.M. Haghiri-Gosnet, F. Carcenac, H. Launois: Electroplating: an alternative transfer technology in the 20 nm range, Microelectron. Eng. **35**, 51–54 (1997)

9.117 K. Pfeiffer, M. Fink, G. Grützner, G. Bleidießel, H. Schulz, H.-C. Scheer: Multistep profiles by mix and match of nanoimprint and UV-lithography, Microelectron. Eng. **57/58**, 381–387 (2001)

9.118 X. Cheng, L.J. Guo: A combined-nanoimprint-and-photolithography patterning technique, Microelectron. Eng. **3/4**, 277–282 (2004)

9.119 X. Cheng, L.J. Guo: One-step lithography for various size patterns with a hybrid mask-mold, Microelectron. Eng. **3/4**, 288–293 (2004)

9.120 N. Kehagias, S. Zankovych, A. Goldschmidt, R. Kian, M. Zelsmann, C.M. Sotomayor Torres, K. Pfeiffer, G. Ahrens, G. Grützner: Embedded polymer waveguides: design and fabrication approaches, Superlattices Microstruct. **36**(1–3), 201–210 (2004)

9.121 W. Zhang, S.Y. Chou: Multilevel nanoimprint lithography with submicron alignment over 4 in. Si wafers, Appl. Phys. Lett. **79**(6), 845–847 (2001)

9.122 H. Schulz, M. Wissen, N. Roos, H.-C. Scheer, K. Pfeiffer, G. Grützner: Low-temperature wafer-scale 'warm' embossing for mix and match with UV-lithography, SPIE Proc. **4688**, 223–231 (2002)

9.123 I. Martini, J. Dechow, M. Kamp, A. Forchel, J. Koeth: GaAs field effect transistors fabricated by imprint lithography, Microelectron. Eng. **60**(3-4), 451–455 (2002)

9.124 A.P. Kam, J. Seekamp, V. Solovyev, C. Clavijo Cedeño, A. Goldschmidt, C.M. Sotomayor Torres: Nanoimprinted organic field-effect transistors: fabrication, transfer mechanism and solvent effects on device characteristics, Microelectron. Eng. **73/74**, 809–813 (2004)

9.125 H. Schulz, A.S. Körbes, H.-C. Scheer, L.J. Balk: Combination of nanoimprint and scanning force lithography for local tailoring of sidewalls of nanometer devices, Microelectron. Eng. **53**, 221–224 (2000)

9.126 M. Tormen, L. Businaro, M. Altissimo, F. Romanato, S. Cabrini, F. Perennes, R. Proietti, H.-B. Sun, S. Kawata, E. Di Fabrizio: 3-D patterning by means of nanoimprinting, x-ray and two-photon lithography, Microelectron. Eng. **73/74**, 535–541 (2004)

9.127 X. Sun, L. Zhuang, W. Zhang, S.Y. Chou: Multilayer resist methods for nanoimprint lithography on non-flat surfaces, J. Vac. Sci. Technol. B **16**(6), 3922–3925 (1998)

9.128 F. van Delft: Bilayer resist used in e-beam lithography for deep narrow structures, Microelectron. Eng. **46**, 369–373 (1999)

9.129 L. Tan, Y.P. Kong, L.-L. Bao, X.D. Huang, L.J. Guo, S.W. Pang, A.F. Yee: Imprinting polymer film on patterned substrates, J. Vac. Sci. Technol. B **21**(6), 2742–2748 (2003)

9.130 B. Faircloth, H. Rohrs, R. Tiberio, R. Ruoff, R.R. Krchnavek: Bilayer nanoimprint lithography, J. Vac. Sci. Technol. B **18**(4), 1866–1873 (2000)

9.131 A. Lebib, M. Natali, S.P. Li, E. Cambril, L. Manin, Y. Chen, H.M. Janssen, R.P. Sijbesma: Control of the critical dimension with a trilayer nanoimprint lithography procedure, Microelectron. Eng. **57/58**, 411–416 (2001)

9.132 Y. Chen, K. Peng, Z. Cui: A lift-off process for high resolution patterns using PMMA/LOR resist stack, Microelectron. Eng. **73/74**, 278–281 (2004)

9.133 P. Carlberg, M. Graczyk, E.-L. Sawe, I. Maximov, M. Beck, L. Montelius: Lift-off process for nanoimprint lithography, Microelectron. Eng. **67/68**, 203–207 (2003)

9.134 W. Li, J.O. Tegenfeldt, L. Chen, R.H. Austin, S.Y. Chou, P.A. Kohl, J. Krotine, J.C. Sturm: Sacrificial polymers for nanofluidic channels in biological applications, Nanotechnology **14**, 578–583 (2003)

9.135 MicroChem Corp.: http://www.microchem.com/ (MicroChem Corp., Newton 2009)

9.136 M.W. Lin, H.-L. Chao, J. Hao, E.K. Kim, F. Palmieri, W.C. Kim, M. Dickey, P.S. Ho, C.G. Willson: Planarization for reverse-tone step and flash imprint lithography, Proc. SPIE **6151**, 688–699 (2006)

9.137 W. Trybula: Sematech, AMRC, and nano, Nanoprint Nanoimpr. Technol. (NNT) Conf., Vienna (2004)

9.138 S. Johnson, D.J. Resnick, D. Mancini, K.J. Nordquist, W.J. Dauksher, K. Gehoski, J.H. Baker, L. Dues, A. Hooper, T.C. Bailey, S.V. Sreenivasan, J.G. Ekerdt, C.G. Willson: Fabrication of multi-tiered structures on step and flash imprint lithography templates, Microelectron. Eng. **67/68**, 221–228 (2003)

9.139 D. Suh, J. Rhee, H.H. Lee: Bilayer reversal imprint lithography: direct metal–polymer transfer, Nanotechnology **15**, 1103–1107 (2004)

9.140 Y.P. Kong, H.Y. Lowa, S.W. Pang, A.F. Yee: Duo-mold imprinting of three-dimensional polymeric structures, J. Vac. Sci. Technol. B **22**(6), 3251–3265 (2004)

9.141 T. Borzenko, M. Tormen, G. Schmidt, L.W. Molenkamp: Polymer bonding process for nanolithography, Appl. Phys. Lett. **79**(14), 2246–2248 (2001)

9.142 X.D. Huang, L.-R. Bao, X. Cheng, L.J. Guo, S.W. Pang, A.F. Yee: Reversal imprinting by transferring polymer from mold to substrate, J. Vac. Sci. Technol. B **20**(6), 2872–2876 (2002)

9.143 N. Kehagias, V. Reboud, G. Chansin, M. Zelsmann, C. Jeppesen, C. Schuster, M. Kubenz, F. Reuther, G. Grützner, C.M. Sotomayor Torres: Reverse-contact UV nanoimprint lithography for multilayered structure fabrication, Nanotechnology **18**, 175303 (2007)

9.144 micro resist technology GmbH: http://www.microresist.de/ (micro resist technology GmbH, Berlin 2009)

9.145 Polysciences Inc.: http://www.polysciences.com (Polysciences Inc., Warrington 2009)

9.146 Allresist GmbH: http://www.allresist.de (Allresist GmbH, Strausberg 2009)

9.147 C.-Y. Chao, L.J. Guo: Polymer microring resonators fabricated by nanoimprint technique, J. Vac. Sci. Technol. B **20**, 2862–2866 (2002)

9.148 Bayer AG: http://plastics.bayer.com (Bayer Material Science, Leverkusen 2009)

9.149 LG Dow Polycarbonate Ltd.: http://www.lg-dow.com (LG Dow Polycarbonate Ltd., Yeosu Chunnam 2009)

9.150 J. Tallal, D. Peyrade, F. Lazzarino, K. Berton, C. Perret, M. Gordon, C. Gourgon, P. Schiavone: Replication of sub-40 nm gap nanoelectrodes over an 8 in. substrate by nanoimprint lithography, Microelectron. Eng. **78/79**, 676–681 (2005)

9.151 Zeon Chemicals L. P.: http://www.zeonchemicals.com (Zeon Chemicals L. P., Louisville 2009)

9.152 Topas Advanced Polymers: http://www.topas.com/ (Topas Advanced Polymers, Florence 2009)

9.153 T. Nielsen, D. Nilsson, F. Bundgaard, P. Shi, P. Szabo, O. Geschke, A. Kristensen: Nanoimprint lithography in the cyclic olefin copolymer, Topas, a highly UV-transparent and chemically resistant thermoplast, J. Vac. Sci. Technol. B **22**, 1770–1775 (2004)

9.154 B. Simmons, B. Lapizco-Encinas, R. Shediac, J. Hachman, J. Chames, J. Brazzle, J. Ceremuga, G. Fiechtner, E. Cummings, Y. Fintschenko: Polymeric insulator-based (electrodeless) dielectrophoresis (iDEP) for the monitoring of water-borne pathogens, Proc. MicroTAS **2**, 171–173 (2004)

9.155 D. Nilsson, S. Balslev, A. Kristensen: A microfluidic dye laser fabricated by nanoimprint lithography in a highly transparent and chemically resistant cyclo-olefin copolymer (COC), J. Micromech. Microeng. **15**, 296–300 (2005)

9.156 K. Pfeiffer, M. Fink, G. Ahrens, G. Grützner, F. Reuther, J. Seekamp, S. Zankovych, C.M. Sotomayor Torres, I. Maximov, M. Beck, M. Graczyk, L. Montelius, H. Schulz, H.-C. Scheer, F. Steingrüber: Polymer stamps for nanoimprinting, Microelectron. Eng. **61/62**, 393–398 (2002)

9.157 M. Wissen, H. Schulz, N. Bogdanski, H.-C. Scheer, Y. Hirai, H. Kikuta, G. Ahrens, F. Reuther, K. Pfeiffer: UV curing of resists for warm embossing, Microelectron. Eng. **73/74**, 184–189 (2004)

9.158 Sumitomo Chemical Corp.: http://www.sumitomo-chem.co.jp/ (Sumitomo Chemical Corp., Sendai 2009)

9.159 S. Landis, N. Chaix, C. Gourgon, C. Perret, T. Leveder: Stamp design effect on 100 nm feature size for 8 inch nanoimprint lithography, Nanotechnology **17**, 2701–2709 (2006)

9.160 N. Chaix, C. Gourgon, S. Landis, C. Perret, M. Fink, F. Reuther, D. Mecerreyes: Influence of the molecular weight and imprint conditions on the formation of capillary bridges in nanoimprint lithography, Nanotechnology **17**, 4082–4087 (2006)

9.161 C.G. Willson, R.A. Dammel, A. Reiser: Photoresist materials: A historical perspective, Proc. SPIE **3049**, 28–41 (1997)

9.162 M.D. Stewart, C.G. Willson: Photoresists. In: *Encyclopedia of Materials: Science and Technology*, ed. by K.H.J. Buschow, R.W. Cahn, M.C. Flemings, B. Ilschner, E.J. Kramer, H.E.H. Meijer, S. Mahajan (Routledge, New York 2001) pp. 6973–6978

9.163 K. Pfeiffer, G. Bleidießel, G. Grützner, H. Schulz, T. Hoffmann, H.-C. Scheer, C.M. Sotomayor Torres, J. Ahopelto: Suitability of new polymer materials with adjustable glass temperature for nanoimprinting, Microelectron. Eng. **46**, 431–434 (1999)

9.164 F. Gaboriau, M.-C. Peignon, A. Barreau, G. Turban, C. Cardinaud, K. Pfeiffer, G. Bleidießel, G. Grutzner: High density fluorocarbon plasma etching of new resists suitable for nanoimprint lithography, Microelectron. Eng. **53**, 501–505 (2000)

9.165 F. Gottschalch, T. Hoffmann, C.M. Sotomayor Torres, H. Schulz, H.-C. Scheer: Polymer issues in nanoimprinting technique, Solid-State Electron. **43**, 1079–1083 (1999)

9.166 H. Schulz, H.-C. Scheer, T. Hoffmann, C.M. Sotomayor Torres, K. Pfeiffer, G. Bleidießel, G. Grützner, C. Cardinaud, F. Gaboriau, M.-C. Peignon, J. Ahopelto, B. Heidari: New polymer materials for nanoimprinting, J. Vac. Sci. Technol. B **18**(4), 1861–1865 (2000)

9.167 D. Lyebyedyev, H. Schulz, H.-C. Scheer: Characterisation of new thermosetting polymer materials for nanoimprint lithography, Mater. Sci. Eng. C **15**(1/2), 241–243 (2001)

9.168 K. Pfeiffer, F. Reuther, M. Fink, G. Grützner, P. Carlberg, I. Maximov, L. Montelius, J. Seekamp, S. Zankovych, C.M. Sotomayor Torres, H. Schulz, H.-C. Scheer: A comparison of thermally and photochemically cross-linked polymers for nanoimprinting, Microelectron. Eng. **67/68**, 266–273 (2003)

9.169 C.D. Schaper, A. Miahnhari: Polyvinyl alcohol templates for low cost, high resolution, complex printing, J. Vac. Sci. Technol. B **22**(6), 3323–3326 (2002)

9.170 R.M. Reano, Y.P. Kong, H.Y. Low, L. Tan, F. Wang, S.W. Pang, A.F. Yee: Stability of functional polymers after plasticizer-assisted imprint lithography, J. Vac. Sci. Technol. B **22**(6), 3294–3299 (2002)

9.171 B.K. Long, B.K. Keitz, C.G. Willson: Materials for step and flash imprint lithography (S-FIL), J. Mater. Chem. **17**, 3575–3580 (2007)

9.172 J. Hao, M. Lin, F. Palmieri, Y. Nishimura, H.-L. Chao, M.D. Stewart, A. Collins, K. Jen, C.G. Willson: Photocurable silicon-base material for imprinting lithography, Proc. SPIE **6517**, 6517–6580 (2007)

9.173 S. Johnson, R. Burns, E.K. Kim, M. Dickey, G. Schmid, J. Meiring, S. Burns, C.G. Willson, D. Convey, Y. Wei, P. Fejes, K. Gehoski, D. Mancini, K. Nordquist, W.J. Dauksher, D.J. Resnick: Effects of etch barrier densification on step and flash imprint lithography, J. Vac. Sci. Technol. B **23**(6), 2553–2556 (2005)

9.174 F. Xu, N. Stacey, M. Watts, V. Truskett, I. McMackin, J. Choi, P. Schumaker, E. Thompson, D. Babbs, S.V. Sreenivasan, G. Willson, N. Schumaker: Development of imprint materials for the step and flash imprint lithography process, Proc. SPIE **5374**, 232–241 (2004)

9.175 M. Vogler, S. Wiedenberg, M. Mühlberger, I. Bergmair, T. Glinsner, H. Schmidt, E.-B. Kley, G. Grützner: Development of a novel, low-viscosity UV-curable polymer system for UV-nanoimprint lithography, Microelectron. Eng. **84**, 984–988 (2007)

9.176 P. Voisin, M. Zelsmann, R. Cluzel, E. Pargon, C. Gourgon, J. Boussey: Characterisation of ultraviolet nanoimprint dedicated resists, Microelectron. Eng. **84**, 967–972 (2007)

9.177 H. Schmitt, L. Frey, H. Ryssel, M. Rommel, C. Lehrer: UV nanoimprint materials: surface energies, residual layers, and imprint quality, J. Vac. Sci. Technol. B **25**(3), 785–790 (2007)

9.178 W.-C. Liao, S.L.-C. Hsu: A novel liquid thermal polymerization resist for nanoimprint lithography with low shrinkage and high flowability, Nanotechnology **18**, 065303 (2007)

9.179 F.A. Houle, C.T. Rettner, D.C. Miller, R. Sooriyakumaran: Antiadhesion considerations for UV nanoimprint lithography, Appl. Phys. Lett. **90**, 213103 (2007)

9.180 F.A. Houle, E. Guyer, D.C. Miller, R. Dauskardt: Adhesion between template materials and UV-cured nanoimprint resists, J. Vac. Sci. Technol. B **25**(4), 1179–1185 (2007)

9.181 M. Köhler: *Etching in Microsystem Technology* (Wiley-VCH, Weinheim 1999)

9.182 H. Schift, J. Gobrecht, B. Satilmis, J. Söchtig, F. Meier, W. Raupach: Nanoreplikation im Verbund: Ein Schweizer Netzwerk, Kunststoffe **94**, 22–26 (2004), in German (English vers.: Nanoreplication in a Network, Kunstst. Plast Eur. **94**, 1–4 (2004))

9.183 S. Park, H. Schift, H.H. Solak, J. Gobrecht: Stamps for nanoimprint lithography by extreme ultraviolet interference lithography, J. Vac. Sci. Technol. B **22**(6), 3246–3250 (2004)

9.184 K.A. Lister, B.G. Casey, P.S. Dobson, S. Thoms, D.S. Macintyre, C.D.W. Wilkinson, J.M.R. Weaver: Pattern transfer of a 23 nm-period grating and sub-15 nm dots into CVD diamond, Microelectron. Eng. **73/74**, 319–322 (2004)

9.185 J. Taniguchi, Y. Tokano, I. Miyamoto, M. Komuro, H. Hiroshima: Diamond nanoimprint lithography, Nanotechnology **13**, 592–596 (2002)

9.186 Y. Hirai, S. Yoshida, N. Takagi, Y. Tanaka, H. Yabe, K. Sasaki, H. Sumitani, K. Yamamoto: High aspect pattern fabrication by nano imprint lithography using fine diamond mold, Jpn. J. Appl. Phys. **42**(6B), 3863–3866 (2003)

9.187 S.W. Pang, T. Tamamura, M. Nakao, A. Ozawa, H. Masuda: Direct nano-printing on Al substrate using SiC mold, J. Vac. Sci. Technol. B **16**, 1145–1149 (1998)

9.188 J. Gao, M.B. Chan-Park, D. Xie, Y. Yan, W. Zhou, B.K.A. Ngoi, C.Y. Yue: UV embossing of submicron patterns on biocompatible polymeric films using a focused ion beam fabricated mold, Chem. Mater. **16**(6), 956–958 (2004)

9.189 M.M. Alkaisi, R.J. Blaikie, S.J. McNab: Low temperature nanoimprint lithography using silicon nitride molds, Microelectron. Eng. **57/58**, 367–373 (2001)

9.190 Y. Hirai, S. Harada, S. Isaka, M. Kobayashi, Y. Tanaka: Nano-imprint lithography using replicated mold by Ni electroforming, Jpn. J. Appl. Phys. **41**(6B), 4186–4189 (2002)

9.191 Z. Yu, L. Chen, W. Wu, H. Ge, S.Y. Chou: Fabrication of nanoscale gratings with reduced line edge roughness using nanoimprint lithography, J. Vac. Sci. Technol. B **21**(5), 2089–2092 (2003)

9.192 N. Roos, H. Schulz, L. Bendfeldt, M. Fink, K. Pfeiffer, H.-C. Scheer: First and second generation purely thermoset stamps for hot embossing, Microelectron. Eng. **61/62**, 399–405 (2002)

9.193 N. Roos, H. Schulz, M. Fink, K. Pfeiffer, F. Osenberg, H.-C. Scheer: Performance of 4″ wafer-scale thermoset working stamps in hot embossing lithography, Proc. SPIE **4688**, 232–239 (2002)

9.194 H. Schift, S. Park, J. Gobrecht, S. Saxer, F. Meier, W. Raupach, K. Vogelsang: Hybrid bendable stamp copies for molding fabricated by nanoimprint, Microelectron. Eng. **78/79**, 605–611 (2005)

9.195 R.W. Jaszewski, H. Schift, B. Schnyder, A. Schneuwly, P. Gröning: The deposition on anti-adhesive ultra-thin teflon-like films and their interaction with polymers during hot embossing, Appl. Surf. Sci. **143**, 301–308 (1999)

9.196 R.W. Jaszewski, H. Schift, P. Gröning, G. Margaritondo: Properties of thin anti-adhesive films used for the replication of microstructures in polymers, Microelectron. Eng. **35**, 381–384 (1997)

9.197 U. Srinivasan, M.R. Houston, R.T. Howe, R. Maboudian: Alkyltrichlorosilane-based self-assembled monolayer films for stiction reduction in silicon micromachines, J. Microelectromech. Syst. **7**, 252–260 (1998)

9.198 H. Schulz, F. Osenberg, J. Engemann, H.-C. Scheer: Mask fabrication by nanoimprint lithography using antisticking layers, Proc. SPIE **3996**, 244–249 (2000)

9.199 M. Beck, M. Graczyk, I. Maximov, E.-L. Sarwe, T.G.I. Ling, M. Keil, L. Montelius: Improving stamps for 10 nm level wafer scale nanoimprint lithography, Microelectron. Eng. **61/62**, 441–448 (2002)

9.200 H. Schift, S. Saxer, S. Park, C. Padeste, U. Pieles, J. Gobrecht: Controlled co-evaporation of silanes for nanoimprint stamps, Nanotechnology **16**, S171–S175 (2005)

9.201 M. Keil, M. Beck, G. Frennesson, E. Theander, E. Bolmsjö, L. Montelius, B. Heidari: Process development and characterization of antisticking layers on nickel-based stamps designed for nanoimprint lithography, J. Vac. Sci. Technol. B **22**(6), 3283–3287 (2002)

9.202 S. Park, H. Schift, C. Padeste, B. Schnyder, R. Kötz, J. Gobrecht: Anti-adhesive layers on nickel stamps for nanoimprint lithography, Microelectron. Eng. **73/74**, 196–201 (2004)

9.203 ABCR GmbH: http://www.abcr.de/ (ABCR GmbH, Karlsruhe 2009)

9.204 B. Heidari, I. Maximov, E.-L. Sarwe, L. Montelius: Large scale nanolithography using imprint lithography, J. Vac. Sci. Technol. B **17**, 2961–2964 (1999)

9.205 B. Heidari, I. Maximov, L. Montelius: Nanoimprint lithography at the 6 in. wafer scale, J. Vac. Sci. Technol. B **18**(6), 3557–3560 (2000)

9.206 N. Roos, T. Luxbacher, T. Glinsner, K. Pfeiffer, H. Schulz, H.-C. Scheer: Nanoimprint lithography with a commercial 4 inch bond system for hot embossing, Proc. SPIE **4343**, 427–436 (2001)

9.207 L. Bendfeldt, H. Schulz, N. Roos, H.-C. Scheer: Groove design of vacuum chucks for hot embossing lithography, Microelectron. Eng. **61/62**, 455–459 (2002)

9.208 T. Haatainen, J. Ahopelto, G. Grützner, M. Fink, K. Pfeiffer: Step and stamp imprint lithography using a commercial flip chip bonder, Proc. SPIE **3997**, 874–879 (2000)

9.209 H. Tana, A. Gilbertson, S.Y. Chou: Roller nanoimprint lithography, J. Vac. Sci. Technol. B **16**(6), 3926–3928 (1998)

9.210 M. Tormen: A nano impression lithographic process which involves the use of a die having a region able to generate heat, European Patent PCT/IB 2004/002120 (2004)

9.211 S.Y. Chou, C. Keimel, J. Gu: Ultrafast and direct imprint of nanostructures in silicon, Nature **417**, 835–837 (2002)

9.212 J.J. Shamaly, V.F. Bunze: I-line to DUV transition for critical levels, Microelectron. Eng. **30**, 87–93 (1996)

9.213 J.E. Bjorkholm: EUV lithography – The successor to optical lithography?, Intel Technol. J. **Q3** (1998), http://www.intel.com/technology/itj/q31998/articles/art_4.htm

9.214 D. Wachenschwanz, W. Jiang, E. Roddick, A. Homola, P. Dorsey, B. Harper, D. Treves, C. Bajorek: Design of a manufacturable discrete track recording medium, IEEE Trans. Mag. **41**, 670–675 (2005)

9.215 G.M. McClelland, M.W. Hart, C.T. Rettner, M.E. Best, K.R. Carter, B.D. Terris: Nanoscale patterning of magnetic islands by imprint lithography using a flexible mold, Appl. Phys. Lett. **81**, 1483–1485 (2002)

9.216 G.F. Cardinale, J.L. Skinner, A.A. Talin, R.W. Brocato, D.W. Palmer, D.P. Mancini, W.J. Dauksher, K. Gehoski, N. Le, K.J. Nordquist, D.J. Resnick: Fabrication of a surface acoustic wave-based correlator using step-and-flash imprint lithography, J. Vac. Sci. Technol. B **22**, 3265–3270 (2004)

9.217 S.-W. Ahn, K.-D. Lee, J.-S. Kim, S.H. Kim, S.H. Lee, J.-D. Park, P.-W. Yoon: Fabrication of subwavelength aluminum wire grating using nanoimprint lithography and reactive ion etching, Microelectron. Eng. **78/79**, 314–318 (2005)

9.218 J. Seekamp, S. Zankovych, A.H. Helfer, P. Maury, C.M. Sotomayor Torres, G. Böttger, C. Liguda, M. Eich, B. Heidari, L. Montelius, J. Ahopelto: Nanoimprinted passive optical devices, Nanotechnology **13**, 581–586 (2002)

9.219 C.M. Sotomayor Torres, S. Zankovych, J. Seekamp, A.P. Kam, C. Clavijo Cedeño, T. Hoffmann, J. Ahopelto, F. Reuther, K. Pfeiffer, G. Bleidießel, G. Grützner, M.V. Maximov, B. Heidari: Nanoimprint lithography: An alternative nanofabrication approach, Mater. Sci. Eng. C **23**, 23–31 (2003)

9.220 J. Wang, X. Sun, L. Chen, S.Y. Chou: Direct nanoimprint of submicron organic light-emitting structures, Appl. Phys. Lett. **75**, 2767–2769 (1999)

9.221 X. Cheng, Y. Hong, J. Kanicki, L.J. Guo: High-resolution organic polymer light-emitting pixels fabricated by imprinting technique, J. Vac. Sci. Technol. B **20**, 2877–2880 (2002)

9.222 D. Pisignano, L. Persano, E. Mele, P. Visconti, R. Cingolani, G. Gigli, G. Barbarella, L. Favaretto: Emission properties of printed organic semiconductor lasers, Opt. Lett. **30**, 260–262 (1995)

9.223 D. Nilsson, T. Nielsen, A. Kristensen: Solid state micro-cavity dye lasers fabricated by nanoimprint lithography, Rev. Sci. Instrum. **75**, 4481–4486 (2004)

9.224 C. Clavijo Cedeño, J. Seekamp, A.P. Kam, T. Hoffmann, S. Zankovych, C.M. Sotomayor Torres, C. Menozzi, M. Cavallini, M. Murgia, G. Ruani, F. Biscarini, M. Behl, R. Zentel, J. Ahopelto: Nanoimprint lithography for organic electronics, Microelectron. Eng. **61/62**, 25–31 (2002)

9.225 A. Manz, N. Graber, H.M. Widmer: Miniaturized total chemical analysis systems: A novel concept for chemical sensing, Sens. Actuators B **1**, 244–248 (1990)

9.226 E. Verpoorte, N.F. De Rooij: Microfluidics meets MEMS, Proc. IEEE **91**, 930–953 (2003)

9.227 A. Pepin, P. Youinou, V. Studer, A. Lebib, Y. Chen: Nanoimprint lithography for the fabrication of DNA electrophoresis chips, Microelectron. Eng. **61/62**, 927–932 (2002)

9.228 J.O. Tegenfeldt, C. Prinz, H. Cao, R.L. Huang, R.H. Austin, S.Y. Chou, E.C. Cox, J.C. Sturm: Micro- and nanofluidics for DNA analysis, Anal. Bioanal. Chem. **378**, 1678–1692 (2004)

9.229 S.Y. Chou: Patterned magnetic nanostructures and quantized magnetic disks, Proc. IEEE **85**, 652–671 (1997)

9.230 M.N. Baibich, J.M. Broto, A. Fert, F. Nguyen Van Dau, F. Petroff, P. Eitenne, G. Creuzet, A. Friederich, J. Chazelas: Giant magnetoresistance of (001)Fe/(001)Cr magnetic superlattices, Phys. Rev. Lett. **61**, 2472–2475 (1988)

9.231 Y. Li, A.K. Menon: Magnetic recording technologies: Overview. In: *Encyclopedia of Materials: Science and Technology*, ed. by K.H.J. Buschow (Elsevier, Amsterdam 2001) pp. 4948–4957

9.232 L.F. Shew: Discrete tracks for saturation magnetic recording, IEEE Trans. Broadcast Telev. Receiv. **9**, 56–62 (1963)

9.233 A.K. Menon: Interface tribology for 100 Gb/in^2, Tribol. Int. **33**, 299–308 (2000)

9.234 Y. Soeno, M. Moriya, K. Ito, K. Hattori, A. Kaizu, T. Aoyama, M. Matsuzaki, H. Sakai: Feasibility of discrete track perpendicular media for high track density recording, IEEE Trans. Magn. **39**, 1967–1971 (2003)

9.235 S.Y. Chou, M. Wei, P.R. Krauss, P.B. Fisher: Study of nanoscale magnetic structures fabricated using electron beam lithography and quantum magnetic disk, J. Vac. Sci. Technol. B **12**, 3695–3698 (1994)

9.236 R.L. White, R.M.H. Newt, R.F.W. Pease: Patterned media: A viable route to 50 Gbit/in^2 and up for magnetic recording?, IEEE Trans. Magn. **33**, 990–995 (1997)

9.237 B.D. Terris, T. Thomson: Nanofabricated and self-assembled magnetic structures as data storage media, J. Phys. D **38**, R199–R222 (2005)

9.238 Z.Z. Bandic, E.A. Dobisz, T.-W. Wu, T.R. Albrecht: Patterning on hard disk drives, Solid State Technol. **Sept**, 57–59 (2006)

9.239 A. Kikitsu, Y. Kamata, M. Sakurai, K. Naito: Recent progress of patterned media, IEEE Trans. Magn. **43**, 3685–3688 (2007)

9.240 M. Natali, A. Lebib, E. Cambril, Y. Chen, I.L. Prejbeanu, K. Ounadjela: Nanoimprint lithography of high-density cobalt dot patterns for fine tuning of dipole interactions, J. Vac. Sci. Technol. B **19**, 2779–2783 (2001)

9.241 J. Moritz, B. Dieny, J.P. Nozieres, S. Landis, A. Lebib, Y. Chen: Domain structure in magnetic dots prepared by nanoimprint and e-beam lithography, J. Appl. Phys. **91**, 7314–7316 (2002)

9.242 P. Lalanne, M. Hutley: Artificial media optical properties-subwavelength scale. In: *Enclopedia of Optical Engineering*, ed. by R.G. Driggers (Dekker, New York 2003) pp. 62–71

9.243 http://www.pcmag.com/ (last accessed December 9, 2009)

9.244 Z. Yu, W. Wu, L. Chen, S. Chou: Fabrication of large area 100 nm pitch grating by spatial frequency doubling an nanoimprint lithography for subwavelength optical applications, J. Vac. Sci. Technol. B **19**, 2816–2819 (2001)

9.245 MOXTEK Inc.: http://www.moxtek.com/ (last accessed December 9, 2009)

9.246 NanoOpto, API Nanotronics Corp.: http://www.nanoopto.com/ (last accessed December 9, 2009)

9.247 A.A. Erchak, D.J. Ripin, S. Fan, P. Rakich, J.D. Joannopoulos, E.P. Ippen, G.S. Petrich, L.A. Kolodziejski: Enhanced coupling to vertical radiation using a two-dimensional photonic crystal in a semiconductor light-emitting diode, Appl. Phys. Lett. **78**, 563–565 (2001)

9.248 S.H. Kim, K.-D. Lee, J.-Y. Kim, M.-K. Kwon, S.-J. Park: Fabrication of photonic crystal structures on light emitting diodes by nanoimprint lithography, Nanotechnology **18**, 055306 (2007)

9.249 L.J. Guo, X. Cheng, C.Y. Chao: Fabrication of photonic nanostructures in nonlinear optical polymers, J. Mod. Opt. **49**, 663–673 (2002)

9.250 C.-Y. Chao, L.J. Guo: reduction of surface scattering loss in polymer microrings using thermal-reflow technique, IEEE Photon. Technol. Lett. **16**, 1498–1500 (2004)

9.251 H.C. Hoch, L.W. Jelinski, H.G. Craighead (Eds.): *Nanofabrication and Biosystems: Integrating Materials Science, Engineering, and Biology* (Cambridge Univ. Press, Cambridge 1996)

9.252 H.G. Craighead: Nanoelectromechanical systems, Science **290**, 1532–1535 (2000)

9.253 L.R. Huang, J.O. Tegenfeldt, J.J. Kraeft, J.C. Sturm, R.H. Austin, E.C. Cox: A DNA prism for high-speed continous frationation of large DNA molecules, Nat. Biotechnol. **20**, 1048–1051 (2002)

9.254 H.G. Craighead: Nanostructure science and technology: Impact and prospects for biology, J. Vac. Sci. Technol. A **21**, S216–S221 (2003)

9.255 J.O. Tegenfeldt, C. Prinz, H. Cao, S. Chou, W.W. Reisner, R. Riehn, Y.M. Wang, E.C. Cox, J.C. Sturm, P. Silberzan, R.H. Austin: The dynamics of genomic-length DNA molecules in 100-nm channels, Proc. Natl. Acad. Sci. USA **101**, 10979–10983 (2004)

9.256 L.J. Guo, X. Cheng, C.-F. Chou: Fabrication of size-controllable nanofluidic channels by nanoimprinting and its application for DNA stretching, Nano Lett. **4**, 69–73 (2004)

9.257 C. Bustamante, J.F. Marko, E.D. Siggia, S. Smith: Entropic elasticity of λ-phage DNA, Science **265**, 1599–1600 (1994)

9.258 W. Kern, D.A. Puotinen: RCA Rev. **31**, 187–206 (1970)

9.259 L.H. Thamdrup, A. Klukowska, A. Kristensen: Stretching DNA in polymer nanochannels fabricated by thermal imprint in PMMA, Nanotechnology **19**, 125301 (2008)

9.260 M.J. Dalby, N. Gadegaard, R. Tare, A. Andar, M.O. Riehle, P. Herzyk, C.D.W. Wilkinson, R.O.C. Oreffo: The control of human mesenchymal cell differentiation using nanoscale symmetry and disorder, Nat. Mater. **6**, 997–1003 (2007)

9.261 K. Seunarine, D.O. Meredith, M.O. Riehle, C.D.W. Wilkinson, N. Gadegaard: Biodegradable polymer tubes with lithographically controlled 3-D micro- and nanotopography, Microelectron. Eng. **85**(5/6), 1350–1354 (2008)

9.262 A. Kapr: *Johann Gutenberg: The Man and His Invention* (Scolar, London 1996), http://www.gutenberg.de/publ.htm

9.263 EVGroup: http://www.evgroup.com/ (EVGroup, St. Florian 2009)

9.264 SÜSS Microtec: http://www.suss.com/ (SÜSS Microtec, Garching 2009)

9.265 Obducat: http://www.obducat.com/ (Obducat, Malmö 2009)

9.266 Smart Equipment Technology S.A.S.: http://www.set-sas.fr/ (Smart Equipment Technology S.A.S., Saint Jeoire 2009)

9.267 Jenoptik: http://www.jenoptik.com (Jenoptik, Jena 2009)

9.268 Molecular Imprints: http://www.molecularimprints.com/ (Molecular Imprints, Austin 2009)

9.269 Nanonex: http://www.nanonex.com/ (Nanonex, Monmouth Junction 2009)

9.270 Eulitha: http://www.eulitha.com/ (Eulitha, Villigen 2009)

9.271 NIL Technology: http://www.nilt.com/ (NIL Technology, Kongens Lyngby 2009)

9.272 Sematech: http://www.sematech.org/ (Sematech, Austin 2009)

9.273 M. Beck, B. Heidari: Nanoimprint lithography for high volume HDI manufacturing, OnBoard Technol. **Sept.**, 52–55 (2006), http://www.onboard-technology.com/

9.274 L. Olsson: Method and device for transferring a pattern, European Patent PCT/SE 2003/001003 (2002)

10. Stamping Techniques for Micro- and Nanofabrication

Etienne Menard, John A. Rogers

Soft-lithographic techniques that use rubber stamps and molds provide simple means to generate patterns with lateral dimensions that can be much smaller than 1 μm and can even extend into the single nanometer regime. These methods rely on the use of soft elastomeric elements typically made out of the polymer poly(dimethylsiloxane). The first section of this chapter presents the fabrication techniques for these elements together with data and experiments that provide insights into the fundamental resolution limits. Next, several representative soft-lithography techniques based on the use of these elements are presented: (i) microcontact printing, which uses molecular *inks* that form self-assembled monolayers, (ii) near- and proximity-field photolithography for producing two- and three-dimensional structures with subwavelength resolution features, and (iii) nanotransfer printing, where soft or hard stamps

10.1	**High-Resolution Stamps** 314
10.2	**Microcontact Printing** 316
10.3	**Nanotransfer Printing** 318
10.4	**Applications** 322
	10.4.1 Unconventional Electronic Systems 322
	10.4.2 Lasers and Waveguide Structures ... 328
10.5	**Conclusions** 329
References 330	

print single or multiple layers of solid inks with feature sizes down to 100 nm. The chapter concludes with descriptions of some device-level applications that highlight the patterning capabilities and potential commercial uses of these techniques.

There is considerable interest in methods that can be used to build structures that have micron or nanometer dimensions. Historically, research and development in this area has been driven mainly by the needs of the microelectronics industry. The spectacularly successful techniques that have emerged from those efforts – such as photolithography and electron beam lithography – are extremely well suited to the tasks for which they were principally designed: forming structures of radiation-sensitive materials (including photoresists or electron beam resists) on ultraflat glass or semiconductor surfaces. Significant challenges exist in adapting these methods for new emerging applications and areas of research that require patterning of unusual systems and materials, (including those in biotechnology and plastic electronics), structures with nanometer dimensions (below 50–100 nm), large areas in a single step (larger than a few square centimeters), or nonplanar (rough or curved) surfaces. These established techniques also have the disadvantage of high capital and operational costs. As a result, some of the oldest and conceptually simplest forms of lithography – embossing, molding, stamping, writing, and so on – are now being reexamined for their potential to serve as the basis for nanofabrication techniques that can avoid these limitations [10.1]. Considerable progress has been made in the last few years, mainly by combining these approaches or variants of them with new materials, chemistries, and processing techniques. This chapter highlights some recent advances in high-resolution printing methods, in which a *stamp* forms a pattern of *ink* on a surface that it contacts. It focuses on approaches whose capabilities, level of development, and demonstrated applications indicate a strong potential for widespread use, especially in areas where conventional methods are unsuitable.

Contact printing involves the use of an element with surface relief (the *stamp*) to transfer material applied to its surface (the *ink*) to locations on a substrate that it contacts. The printing press, one of the earliest manufacturable implementations of this approach, was introduced by Gutenberg in the fifteenth century. Since then, this general approach has been used almost exclusively for producing printed text or images with features that are 100 μm or larger in their smallest dimension. The resolution is determined by the nature of the ink and its interaction with the stamp and/or substrate, the resolution of the stamp, and the processing conditions that are used for printing or to convert the pattern of ink into a pattern of functional material. This chapter focuses on (1) printing techniques that are capable of micron and nanometer resolution, and (2) their use for fabricating key elements of active electronic or optical devices and subsystems. It begins with an overview of some methods for fabricating high-resolution stamps and then illustrates two different ways that these stamps can be used to print patterns of functional materials. Applications that highlight the capabilities of these techniques and the performances of systems that are constructed with them are also presented.

10.1 High-Resolution Stamps

The printing process can be separated into two parts: fabrication of the stamp and the use of this stamp to pattern features defined by the relief on its surface. These two processes are typically quite different, although it is possible in some cases to use patterns generated by a stamp to produce a replica of that stamp. The structure from which the stamp is derived, which is known as the *master*, can be fabricated with any technique that is capable of producing well-defined structures of relief on a surface. This master can then be used directly as the stamp, or to produce stamps via molding or printing procedures. It is important to note that the technique for producing the master does not need to be fast or low in cost. It also does not need to possess many other characteristics that might be desirable for a given patterning task: it is used just once to produce a master, which is directly or indirectly used to fabricate stamps. Each one of these stamps can then be used many times for printing.

In a common approach for the high-resolution techniques that are the focus of this chapter, an established lithographic technique, such as one of those developed for the microelectronics industry, defines the master. Figure 10.1 schematically illustrates typical processes. Here, photolithography patterns a thin layer of resist onto a silicon wafer. Stamps are generated from this structure in one of two ways: by casting against this master, or by etching the substrate with the patterned resist as a mask. In the first approach, the master itself can be used multiple times to produce many stamps, typically using a light or heat-curable prepolymer. In the second, the etched substrate serves as the stamp. Additional stamps can be generated either by repeating the lithography and etching, or by using the original stamp to print replica stamps. For minimum lateral feature sizes that are greater than ≈ 1−2 μm, contact-

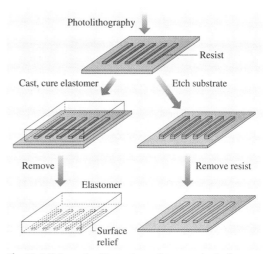

Fig. 10.1 Schematic illustration of two methods for producing high-resolution stamps. The first step involves patterning a thin layer of some radiation-sensitive material, known as the resist, on a flat substrate, such as a silicon wafer. It is convenient to use an established technique, such as photolithography or electron beam lithography, for this purpose. This structure, known as the *master*, is converted to a stamp either by etching or by molding. In the first case, the resist acts as a mask for etching the underlying substrate. Removing the resist yields a stamp. This structure can be used directly as a stamp to print patterns or to produce additional stamps. In the molding approach, a prepolymer is cast against the relief structure formed by the patterned resist on the substrate. Curing (thermally or optically) and then peeling the resulting polymer away from the substrate yields a stamp. In this approach, many stamps can be made with a single *master* and each stamp can be used many times

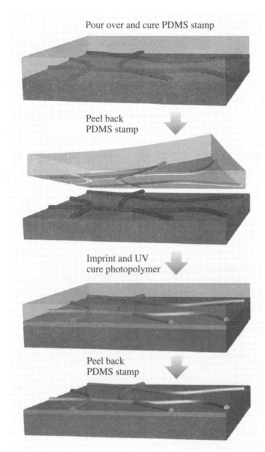

Fig. 10.2 Schematically illustrates a process for examining the ultimate limits in resolution of soft lithographic methods. The approach uses a SWNT master to create a PDMS mold with nanoscale relief features. Soft nanoimprint lithography transfers the relief on the PDMS to that on the surface of an ultraviolet curable photopolymer film

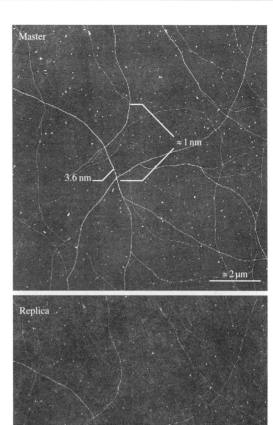

Fig. 10.3 Atomic force micrographs (*top* picture) of a *master* that consists of a submonolayer of single-walled carbon nanotubes (SWNTs; diameter between 0.5 and 5 nm) grown on a SiO_2/Si wafer. The *bottom* atomic force micrograph shows a replica of the relief structures in poly(urethane). These results indicate effective operation of a PDMS stamp for soft imprint lithography at the single nanometer scale

or proximity-mode photolithography with a mask produced by direct write photolithography represents a convenient method of fabricating the master. For features smaller than $\approx 2\,\mu m$, several different techniques can be used [10.2], including: (1) projection mode photolithography [10.3], (2) direct write electron beam (or focused ion beam) lithography [10.4, 5], (3) scanning probe lithography [10.6–9] or (4) laser interference lithography [10.10]. The first approach requires a photomask generated by some other method, such as direct write photolithography or electron beam lithography. The reduction (typically 4×) provided by the projection optics relaxes the resolution requirements on the mask and enables features as small as $\approx 90\,nm$ when deep ultraviolet radiation and phase shifting masks are used. The costs for these systems are, however, very high and their availability for general research purposes is limited. The second method is flexible in the geometry of patterns that can be produced, and the writing systems are highly developed: 30–50 nm features can be achieved with commercial systems [10.11], and < 10 nm features are possible with research tools, as first demonstrated more than 25 years ago by *Broers* [10.12]. The main drawbacks of this method are that it is relatively slow and it is difficult to pattern large areas. Like projection-mode photolithography, it can be expensive. The third method, scanning probe lithography, is quite powerful in principle, but the tools are not as well established as those for other approaches. This technique has atomic resolution, but its writing speed can be lower and the areas that can be patterned are smaller than electron beam systems. Interference lithography provides a powerful, low-cost tool for generating periodic arrays of features with dimensions down to 100–200 nm; smaller sizes demand ultraviolet lasers, and patterns with aperiodic or nonregular features are difficult to produce.

In order to evaluate the ultimate resolution limit of the soft lithography methods, masters with relief structures in the single nanometer range must be fabricated. A simple method, presented in Fig. 10.2, uses submonolayer coverage of single-walled carbon nanotubes (SWNT grown, by established chemical vapor deposition techniques, on an ultraflat silicon wafer. The SWNT, which have diameters (heights and widths) in the 0.5–5 nm range, are molded on the bottom surface of a PDMS stamp generated by casting and curing against this master. Such a mold can be used to replicate the relief structure into a variety of photocurable polymers in a kind of soft nanoimprinting technique [10.13–15]. A single mold can give exceedingly high resolution, approaching the single nanometer scale range (comparable to a few bond lengths in the polymer backbone), as can be seen in Fig. 10.3 [10.16]. These results demonstrate the extreme efficiency of the basic soft lithographic procedure for generating and using elastomeric elements. The ultimate limits are difficult to predict, due to substantial uncertainties surrounding the polymer physics and chemistry that dominates in the nanometer regime.

10.2 Microcontact Printing

Microcontact printing (μCP) [10.17] is one of several soft lithographic techniques – replica molding, micromolding in capillaries, microtransfer molding, near-field conformal photolithography using an elastomeric phase-shifting mask, and so on – that have been developed as alternatives to established methods for micro- and nanofabrication [10.18–22]. μCP uses an elastomeric element (usually polydimethylsiloxane – PDMS) with high-resolution features of relief as a stamp to print patterns of chemical inks. It was mainly developed for use with inks that form self-assembled monolayers (SAMs) of alkanethiolates on gold and silver. The procedure for carrying out μCP in these systems is remarkably simple: a stamp, inked with a solution of alkanethiol, is brought into contact with the surface of a substrate in order to transfer ink molecules to regions where the stamp and substrate contact. The resolution and effectiveness of μCP rely on conformal contact between the stamp and the surface of the substrate, rapid formation of highly ordered monolayers [10.23], and the autophobicity of the SAM, which effectively blocks the reactive spreading of the ink across the surface [10.24]. It can pattern SAMs over relatively large areas ($\approx$ up to 0.25 ft^2 has been demonstrated in prototype electronic devices) in a single impression [10.25]. The edge resolution of SAMs printed onto thermally evaporated gold films is on the order of 50 nm, as determined by lateral force microscopy [10.26]. Microcontact printing has been used with a range of different SAMs on various substrates [10.18]. Of these, alkanethiolates on gold, silver, and palladium [10.27] presently give the highest resolution. In many cases, the mechanical properties of the stamp limit the sizes of the smallest features that can be achieved: the most commonly used elastomer (Sylgard 184, Dow Corning) has a low modulus, which can lead to mechanical collapse or sagging for features of relief with aspect ratios greater than $\approx$ 2 or less than $\approx$ 0.05. Stamps fabricated with high modulus elastomers avoid some of these problems [10.28, 29]. Conventional stamps are also susceptible to in-plane mechanical strains that can cause distortions in the printed patterns. Composite stamps that use thin elastomer layers on stiff supports are effective at minimizing this source of distortion [10.30]. Methods for printing that avoid direct mechanical manipulation of the stamp can reduce distortions with conventional and composite stamps [10.25]. This approach has proven

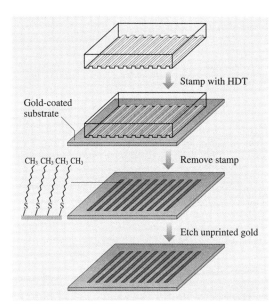

Fig. 10.4 Schematic illustration of microcontact printing. The first step involves *inking* a *stamp* with a solution of a material that is capable of forming a self-assembled monolayer (SAM) on a substrate that will be printed. In the case illustrated here, the ink is a millimolar concentration of hexadecanethiol (HDT) in ethanol. Directly applying the ink to the surface of the stamp with a pipette prepares the stamp for printing. Blowing the surface of the stamp dry and contacting it to a substrate delivers the ink to areas where the stamp contacts the substrate. The substrate consists of a thin layer of Au on a flat support. Removing the stamp after a few seconds of contact leaves a patterned SAM of HDT on the surface of the Au film. The printed SAM can act as a resist for the aqueous-based wet etching of the exposed regions of the Au. The resulting pattern of conducting gold can be used to build devices of various types

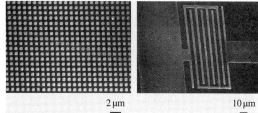

Fig. 10.5 Scanning electron micrographs of typical structures formed by microcontact printing a self-assembled monolayer ink of hexadecanethiol onto a thin metal film followed by etching of the unprinted areas of the film. The *left frame* shows an array of Au dots (20 nm thick) with ≈ 500 nm diameters. The *right frame* shows a printed structure of Ag (100 nm thick) in the geometry of interdigitated source/drain electrodes for a transistor in a simple inverter circuit. The edge resolution of patterns that can be easily achieved with microcontact printing is 50–100 nm

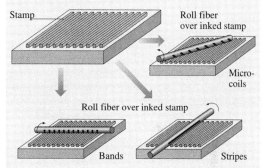

Fig. 10.6 Schematic illustration of a simple method to print lines on the surfaces of optical fibers. Rolling a fiber over the inked stamp prints a pattern onto the fiber surface. Depending on the orientation of the fiber axis with the line stamp illustrated here, it is possible, in a single rotation of the fiber, to produce a continuous microcoil, or arrays of bands or stripes

effective in large-area flexible circuit applications that require accurate multilevel registration.

The patterned SAM can be used either as a resist in selective wet etching or as a template in selective deposition to form structures of a variety of materials: metals, silicon, liquids, organic polymers and even biological species. Figure 10.4 schematically illustrates the use of μCP and wet etching to pattern a thin film of Au. Figure 10.5 shows SEM images of nanostructures of gold (20 nm thick, thermally evaporated with a 2.5 nm layer of Ti as an adhesion promoter) and silver (≈ 100 nm thick formed by electroless deposition using commercially available plating baths) [10.31] that were fabricated using this approach. In the first and second examples, the masters for the stamps consisted of photoresist patterned on silicon wafers with projection and contact mode photolithography, respectively. Placing these masters in a desiccator for ≈ 1 h with a few drops of tridecafluoro-1,1,2,2-tetrahydrooctyl-1-trichlorosilane forms a silane monolayer on the exposed native oxide of the silicon. This monolayer prevents adhesion of the master to PDMS (Sylgard 184), which is cast and cured from a 10 : 1 mixture of prepolymer and curing agent. Placing a few drops of a ≈ 1 mM solution of hexadecanethiol (HDT) in ethanol on the surface of the stamps and then blowing them dry with a stream of nitrogen prepares them for printing. Contacting the metal film for a few seconds with the stamp produces a patterned self-assembled monolayer (SAM) of HDT. An aqueous etchant (1 mM $K_4Fe(CN)_6$, 10 mM $K_3Fe(CN)_6$, and 0.1 M $Na_2S_2O_3$) removes the unprinted regions of the silver [10.32]. A similar solution (1 mM $K_4Fe(CN)_6$, 10 mM $K_3Fe(CN)_6$, 1.0 M KOH, and 0.1 M $Na_2S_2O_3$) can be used to etch the bare gold [10.33]. The results in Fig. 10.5 show that the roughness on the edges of the patterns is ≈ 50–100 nm. The resolution is determined by the grain size of the metal films, the isotropic etching process, slight reactive spreading of the inks, and edge disorder in the patterned SAMs.

The structures of Fig. 10.5 were formed on the flat surfaces of silicon wafers (left image) and glass slides (right image). An attractive feature of µCP and certain other contact printing techniques is their ability to pattern features with high resolution on highly curved or rough surfaces [10.22, 34, 35]. This type of patterning task is difficult or impossible to accomplish with photolithography due to its limited depth of focus and the difficulty involved with casting uniform films of photoresist on nonflat surfaces. Figure 10.6 shows, as an example, a straightforward approach for high-resolution printing on the highly curved surfaces of optical fibers. Here, simply rolling the fiber over an inked stamp prints a pattern on the entire outer surface of the fiber. Simple staging systems allow alignment of features to the fiber axis; they also ensure registration of the pattern from one side of the fiber to the other [10.20]. Figure 10.7 shows 3 µm wide lines and spaces printed onto the surface of a single mode optical fiber (diameter 125 µm). The bottom frame shows a freestanding metallic structure with the geometry and mechanical properties of an

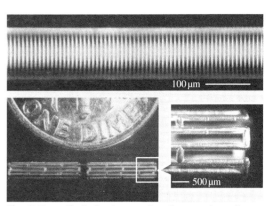

Fig. 10.7a–c Optical micrographs of some three-dimensional microstructures formed by microcontact printing on curved surfaces. The *top* frame shows an array of 3 µm lines of Au (20 nm)/Ti (1.5 nm) printed onto the surface of an optical fiber. This type of structure can be used as an integrated photomask for producing mode-coupling gratings in the core of the fiber. The *bottom* frames show a free-standing metallic microstructure formed by (**a**) microcontact printing and etching a thin film (100 nm thick) of Ag on the surface of a glass microcapillary tube, (**b**) electroplating the Ag to increase its thickness (to tens of micrometer) and (**c**) etching away the glass microcapillary with concentrated hydrofluoric acid. The structure shown here has the geometry and mechanical properties of an intravascular stent, which is a biomedical device commonly used in balloon angioplasty

intravascular stent, which is a biomedical device that is commonly used in balloon angioplasty procedures. In this latter case µCP followed by electroplating generated the Ag microstructure on a sacrificial glass cylinder that was subsequently etched away with concentrated hydrofluoric acid [10.36]. Other examples of microcontact printing on nonflat surfaces (low cost plastic sheets and optical ridge waveguides) appear in the Sect. 10.4.

10.3 Nanotransfer Printing

Nanotransfer printing (nTP) is a more recent high-resolution printing technique, which uses surface chemistries as interfacial *glues* and *release* layers (rather than *inks*, as in µCP) to control the transfer of solid material layers from relief features on a stamp to a substrate [10.37–39]. This approach is purely additive (material is only deposited in locations where it is needed), and it can generate complex patterns of single or multiple layers of materials with nanometer resolution over large areas in a single process step. It does not suffer from surface diffusion or edge disorder in the patterned inks of µCP, nor does it require post-printing etching or deposition steps to produce structures of functional materials. The method involves four components: (1) a stamp (rigid, flexible, or elastomeric) with relief features in the geometry of the desired pattern, (2) a method for depositing a thin layer of solid material onto the raised features of this stamp, (3) a means of bringing the stamp into intimate physical contact with a substrate, and (4) surface chemistries that prevent adhesion of the deposited material to the stamp and promote its strong adhesion to the substrate. nTP has been demonstrated with SAMs and other surface chemistries for printing onto flexible and rigid substrates with hard inorganic and soft polymer stamps. Figure 10.8 presents a set of procedures for using nTP to pattern a thin metal bilayer of Au/Ti with a surface transfer chemistry that relies on a dehydration reaction [10.37]. The process begins with fabrication of a suitable stamp. Elastomeric stamps can be built using the same casting and curing procedures described for μCP. Rigid stamps can be fabricated by (1) patterning resist (such as electron beam resist or photoresist) on a substrate (such as Si or GaAs), (2) etching the exposed regions of the substrate with an anisotropic reactive ion etch, and (3) removing the resist, as illustrated in Fig. 10.1. For both types of stamps, careful control of the lithography and the etching steps yields features of relief with nearly vertical or slightly reentrant sidewalls. The stamps typically have depths of relief $> 0.2\,\mu m$ for patterning metal films with thicknesses $< 50\,nm$.

Electron beam evaporation of Au (20 nm; 1 nm/s) and Ti (5 nm; 0.3 nm/s) generates uniform metal bilayers on the surfaces of the stamp. A vertical, collimated flux of metal from the source ensures uniform deposition only on the raised and recessed regions of relief. The gold adheres poorly to the surfaces of stamps made of GaAs, PDMS, glass, or Si. In the process of Fig. 10.8, a fluorinated silane monolayer acts to reduce the adhesion further when a Si stamp (with native oxide) is used. The Ti layer serves two purposes: (1) it promotes adhesion between the Au layer and the substrate after pattern transfer, and (2) it readily forms a ≈ 3 nm oxide layer at ambient conditions, which provides a surface where the dehydration reaction can take place. Exposing the titanium oxide (TiO_x) surface to an oxygen plasma breaks bridging oxygen bonds, thus creating defect sites where water molecules can adsorb. The result is a titanium oxide surface with some fractional coverage of hydroxyl (−OH) groups (titanol).

In the case of Fig. 10.8, the substrate is a thin film of PDMS (10−50 μm thick) cast onto a sheet of poly(ethylene terephthalate) (PET; 175 μm thick). Exposing the PDMS to an oxygen plasma produces surface (−OH) groups (silanol). Placing the plasma-oxidized,

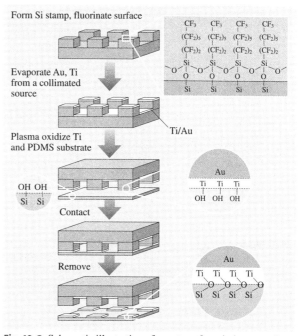

Fig. 10.8 Schematic illustration of nanotransfer printing procedure. Here, interfacial dehydration chemistries control the transfer of a thin metal film from a hard inorganic stamp to a conformable elastomeric substrate (thin film of polydimethylsiloxane (PDMS) on a plastic sheet). The process begins with fabrication of a silicon stamp (by conventional lithography and etching) followed by surface functionalization of the native oxide with a fluorinated silane monolayer. This layer ensures poor adhesion between the stamp and a bilayer metal film (Au and Ti) deposited by electron beam evaporation. A collimated flux of metal oriented perpendicular to the surface of the stamp avoids deposition on the sidewalls of the relief. Exposing the surface Ti layer to an oxygen plasma produces titanol groups. A similar exposure for the PDMS produces silanol groups. Contacting the metal-coated stamp to the PDMS results in a dehydration reaction that links the metal to the PDMS. Removing the stamp leaves a pattern of metal in the geometry of the relief features

Au/Ti-coated stamp on top of these substrates leads to intimate, conformal contact between the raised regions of the stamp and the substrate, without the application of any external pressure. (The soft, conformable PDMS is important in this regard.) It is likely that a dehydration reaction takes place at the (−OH)-bearing interfaces during contact; this reaction results in permanent Ti−O−Si bonds that produce strong adhesion between the two surfaces. Peeling the substrate and stamp apart transfers the Au/Ti bilayer from the raised

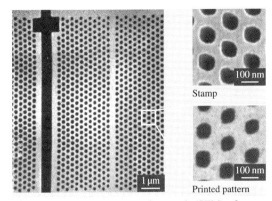

Fig. 10.9 Scanning electron micrograph (SEM) of a pattern produced by nanotransfer printing. The structure consists of a bilayer of Au (20 nm)/Ti (1 nm) (*white*) in the geometry of a photonic bandgap waveguide printed onto a thin layer of polydimethylsiloxane on a sheet of plastic (*black*). Electron beam lithography and etching of a GaAs wafer produced the stamp that was used in this case. The transfer chemistry relied on condensation reactions between titanol groups on the surface of the Ti and silanol groups on the surface of the PDMS. The *frames* on the *right* show SEMs of the Au/Ti-coated stamp (*top*) before printing and on the substrate (*bottom*) after printing. The electron beam lithography and etching used to fabricate the stamp limit the minimum feature size (≈ 70 nm) and the edge resolution ($\approx 5–10$ nm) of this pattern

regions of the stamp (to which the metal has extremely poor adhesion) to the substrate. Complete pattern transfer from an elastomeric stamp to a thin elastomeric substrate occurs readily at room temperature in open air with contact times of less than 15 s. When a rigid stamp is employed, slight heating is needed to induce transfer. While the origin of this difference is unclear, it may reflect the comparatively poor contact when rigid stamps are used; similar differences are also observed in cold welding of gold films [10.40].

Figure 10.9 shows scanning electron micrographs of a pattern produced using a GaAs stamp generated by electron beam lithography and etching. The frames on the right show images of the metal-coated stamp before printing (top) and the transferred pattern (bottom). The resolution appears to be limited only by the resolution of the stamp itself, and perhaps by the grain size of the metal films. Although the accuracy in multilevel registration that is possible with nTP has not yet been quantified, its performance is likely similar to that of embossing techniques when rigid stamps are used [10.41].

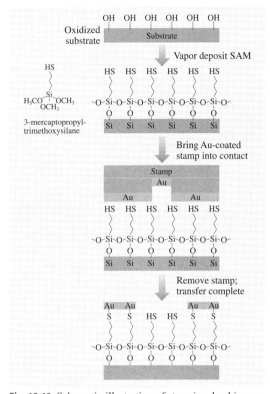

Fig. 10.10 Schematic illustration of steps involved in nanotransfer printing a pattern of a thin layer of Au onto a silicon wafer using a self-assembled monolayer (SAM) surface chemistry. Plasma oxidizing the surface of the wafer generates OH groups. Solution or vapor phase exposure of the wafer to 3-mercaptopropyltrimethoxysilane yields a SAM with exposed thiol groups. Contacting an Au-coated stamp to this surface produces thiol linkages that bond the gold to the substrate. Removing the stamp completes the transfer printing process

A wide range of surface chemistries can be used for the transfer. SAMs are particularly attractive due to their chemical flexibility. Figure 10.10 illustrates the use of a thiol-terminated SAM and nTP for forming patterns of Au on a silicon wafer [10.38]. Here, the vapor phase cocondensation of the methoxy groups of molecules of 3-mercaptopropyltrimethoxysilane (MPTMS) with the —OH-terminated surface of the wafer produces a SAM of MPTMS with exposed thiol (—SH) groups. PDMS stamps can be prepared for printing on this surface by coating them with a thin film (≈ 15 nm) of Au using conditions (thermal evaporation 1.0 nm/s; $\approx 10^{-7}$ Torr

Fig. 10.11 Optical micrographs of patterns of Au (15 nm thick) formed on plastic (*left* frame) and silicon (*right* frame) substrates with nanotransfer printing. The transfer chemistries in both cases rely on self-assembled monolayers with exposed thiol groups. The minimum feature sizes and the edge resolution are both limited by the photolithography used to fabricate the stamps

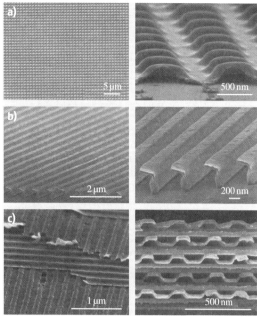

Fig. 10.13a–c Scanning electron micrographs of three-dimensional metal structures obtained by nanotransfer printing gold metal films. Part (**a**) shows closed gold nanocapsules. Part (**b**) shows free-standing *L* structures obtained using a stamp coated with a steeply angled flux of metal. Part (**c**) shows a multilayer 3-D structure obtained by the successive transfer and cold welding of continuous gold nanocorrugated films

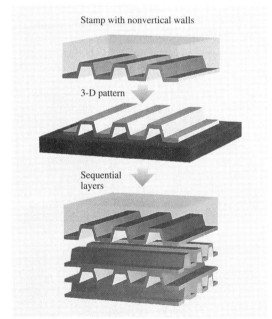

Fig. 10.12 Schematic illustration of the nanotransfer printing (nTP) process for generating continuous 3-D structures when the stamp relief side walls are not vertical. Successive transfer by cold welding the gold films on top of each other yields complex multilayer structures

base pressure) that yield optically smooth, uniform films without the buckling that has been observed in the past with similar systems [10.42]. Nanocracking that sometimes occurs in the films deposited in this way can be reduced or eliminated by evaporating a small amount of Ti onto the PDMS before Au deposition and/or by exposing the PDMS surface briefly to an oxygen plasma. Bringing this coated stamp into contact with the MPTMS SAM leads to the formation of sulfur–gold bonds in the regions of contact. Removing the stamp after a few seconds efficiently transfers the gold from the raised regions of the stamp (Au does not adhere to the PDMS) to the substrate. Covalent bonding of the SAM glue to both the substrate and the gold leads to good adhesion of the printed patterns: They easily pass Scotch tape adhesion tests. Similar results can be obtained with other substrates containing surface −OH groups. For example, Au patterns can be printed onto ≈ 250 μm-thick sheets of poly(ethylene terephthalate) (PET) by first spin-casting and curing (130 °C for 24 h) a thin film of an organosilsesquioxane on the PET. Exposing the cured film (≈ 1 μm thick) to an oxygen plasma and then to air produces the necessary surface (−OH) groups.

Figure 10.11 shows some optical micrographs of typical printed patterns in this case [10.38].

Similar surface chemistries can guide transfer to other substrates. Alkanedithiols, for example, are useful for printing Au onto GaAs wafers [10.39]. Immersing these substrates (freshly etched with 37% HCl for ≈ 2 min to remove the surface oxide) in a 0.05 M solution of 1,8-octanedithiol in ethanol for 3 h produces a monolayer of dithiol on the surface. Although the chemistry of this system is not completely clear, it is generally believed that the thiol end groups bond chemically to the surface. Surface spectroscopy suggests the formation of Ga−S and As−S bonds. Contacting an Au-coated PDMS stamp with the treated substrate causes the exposed thiol endgroups to react with Au in the regions of contact. This reaction produces permanent Au−S bonds at the stamp/substrate interface (see insets in Fig. 10.4 for idealized chemical reaction schemes). Figure 10.12 schematically illustrates how this procedure can generate continuous 3-D metal patterns using stamps with nonvertical side walls. Several layers can be transferred on top of each other by successively cold-welding the different gold metal layers. Figure 10.13 shows high- and low-magnification scanning electron microscope images of nanotransfer printed single- and multilayer 3-D metal structures [10.43, 44]. The integrity of these free-standing 3-D metal structures is remarkable but depends critically on the careful optimization of the metal evaporation conditions and stamp & substrate surface chemistries.

10.4 Applications

Although conventional patterning techniques, such as photolithography or electron beam lithography, have the required resolution, they are not appropriate because they are expensive and generally require multiple processing steps with resists, solvents and developers that can be difficult to use with organic active materials and plastic substrates. Microcontact and nanotransfer printing are both particularly well suited for this application. They can be combined and matched with other techniques, such as ink-jet or screen printing, to form a complete system for patterning all layers in practical plastic electronic devices [10.45]. We have focused our efforts partly on unusual electronic systems such as flexible plastic circuits and devices that rely on electrodes patterned on curved objects such as microcapillaries and optical fibers. We have also explored photonic systems such as distributed feedback structures for lasers and other integrated optical elements that demand submicron features. The sections below highlight several examples in each of these areas.

Fig. 10.14 Schematic cross-sectional view (*left*) and electrical performance (*right*) of an organic thin film transistor with microcontact printed source and drain electrodes. The structure consists of a substrate (PET), a gate electrode (indium tin oxide), a gate dielectric (spin-cast layer of organosilsesquioxane), source and drain electrodes (20 nm Au and 1.5 nm Ti), and a layer of the organic semiconductor pentacene. The electrical properties of this device are comparable to or better than those that use pentacene with photolithographically defined source/drain electrodes and inorganic dielectrics, gates and substrates ▶

10.4.1 Unconventional Electronic Systems

A relatively new direction in electronics research seeks to establish low-cost plastic materials, substrates and printing techniques for large-area flexible electronic devices, such as paperlike displays. These types of novel devices can complement those (including high-density memories and high-speed microprocessors) that are well suited to existing inorganic (such as

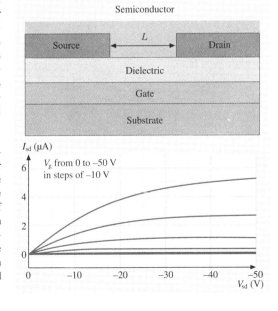

silicon) electronics technologies. High-resolution patterning methods for defining the separation between the source and drain electrodes (the channel length) of transistors in these plastic circuits are particularly important because this dimension determines current output and other important characteristics [10.46].

Figure 10.14 illustrates schematically a cross-sectional view of a typical organic transistor. The frame on the right shows the electrical switching characteristics of a device that uses source/drain electrodes of Au patterned by μCP, a dielectric layer of an organosilsesquioxane, a gate of indium tin oxide (ITO), and a PET substrate. The effective semiconductor mobility extracted from these data is comparable to those measured in devices that use the same semiconductor (pentacene in this case) with inorganic substrates and dielectrics, and gold source/drain electrodes defined by photolithography. Our recent work [10.1, 31, 47] with μCP in the area of plastic electronics demonstrates: (1) methods for using cylindrical *roller* stamps mounted on fixed axles for printing in a continuous reel-to-reel fashion, high-resolution source/drain electrodes in ultrathin gold and silver deposited from solution at room temperature using electroless deposition, (2) techniques for performing registration and alignment of the printed features with other elements of a circuit over large areas, (3) strategies for achieving densities of defects that are as good as those observed with photolithography when the patterning is performed outside of clean room facilities, (4) methods for removing the printed SAMs to allow good electrical contact of the electrodes

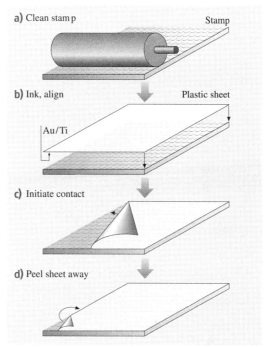

Fig. 10.16a–d Schematic illustration of fabrication steps for microcontact printing over large areas onto plastic sheets. The process begins with cleaning the stamp using a conventional adhesive roller lint remover. This procedure effectively removes dust particles. To minimize distortions, the stamp rests face-up on a flat surface and it is not manipulated directly during the printing. Alignment and registration are achieved with alignment marks on one side of the substrate and the stamp. By bending the plastic sheet, contact is initiated on one side of the stamp; the contact line is then allowed to progress gradually across the stamp. This approach avoids formation of air bubbles that can frustrate good contact. After the substrate is in contact with the stamp for a few seconds, the plastic substrate is separated from the stamp by peeling it away beginning in one corner. Good registration (maximum cumulative distortions of less than 50 μm over an area of 130 cm^2) and low defect density can be achieved with this simple approach. It is also well suited for use with rigid composite stamps designed to reduce the level of distortions even further

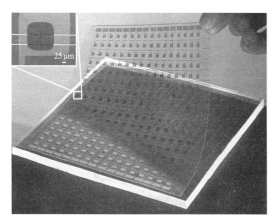

Fig. 10.15 Image of a flexible plastic active matrix backplane circuit whose finest features (transistor source/drain electrodes and related interconnects) are patterned by microcontact printing. The circuit rests partly on the elastomeric stamp that was used for printing. The circuit consists of a square array of interconnected organic transistors, each of which acts locally as a voltage-controlled switch to control the color of an element in the display. The *inset* shows an optical micrograph of one of the transistors

with organic semiconductors deposited on top of them, and (5) materials and fabrication sequences that can efficiently exploit these printed electrodes for working organic TFTs in large-scale circuits.

Figure 10.15 provides an image of a large-area plastic circuit with critical features defined by μCP. This circuit is a flexible active matrix backplane for a display. It consists of a square array of interconnected transistors, each of which serves as a switching element that controls the color of a display pixel [10.25, 48]. The transistors themselves have the layout illustrated in Fig. 10.11, and they use similar materials. The semiconductor in this image is blue (pentacene), the source/drain level is Au, the ITO appears green in the optical micrograph in the inset. Part of the circuit rests on the stamp that was used for μCP. The smallest features are the source and drain electrodes ($\approx 15\,\mu m$ lines), the interconnecting lines ($\approx 15\,\mu m$ lines), and the channel length of the transistor ($\approx 15\,\mu m$). This circuit incorporates five layers of material patterned with good registration of the source/drain, gate, and semiconductor levels. The simple printing approach is illustrated in Fig. 10.16 [10.25]. Just before use, the surface of the stamp is cleaned using a conventional adhesive roller lint remover; this procedure removes dust from the stamp in such a way that does not contaminate or damage its surface. Inking the stamp and placing it face-up on a flat surface prepares it for printing. Matching the cross-hair alignment marks on the corners of one edge of the stamp with those patterned in the ITO brings the substrate into registration with the stamp. During this alignment, features on the stamp are viewed directly through the semitransparent substrate. By bending the PET sheet, contact with the stamp is initiated on the edge of the substrate that contains the cross-hair marks. Gradually unbending the sheet allows contact to progress across the rest of the surface. This printing procedure is attractive because it avoids distortions that can arise when directly manipulating the flexible rubber stamp. It also minimizes the number and size of trapped air pockets that can form between the stamp and substrate. Careful measurements performed after etching the unprinted areas of the gold show that over the entire $6'' \times 6''$ area of the circuit, (1) the overall alignment accuracy for positioning the stamp relative to the substrate (the offset of the center of the distribution of registration errors) is $\approx 50-100\,\mu m$, even with the simple approach used here, and (2) the distortion in the positions of features in the source/drain level, when referenced to the gate level, can be as small as $\approx 50\,\mu m$ (the full width at half maximum of the distribution of registration errors). These distortions represent the cumulative effects of deformations in the stamp and distortions in the gate and column electrodes that may arise during the patterning and processing of the flexible PET sheet. The density of defects in the printed patterns is comparable to (or smaller than) that in resist patterned by contact-mode photolithography when both procedures are performed outside of a clean-room facility (when dust is the dominant source of defects).

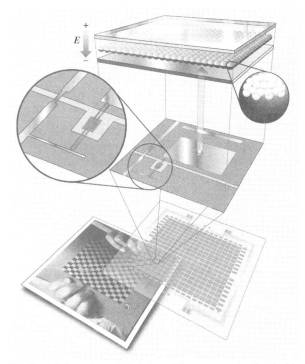

Fig. 10.17 Schematic exploded view of the components of a pixel in an electronic paperlike display (*bottom frame*) that uses a microcontact printed flexible active matrix backplane circuit (illustration near the *bottom frame*). The circuit is laminated against an unpatterned thin sheet of electronic ink (*top frame*) that consists of a monolayer of transparent polymer microcapsules (diameter $\approx 100\,\mu m$). These capsules contain a heavily dyed black fluid and a suspension of charged white pigment particles (see *right inset*). When one of the transistors turns on, electric fields develop between an unpatterned transparent frontplane electrode (indium tin oxide) and a backplane electrode that connects to the transistor. Electrophoretic flow drives the pigment particles to the front or the back of the display, depending on the polarity of the field. This flow changes the color of the pixel, as viewed from the front of the display, from black to white or vica versa

Fig. 10.18 Electronic paperlike display showing two different images. The device consists of several hundred pixels controlled by a flexible active matrix backplane circuit formed by microcontact printing. The relatively coarse resolution of the display is not limited by material properties or by the printing techniques. Instead, it is set by practical considerations for achieving high pixel yields in the relatively uncontrolled environment of the chemistry laboratory in which the circuits were fabricated

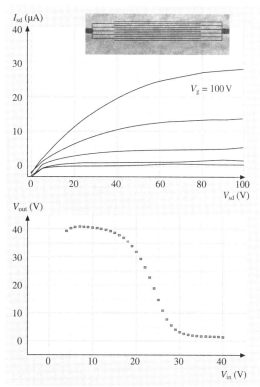

Fig. 10.19 The *upper* frame shows current–voltage characteristics of an n-channel transistor formed with electrodes patterned by nanotransfer printing that are laminated against a substrate that supports an organic semiconductor, a gate dielectric and a gate. The *inset* shows an optical micrograph of the interdigitated electrodes. The *lower* frame shows the transfer characteristics of a simple CMOS inverter circuit that uses this device and a similar one for the p-channel transistor

Figure 10.17 shows an *exploded* view of a paperlike display that consists of a printed flexible plastic backplane circuit, like the one illustrated in Fig. 10.16, laminated against a thin layer of *electronic ink* [10.25, 49]. The electronic ink is composed of a monolayer of transparent polymer microcapsules that contain a suspension of charged white pigment particles suspended in a black liquid. The printed transistors in the backplane circuit act as local switches, which control electric fields that drive the pigments to the front or back of the display. When the particles flow to the front of a microcapsule, it appears white; when they flow to the back, it appears black. Figure 10.18 shows a working sheet of active matrix electronic paper that uses this design. This prototype display has several hundred pixels and an optical contrast that is both independent of the viewing angle and significantly better than newsprint. The device is ≈ 1 mm thick, is mechanically flexible, and weighs $\approx 80\%$ less than a conventional liquid crystal display of similar size. Although these displays have only a relatively coarse resolution, all of the processing techniques, the μCP method, the materials, and the electronic inks, are suitable for the large numbers of pixels required

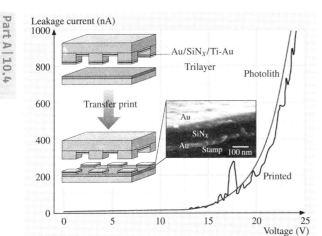

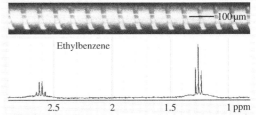

Fig. 10.20 Multilayer thin film capacitor structure printed in a single step onto a plastic substrate using the nanotransfer printing technique. A multilayer of Au/SiN$_x$/Ti/Au was first deposited onto a silicon stamp formed by photolithography and etching. Contacting this stamp to a substrate of Au/PDMS/PET forms a cold weld that bonds the exposed Au on the stamp to the Au-coating on the substrate. Removing the stamp produces arrays of square (250× 250 mm^2) metal/insulator/metal capacitors on the plastic support. The *dashed line* shows the measured current–voltage characteristics of one of these printed capacitors. The *solid line* corresponds to a similar structure formed on a rigid glass substrate using conventional photolithographic procedures. The characteristics are the same for these two cases. The slightly higher level of noise in the printed devices results, at least partly, from the difficulties involved with making good electrical contacts to structures on the flexible plastic substrate

Fig. 10.21 The *top frame* shows an optical micrograph of a continuous conducting microcoil formed by microcontact printing onto a microcapillary tube. This type of printed microcoil is well suited for excitation and detection of nuclear magnetic resonance spectra from nanoliter volumes of fluid housed in the bore of the microcapillary. The *bottom frame* shows a spectra trace collected from an ≈ 8 nL volume of ethyl benzene using a structure similar to the one shown in the *top frame*

for high-information content electronic newspapers and other systems.

Like μCP, nTP is well suited to forming high-resolution source/drain electrodes for plastic electronics. nTP of Au/Ti features in the geometry of the drain and source level of organic transistors, and with appropriate interconnects on a thin layer of PDMS on PET it yields a substrate that can be used in an unusual but powerful way for building circuits: soft, room temperature lamination of such a structure against a plastic substrate that supports the semiconductor, gate dielectric, and gate levels yields a high performance circuit embedded between two plastic sheets [10.37, 50]. (Details of this lamination procedure are presented elsewhere.) The left frame of Fig. 10.19 shows the current–voltage characteristics of a laminated n-channel transistor that uses the organic semiconductor copper hexadecafluorophthalocyanine (n-type) and source/drain electrodes patterned with nTP. The inset shows an optical micrograph of the printed interdigitated source/drain electrodes of this device. The bottom frame of Fig. 10.16 shows the transfer characteristics of a laminated complementary organic inverter circuit whose electrodes and connecting lines are defined by nTP. The p-channel transistor in this circuit used pentacene for the semiconductor [10.37].

In addition to high-resolution source/drain electrodes, it is possible to use nTP to form complex multilayer devices with electrical functionality on plastic substrates [10.38]. Figure 10.20 shows a metal/insulator/metal (MIM) structure of Au (50 nm), SiN$_x$ (100 nm; by plasma enhanced vapor deposition, PECVD), Ti (5 nm) and Au (50 nm) formed by transfer printing with a silicon stamp that is coated sequentially with these layers. In this case, a short reactive ion etch (with CF$_4$) after the second Au deposition removes the SiN$_x$ from the sidewalls of the stamp. nTP transfers these layers in a patterned geometry to a substrate of Au (15 nm)/Ti (1 nm)-coated PDMS (50 μm)/PET (250 μm). Interfacial cold-welding between the Au on the surfaces of the stamp and substrate bonds the multilayers to the substrate. Figure 10.8 illustrates the procedures, the structures (lateral dimensions of 250× 250 μm^2, for ease of electrical probing), and their electrical characteristics. These MIM capacitors have performances similar to devices fabricated on silicon wafers by photolithography and lift-off. This example illustrates the ability of nTP to print patterns of materials whose growth conditions (high-temperature SiN$_x$

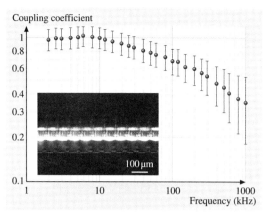

Fig. 10.22 The *inset* shows a concentric microtransformer formed using microcoils printed onto two different microcapillary tubes. The smaller of the tubes (outer diameter 135 μm) has a ferromagnetic wire threaded through its core. The larger one (outer diameter 350 μm) has the smaller tube threaded through its core. The resulting structure is a microtransformer that shows good coupling coefficients at frequencies up to ≈ 1 MHz. The *graph* shows its performance

by PECVD, in this case) prevent their direct deposition or processing on the substrate of interest (PET, in this case). The cold-welding transfer approach has also been exploited in other ways for patterning components for plastic electronics [10.51, 52].

Another class of unusual electronic/optoelectronic devices relies on circuits or circuit elements on curved surfaces. This emerging area of research was stimulated primarily by the ability of μCP to print high-resolution features on fibers and cylinders. Figure 10.21 shows a conducting microcoil printed with μCP on a microcapillary tube using the approach illustrated in Fig. 10.4. The coil serves as the excitation and detection element for high-resolution proton nuclear magnetic resonance of nanoliter volumes of fluid that are housed in the bore of the microcapillary [10.53]. The high fill factor and other considerations lead to extremely high sensitivity with such printed coils. The bottom frame of Fig. 10.21 shows the spectrum of an ≈ 8 nL volume of ethylbenzene. The narrow lines demonstrate the high resolution that is possible with this approach. Similar coils can be used as magnets [10.54], springs [10.36], and electrical transformers [10.55]. Figure 10.22 shows an optical micrograph and the electrical measurements from a concentric cylindrical microtransformer that uses a microcoil printed on a microcapillary tube with

a ferromagnetic wire threaded through its core. Inserting this structure into the core of a larger microcapillary that also supports a printed microcoil completes the transformer [10.55]. This type of device shows good coupling coefficients up to relatively high frequencies. Examples of other optoelectronic components appear in fiber optics where microfabricated on-fiber structures

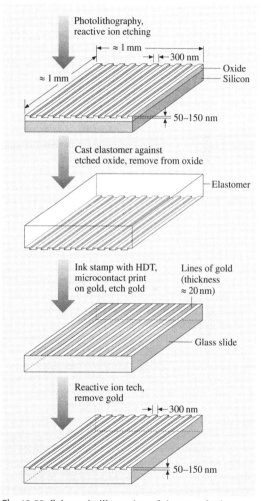

Fig. 10.23 Schematic illustration of the use of microcontact printing (μCP) for fabricating high-resolution gratings that can be incorporated into distributed planar laser structures or other components for integrated optics. The geometries illustrated here are suitable for third- order distributed feedback (DFB) lasers that operate in the red

serve as integrated photomasks [10.20] and distributed thermal actuators [10.22].

10.4.2 Lasers and Waveguide Structures

In addition to integral components of unconventional electronic systems, useful structures for integrated optics can be built by using μCP and nTP to print sacrificial resist layers for etching glass waveguides. These printing techniques offer the most significant potential value for this area when they are used to pattern features that are smaller than those that can be achieved with contact mode photolithography. Mode coupling gratings and distributed laser resonators are two such classes of structures. We have demonstrated μCP for forming distributed feedback (DFB) and distributed Bragg reflector (DBR) lasers that have narrow emission line widths [10.56]. This challenging fabrication demonstrates the suitability of μCP for building structures that have (1) feature sizes of significantly less than 1 μm ($\approx$ 300 nm), *and* (2) long-range spatial coherence ($\approx$ 1 mm). The lasers employ optically pumped gain material deposited onto DFB or DBR resonators formed from periodic relief on a transparent substrate. The gain media confines light to the surface of the structure; its thickness is chosen to support a single transverse mode. To generate the required relief, lines of gold formed by μCP on a glass slide act as resists for reactive ion etching of the glass. Removing the gold leaves a periodic pattern of relief (600 nm period, 50 nm depth) on the surface of the glass (Fig. 10.23). Figure 10.24 shows the performance of plastic lasers that use printed DFB and DBR resonators with gain media consisting of thin films of PBD doped with 1 wt. % of coumarin 490 and DCMII, photopumped with 2 ns pulses from a nitrogen laser with intensities $> 5\,\text{kW}/\text{cm}^2$ [10.56]. Multimode lasing at resolution-limited line widths was

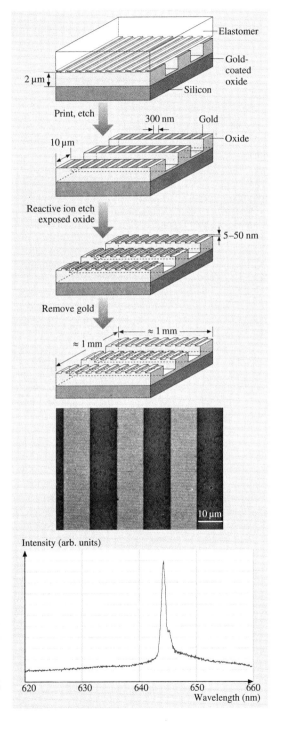

Fig. 10.24 The *top frames* gives a schematic illustration of steps for microcontact printing high-resolution gratings directly onto the top surfaces of ridge waveguides. The printing defines a sacrificial etch mask of gold which is subsequently removed. Producing this type of structure with photolithography is difficult because of severe thickness nonuniformities that appear in photoresist spin-cast on this type of nonplanar substrate. The *upper bottom* frame shows a top view optical micrograph of printed gold lines on the ridge waveguides. The *lower bottom* frame shows the emission output of a plastic photopumped laser that uses the printed structure and a thin evaporated layer of gain media ▶

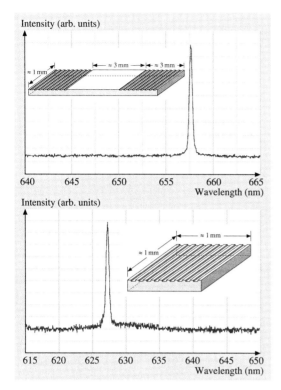

Fig. 10.25 Schematic illustrations and lasing spectra of plastic lasers that use microcontact printed resonators based on surface relief distributed Bragg reflectors (DBRs) and distributed feedback gratings (DFBs) on glass substrates. The grating periods are ≈ 600 nm in both cases. The lasers use thin film plastic gain media deposited onto the printed gratings. This layer forms a planar waveguide that confines the light to the surface of the substrate. The laser shows emission over a narrow wavelength range, with a width that is limited by the resolution of the spectrometer used to characterize the output. In both cases, the emission profiles, the lasing thresholds and other characteristics of the devices are comparable to similar lasers that use resonators formed by high-resolution projection-mode photolithography ◄

with photolithography. For example, μCP can be used to form DFB resonators directly on the top surfaces of ridge waveguides [10.58]. Figure 10.25 illustrates the procedures. The bottom left frame shows an optical micrograph of the printed gold lines. Sublimation of a ≈ 200 nm film of tris(8-hydroxyquinoline) aluminum (Al) doped with 0.5–5.0 wt.% of the laser dye DCMII onto the resonators produces waveguide DFB lasers. The layer of gain material itself provides a planar waveguide with air and polymer as the cladding layers. The relief waveguide provides lateral confinement of the light. Photopumping these devices with the output of a pulsed nitrogen laser (≈ 2 ns, 337 nm) causes lasing due to Bragg reflections induced by the DFB structures on the top surfaces of the ridge waveguides. Some of the laser emission scatters out of the plane of the waveguide at an angle allowed by phase matching conditions. In this way, the grating also functions as an output coupler and offers a convenient way to characterize the laser emission. The bottom right frame of Fig. 10.22 shows the emission profile.

observed at wavelengths corresponding to the third harmonic of the gratings. These characteristics are similar to those observed in lasers that use resonators generated with photolithography and are better than those that use imprinted polymers [10.57].

Contact printing not only provides a route to low-cost equivalents of gratings fabricated with other approaches, but also allows the fabrication of structures that would be difficult or impossible to generate

10.5 Conclusions

This chapter provides an overview of two contact printing techniques that are capable of micrometer and submicrometer resolution. It also illustrates some applications of these methods that may provide attractive alternatives to more established lithographic methods. The growing interest in nanoscience and technology makes it crucial to develop new methods for fabricating the relevant test structures and devices. The simplicity of these techniques together with the interesting and subtle materials science, chemistry, and physics associated with them make this a promising area for basic and applied study.

References

10.1 C.A. Mirkin, J.A. Rogers: Emerging methods for micro- and nanofabrication, MRS Bulletin **26**, 506–507 (2001)

10.2 H.I. Smith, H.G. Craighead: Nanofabrication, Phys. Today **43**, 24–43 (1990)

10.3 W.M. Moreau (Ed.): *Semiconductor Lithography: Principles and Materials* (Plenum, New York 1988)

10.4 S. Matsui, Y. Ochiai: Focused ion beam applications to solid state devices, Nanotechnology **7**, 247–258 (1996)

10.5 J.M. Gibson: Reading and writing with electron beams, Phys. Today **50**, 56–61 (1997)

10.6 L.L. Sohn, R.L. Willett: Fabrication of nanostructures using atomic-force microscope-based lithography, Appl. Phys. Lett. **67**, 1552–1554 (1995)

10.7 E. Betzig, K. Trautman: Near-field optics – Microscopy, spectroscopy, and surface modification beyond the diffraction limit, Science **257**, 189–195 (1992)

10.8 A.J. Bard, G. Denault, C. Lee, D. Mandler, D.O. Wipf: Scanning electrochemical microscopy: A new technique for the characterization and modification of surfaces, Acc. Chem. Res. **23**, 357 (1990)

10.9 J.A. Stroscio, D.M. Eigler: Atomic and molecular manipulation with the scanning tunneling microscope, Science **254**, 1319–1326 (1991)

10.10 J. Nole: Holographic lithography needs no mask, Laser Focus World **33**, 209–212 (1997)

10.11 A.N. Broers, A.C.F. Hoole, J.M. Ryan: Electron beam lithography – Resolution limits, Microelectron. Eng. **32**, 131–142 (1996)

10.12 A.N. Broers, W. Molzen, J. Cuomo, N. Wittels: Electron-beam fabrication of 80 metal structures, Appl. Phys. Lett. **29**, 596 (1976)

10.13 G.D. Aumiller, E.A. Chandross, W.J. Tomlinson, H.P. Weber: Submicrometer resolution replication of relief patterns for integrated optics, J. Appl. Phys. **45**, 4557–4562 (1974)

10.14 Y. Xia, J.J. McClelland, R. Gupta, D. Qin, X.-M. Zhao, L.L. Sohn, R.J. Celotta, G.M. Whiteside: Replica molding using polymeric materials: A practical step toward nanomanufacturing, Adv. Mater. **9**, 147–149 (1997)

10.15 T. Borzenko, M. Tormen, G. Schmidt, L.W. Molenkamp, H. Janssen: Polymer bonding process for nanolithography, Appl. Phys. Lett. **79**, 2246–2248 (2001)

10.16 H. Hua, Y. Sun, A. Gaur, M.A. Meitl, L. Bilhaut, L. Rotinka, J. Wang, P. Geil, M. Shim, J.A. Rogers: Polymer imprint lithography with molecular-scale resolution, Nano Lett. **4**(12), 2467–2471 (2004)

10.17 A. Kumar, G.M. Whitesides: Features of gold having micrometer to centimeter dimensions can be formed through a combination of stamping with an elastomeric stamp and an alkanethiol ink followed by chemical etching, Appl. Phys. Lett. **63**, 2002–2004 (1993)

10.18 Y. Xia, G.M. Whitesides: Soft lithography, Angew. Chem. Int. Ed. **37**, 550–575 (1998)

10.19 Y. Xia, J.A. Rogers, K.E. Paul, G.M. Whitesides: Unconventional methods for fabricating and patterning nanostructures, Chem. Rev. **99**, 1823–1848 (1999)

10.20 J.A. Rogers, R.J. Jackman, J.L. Wagener, A.M. Vengsarkar, G.M. Whitesides: Using microcontact printing to generate photomasks on the surface of optical fibers: A new method for producing in-fiber gratings, Appl. Phys. Lett. **70**, 7–9 (1997)

10.21 B. Michel, A. Bernard, A. Bietsch, E. Delamarche, M. Geissler, D. Juncker, H. Kind, J.P. Renault, H. Rothuizen, H. Schmid, P. Schmidt-Winkel, R. Stutz, H. Wolf: Printing meets lithography: soft approaches to high-resolution printing, IBM J. Res. Dev. **45**, 697–719 (2001)

10.22 J.A. Rogers: Rubber stamping for plastic electronics and fiber optics, MRS Bulletin **26**, 530–534 (2001)

10.23 N.B. Larsen, H. Biebuyck, E. Delamarche, B. Michel: Order in microcontact printed self-assembled monolayers, J. Am. Chem. Soc. **119**, 3017–3026 (1997)

10.24 H.A. Biebuyck, G.M. Whitesides: Self-organization of organic liquids on patterned self-assembled monolayers of alkanethiolates on gold, Langmuir **10**, 2790–2793 (1994)

10.25 J.A. Rogers, Z. Bao, K. Baldwin, A. Dodabalapur, B. Crone, V.R. Raju, V. Kuck, H. Katz, K. Amundson, J. Ewing, P. Drzaic: Paper-like electronic displays: Large area, rubber stamped plastic sheets of electronics and electrophoretic inks, Proc. Natl. Acad. Sci. USA **98**, 4835–4840 (2001)

10.26 J.L. Wilbur, H.A. Biebuyck, J.C. MacDonald, G.M. Whitesides: Scanning force microscopies can image patterned self-assembled monolayers, Langmuir **11**, 825–831 (1995)

10.27 J.C. Love, D.B. Wolfe, M.L. Chabinyc, K.E. Paul, G.M. Whitesides: Self-assembled monolayers of alkanethiolates on palladium are good etch resists, J. Am. Chem. Soc. **124**, 1576–1577 (2002)

10.28 H. Schmid, B. Michel: Siloxane polymers for high-resolution, high-accuracy soft lithography, Macromolecules **33**, 3042–3049 (2000)

10.29 K. Choi, J.A. Rogers: A photocurable poly(dimethylsiloxane) chemistry for soft lithography in the nanometer regime, J. Am. Chem. Soc. **125**, 4060–4061 (2003)

10.30 J.A. Rogers, K.E. Paul, G.M. Whitesides: Quantifying distortions in soft lithography, J. Vac. Sci. Technol. B **16**, 88–97 (1998)

10.31 J. Tate, J.A. Rogers, C.D.W. Jones, W. Li, Z. Bao, D.W. Murphy, R.E. Slusher, A. Dodabalapur, H.E. Katz, A.J. Lovinger: Anodization and microcontact printing on electroless silver: solution-based fabrication procedures for low voltage organic electronic systems, Langmuir **16**, 6054–6060 (2000)

10.32 Y. Xia, E. Kim, G.M. Whitesides: Microcontact printing of alkanethiols on silver and its application to microfabrication, J. Electrochem. Soc. **143**, 1070–1079 (1996)

10.33 Y.N. Xia, X.M. Zhao, E. Kim, G.M. Whitesides: A selective etching solution for use with patterned self-assembled monolayers of alkanethiolates on gold, Chem. Mater. **7**, 2332–2337 (1995)

10.34 R.J. Jackman, J. Wilbur, G.M. Whitesides: Fabrication of submicrometer features on curved substrates by microcontact printing, Science **269**, 664–666 (1995)

10.35 R.J. Jackman, S.T. Brittain, A. Adams, M.G. Prentiss, G.M. Whitesides: Design and fabrication of topologically complex, three-dimensional microstructures, Science **280**, 2089–2091 (1998)

10.36 J.A. Rogers, R.J. Jackman, G.M. Whitesides: Microcontact printing and electroplating on curved substrates: A new means for producing freestanding three-dimensional microstructures with possible applications ranging from micro-coil springs to coronary stents, Adv. Mater. **9**, 475–477 (1997)

10.37 Y.-L. Loo, R.W. Willett, K. Baldwin, J.A. Rogers: Additive, nanoscale patterning of metal films with a stamp and a surface chemistry mediated transfer process: applications in plastic electronics, Appl. Phys. Lett. **81**, 562–564 (2002)

10.38 Y.-L. Loo, R.W. Willett, K. Baldwin, J.A. Rogers: Interfacial chemistries for nanoscale transfer printing, J. Am. Chem. Soc. **124**, 7654–7655 (2002)

10.39 Y.-L. Loo, J.W.P. Hsu, R.L. Willett, K.W. Baldwin, K.W. West, J.A. Rogers: High-resolution transfer printing on GaAs surfaces using alkane dithiol self-assembled monolayers, J. Vac. Sci. Technol. B **20**, 2853–2856 (2002)

10.40 G.S. Ferguson, M.K. Chaudhury, G.B. Sigal, G.M. Whitesides: Contact adhesion of thin gold-films on elastomeric supports – cold welding under ambient conditions, Science **253**, 776–778 (1991)

10.41 W. Zhang, S.Y. Chou: Multilevel nanoimprint lithography with submicron alignment over 4 in Si wafers, Appl. Phys. Lett. **79**, 845–847 (2001)

10.42 N. Bowden, S. Brittain, A.G. Evans, J.W. Hutchinson, G.M. Whitesides: Spontaneous formation of ordered structures in thin films of metals supported on an elastomeric polymer, Nature **393**, 146–149 (1998)

10.43 E. Menard, L. Bilhaut, J. Zaumseil, J.A. Rogers: Improved surface chemistries, thin film deposition techniques, and stamp designs for nanotransfer printing, Langmuir **20**, 6871–6878 (2004)

10.44 J. Zaumseil, M.A. Meitl, J.W.P. Hsu, B. Acharya, K.W. Baldwin, Y.-L. Loo, J.A. Rogers: Three-dimensional and multilayer nanostructures formed by nanotransfer printing, Nano Lett. **3**, 1223–1227 (2003)

10.45 Z. Bao, J.A. Rogers, H.E. Katz: Printable organic and polymeric semiconducting materials and devices, J. Mater. Chem. **9**, 1895–1904 (1999)

10.46 J.A. Rogers, Z. Bao, A. Dodabalapur, A. Makhija: Organic smart pixels and complementary inverter circuits formed on plastic substrates by casting, printing and molding, IEEE Electron Dev. Lett. **21**, 100–103 (2000)

10.47 J.A. Rogers, Z. Bao, A. Makhija: Non-photolithographic fabrication sequence suitable for reel-to-reel production of high performance organic transistors and circuits that incorporate them, Adv. Mater. **11**, 741–745 (1999)

10.48 P. Mach, S. Rodriguez, R. Nortrup, P. Wiltzius, J.A. Rogers: Active matrix displays that use printed organic transistors and polymer dispersed liquid crystals on flexible substrates, Appl. Phys. Lett. **78**, 3592–3594 (2001)

10.49 J.A. Rogers: Toward paperlike displays, Science **291**, 1502–1503 (2001)

10.50 Y.-L. Loo, T. Someya, K.W. Baldwin, P. Ho, Z. Bao, A. Dodabalapur, H.E. Katz, J.A. Rogers: Soft, conformable electrical contacts for organic transistors: High resolution circuits by lamination, Proc. Natl. Acad. Sci. USA **99**, 10252–10256 (2002)

10.51 C. Kim, P.E. Burrows, S.R. Forrest: Micropatterning of organic electronic devices by cold-welding, Science **288**, 831–833 (2000)

10.52 C. Kim, M. Shtein, S.R. Forrest: Nanolithography based on patterned metal transfer and its application to organic electronic devices, Appl. Phys. Lett. **80**, 4051–4053 (2002)

10.53 J.A. Rogers, R.J. Jackman, G.M. Whitesides, D.L. Olson, J.V. Sweedler: Using microcontact printing to fabricate microcoils on capillaries for high resolution ^{1}H-NMR on nanoliter volumes, Appl. Phys. Lett. **70**, 2464–2466 (1997)

10.54 J.A. Rogers, R.J. Jackman, G.M. Whitesides: Constructing single and multiple helical microcoils and characterizing their performance as components of microinductors and microelectromagnets, J. Microelectromech. Syst. **6**, 184–192 (1997)

10.55 R.J. Jackman, J.A. Rogers, G.M. Whitesides: Fabrication and characterization of a concentric, cylindrical microtransformer, IEEE Trans. Magn. **33**, 2501–2503 (1997)

10.56 J.A. Rogers, M. Meier, A. Dodabalapur: Using stamping and molding techniques to produce distributed feedback and Bragg reflector resonators for plastic lasers, Appl. Phys. Lett. **73**, 1766–1768 (1998)

10.57 M. Berggren, A. Dodabalapur, R.E. Slusher, A. Timko, O. Nalamasu: Organic solid-state lasers with imprinted gratings on plastic substrates, Appl. Phys. Lett. **72**, 410–411 (1998)

10.58 J.A. Rogers, M. Meier, A. Dodabalapur: Distributed feedback ridge waveguide lasers fabricated by nanoscale printing and molding on non-planar substrates, Appl. Phys. Lett. **74**, 3257–3259 (1999)

11. Material Aspects of Micro- and Nanoelectromechanical Systems

Christian A. Zorman, Mehran Mehregany

One of the more significant technological achievements during the last 20 years has been the development of MEMS and its new offshoot, NEMS. These developments were made possible by significant advancements in the materials and processing technologies used in the fabrication of MEMS and NEMS devices. While initial developments capitalized on a mature Si infrastructure built for the integrated circuit (IC) industry, recent advances have come about using materials and processes not associated with IC fabrication, a trend that is likely to continue as new application areas emerge.

A well-rounded understanding of MEMS and NEMS technology requires a basic knowledge of the materials used to construct the devices, since material properties often govern device performance and dictate fabrication approaches. An understanding of the materials used in MEMS and NEMS involves an understanding of material systems, since such devices are rarely constructed of a single material but rather a collection of materials working in conjunction with each other to provide critical functions. It is from this perspective that the following chapter is constructed. A preview of the materials selected for inclusion in this chapter is presented in Table 11.1. It should be clear from this table that this chapter is not a summary of all materials used in MEMS and NEMS, as such a work would itself constitute a text of

11.1	**Silicon**	333
	11.1.1 Single-Crystal Silicon	333
	11.1.2 Polycrystalline and Amorphous Silicon	336
	11.1.3 Porous Silicon	338
	11.1.4 Silicon Dioxide	339
	11.1.5 Silicon Nitride	340
11.2	**Germanium-Based Materials**	340
	11.2.1 Polycrystalline Ge	340
	11.2.2 Polycrystalline SiGe	341
11.3	**Metals**	341
11.4	**Harsh-Environment Semiconductors**	343
	11.4.1 Silicon Carbide	343
	11.4.2 Diamond	346
11.5	**GaAs, InP, and Related III–V Materials**	349
11.6	**Ferroelectric Materials**	350
11.7	**Polymer Materials**	351
	11.7.1 Polyimide	351
	11.7.2 SU-8	351
	11.7.3 Parylene	352
	11.7.4 Liquid Crystal Polymer	352
11.8	**Future Trends**	352
References		353

significant size. It does, however, present a selection of some of the more important material systems, and especially those that illustrate the importance of viewing MEMS and NEMS in terms of material systems.

11.1 Silicon

11.1.1 Single-Crystal Silicon

Use of silicon (Si) as a material for microfabricated sensors dates back to the middle of the 20th century when the piezoresistive effect in germanium (Ge) and Si was first identified [11.1]. It was discovered that the piezoresistive coefficients of Si were significantly higher than those associated with metals used in conventional strain

Table 11.1 Distinguishing characteristics and application examples of selected materials for MEMS and NEMS

Material	Distinguishing characteristics	Application examples
Single-crystal silicon (Si)	High-quality electronic material, selective anisotropic etching	Bulk micromachining, piezoresistive sensing
Polycrystalline Si (polysilicon)	Doped Si films on sacrificial layers	Surface micromachining, electrostatic actuation
Silicon dioxide (SiO_2)	Insulating, etched by HF, compatible with polysilicon	Sacrificial layer in polysilicon surface micromachining, passivation layer for devices
Silicon nitride (Si_3N_4, Si_xN_y)	Insulating, chemically resistant, mechanically durable	Isolation layer for electrostatic devices, membrane and bridge material
Polycrystalline germanium (polyGe), Polycrystalline silicon-germanium (poly SiGe)	Deposited at low temperatures	Integrated surface micromachined MEMS
Gold (Au), aluminum (Al)	Conductive thin films, flexible deposition techniques	Innerconnect layers, masking layers, electromechanical switches
Bulk Ti	High strength, corrosion resistant	Optical MEMS
Nickel-iron (NiFe)	Magnetic alloy	Magnetic actuation
Titanium-nickel (TiNi)	Shape-memory alloy	Thermal actuation
Silicon carbide (SiC) diamond	Electrically and mechanically stable at high temperatures, chemically inert, high Young's modulus to density ratio	Harsh-environment MEMS, high-frequency MEMS/NEMS
Gallium arsenide (GaAs), indium phosphide (InP), indium arsenide (InAs) and related materials	Wide bandgap, epitaxial growth on related ternary compounds	RF MEMS, optoelectronic devices, single-crystal bulk and surface micromachining
Lead zirconate titanate (PZT)	Piezoelectric material	Mechanical sensors and actuators
Polyimide	Chemically resistant, high-temperature polymer	Mechanically flexible MEMS, bioMEMS
SU-8	Thick, photodefinable resist	Micromolding, High-aspect-ratio structures
Parylene	Biocompatible polymer, deposited at room temperature by CVD	Protective coatings, molded polymer structures
Liquid crystal polymer	Chemically resistant, low moisture permeability, insulating	bioMEMS, RF MEMS

gauges; and this finding initiated the development of Si-based strain gauge devices, and along with Si bulk micromachining techniques, piezoresistive Si pressure sensors during the 1960s and 1970s. The subsequent development of Si surface micromachining techniques along with the recognition that micromachined Si structures could be integrated with Si IC devices marked the advent of MEMS with Si firmly positioned as the primary MEMS material.

For MEMS applications, single-crystal Si serves several key functions. Single-crystal Si is one of the most versatile materials for bulk micromachining due to the availability of anisotropic etching processes in conjunction with good mechanical properties. Single-crystal Si has favorable mechanical properties (i.e., a Young's modulus of about 190 GPa), enabling its use as a material for membranes, resonant beams, and other such structures. For surface micromachining applications, single-crystal Si substrates are used primarily as mechanical platforms on which device structures are fabricated, although the advent of silicon-on-insulator (SOI) substrates enables the fabrication of single-crystal Si surface micromachined structures by using the buried oxide as a sacrificial layer. Use of high-quality single-crystal wafers enables the fabrication of integrated MEMS devices, at least for materials and processes that are compatible with Si ICs.

From the materials perspective, single-crystal Si is a relatively easy material to bulk micromachine due to the availability of anisotropic etchants such as potassium hydroxide (KOH) and tetramethyl-aluminum hydroxide (TMAH) that attack the (100) and (110) Si crystal planes significantly faster than the (111) crystal planes. For example, the etching rate ratio of (100) to (111) planes in Si is about 400:1 for a typical KOH/water etching solution. Silicon dioxide (SiO_2), silicon nitride (Si_3N_4), and some metallic thin films (e.g., Cr, Au, etc.) provide good etch masks for most Si anisotropic etchants. Heavily boron-doped Si is an effective etch stop for some liquid reagents. Boron-doped etch stops are often less than $10\,\mu m$ thick, since the boron concentration in Si must exceed $7 \times 10^{19}\,cm^3$ for the etch stop to be effective and the doping is done by thermal diffusion. Ion implantation can be used to create a subsurface etch stop layer; however, the practical limit is a few micrometer.

In contrast to anisotropic etching, isotropic etching exhibits no selectivity to the various crystal planes. Commonly used isotropic Si etchants consist of hydrofluoric (HF) and nitric (HNO_3) acid mixtures in water or acetic acid (CH_3COOH), with the etch rate dependent on the ratio of HF to HNO_3. From a processing perspective, isotropic etching of Si is commonly used for removal of work-damaged surfaces, creation of structures in single-crystal slices, and patterning of single-crystal or polycrystalline films.

Well-established dry etching processes are routinely used to pattern single-crystal Si. The process spectrum ranges from physical techniques such as sputtering and ion milling to chemical techniques such as plasma etching. Reactive ion etching (RIE) is the most commonly used dry etching technique for Si patterning. By combining both physical and chemical processes, RIE is a highly effective anisotropic Si etching technique that can be used to generate patterns that are independent of crystalline orientation. Fluorinated compounds such as CF_4, SF_6, and NF_3, or chlorinated compounds such as CCl_4 or Cl_2, sometimes mixed with He, O_2, or H_2, are commonly used in Si RIE. The RIE process is highly directional, which enables direct lateral pattern transfer from an overlying masking material to the etched Si surface. SiO_2 thin films are often used as masking and sacrificial layers owing to its chemical durability under these plasma conditions. Process limitations (i.e., etch rates) restrict the etch depths of conventional Si RIE to less than $10\,\mu m$; however, a process called deep reactive ion etching (DRIE) has extended the use of anisotropic dry etching to depths well beyond several hundred micrometer.

Using the aforementioned processes and techniques, a wide variety of microfabricated devices have been made from single-crystal Si, such as piezoresistive pressure sensors, accelerometers, and mechanical resonators, to name a few. Using nearly the same approaches but on a smaller scale, *top-down* nanomachining techniques have been used to fabricate nanoelectromechanical devices from single-crystal Si. Single-crystal Si is particularly well suited for nanofabrication because high crystal quality substrates with very smooth surfaces are readily available. By coupling electron-beam (e-beam) lithographic techniques with conventional Si etching, device structures with submicrometer dimensions have been fabricated. Submicrometer, single-crystal Si nanomechanical structures have been successfully micromachined from bulk Si wafers [11.2]

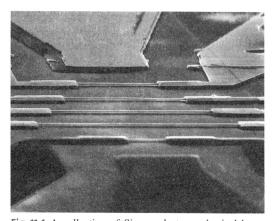

Fig. 11.1 A collection of Si nanoelectromechanical beam resonators fabricated from a single-crystal Si substrate (courtesy M. Roukes, Caltech)

and silicon-on-insulator (SOI) wafers [11.3]. In the former case, an isotropic Si etch was performed to release the device structures, whereas in the latter case, the 50–200 nm structures were released by dissolving the underlying oxide layer in HF. An example of nanoelectromechanical beam structures fabricated from a single-crystal Si substrate is shown in Fig. 11.1.

11.1.2 Polycrystalline and Amorphous Silicon

Surface micromachining is a process where a sequence of thin films, often of different materials, is deposited and selectively etched to form the desired micromechanical (or microelectromechanical) structure. In contrast to bulk micromachining, the substrate serves primarily as a device-supporting platform. For Si-based surface micromachined MEMS, polycrystalline Si (polysilicon) is most often used as the structural material, SiO_2 as the sacrificial material, silicon nitride (Si_3N_4) for electrical isolation of device structures, and single-crystal Si as the substrate. Like single-crystal Si, polysilicon can be doped during or after film deposition. SiO_2 can be thermally grown or deposited on polysilicon over a broad temperature range (e.g., 200–1150 °C) to meet various process and material requirements. SiO_2 is readily dissolvable in hydrofluoric acid (HF), which does not etch polysilicon and thus can be used to dissolve SiO_2 sacrificial layers. Si_3N_4 is an insulating film that is highly resistant to oxide etchants. The polysilicon micromotor shown in Fig. 11.2 was surface micromachined using a process that included these materials.

Fig. 11.2 SEM micrograph of a surface micromachined polysilicon micromotor fabricated using a SiO_2 sacrificial layer

For MEMS and IC applications, polysilicon films are commonly deposited using a process known as low-pressure chemical vapor deposition (LPCVD). The typical polysilicon LPCVD reactor is based on a hot-wall, resistance-heated furnace. Typical processes are performed at temperatures ranging from 580 to 650 °C and pressures from 100 to 400 mtorr. The most commonly used source gas is silane (SiH_4). The microstructure of polysilicon thin films consist of a collection of small grains whose microstructure and orientation is a function of the deposition conditions [11.4]. For typical LPCVD processes (e.g., 200 mtorr), the amorphous-to-polycrystalline transition temperature is about 570 °C, with polycrystalline films deposited above the transition temperature. At 600 °C, the grains are small and equiaxed, while at 625 °C, the grains are large and columnar [11.4]. The crystal orientation is predominantly (110) Si for temperatures between 600 and 650 °C, while the (100) orientation is dominant for temperatures between 650 and 700 °C.

The resistivity of polysilicon can be modified using the doping methods developed for single-crystal Si. Diffusion is an effective method for doping polysilicon films, especially for heavy doping of thick films. Phosphorus, which is the most commonly used dopant in polysilicon MEMS, diffuses significantly faster in polysilicon than in single-crystal Si due primarily to enhanced diffusion rates along grain boundaries. The diffusivity of phosphorus in polysilicon thin films with small equiaxed grains is about $1 \times 10^{12}\,cm^2/s$. Ion implantation is also used to dope polysilicon films. A high-temperature annealing step is usually required to electrically activate the implanted dopants as well as to repair implant-related damage in the polysilicon films. In general, the conductivity of implanted polysilicon films is not as high as films doped by diffusion.

In situ doping of polysilicon is performed by simply including a dopant gas, usually diborane (B_2H_6) or phosphine (PH_3), in the CVD process. The addition of dopants during the deposition process not only modifies the conductivity but also affects the deposition rate of the polysilicon films. As shown in Fig. 11.3, the inclusion of boron generally increases the deposition rate of polysilicon relative to undoped films [11.5], while phosphorus (not shown) reduces the rate. In situ doping can be used to produce conductive films with uniform doping profiles without requiring the high-temperature steps commonly associated with diffusion or ion implantation. Although commonly used to produce doped polysilicon for electrostatic devices, *Cao* et al. [11.6] have used in situ phosphorus-doped polysilicon films in piezoresistive strain gauges, achieving gauge factors as high as 15 for a single strip sensor.

The thermal conductivity of polysilicon is a strong function of its microstructure, and therefore the conditions used during deposition [11.4]. For fine-grained films, the thermal conductivity is about 25% of the value of single-crystal Si. For thick films with large grains, the thermal conductivity ranges between 50% and 85% of the single-crystal value.

Like the electrical and thermal properties of polysilicon, the as-deposited residual stress in polysilicon films depends on microstructure. For films deposited under typical conditions (200 mtorr, 625 °C), the as-deposited polysilicon films have compressive residual stresses. The highest compressive stresses are found in amorphous Si films and polysilicon films with a strong, columnar (110) texture. For films with fine-grained microstructures, the stress tends to be tensile. Annealing can be used to reduce the compressive stress in as-deposited polysilicon films. For instance, compressive residual stresses on the order of 500 MPa can be reduced to less than 10 MPa by annealing the as-deposited films at 1000 °C in a N_2 ambient [11.7, 8]. Rapid thermal annealing (RTA) provides an effective method of stress reduction in polysilicon films on temperature-sensitive substrates. *Zhang* et al. [11.9] reported that a 10 s anneal at 1100 °C was sufficient to completely relieve the stress in films that originally had a compressive stress of about 340 MPa. RTA is particularly attractive in situations where the process parameters require a low thermal budget.

As an alternative to high-temperature annealing, *Yang* et al. [11.10] have developed an approach that actually utilizes the residual stress characteristics of polysilicon deposited under various conditions to construct polysilicon multilayers that have the desired thickness and stress values. The multilayers are comprised of alternating tensile and compressive polysilicon layers that are deposited in a sequential manner. The tensile layers consist of fine-grained polysilicon grown at a temperature of 570 °C, while the compressive layers are made up of columnar polysilicon deposited at 615 °C. The overall stress in the composite film depends on the number of alternating layers and the thickness of each layer. With the proper set of parameters, a composite polysilicon multilayer can be deposited with near zero residual stress and no stress gradient. The process achieves stress reduction without high-temperature annealing, a considerable advantage for integrated MEMS processes.

Many device designs require polysilicon thicknesses that are not readily achievable using conventional LPCVD polysilicon due to the low deposition rates associated with such systems. For these applications, epitaxial Si reactors can be used to grow polysilicon films. Unlike conventional LPCVD processes with deposition rates of less than 100 Å/min, epitaxial processes have deposition rates on the order of 1 μm/min [11.11]. The high deposition rates result from the much higher substrate temperatures (> 1000 °C) and deposition pressures (> 50 torr) used in these processes. The polysilicon films are usually deposited on SiO_2 sacrificial layers to enable surface micromachining. An LPCVD polysilicon seed layer is sometimes used in order to control nucleation, grain size, and surface roughness. As with conventional polysilicon, the microstructure and residual stress of the epi-poly films, as they are known, are related to deposition conditions. Compressive films generally have a mixture of [110] and [311] grains [11.12, 13], while tensile films have a random mix of [110], [100], [111], and [311] grains [11.12]. The Young's modulus of epi-poly measured from micromachined test structures is comparable with LPCVD polysilicon [11.13]. Mechanical properties test structures [11.11–13], thermal actuators [11.11], electrostatically actuated accelerometers [11.11], and gryoscopes [11.14] have been fabricated from these films.

As a low-temperature alternative to LPCVD polysilicon, physical vapor deposition (PVD) techniques have been developed to produce Si thin films on temperature-sensitive substrates. *Abe* et al. [11.15] and *Honer* et al. [11.16] have developed sputtering pro-

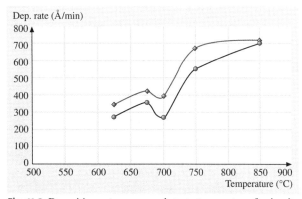

Fig. 11.3 Deposition rate versus substrate temperature for in situ boron-doped (♦) and undoped (○) polysilicon films grown by atmospheric pressure chemical vapor deposition (after [11.5])

cesses for polysilicon. Early work [11.15] emphasized the ability to deposit very smooth (2.5 nm) polysilicon films on thermally oxidized wafers at reasonable deposition rates (19.1 nm/min) and with low residual compressive stresses. The process involved DC magnitron sputtering from a Si target using an Ar sputtering gas, a chamber pressure of 5 mtorr, and a power of 100 W. The authors reported that a post-deposition anneal at 700 °C in N_2 for 2 h was needed to crystallize the deposited film and perhaps lower the stress. *Honer* et al. [11.16] sought to develop a polymer-friendly, Si-based surface micromachining process based on polysilicon sputtered onto polyimide and PSG sacrificial layers. To improve the conductivity of the micromachined Si structures, the sputtered Si films were sandwiched between two TiW cladding layers. The device structures on polyimide were released using oxygen plasma etching. The processing step with the highest temperature was, in fact, the polyimide cure at 350 °C. To test the robustness of the process, sputter-deposited Si microstructures were fabricated on substrates containing CMOS devices. As expected from thermal budget considerations, the authors reported no measurable degradation of device performance.

PECVD has emerged as an alternative to LPCVD for the production of Si-based surface micromachined structures on temperature-sensitive substrates. *Gaspar* et al. [11.17] recently reported on the development of surface micromachined microresonators fabricated from hydrogenated amorphous Si (a-Si:H) thin films deposited by PECVD. The vertically actuated resonators consisted of doubly-clamped microbridges suspended over fixed Al electrodes. The a-Si:H films were deposited using SiH_4 and H_2 precursors and PH_3 as a doping gas. The substrate temperature was held to around 100 °C, which enabled the use of photoresist as a sacrificial layer. The microbridges consisted of a large paddle suspended by two thin paddle supports, with the paddle providing a large reflective surface for optical detection of resonant frequency. The megahertz-frequency resonators exhibited quality factors in the 1×10^5 range when tested in vacuum.

11.1.3 Porous Silicon

Porous Si is produced by room temperature electrochemical etching of Si in HF. If configured as an electrode in an HF-based electrochemical circuit, positive charge carriers (holes) at the Si surface facilitate the exchange of F atoms with H atoms terminating the Si surface. The exchange continues in the subsurface region, leading to the eventual removal of the fluorinated Si. The quality of the etched surface is related to the density of holes at the surface, which is controlled by the applied current density. For high current densities, the density of holes is high and the etched surface is smooth. For low current densities, the hole density is low and clustered in highly localized regions associated with surface defects. Surface defects become enlarged by etching, which leads to the formation of pores. Pore size and density are related to the type of Si used and the conditions of the electrochemical cell. Both single-crystal and polycrystalline Si can be converted to porous Si.

The large surface-to-volume ratios make porous Si attractive for gaseous and liquid applications, including filter membranes and absorbing layers for chemical and mass sensing [11.18]. When single-crystal substrates are used, the unetched porous layer remains single crystalline and is suitable for epitaxial Si growth. It has been shown that CVD coatings do not generally penetrate the porous regions, but rather overcoat the pores at the surface of the substrate [11.19]. The formation of localized Si-on-insulator structures is therefore possible by simply combining pore formation with epitaxial growth, followed by dry etching to create access holes to the porous region and thermal oxidation of the underlying porous region. A third application uses porous Si as a sacrificial layer for polysilicon and single-crystalline Si surface micromachining. As shown by *Lang* et al. [11.19], the process involves the electrical isolation of the solid structural Si layer by either pn-junction formation through selective doping or use of electrically insulating thin films since the formation of pores only occurs on electrically charged surfaces. A weak Si etchant will aggressively attack the porous regions with little damage to the structural Si layers and can be used to release the devices.

Porous polysilicon is currently being developed as a structural material for chip-level vacuum packaging [11.20]. In this example, a 1.5 μm thick polysilicon is deposited onto a supporting PSG sacrificial layer, electrochemically etched in an HF solution to render it porous, and then annealed by RTA to reduce stress in the porous layer. When fabricated locally over a prefabricated device structure (prior to release), the porous Si forms a localized shell that will serve as a mechanical support for the main packaging structure. The porous structure enables an HF etch to remove the supporting PSG layer as well as any sacrificial oxide layers associated with the prefabricated MEMS device. After the sacrificial etch, the packaging sequence is completed by depositing a polysilicon film by LPCVD at 179 mtorr on the porous shell, thus fully encapsulating the device under vacuum conditions. This technique was used to package a microfabricated Pirani vacuum gauge, which enabled an in situ measurement of pressure versus time. The authors found no detectable change in pressure over a 3-month period.

11.1.4 Silicon Dioxide

Silicon dioxide (SiO_2) is one of the most widely used materials in the fabrication of MEMS. In polysilicon surface micromachining, SiO_2 is used as a sacrificial material since it can be easily dissolved using etchants that do not attack polysilicon. SiO_2 is widely used as an etch mask for dry etching of thick polysilicon films since it is chemically resistant to dry etching processes for polysilicon. SiO_2 films are also used as passivation layers on the surfaces of environmentally sensitive devices.

The most common processes used to produce SiO_2 films for polysilicon surface micromachining are thermal oxidation and LPCVD. Thermal oxidation of Si is performed at temperatures of 900–1200 °C in the presence of oxygen or steam. Since thermal oxidation is a self-limiting process, the maximum practical film thickness that can be obtained is about 2 μm, which is sufficient for many sacrificial applications. As noted by its name, thermal oxidation of Si can only be performed on Si surfaces.

SiO_2 films can be deposited on a wide variety of substrate materials by LPCVD. In general, LPCVD provides a means for depositing thick ($> 2\,\mu$m) SiO_2 films at temperatures much lower than thermal oxidation. Known as low-temperature oxides, or LTO for short, these films have a higher etch rate in HF than thermal oxides, which translates to significantly faster release times when LTO films are used as sacrificial layers. Phosphosilicate glass (PSG) can be formed using nearly the same deposition process as LTO by adding a phosphorus-containing gas to the precursor flows. PSG films are useful as sacrificial layers since they generally have higher etching rates in HF than LTO films.

PSG and LTO films are deposited in hot-wall, low-pressure, fused-silica furnaces in systems similar to those described previously for polysilicon. Precursor gases include SiH_4 as a Si source, O_2 as an oxygen source, and, in the case of PSG, PH_3 as a source of phosphorus. LTO and PSG films are typically deposited at temperatures of 425–450 °C and pressures ranging from 200 to 400 mtorr. The low deposition temperatures result in LTO and PSG films that are slightly less dense than thermal oxides due to the incorporation of hydrogen in the films. LTO films can, however, be densified by an annealing step at high temperature (1000 °C). The low density of LTO and PSG films is partially responsible for the increased etch rate in HF.

Thermal SiO_2 and LTO are electrical insulators used in numerous MEMS applications. The dielectric constants of thermal oxide and LTO are 3.9 and 4.3, respectively. The dielectric strength of thermal SiO_2 is 1.1×10^6 V/cm, and for LTO it is about 80% of that value [11.21]. The stress in thermal SiO_2 is compressive with a magnitude of about 300 MPa [11.21]. For LTO, however, the typical as-deposited residual stress is tensile, with a magnitude of about 100–400 MPa [11.21]. The addition of phosphorus to LTO decreases the tensile residual stress to about 10 MPa for phosphorus concentrations of 8% [11.22]. As with polysilicon, the properties of LTO and PSG are dependent on processing conditions.

Plasma enhanced chemical vapor deposition (PECVD) is another common method to produce oxides of silicon. Using a plasma to dissociate the gaseous precursors, the deposition temperatures needed to deposit PECVD oxide films is lower than for LPCVD films. For this reason, PECVD oxides are quite commonly used as masking, passivation, and protective layers, especially on devices that have been coated with metals.

Quartz is the crystalline form of SiO_2 and has interesting properties for MEMS. Quartz is optically transparent, piezoelectric, and electrically insulating. Like single-crystal Si, quartz substrates are available as high-quality, large-area wafers that can be bulk micromachined using anisotropic etchants. A short review of the basics of quartz etching was written by *Danel* et al. [11.23] and is recommended for those interested in the subject. Quartz has recently become a popular substrate material for microfluidic devices due to its optical, electronic, and chemical properties.

Another SiO_2-related material that has recently found uses in MEMS is spin-on-glass (SOG). SOG is a polymeric material with a viscosity suitable for spin coating. Two recent publications illustrate the potential for SOG in MEMS fabrication. In the first example, *Yasseen* et al. [11.24] detailed the development of SOG as a thick-film sacrificial molding material for thick polysilicon films. The authors reported a process to deposit, polish, and etch SOG films that were 20 μm thick. The thick SOG films were patterned into molds and filled with 10 μm thick LPCVD polysilicon films, planarized by selective CMP, and subsequently dissolved in a wet etchant containing HCl, HF, and H_2O to reveal the patterned polysilicon structures. The cured SOG films were completely compatible with the polysilicon deposition process. In the second example, *Liu* et al. [11.25] fabricated high aspect ratio channel plate microstructures from SOG. Electroplated nickel (Ni) was used as a molding material, with Ni channel plate molds fabricated using a conventional LIGA process. The Ni molds were then filled with SOG, and the sacrificial Ni molds were removed in a reverse electroplating process. In this case, the fabricated SOG structures (over 100 μm tall) were micromachined glass structures fabricated using a molding material more commonly used for structural components.

11.1.5 Silicon Nitride

Silicon nitride (Si_3N_4) is widely used in MEMS for electrical isolation, surface passivation, etch masking, and as a mechanical material typically for membranes and other suspended structures. Two deposition methods are commonly used to deposit Si_3N_4 thin films, LPCVD, and PECVD. PECVD silicon nitride is generally nonstoichiometric (sometimes denoted as Si_xN_y:H) and may contain significant concentrations of hydrogen. Use of PECVD silicon nitride in micromachining applications is somewhat limited because it has a high etch rate in HF (e.g., often higher than that of thermally grown SiO_2). However, PECVD offers the ability to deposit nearly stress-free silicon nitride films, an attractive property for encapsulation and packaging.

Unlike its PECVD counterpart, LPCVD Si_3N_4 is extremely resistant to chemical attack, thereby making it the material of choice for many Si bulk and surface micromachining applications. LPCVD Si_3N_4 is commonly used as an insulating layer because it has a resistivity of 10^{16} Ω cm and field breakdown limit of 10^7 V/cm. LPCVD Si_3N_4 films are deposited in horizontal furnaces similar to those used for polysilicon deposition. Typical deposition temperatures and pressures range between 700 and 900 °C and 200 and 500 mtorr, respectively. The standard source gases are dichlorosilane (SiH_2Cl_2) and ammonia (NH_3). To produce stoichiometric Si_3N_4 a NH_3 to SiH_2Cl_2 ratio 10:1 is commonly used. The microstructure of films deposited under these conditions is amorphous.

The residual stress in stoichiometric Si_3N_4 is large and tensile, with a magnitude of about 1 GPa. Such a large residual stress causes films thicker than a few thousand angstroms to crack. Nonetheless thin stoichiometric Si_3N_4 films have been used as mechanical support structures and electrical insulating layers in piezoresistive pressure sensors [11.26]. To enable the use of Si_3N_4 films for applications that require micrometer-thick, durable, and chemically resistant membranes, Si_xN_y films can be deposited by LPCVD. These films, often referred to as Si-rich or low-stress nitride, are intentionally deposited with an excess of Si by simply decreasing the ratio of NH_3 to SiH_2Cl_2 during deposition. Nearly stress-free films can be deposited using a NH_3-to-SiH_2Cl_2 ratio of 1/6, a deposition temperature of 850 °C, and a pressure of 500 mtorr [11.27]. The increase in Si content not only leads to a reduction in tensile stress, but also a decrease in the etch rate in HF. Such properties have enabled the development of fabrication techniques that would otherwise not be feasible with stoichiometric Si_3N_4. For example, low-stress silicon nitride has been surface micromachined using polysilicon as the sacrificial material [11.28]. In this case, Si anisotropic etchants such as KOH and EDP were used for dissolving the sacrificial polysilicon. *French* et al. [11.29] used PSG as a sacrificial layer to surface micromachine low-stress nitride, capitalizing on the HF resistance of the nitride films.

11.2 Germanium-Based Materials

11.2.1 Polycrystalline Ge

Like Si, Ge has a long history as a semiconductor device material, dating back to the development of the earliest transistors and semiconductor strain gauges. Issues related to germanium oxide, however, stymied the development of Ge for microelectronic devices. Nonetheless, there is a renewed interest in using Ge in surface micromachined devices due to the relatively low processing temperatures required to deposit the material and its compatibility with Si.

Thin polycrystalline Ge (poly-Ge) films can be deposited by LPCVD at temperatures as low as 325 °C on Si, Ge, and SiGe substrates [11.30]. Ge does not nucleate on SiO_2 surfaces, which prohibits the use of thermal oxides and LTO films as sacrificial layers but enables the use of these films as sacrificial molds. Residual stress in poly-Ge films deposited on Si substrates can be reduced to nearly zero after short anneals at modest temperatures (30 s at 600 °C). Poly-Ge is essentially impervious to KOH, TMAH, and BOE, enabling the fabrication of Ge membranes on Si substrates [11.30]. The mechanical properties of poly-Ge are comparable to those of polysilicon, having a Young's modulus of 132 GPa and a fracture stress ranging between 1.5 and 3.0 GPa [11.31]. Mixtures of HNO_3, H_2O, and HCl and H_2O, H_2O_2, and HCl, as well as the RCA SC-1 cleaning solution, isotropically etch Ge. Since these mixtures do not etch Si, SiO_2, Si_3N_4, and SiN, poly-Ge can be used as a sacrificial substrate layer in polysilicon surface micromachining. Using these techniques, devices such as poly-Ge-based thermistors and Si_3N_4 membrane-based pressure sensors made using poly-Ge sacrificial layers have been fabricated [11.30]. *Franke* et al. [11.31] found no performance degradation in Si CMOS devices following the fabrication of surface micromachined poly-Ge structures, thus demonstrating the potential for on-chip integration of Ge electromechanical devices with Si circuitry.

11.2.2 Polycrystalline SiGe

Like poly-Ge, polycrystalline SiGe (poly-SiGe) is a material that can be deposited at temperatures lower than polysilicon. Deposition processes include LPCVD, APCVD, and RTCVD (rapid thermal CVD) using SiH_4 and GeH_4 as precursor gases. Deposition temperatures range between 450 °C for LPCVD [11.32] and 625 °C by rapid thermal CVD (RTCVD) [11.33]. In general, the deposition temperature is related to the concentration of Ge in the films, with higher Ge concentrations resulting in lower deposition temperatures. Like polysilicon, poly-SiGe can be doped with boron and phosphorus to modify its conductivity. In situ boron doping can be performed at temperatures as low as 450 °C [11.32]. *Sedky* et al. [11.33] showed that the deposition temperature of conductive films doped with boron could be further reduced to 400 °C if the Ge content was kept at or above 70%.

Unlike poly-Ge, poly-SiGe can be deposited on a number of sacrificial substrates, including SiO_2 [11.33], PSG [11.31], and poly-Ge [11.31]. For Ge-rich films, a thin polysilicon seed layer is sometimes used on SiO_2 surfaces since Ge does not readily nucleate on oxide surfaces. Like many compound materials, variations in film composition can change the physical properties of the material. For instance, etching of poly-SiGe by H_2O_2 becomes significant for Ge concentrations over 70%. *Sedky* et al. [11.33] has shown that the microstructure, film conductivity, residual stress, and residual stress gradient are related to the concentration of Ge in the material. With respect to residual stress, *Franke* et al. [11.32] produced in situ boron-doped films with residual compressive stresses as low as 10 MPa.

The poly-SiGe, poly-Ge material system is particularly attractive for surface micromachining since H_2O_2 can be used as a release agent. It has been reported that poly-Ge etches at a rate of $0.4\,\mu m/min$ in H_2O_2, while poly-SiGe with Ge concentrations below 80% have no observable etch rate after 40 h [11.34]. The ability to use H_2O_2 as a sacrificial etchant makes the combination of poly-SiGe and poly-Ge extremely attractive for surface micromachining from processing, safety, and materials compatibility points of view. Due to the conformal nature of LPCVD processing, poly-SiGe structural elements, such as gimbal-based microactuator structures have been made by high-aspect-ratio micromolding [11.34]. Capitalizing on the low deposition temperatures, poly-SiGe MEMS integrated with Si ICs has been demonstrated [11.32]. In this process, CMOS structures are first fabricated on Si wafers. Poly-SiGe mechanical structures are then surface micromachined using a poly-Ge sacrificial layer. A significant advantage of this design lies in the fact that the MEMS structure is positioned directly above the CMOS structure, thus reducing the parasitic capacitance and contact resistance characteristic of interconnects associated with side-by-side integration schemes. Use of H_2O_2 as the sacrificial etchant eliminates the need for layers to protect the underlying CMOS structure during release. In addition to its utility as a material for integrated MEMS devices, poly-SiGe has been identified as a material well suited for micromachined thermopiles [11.35] to its lower thermal conductivity relative to Si.

11.3 Metals

It can be argued that of all the material categories associated with MEMS, metals may be among the most enabling, since metallic thin films are used in many different capacities, from etch masks used in device fabrication to interconnects and structural elements in microsensors and microactuators. Metallic thin films can be deposited using a wide range of techniques, including evaporation, sputtering, CVD, and electroplating. Since a complete review of the metals used in MEMS is far beyond the scope of this chapter, the examples presented in this section were selected to represent a broad cross section where metals have found uses in MEMS.

Aluminum (Al) and gold (Au) are among the most widely employed metals in microfabricated electronic and electromechanical devices as a result of their use as innerconnect and packaging materials. In addition to these critical electrical functions, Al and Au are also desirable as electromechanical materials. One such example is the use of Au micromechanical switches for RF MEMS. For conventional RF applications, chip level switching is currently performed using FET and PIN diode-based solid state devices fabricated from gallium arsenide (GaAs) substrates. Unfortunately, these devices suffer from insertion losses and poor electrical isolation. In an effort to develop replacements for GaAs-based solid state switches, *Hyman* et al. [11.36] reported the development of an electrostatically actuated, cantilever-based micromechanical switch fabricated on GaAs substrates. The device consisted of a silicon-nitride-encased Au cantilever constructed on a sacrificial silicon dioxide layer. The silicon nitride and silicon dioxide layers were deposited by PECVD, and the Au beam was electroplated from a sodium sulfite solution inside a photoresist mold. A thin multilayer of Ti and Au was sputter deposited in the mold prior to electroplating. The trilayer cantilever structure was chosen to minimize the deleterious effects of thermal- and process-related stress gradients in order to produce unbent and thermally stable beams. After deposition and patterning, the cantilevers were released in HF. The processing steps proved to be completely compatible with GaAs substrates. The released cantilevers demonstrated switching speeds of better than $50\,\mu s$ at $25\,V$ with contact lifetimes exceeding 10^9 cycles.

In a second example from RF MEMS, *Chang* et al. [11.37] reported the fabrication of an Al-based micromachined switch as an alternative to GaAs FETs and PIN diodes. In contrast to the work by *Hyman* et al. [11.36], this switch utilizes the differences in the residual stresses in Al and Cr thin films to create bent cantilever switches that capitalize on the stress differences in the materials. Each switch is comprised of a series of linked bimorph cantilevers designed in such a way that the resulting structure bends significantly out of the plane of the wafer due to the stress differences in the bimorph. The switch is drawn closed by electrostatic attraction. The bimorph consists of metals that can easily be processed with GaAs wafers, thus making integration with GaAs devices possible. The released switches were relatively slow, at $10\,ms$, but an actuation voltage of only $26\,V$ was needed to close the switch.

Direct bulk micromachining of metal substrates is being developed for MEMS applications requiring structures with the dimensional complexity associated with Si DRIE and the physical properties of metals. One such example is Ti, which has a higher fracture toughness, a greater biocompatibility, and a more stable passivating oxide than Si. A process to fabricate high-aspect-ratio, three-dimensional structures from bulk Ti substrates has recently been developed [11.38]. This process involves inductively coupled plasma etching of a TiO_2-capped Ti substrate. The TiO_2 capping layer is deposited by DC reactive sputtering and photolithographically patterned using a CHF_3-based dry etch. The deep Ti etch is then performed using a Cl/Ar-based plasma that exhibits a selectivity of $40:1$ with the masking TiO_2 layer. The etch process consists of a series of two-step sequences, where the first step involves Ti removal by the Cl/Ar plasma while the second step involves sidewall passivation using an oxygen plasma. After the prescribed etch period, the masking thin film can be removed by HF etching. High-aspect-ratio comb-drive actuators and other beam-based structures have been fabricated directly from bulk Ti using this method.

Thin-film metallic alloys that exhibit the shape-memory effect are of particular interest to the MEMS community for their potential in microactuators. The shape-memory effect relies on the reversible transformation from a ductile martensite phase to a stiff austenite phase in the material with the application of heat. The reversible phase change allows the shape-memory effect to be used as an actuation mechanism since the material changes shape during the transition. It has been found that high forces and strains can be generated from shape-memory thin films at reasonable power inputs, thus enabling shape memory actuation to be used in MEMS based microfluidic devices such as microvalves and micropumps. Titanium-nickel (TiNi) is among the most popular of the shape-memory alloys owing to its high actuation work density, ($50\,MJ/m^3$), and large bandwidth (up to $0.1\,kHz$) [11.39]. TiNi is also attractive because conventional sputtering techniques can be employed to deposit thin films, as detailed in a recent report by *Shih* et al. [11.39]. In this study, TiNi films were deposited by cosputtering elemental Ti and Ni targets and cosputtering TiNi alloy and elemental Ti targets. It was reported that cosputtering from TiNi and Ti targets produced better films due to process variations related to roughening of the Ni target in the case of Ti and Ni cosputtering. The TiNi/Ti cosputtering process has been used to produce shape-memory material for a silicon spring-based microvalve [11.40].

Use of thin-film metal alloys in magnetic actuator systems is another example of the versatility of metallic materials in MEMS. Magnetic actuation in microdevices generally requires the magnetic layers to be relatively thick (tens to hundreds of micrometer) to generate magnetic fields of sufficient strength to generate the desired actuation. To this end, magnetic materials are often deposited by thick-film methods such as electroplating. The thicknesses of these layers exceeds what can feasibly be patterned by etching, so plating is often performed in microfabricated molds made from materials such as polymethylmethacrylate (PMMA). The PMMA mold thickness can exceed several hundred micrometer, so x-rays are used as the exposure source during the patterning steps. When necessary a metallic thin-film seed layer is deposited prior to plating. After plating, the mold is dissolved, which frees the metallic component. Known as LIGA (short for lithography, galvanoforming, and abformung), this process has been used to produce a wide variety of high-aspect-ratio structures from plateable materials, such as nickel-iron (NiFe) magnetic alloys [11.41] and Ni [11.42].

In addition to elemental metals and simple compound alloys, more complex metallic alloys commonly used in commerical macroscopic applications are finding their way into MEMS applications. One such example is an alloy of titanium known as Ti-6Al-4V. Composed of 88% titanium, 6% aluminum, and 4% vanadium, this alloy is widely used in commercial avation due to its weight, strength, and temperature tolerance. *Pornsin-Sirirak* et al. [11.43] have explored the use of this alloy in the manufacture of MEMS-based winged structures for micro aerial vehicles. The authors considered this alloy not only because of its weight and strength, but also because of its ductility and its etching rate at room temperature. The designs for the wing prototype were modeled after the wings of bats and various flying insects. For this application, Ti-alloy structures patterned from bulk (250 μm thick) material by an $HF/HO_3/H_2O$ etching solution were used rather than thin films. Parylene-C (detailed in a later section) was deposited on the patterned alloy to serve as the wing membrane. The miniature micromachined wings were integrated into a test setup, and several prototypes actually demonstrated short duration flight.

11.4 Harsh-Environment Semiconductors

11.4.1 Silicon Carbide

Silicon carbide (SiC) has long been recognized as the leading semiconductor for use in high-temperature and high-power electronics and is currently being developed as a material for harsh-environment MEMS. SiC is a polymorphic material that exists in cubic, hexagonal, and rhombehedral polytypes. The cubic polytype, called 3C-SiC, has an electronic bandgap of 2.3 eV, which is over twice that of Si. Numerous hexagonal and rhombehedral polytypes have been identified, with the two most common being 4H-SiC and 6H-SiC. The electronic bandgaps of 4H- and 6H-SiC are even higher than 3C-SiC, being 2.9 and 3.2 eV, respectively. SiC films can be doped to create n-type and p-type materials. The Young's modulus of SiC is still the subject of research, but most reported values range from 300 to 450 GPa, depending on the microstructure and measurement technique. SiC is not etched in any wet Si etchants and is not attacked by XeF_2, a popular dry Si etchant used for releasing device structures [11.44]. SiC is a material that does not melt, but rather sublimes at temperatures in excess of 1800 °C. Single-crystal 4H- and 6H-SiC wafers are commercially available, but they are smaller in diameter (3 inch) and much more expensive than Si wafers.

SiC thin films can be grown or deposited using a number of different techniques. For high-quality single-crystal films, APCVD and LPCVD processes are most commonly employed. Homoepitaxial growth of 4H- and 6H-SiC yields high-quality films suitable for microelectronic applications but typically only on substrates of the same polytype. These processes usually employ dual precursors, such as SiH_4 and C_3H_8, and are performed at temperatures ranging from 1500 to 1700 °C. Epitaxial films with p-type or n-type conductivity can be grown using Al and B for p-type films and N and P for n-type films. Nitrogen is so effective at modifying the conductivity of SiC that growth of undoped SiC films is extremely challenging because the concentrations of residual nitrogen in typical deposition systems are sufficient for n-type doping.

APCVD and LPCVD can also be used to deposit 3C-SiC on Si substrates. Heteroepitaxy is possible despite a 20% lattice mismatch because 3C-SiC and Si have the same lattice structure. The growth process involves two key steps. The first step, called carbonization, converts the near surface region of the Si substrate to 3C-SiC by simply exposing it to a hydrocarbon/hydrogen mixture at high substrate temperatures ($> 1200\,°C$). The carbonized layer forms a crystalline template on which a 3C-SiC film can be grown by adding a silicon-containing gas to the hydrogen/hydrocarbon mix. The lattice mismatch between Si and 3C-SiC results in the formation of crystalline defects in the 3C-SiC film, with the density being highest in the carbonization layer and decreasing with increasing thickness. The crystal quality of 3C-SiC films is nowhere near that of epitaxially grown 4H- and 6H-SiC films; however, the fact that 3C-SiC can be grown on Si substrates enables the use of Si bulk micromachining techniques for fabrication of a host of 3C-SiC-based mechanical devices. These include microfabricated pressure sensors [11.45] and nanoelectromechanical resonant structures [11.46]. For designs that require electrical isolation from the substrate, 3C-SiC devices can be made directly on SOI substrates [11.45] or by wafer bonding and etchback, such as the capacitive pressure sensor developed by *Young* et al. [11.47].

Polycrystalline SiC (poly-SiC) is a more versatile material for SiC MEMS than its single-crystal counterparts. Unlike single-crystal versions of SiC, poly-SiC can be deposited on a variety of substrate types, including common surface micromachining materials such as polysilicon, SiO_2, and Si_3N_4. Commonly used deposition techniques include LPCVD [11.44, 48, 49] and APCVD [11.50, 51]. The deposition of poly-SiC requires much lower substrate temperatures than epitaxial films, ranging from roughly 700 to 1200 °C. Amorphous SiC can be deposited at even lower temperatures (25–400 °C) by PECVD [11.52] and sputtering [11.53]. The microstructure of poly-SiC films is temperature, substrate, and process dependent. For amorphous substrates such as SiO_2 and Si_3N_4, APCVD poly-SiC films deposited from SiH_4 and C_3H_8 are randomly oriented with equiaxed grains [11.51], whereas for oriented substrates such as polysilicon, the texture of the poly-SiC film matches that of the substrate itself [11.50]. By comparison, poly-SiC films deposited by LPCVD from SiH_2Cl_2 and C_2H_2 are highly textured (111) films with a columnar microstructure [11.48], while films deposited from disilabutane have a distribution of orientations [11.44]. This variation suggests that device performance can be tailored by selecting the proper substrate and deposition conditions.

SiC films deposited by AP- and LPCVD generally suffer from large tensile stresses on the order of several hundred MPa. Moreover, the residual stress gradients in these films tend to be large, leading to significant out-of-plane bending of structures that are anchored at a single location. The thermal stability of SiC makes a postdeposition annealing step impractical for films deposited on Si substrates, since the temperatures needed to significantly modify the film are likely to exceed the melting temperature of the wafer. For LPCVD processes using SiH_2Cl_2 and C_2H_2 precursors, *Fu* et al. [11.54] has described a relationship between deposition pressure and residual stress that enables the deposition of undoped poly-SiC films with nearly zero residual stresses and negligible stress gradients. This work has recently been extended to include films doped with nitrogen [11.55].

Direct bulk micromachining of SiC is very difficult, due to its chemical inertness. Although conventional wet chemical techniques are not effective, several electrochemical etch processes have been demonstrated and used in the fabrication of 6H-SiC pressure sensors [11.56]. The etching processes are selective to the conductivity of the material, so dimensional control of the etched structures depends on the ability to form doped layers, which can only be formed by in situ or ion-implantation processes since solid source diffusion is not possible at reasonable processing temperatures. This constraint somewhat limits the geometrical complexity of the patterned structures as compared with conventional plasma-based etching. To fabricate thick (hundreds of micrometer), 3-D, high-aspect-ratio SiC structures, a molding technique has been developed [11.42]. The molds are fabricated from Si substrates using deep reactive ion etching and then filled with SiC using a combination of thin epitaxial and thick polycrystalline film CVD processes. The thin-film process is used to protect the mold from pitting during the more aggressive mold-filling SiC growth step. The mold-filling process coats all surfaces of the mold with a SiC film as thick as the mold is deep. To release the SiC structure, the substrate is first mechanically polished to expose sections of the Si mold; then the substrate is immersed in a Si etchant to completely dissolve the mold. This process has been used to fabricate solid SiC fuel atomizers [11.42], and a variant has been used to fabricate SiC structures for micropower systems [11.57]. Recently, *Min* et al. [11.58] reported a process to fabricate reusable glass press molds made from SiC structures that were patterned using Si molding masters. SiC was selected as the material for the glass press mold because the application requires a hard, mechanically strong, and chemically stable material that can withstand and maintain its properties at temperatures between 600 and 1400 °C.

In addition to CVD processes, bulk micromachined SiC structures can be fabricated using sintered SiC powders. *Tanaka* et al. [11.59] describe a process where SiC components, such as micro gas turbine engine rotors, can be fabricated from SiC powders using a microreaction-sintering process. The molds are microfabricated from Si using DRIE and filled with SiC and graphite powders mixed with a phenol resin. The molds are then reaction-sintered using a hot isostatic pressing technique. The SiC components are then released from the Si mold by wet chemical etching. The authors reported that the component shrinkage was less than 3%. The bending strength and Vickers hardness of the microreaction-sintered material was roughly 70 to 80% of commercially available reaction-sintered SiC, the difference being attributed to the presence of unreacted Si in the microscale components.

In a related process, *Liew* et al. [11.60] detail a technique to create silicon carbon nitride (SiCN) MEMS structures by molding injectable polymer precursors. Unlike the aforementioned processes, this technique uses SU-8 photoresists for the molds. To be detailed later in this chapter, SU-8 is a versatile photodefinable polymer in which thick films (hundreds of micrometer) can be patterned using conventional UV photolithographic techniques. After patterning, the molds are filled with the SiCN-containing polymer precursor, lightly polished, and then subjected to a multistep heat-treating process. During the thermal processing steps, the SU-8 mold decomposes and the SiCN structure is released. The resulting SiCN structures retain many of the same properties of stoichiometric SiC.

Although SiC cannot be etched using conventional wet etch techniques, SiC can be patterned using conventional dry etching techniques. RIE processes using fluorinated compounds such as CHF_3 and SF_6 combined with O_2 and sometimes with an inert gas or H_2 are used to pattern thin films. The high oxygen content in these plasmas generally prohibits the use of photoresist as a masking material; therefore, hard masks made of Al, Ni, and ITO are often used. RIE-based SiC surface micromachining processes with polysilicon and SiO_2 sacrificial layers have been developed for single-layer devices [11.61, 62]. ICP RIE of SiC using SF_6 plasmas and Ni or ITO etch masks has been developed for bulk micromachining SiC substrates, with structural depths in excess of 100 μm reported [11.63].

Until recently, multilayer thin-film structures were very difficult to fabricate by direct RIE because the etch rates of the sacrificial layers were much higher than the SiC structural layers, making dimensional control very difficult. To address this issue, a micromolding process for patterning SiC films on sacrificial-layer substrates was developed [11.64]. In essence, the micromolding technique is the thin-film analog to the molding-based, bulk micromachining technique presented earlier. The micromolding process utilizes polysilicon and SiO_2 films as both molds and sacrificial substrate layers, with SiO_2 molds used with polysilicon sacrificial layers and vice versa. These films are deposited and patterned using conventional methods, thus leveraging the well-characterized and highly selective processes developed for polysilicon MEMS. Poly-SiC films are simply deposited into the micromolds and mechanical polishing is used to remove poly-SiC from atop the molds. Appropriate etchants are then used to dissolve the molds and sacrificial layers. The micromolding method utilizes the differences in chemical properties of the three materials in this system in a way that bypasses the difficulties associated with chemical etching of SiC. This technique has been developed specifically for multilayer processing and has been used successfully to fabricate SiC micromotors [11.64] and the lateral resonant structure shown in Fig. 11.4 [11.65].

Recent advancements in the area of SiC RIE show that significant progress has been made in developing etch recipes with selectivities to nonmetal mask and sacrificial layers that are suitable for multilayer SiC surface micromachining. For instance, *Gao* et al. [11.66] have developed a transformer-coupled RIE process using a HBr-based chemistry for thin-film poly-SiC etching. The recipe exhibits a SiC-to-SiO_2 selectivity of 20:1 and a SiC-to-Si_3N_4 selectivity of 22:1, which are the highest reported thus far. In addition, the anisotropy of the etch was quite high, and micromasking, a common problem when metal masks are used, was not an issue. This process has since been used to fabricate multilayered lateral resonant structures that utilize poly-SiC as the main structural material and polysilicon as

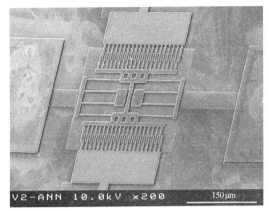

Fig. 11.4 SEM micrograph of a poly-SiC lateral resonant structure fabricated using a multilayer, micromolding-based micromachining process (after [11.65])

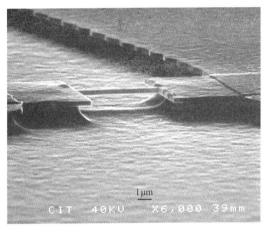

Fig. 11.5 SEM micrograph of a 3C-SiC nanomechanical beam resonator fabricated by electron-beam lithography and dry etching processes (courtesy of M. Roukes, Caltech)

After patterning, the beams were released by etching the underlying Si isotropically using a NF_3/Ar plasma. The inertness of the SiC film to the Si etchant enables the dry release of the nanomechanical beams. An example of a 3C-SiC nanomechanical beam is shown in Fig. 11.5.

11.4.2 Diamond

Diamond is commonly known as nature's hardest material, making it ideal for high wear environments. Diamond has a very large electronic bandgap (5.5 eV), which makes it attractive for high temperature electronics. Undoped diamond is a high-quality insulator with a dielectric constant of 5.5; however, it can be relatively easily doped with boron to create p-type conductivity. Diamond has a very high Young's modulus (1035 GPa), making it suitable for high-frequency micromachined resonators, and it is among nature's most chemically inert materials, making it well suited for harsh chemical environments.

Unlike SiC, fabrication of diamond MEMS is currently restricted to polycrystalline and amorphous material, since single-crystal diamond wafers are not yet commercially available. Polycrystalline diamond films can be deposited on Si and SiO_2 substrates by CVD methods, but the surfaces must often be seeded by diamond powders or biased with a negative charge to initiate growth. In general, diamond nucleates much more readily on Si surfaces than on SiO_2 surfaces, an effect that has been used to selectively pattern diamond films into micromachined AFM cantilever probes using SiO_2 molding masks [11.67].

Bulk micromachining of diamond using wet and dry etching is extremely difficult given its extreme chemical inertness. Diamond structures have nevertheless been fabricated using bulk micromachined Si molds to pattern the structures [11.68]. The Si molds were fabricated using conventional micromachining techniques and filled with polycrystalline diamond deposited by hot filament chemical vapor deposition (HFCVD). The HFCVD process uses H_2 and CH_4 precursors. The process was performed at a substrate temperature of 850–900 °C and a pressure of 50 mtorr. The Si substrate was seeded prior to deposition using a diamond particle/ethanol solution. After deposition, the top surface of the structure was polished using a hot iron plate. After polishing, the Si mold was removed in a Si etchant, leaving behind the micromachined diamond structure. This process was used to produce high-aspect-ratio capillary channels for microfluidic applications [11.69] and components for diffractive op-

a conducting plane that underlies the resonating shuttle [11.66].

Yang et al. [11.46] have recently shown that the chemical inertness of SiC facilitates the fabrication of NEMS devices. In this work, the authors present a fabrication method to realize SiC mechanical resonators with submicrometer thickness and width dimensions. The resonators were fabricated from ≈ 260 nm thick 3C-SiC films epitaxially grown on (100) Si wafers. The films were patterned into 150 nm wide beams ranging in length from 2 to 8 μm. The beams were etched in a $NF_3/O_2/Ar$ plasma using an evaporated Cr etch mask.

tics, laser-to-fiber alignment, and power device cooling structures [11.70].

Due to the nucleation processes associated with diamond film growth, surface micromachining of polycrystalline diamond thin films requires modifications to conventional micromachining to facilitate film growth on sacrificial substrates. Initially, conventional RIE methods were generally ineffective, so work was focused on developing selective deposition techniques. One early method used selective seeding to form patterned templates for diamond nucleation. The selective seeding process employed the lithographic patterning of photoresist that contained diamond powders [11.71]. The diamond-loaded photoresist was deposited and patterned onto a Cr-coated Si wafer. During the onset of diamond growth, the patterned photoresist rapidly evaporates, leaving behind the diamond seed particles in the desired locations. A patterned diamond film is then selectively grown on these locations.

A second process utilized selective deposition directly on sacrificial substrate layers. This process combined conventional diamond seeding with photolithographic patterning and etching to fabricate micromachined diamond structures on SiO_2 sacrificial layers [11.72]. The process was performed in one of two ways. The first approach begins with the seeding of an oxidized Si wafer. The wafer is coated with a photoresist and photolithographically patterned. Unmasked regions of the seeded SiO_2 film are then partially etched, forming a surface unfavorable for diamond growth. The photoresist is then removed and a diamond film is deposited on the seeded regions. The second approach also begins with an oxidized Si wafer. The wafer is coated with a photoresist, photolithographically patterned, and then seeded with diamond particles. The photoresist is removed, leaving behind a patterned seed layer suitable for selective growth. These techniques have been successfully used to fabricate cantilever beams and bridge structures.

A third method to surface micromachine polycrystalline diamond films follows the conventional approach of film deposition, dry etching, and release. The chemical inertness of diamond renders most conventional plasma chemistries useless; however, oxygen-based ion-beam plasmas can be used to etch diamond thin films [11.73]. A simple surface micromachining process begins with the deposition of a polysilicon sacrificial layer on a Si_3N_4-coated Si wafer. The polysilicon layer is seeded using diamond slurry, and a diamond film is deposited by HFCVD. Since photoresists are not resistant to O_2 plasmas, an Al masking film is deposited and patterned. The diamond films are then etched in the O_2 ion-beam plasma, and the structures are released by etching the polysilicon with KOH. This process has been used to create lateral resonant structures, but a significant stress gradient in the films rendered the devices inoperable.

In general, conventional HFCVD requires that the substrate be pretreated with a seeding layer prior to diamond film growth. However, a method called biased enhanced nucleation (BEN) has been developed that enables the growth of diamond on unseeded Si surfaces. *Wang* et al. [11.74] have shown that if Si substrates are masked with patterned SiO_2 films, selective diamond growth will occur primarily on the exposed Si surfaces, and a slight HF etch is sufficient to remove the adventitious diamond from the SiO_2 mask. This group was able to use this method to fabricate diamond micromotor rotors and stators on Si surfaces.

Diamond is a difficult, but not impossible, material to etch using conventional RIE techniques. It is well known that diamond can be etched in oxygen plasmas, but these plasmas can be problematic for device fabrication because the etching tends to be isotropic. A recent development, however, suggests that RIE processes for diamond are close at hand. *Wang* et al. [11.74] describe a process to fabricate a vertically actuated, doubly clamped micromechanical diamond beam resonator using RIE. The process outlined in this paper addresses two key issues related to diamond surface micromachining, namely, residual stress gradients in the diamond films and diamond patterning techniques. A microwave plasma CVD (MPCVD) reactor was used to grow the diamond films on sacrificial SiO_2 layers pretreated with a nanocrystalline diamond powder, resulting in a uniform nucleation density at the diamond/SiO_2 interface. The diamond films were etched in a CF_4/O_2 plasma using Al as a hard mask. Reasonably straight sidewalls were created, with roughness attributable to the surface roughness of the faceted diamond film. An Au/Cr drive electrode beneath the sacrificial oxide remained covered throughout the diamond-patterning steps and thus was undamaged during the diamond-etching process. This work has since been extended to develop a 1.51 GHz diamond micromechanical disk resonator [11.74]. In this instance, the nanocrystalline diamond film was deposited my MPCVD, coated with an oxide film that had been patterned into an etch mask, and then etched in a O_2/CF_4 RIE plasma under conditions that yielded a fairly anisotropic etch with a diamond-to-oxide selectivity of 15:1. The disk was suspended over the substrate on a polysilicon stem using an oxide sacrificial layer. Polysilicon was also used as the drive and sense electrodes. The material mismatch between the step and the resonating disk substantially reduced anchor losses, thus allowing for very high-quality factors (11, 500) for 1.5 GHz resonators tested in a vacuum.

In conjunction with recent advances in RIE and micromachining techniques, work is being performed to develop diamond-deposition processes specifically for MEMS applications. Diamond films grown using conventional techniques, especially processes that require pregrowth seeding, tend to have high residual stress gradients and roughened surface morphologies as a result of the highly faceted, large-grain polycrystalline films that are produced by these methods (Fig. 11.6). The rough surface morphology degrades the patterning process, resulting in roughened sidewalls in etched structures and roughened surfaces of films deposited over these layers. Unlike polysilicon and SiC, a postdeposition polishing process is not technically feasible for diamond due to its extreme hardness. For the fabrication of multilayer diamond devices, methods to reduce the surface roughness of the as-deposited films are highly desirable. Along these lines, *Krauss* et al. [11.75] have reported on the development of an ultrananocrystalline diamond (UCND) film that exhibits a much smoother surface morphology than comparable diamond films grown using conventional methods. Unlike conventional CVD diamond films that are grown using a mixture of H_2 and CH_4, the ultrananocrystalline diamond films are grown from mixtures of Ar, H_2, and C_{60} or Ar, H_2, and CH_4. Films produced by this method have proven to be effective as conformal coatings on Si surfaces and have been used successfully in several surface micromachining processes. Recently, this group has extended the UCND deposition technology to low deposition temperatures, with high-quality nanocrystalline diamond films being deposited at rates of 0.2 μm/h at substrate temperatures of 400 °C, making these films compatible from a thermal budget perspective with Si IC technology [11.76].

Another alternative deposition method that is proving to be well suited for diamond MEMS is based on pulsed laser deposition [11.77]. The process is performed in a high vacuum chamber and uses a pulsed eximer laser to ablate a pyrolytic graphite target. Material from the ejection plume deposits on a substrate,

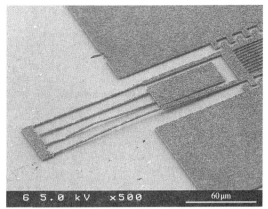

Fig. 11.6 SEM micrograph of the folded beam truss of diamond lateral resonator. The diamond film was deposited using a seeding-based hot filament CVD process. The micrograph illustrates the challenges facing MEMS structures made from polycrystalline material, namely roughened surfaces and residual stress gradients

which is kept at room temperature. Background gases composed of N_2, H_2, and Ar can be introduced to adjust the deposition pressure and film properties. The as-deposited films consist of tetrahedrally bonded carbon that is amorphous in microstructure, hence the name amorphous diamond. Nominally stress-free films can be deposited by proper selection of deposition parameters [11.78] or by a short postdeposition annealing step [11.77]. The amorphous diamond films exhibit many of the properties of single-crystal diamond, such as a high hardness (88 GPa), a high Young's modulus (1100 GPa), and chemical inertness. Many single-layer surface micromachined structures have been fabricated using these films, in part because the films can be readily deposited on oxide sacrificial layers and etched in an oxygen plasma. Recently, amorphous diamond films have been used as a dielectric isolation layer in vertically actuated microbridges in micromachined RF capacitive switches [11.79]. The diamond films sit atop fixed tungsten electrodes to provide dielectric isolation from an Au microbridge that spans the fixed electrode structure. The diamond films are particularly attractive for such applications since the surfaces are hydrophobic and thus do not suffer from stiction and are highly resistant to wear over repeated use.

11.5 GaAs, InP, and Related III–V Materials

Gallium arsenide (GaAs), indium phosphide (InP), and related III–V compounds have favorable piezoelectric and optoelectric properties, high piezoresistive constants, and wide electronic bandgaps relative to Si, making them attractive for various sensor and optoelectronic applications. Like Si, significant research in bulk crystal growth has led to the development of GaAs and InP substrates that are commercially available as high-quality, single-crystal wafers. Unlike compound semiconductors such as SiC, III–V materials can be deposited as ternary and quaternary alloys with lattice constants that closely match the binary compounds from which they are derived (i.e., $Al_xGa_{1-x}As$ and GaAs), thus permitting the fabrication of a wide variety of heterostructures that facilitate device performance.

Crystalline GaAs has a zinc blend crystal structure with an electronic bandgap of 1.4 eV, enabling GaAs electronic devices to function at temperatures as high as 350 °C [11.80]. High-quality, single-crystal wafers are commercially available, as are well-developed metalorganic chemical vapor deposition (MOCVD) and molecular beam epitaxy (MBE) growth processes for epitaxial layers of GaAs and its alloys. GaAs does not outperform Si in terms of mechanical properties; however, its stiffness and fracture toughness are still suitable for micromechanical devices.

Micromachining of GaAs is relatively straightforward, since many of its lattice-matched ternary and quaternary alloys have sufficiently different chemical properties to allow their use as sacrificial layers [11.81]. For example, the most common ternary alloy for GaAs is $Al_xGa_{1-x}As$. For values of x less than or equal to 0.5, etchants containing mixtures of HF and H_2O will etch $Al_xGa_{1-x}As$ without attacking GaAs, while etchants containing NH_4OH and H_2O_2 attack GaAs isotropically but do not etch $Al_xGa_{1-x}As$. Such selectivity enables the micromachining of GaAs wafers using lattice-matched etch stops and sacrificial layers. Devices fabricated using these methods include comb drive lateral resonant structures [11.81], pressure sensors [11.82, 83], thermopile sensors [11.83], Fabry–Perot detectors [11.84], and cantilever-based sensors and actuators [11.85, 86]. In addition, nanoelectromechanical devices, such as suspended micromechanical resonators [11.87] and tethered membranes [11.88], have been fabricated using these techniques. An example of a nanoelectromechanical beam structure fabricated from GaAs is shown in Fig. 11.7.

In addition to using epitaxial layers as etch stops, ion-implantation methods can also be used to produce etch stops in GaAs layers. *Miao* et al. [11.89] describe a process that uses electrochemical etching to selectively remove n-type GaAs layers. The process relies on the creation of a highly resistive near-surface GaAs layer on an n-type GaAs substrate by low-dose nitrogen implantation in the MeV energy range. A pulsed electrochemical etch method using an H_2PtCl_6, H_3PO_4, H_2SO_4 platinum electrolytic solution at 40 °C with 17 V, 100 ms pulses is sufficient to selectively remove n-type GaAs at about 3 μm/min. Using this method, stress-free, tethered membranes could readily be fabricated from the highly resistive GaAs layer. The high implant energies enable the fabrication of membranes several micrometer thick. Moreover, the authors demonstrated that if the GaAs wafer were etched in such a way as to create an undulating surface prior to ion implantation, corrugated membranes could be fabricated. These structures can sustain much higher deflection amplitudes than flat structures.

Micromachining of InP closely resembles the techniques used for GaAs. Many of the properties of InP are similar to GaAs in terms of crystal structure, mechanical stiffness, and hardness; however, the optical properties of InP make it particularly attractive for microoptomechanical devices to be used in the 1.3–1.55 μm wavelength range [11.90]. Like GaAs, single-crystal

Fig. 11.7 SEM micrograph of a GaAs nanomechanical beam resonator fabricated by epitaxial growth, electron-beam lithography, and selective etching (courtesy of M. Roukes, Caltech)

wafers of InP are readily available, and ternary and quaternary lattice-matched alloys, such as InGaAs, InAlAs, InGaAsP, and InGaAlAs, can be used as either etch stop and/or sacrificial layers depending on the etch chemistry [11.81]. For instance, InP structural layers deposited on $In_{0.53}Al_{0.47}As$ sacrificial layers can be released using etchants containing $C_6H_8O_7$, H_2O_2, and H_2O. In addition, InP films and substrates can be etched in solutions containing HCl and H_2O using $In_{0.53}Ga_{0.47}As$ films as etch stops. Using InP-based micromachining techniques, multiair gap filters [11.91] bridge structures [11.90], and torsional membranes [11.84] have been fabricated from InP and its related alloys.

In addition to GaAs and InP, materials such as indium arsenide (InAs) can be micromachined into device structures. Despite a 7% lattice mismatch between InAs and (111) GaAs, high-quality epitaxial layers can be grown on GaAs substrates. As described by *Yamaguchi* et al. [11.92], the surface Fermi level of InAs/GaAs structures is pinned in the conduction band, enabling the fabrication of very thin conductive membranes. In fact, the authors have successfully fabricated free-standing InAs structures that range in thickness from 30 to 300 nm. The thin InAs films were grown directly on GaAs substrates by MBE and etched using a solution containing H_2O, H_2O_2, and H_2SO_4. The structures, mainly doubly clamped cantilevers, were released by etching the GaAs substrate using an $H_2O/H_2O_2/NH_4OH$ solution.

11.6 Ferroelectric Materials

Piezoelectric materials play an important role in MEMS technology for sensing and mechanical actuation applications. In a piezoelectric material, mechanical stress produces a polarization, and conversely a voltage-induced polarization produces a mechanical stress. Many asymmetric materials, such as quartz, GaAs, and zinc oxide (ZnO), exhibit some piezoelectric behavior. Recent work in MEMS has focused on the development of ferroelectric compounds such as lead zirconate titanate, $Pb(Zr_xTi_{1-x})O_3$, or PZT for short, because such compounds have high piezoelectric constants that result in high mechanical transduction. It is relatively straightforward to fabricate a PZT structure on top of a thin free-standing structural layer (i.e., cantilever, diaphragm). Such a capability enables the piezoelectric material to be used in sensor applications or actuator applications where piezoelectric materials are particularly well suited. Like Si, PZT films can be patterned using dry etch techniques based on chlorine chemistries, such as Cl_2/CCl_4, as well as ion-beam milling using inert gases like Ar.

PZT has been successfully deposited in thin-film form using cosputtering, CVD, and sol-gel processing. So-gel processing is particularly attractive because the composition and homogeneity of the deposited material over large surface areas can be readily controlled. The sol gel process outlined by *Lee* et al. [11.93] uses PZT solutions made from liquid precursors containing Pb, Ti, Zr, and O. The solution is deposited by spin coating on a Si wafer that has been coated with a $Pt/Ti/SiO_2$ thin-film multilayer. The process is executed to produce a PZT film in layers, with each layer consisting of a spin-coated layer that is dried at 110 °C for 5 min and then heat-treated at 600 °C for 20 min. After building up the PZT layer to the desired thickness, the multilayer was heated at 600 °C for up to 6 h. Prior to this anneal, a PbO top layer was deposited on the PZT surface. An Au/Cr electrode was then sputter-deposited on the surface of the piezoelectric stack. This process was used to fabricate a PZT-based force sensor. *Xu* et al. [11.94] describe a similar sol-gel process to produce 12 μm thick, crack-free PZT films on Pt-coated Si wafers and 5 μm thick films on insulating ZrO_2 layers to produce micromachined MHz-range two-dimensional transducer arrays for acoustic imaging.

Thick-film printing techniques for PZT have been developed to produce thick films in excess of 100 μm. Such thicknesses are desired for applications that require actuation forces that cannot be achieved with the much thinner sol-gel films. *Beeby* et al. [11.95] describe a thick-film printing process whereby a PZT paste is made from a mixture of 95% PZT powder, 5% lead borosilicate powder, and an organic carrier. The paste was then printed through a stainless steel screen using a thick-film printer. Printing was performed on an oxidized Si substrate that is capped with a Pt electrode. After printing, the paste was dried and then fired at 850–950 °C. Printing could be repeated to achieve the desired thickness. The top electrode consisted of an evaporated Al film. The authors found that it was possible to perform plasma-based processing on the printed substrates but that the porous nature of the printed PZT films made them unsuitable for wet chemical processing.

11.7 Polymer Materials

11.7.1 Polyimide

Polyimides comprise an important class of durable polymers that are well suited for many of the techniques used in conventional MEMS processing. In general, polyimides can be acquired in bulk or deposited as thin films by spin coating, and they can be patterned using conventional dry etching techniques and processed at relatively high temperatures. These attributes make polyimides an attractive group of polymers for MEMS that require polymer structural and/or substrate layers, such as microfabricated biomedical devices where inertness and flexibility are important parameters.

Shearwood et al. [11.96] explored the use of polyimides as a robust mechanical material for microfabricated audio membranes. The authors fabricated 7 μm thick, 8 mm diameter membranes on GaAs substrates by bulk micromachining the GaAs substrate using a NH_3/H_2O_2 solution. They realized 100% yield and, despite a low Young's modulus (≈ 3 GPa), observed flat membranes to within 1 nm after fabrication.

Jiang et al.[11.97] capitalized on the strength and flexibility of polyimides to fabricate a flexible sheer-stress sensor array based on Si sensors. The sensor array consisted of a collection of Si islands linked by two polyimide layers. Each Si sensor island was $250 \times 250\,\mu m^2$ in area and 80 μm in thickness. Al was used as an electrical innerconnect layer. The two polyimide layers served as highly flexible hinges, making it possible to mount the sensor array on curved surfaces. The sensor array was successful in profiling the shear-stress distribution along the leading edge of a rounded delta wing.

The chemical and temperature durability of polyimides enables their use as a sacrificial layer for a number of commonly used materials, such as evaporated or sputter-deposited metals. *Memmi* et al. [11.98] developed a fabrication process for capacitive micromechanical ultrasonic transducers using a polyimide as a sacrificial layer. The authors showed that the polyimide could withstand the conditions used to deposit silicon monoxide by evaporation and silicon nitride by PECVD at 400 °C. Recent work by *Bagolini* et al. [11.99] has shown that polyimides can even be used as sacrificial layers for PECVD SiC.

In the area of microfabricated biomedical devices, polyimides are receiving attention as a substrate material for implantable devices, owing to their potential biocompatiblity and mechanical flexibility.

Stieglitz [11.100] reported on the fabrication of multichannel microelectrodes on polyimide substrates. Instead of using polyimide sheets as starting substrates, Si carrier wafers coated with a 5 μm thick polyimide film were used. Pt microelectrodes were then fabricated on these substrates using conventional techniques. Thin polyimide layers were deposited between various metal layers to serve as insulating layers. A capping polyimide layer was then deposited on the top of the substrates, and then the entire polyimide/metal structure was peeled off the Si carrier wafers. Backside processing was then performed on the free-standing polyimide structures to create devices that have exposed electrodes on both surfaces. In a later paper, *Stieglitz* et al. [11.101] describe a variation of this process for neural prostheses.

11.7.2 SU-8

SU-8 is a negative-tone epoxylike photoresist that is receiving much attention for its versatility in MEMS processing. It is a high-aspect-ratio, UV-sensitive resist designed for applications requiring single-coat resists with thicknesses on the order of 500 μm [11.102]. SU-8 has favorable chemical properties that enable it to be used as a molding material for high-aspect-ratio electroplated structures (as an alternative to LIGA) and as a structural material for microfluidics [11.102]. In terms of mechanical properties, *Lorenz* et al. [11.103] reported that SU-8 has a modulus of elasticity of 4.02 GPa, which compares favorably with a commonly-used polyamid (3.4 GPa).

In addition to the above-mentioned conventional uses for SU-8, several interesting alternative uses are beginning to appear in the literature. *Conradie* et al. [11.104] have used SU-8 to trim the mass of silicon paddle oscillators as a means to adjust the resonant frequency of the beams. The trimming process involves the patterning of SU-8 posts on Si paddles. The process capitalizes on the relative chemical stability of the SU-8 resin in conjunction with the relatively large masses that can be patterned using standard UV exposure processes.

SU-8 is also of interest as a bonding layer material for wafer bonding processes using patterned bonding layers. *Pan* et al. [11.105] compared several UV photodefinable polymeric materials and found that SU-8 exhibited the highest bonding strength (20.6 MPa) for layer thicknesses up to 100 μm.

11.7.3 Parylene

Parylene (poly-paraxylylene) is another emerging polymeric MEMS material due in large part to its biocompatibility. It is particularly attractive from the fabrication point of view because it can be deposited by CVD at room temperature. Moreover, the deposition process is conformal, which enables parylene coatings to be applied to prefabricated structures, such as Si microneedles [11.106], low-stress silicon nitride membrane particle filters [11.73], and micromachined polyimide/Au optical scanners [11.107]. In the former cases, the parylene coating served to strengthen the microfabricated structures, while in the latter case it served to protect the structure from condensing water vapor.

In addition to its function as a protective coating, parylene can actually be micromachined into free-standing components. *Noh* et al. [11.108] demonstrated a method to create bulk micromachined parylene microcolumns for miniature gas chromatographs. The structure is fabricated using a micromolding technique where Si molds are fabricated by DRIE and coated with parylene to form three sides of the microcolumn. A second wafer is coated with parylene, and the two are bonded together via a fusion bonding process. After bonding, the structure is released from the Si mold by KOH etching. In a second example, *Yao* et al. [11.109] describe a dry release process for parylene surface micromachining. In this process, sputtered Si is used as a sacrificial layer onto which a thick sacrificial photoresist is deposited. Parylene is then deposited on the photoresist and patterned into the desired structural shape. The release procedure is a two-step process. First the photoresist is dissolved in acetone. This results in the parylene structure sticking to the sputtered Si. Next, a dry BrF_3 etch is performed that dissolves the Si and releases the parylene structures. Parylene beams that were 1 mm long and 4.5 μm thick were successfully fabricated using this technique.

11.7.4 Liquid Crystal Polymer

Liquid crystal polymer (LCP) is a high-performance thermoplastic currently being used in printed circuit board and electronics packaging applications and has recently been investigated for use in MEMS applications requiring a material that is mechanically flexible, electrically insulating, chemically durable, and impermeable to moisture. LCP can be bonded to itself and other substrate materials such as glass and Si by thermal lamination. It can be micromachined using an oxygen plasma and yet is highly resistant to HF and many metal etchants [11.110]. The moisture absorption is less than 0.02% as compared with about 1% for polyimide [11.111], making it well suited as a packaging material.

Applications where LCPs are used as a key component in a MEMS device are beginning to emerge. *Faheem* et al. [11.112] reported on the use of LCP for encapsulation of variable RF MEMS capacitors. In this example, LCP, dispensed in liquid form, was used to join and seal a glass microcap to a prefabricated, microbridge capacitor. LCP was chosen in part because in addition to the aforementioned properties, it has very low RF loss characteristics, making it very well suited as an RF MEMS packaging material. *Wang* et al. [11.110] showed that LCP is a very versatile material that is highly compatible with many standard Si-based processing techniques. They also showed that micromachining techniques can be used to make LCP cantilever flow sensors that incorporate metal strain gauges and LCP membrane tactile sensors using NiCr strain gauges. *Lee* et al. [11.111] has developed a LCP-based, mechanically flexible, multichannel microelectrode array structure for neural stimulation and recording.

11.8 Future Trends

The rapid expansion of MEMS in recent years is due in large part to the inclusion of new materials that have expanded the functionality of microfabricated devices beyond what is achievable in silicon. This trend will certainly continue as new application areas for micro- and nanofabricated devices are identified. Many of these applications will likely require both new materials and new processes to fabricate the micro- and nanomachined devices for these yet-to-be-identified applications. Currently, conventional micromachining techniques employ a *top-down* approach that begins with either bulk substrates or thin films. Future MEMS and NEMS will likely incorporate materials that are created using a *bottom-up* approach. A significant challenge facing device design and fabrication engineers alike will be how to marry top-down and bottom-up approaches to create devices and systems that cannot be made using either process alone.

References

11.1 C.S. Smith: Piezoresistive effect in germanium and silicon, Phys. Rev. **94**, 1–10 (1954)
11.2 A.N. Cleland, M.L. Roukes: Fabrication of high frequency nanometer scale mechanical resonators from bulk Si crystals, Appl. Phys. Lett. **69**, 2653–2655 (1996)
11.3 D.W. Carr, H.G. Craighead: Fabrication of nanoelectromechanical systems in single crystal silicon using silicon on insulator substrates and electron beam lithography, J. Vac. Sci. Technol. B **15**, 2760–2763 (1997)
11.4 T. Kamins: *Polycrystalline Silicon for Integrated Circuits and Displays*, 2nd edn. (Kluwer, Boston 1988)
11.5 J.J. McMahon, J.M. Melzak, C.A. Zorman, J. Chung, M. Mehregany: Deposition and characterization of in-situ boron doped polycrystalline silicon films for microelectromechanical systems applications, Mater. Res. Symp. Proc. **605**, 31–36 (2000)
11.6 L. Cao, T.S. Kin, S.C. Mantell, D. Polla: Simulation and fabrication of piezoresistive membrane type MEMS strain sensors, Sens. Actuators **80**, 273–279 (2000)
11.7 H. Guckel, T. Randazzo, D.W. Burns: A simple technique for the determination of mechanical strain in thin films with application to polysilicon, J. Appl. Phys. **57**, 1671–1675 (1983)
11.8 R.T. Howe, R.S. Muller: Stress in polysilicon and amorphous silicon thin films, J. Appl. Phys. **54**, 4674–4675 (1983)
11.9 X. Zhang, T.Y. Zhang, M. Wong, Y. Zohar: Rapid thermal annealing of polysilicon thin films, J. Microelectromech. Syst. **7**, 356–364 (1998)
11.10 J. Yang, H. Kahn, A.-Q. He, S.M. Phillips, A.H. Heuer: A new technique for producing large-area as-deposited zero-stress LPCVD polysilicon films: The multipoly process, J. Microelectromech. Syst. **9**, 485–494 (2000)
11.11 P. Gennissen, M. Bartek, P.J. French, P.M. Sarro: Bipolar-compatible epitaxial poly for smart sensors: Stress minimization and applications, Sens. Actuators A **62**, 636–645 (1997)
11.12 P. Lange, M. Kirsten, W. Riethmuller, B. Wenk, G. Zwicker, J.R. Morante, F. Ericson, J.A. Schweitz: Thick polycrystalline silicon for surface-micromechanical applications: deposition, structuring, and mechanical characterization, Sens. Actuators A **54**, 674–678 (1996)
11.13 S. Greek, F. Ericson, S. Johansson, M. Furtsch, A. Rump: Mechanical characterization of thick polysilicon films: Young's modulus and fracture strength evaluated with microstructures, J. Micromech. Microeng. **9**, 245–251 (1999)
11.14 K. Funk, H. Emmerich, A. Schilp, M. Offenberg, R. Neul, F. Larmer: A surface micromachined silicon gyroscope using a thick polysilicon layer, Proc. 12th Int. Conf. Microelectromech. Syst. (IEEE, Piscataway 1999) pp. 57–60
11.15 T. Abe, M.L. Reed: Low strain sputtered polysilicon for micromechanical structures, Proc. 9th Int. Workshop Microelectromech. Syst. (IEEE, Piscataway 1996) pp. 258–262
11.16 K. Honer, G.T.A. Kovacs: Integration of sputtered silicon microstructures with pre-fabricated CMOS circuitry, Sens. Actuators A **91**, 392–403 (2001)
11.17 J. Gaspar, T. Adrega, V. Chu, J.P. Conde: Thin-film paddle microresonators with high quality factors fabricated at temperatures below 110 °C, Proc. 18th Int. Conf. Microelectromech. Syst. (IEEE, Piscataway 2005) pp. 125–128
11.18 R. Anderson, R.S. Muller, C.W. Tobias: Porous polycrystalline silicon: A new material for MEMS, J. Microelectromech. Syst. **3**, 10–18 (1994)
11.19 W. Lang, P. Steiner, H. Sandmaier: Porous silicon: A novel material for microsystems, Sens. Actuators A **51**, 31–36 (1995)
11.20 R. He, C.J. Kim: On-chip hermetic packaging enabled by post-deposition electrochemical etching of polysilicon, Proc. 18th Int. Conf. Microelectromech. Syst. (IEEE, Piscataway 2005) pp. 544–547
11.21 S.K. Ghandhi: *VLSI Fabrication Principles – Silicon and Gallium Arsenide* (Wiley, New York 1983)
11.22 W.A. Pilskin: Comparison of properties of dielectric films deposited by various methods, J. Vac. Sci. Technol. **21**, 1064–1081 (1977)
11.23 J.S. Danel, F. Michel, G. Delapierre: Micromachining of quartz and its application to an acceleration sensor, Sens. Actuators A **21–23**, 971–977 (1990)
11.24 A. Yasseen, J.D. Cawley, M. Mehregany: Thick glass film technology for polysilicon surface micromachining, J. Microelectromech. Syst. **8**, 172–179 (1999)
11.25 R. Liu, M.J. Vasile, D.J. Beebe: The fabrication of nonplanar spin-on glass microstructures, J. Microelectromech. Syst. **8**, 146–151 (1999)
11.26 B. Folkmer, P. Steiner, W. Lang: Silicon nitride membrane sensors with monocrystalline transducers, Sens. Actuators A **51**, 71–75 (1995)
11.27 M. Sekimoto, H. Yoshihara, T. Ohkubo: Silicon nitride single-layer x-ray mask, J. Vac. Sci. Technol. **21**, 1017–1021 (1982)

11.28 D.J. Monk, D.S. Soane, R.T. Howe: Enhanced removal of sacrificial layers for silicon surface micromachining, 7th Int. Conf. Solid State Sens. Actuators, Technical Digest (Institute of Electrical Engineers of Japan, Tokyo 1993) pp. 280–283

11.29 P.J. French, P.M. Sarro, R. Mallee, E.J.M. Fakkeldij, R.F. Wolffenbuttel: Optimization of a low-stress silicon nitride process for surface micromachining applications, Sens. Actuators A **58**, 149–157 (1997)

11.30 B. Li, B. Xiong, L. Jiang, Y. Zohar, M. Wong: Germanium as a versatile material for low-temperature micromachining, J. Microelectromech. Syst. **8**, 366–372 (1999)

11.31 A. Franke, D. Bilic, D.T. Chang, P.T. Jones, T.J. King, R.T. Howe, C.G. Johnson: Post-CMOS integration of germanium microstructures, Proc. 12th Int. Conf. Microelectromech. Syst. (IEEE, Piscataway 1999) pp. 630–637

11.32 A.E. Franke, Y. Jiao, M.T. Wu, T.J. King, R.T. Howe: Post-CMOS modular integration of poly-SiGe microstructures using poly-Ge sacrificial layers, Solid State Sens. Actuator Workshop, Technical Digest (Transducers Research Foundation, Hilton Head 2000) pp. 18–21

11.33 S. Sedky, P. Fiorini, M. Caymax, S. Loreti, K. Baert, L. Hermans, R. Mertens: Structural and mechanical properties of polycrystalline silicon germanium for micromachining applications, J. Microelectromech. Syst. **7**, 365–372 (1998)

11.34 J.M. Heck, C.G. Keller, A.E. Franke, L. Muller, T.-J. King, R.T. Howe: High aspect ratio polysilicon-germanium microstructures, Proc. 10th Int. Conf. Solid State Sens. Actuators (Institute of Electrical Engineers of Japan, Tokyo 1999) pp. 328–334

11.35 P. Van Gerwen, T. Slater, J.B. Chevrier, K. Baert, R. Mertens: Thin-film boron-doped polycrystalline silicon$_{70\%}$-germanium$_{30\%}$ for thermopiles, Sens. Actuators A **53**, 325–329 (1996)

11.36 D. Hyman, J. Lam, B. Warneke, A. Schmitz, T.Y. Hsu, J. Brown, J. Schaffner, A. Walson, R.Y. Loo, M. Mehregany, J. Lee: Surface micromachined RF MEMS switches on GaAs substrates, Int. J. Radio Freq. Microw. Commun. Eng. **9**, 348–361 (1999)

11.37 C. Chang, P. Chang: Innovative micromachined microwave switch with very low insertion loss, Sens. Actuators **79**, 71–75 (2000)

11.38 M.F. Aimi, M.P. Rao, N.C. MacDonald, A.S. Zuruzi, D.P. Bothman: High-aspect-ratio bulk micromachining of Ti, Nat. Mater. **3**, 103–105 (2004)

11.39 C.L. Shih, B.K. Lai, H. Kahn, S.M. Phillips, A.H. Heuer: A robust co-sputtering fabrication procedure for TiNi shape memory alloys for MEMS, J. Microelectromech. Syst. **10**, 69–79 (2001)

11.40 G. Hahm, H. Kahn, S.M. Phillips, A.H. Heuer: Fully microfabricated silicon spring biased shape memory actuated microvalve, Solid State Sens. Actuator Workshop, Technical Digest (Transducers Research Foundation, Hilton Head Island 2000) pp. 230–233

11.41 S.D. Leith, D.T. Schwartz: High-rate through-mold electrodeposition of thick (> 200 micron) NiFe MEMS components with uniform composition, J. Microelectromech. Syst. **8**, 384–392 (1999)

11.42 N. Rajan, M. Mehregany, C.A. Zorman, S. Stefanescu, T. Kicher: Fabrication and testing of micromachined silicon carbide and nickel fuel atomizers for gas turbine engines, J. Microelectromech. Syst. **8**, 251–257 (1999)

11.43 T. Pornsin-Sirirak, Y.C. Tai, H. Nassef, C.M. Ho: Titanium-alloy MEMS wing technology for a microaerial vehicle application, Sens. Actuators A **89**, 95–103 (2001)

11.44 C.R. Stoldt, C. Carraro, W.R. Ashurst, D. Gao, R.T. Howe, R. Maboudian: A low temperature CVD process for silicon carbide MEMS, Sens. Actuators A **97/98**, 410–415 (2002)

11.45 M. Eickhoff, H. Moller, G. Kroetz, J. von Berg, R. Ziermann: A high temperature pressure sensor prepared by selective deposition of cubic silicon carbide on SOI substrates, Sens. Actuators **74**, 56–59 (1999)

11.46 Y.T. Yang, K.L. Ekinci, X.M.H. Huang, L.M. Schiavone, M.L. Roukes, C.A. Zorman, M. Mehregany: Monocrystalline silicon carbide nanoelectromechanical systems, Appl. Phys. Lett. **78**, 162–164 (2001)

11.47 D. Young, J. Du, C.A. Zorman, W.H. Ko: High-temperature single crystal 3C-SiC capacitive pressure sensor, IEEE Sens. J. **4**, 464–470 (2004)

11.48 C.A. Zorman, S. Rajgolpal, X.A. Fu, R. Jezeski, J. Melzak, M. Mehregany: Deposition of polycrystalline 3C-SiC films on 100 mm-diameter (100) Si wafers in a large-volume LPCVD furnace, Electrochem. Solid State Lett. **5**, G99–G101 (2002)

11.49 I. Behrens, E. Peiner, A.S. Bakin, A. Schlachetzski: Micromachining of silicon carbide on silicon fabricated by low-pressure chemical vapor deposition, J. Micromech. Microeng. **12**, 380–384 (2002)

11.50 C.A. Zorman, S. Roy, C.H. Wu, A.J. Fleischman, M. Mehregany: Characterization of polycrystalline silicon carbide films grown by atmospheric pressure chemical vapor deposition on polycrystalline silicon, J. Mater. Res. **13**, 406–412 (1996)

11.51 C.H. Wu, C.H. Zorman, M. Mehregany: Growth of polycrystalline SiC films on SiO$_2$ and Si$_3$N$_4$ by APCVD, Thin Solid Films **355/356**, 179–183 (1999)

11.52 P. Sarro: Silicon carbide as a new MEMS technologytuators, Sens. Actuators A **82**, 210–218 (2000)

11.53 N. Ledermann, J. Baborowski, P. Muralt, N. Xantopoulos, J.M. Tellenbach: Sputtered silicon carbide thin films as protective coatings for MEMS applications, Surf. Coat. Technol. **125**, 246–250 (2000)

11.54 X.A. Fu, R. Jezeski, C.A. Zorman, M. Mehregany: Use of deposition pressure to control the residual stress in polycrystalline SiC films, Appl. Phys. Lett. **84**, 341–343 (2004)

11.55 J. Trevino, X.A. Fu, M. Mehregany, C. Zorman: Low-stress, heavily-doped polycrystalline silicon carbide for MEMS applications, Proc. 18th Int. Conf. Microelectromech. Syst. (IEEE, Piscataway 2005) pp. 451–454

11.56 R.S. Okojie, A.A. Ned, A.D. Kurtz: Operation of a 6H-SiC pressure sensor at 500 °C, Sens. Actuators A **66**, 200–204 (1998)

11.57 K. Lohner, K.S. Chen, A.A. Ayon, M.S. Spearing: Microfabricated silicon carbide microengine structures, Mater. Res. Soc. Symp. Proc. **546**, 85–90 (1999)

11.58 K.O. Min, S. Tanaka, M. Esashi: Micro/nano glass press molding using silicon carbide molds fabricated by silicon lost molding, Proc. 18th Int. Conf. Microelectromech. Syst. (IEEE, Miami 2005) pp. 475–478

11.59 S. Tanaka, S. Sugimoto, J.-F. Li, R. Watanabe, M. Esashi: Silicon carbide micro-reaction-sintering using micromachined silicon molds, J. Microelectromech. Syst. **10**, 55–61 (2001)

11.60 L.A. Liew, W. Zhang, V.M. Bright, A. Linan, M.L. Dunn, R. Raj: Fabrication of SiCN ceramic MEMS using injectable polymer-precursor technique, Sens. Actuators A **89**, 64–70 (2001)

11.61 A.J. Fleischman, S. Roy, C.A. Zorman, M. Mehregany: Polycrystalline silicon carbide for surface micromachining, Proc. 9th Int. Workshop Microelectromech. Syst. (IEEE, San Diego 1996) pp. 234–238

11.62 A.J. Fleischman, X. Wei, C.A. Zorman, M. Mehregany: Surface micromachining of polycrystalline SiC deposited on SiO2 by APCVD, Mater. Sci. Forum **264-268**, 885–888 (1998)

11.63 G. Beheim, C.S. Salupo: Deep RIE process for silicon carbide power electronics and MEMS, Mater. Res. Soc. Symp. Proc. **622**, T8.8.1–T8.8.6 (2000)

11.64 A. Yasseen, C.H. Wu, C.A. Zorman, M. Mehregany: Fabrication and testing of surface micromachined polycrystalline SiC micromotors, Electron. Device Lett. **21**, 164–166 (2000)

11.65 X. Song, S. Rajgolpal, J.M. Melzak, C.A. Zorman, M. Mehregany: Development of a multilayer SiC surface micromachining process with capabilities and design rules comparable with conventional polysilicon surface micromachining, Mater. Sci. Forum **389-393**, 755–758 (2001)

11.66 D. Gao, M.B. Wijesundara, C. Carraro, R.T. Howe, R. Maboudian: Recent progress toward and manufacturable polycrystalline SiC surface micromachining technology, IEEE Sens. J. **4**, 441–448 (2004)

11.67 T. Shibata, Y. Kitamoto, K. Unno, E. Makino: Micromachining of diamond film for MEMS applications, J. Microelectromech. Syst. **9**, 47–51 (2000)

11.68 H. Bjorkman, P. Rangsten, P. Hollman, K. Hjort: Diamond replicas from microstructured silicon masters, Sens. Actuators **73**, 24–29 (1999)

11.69 P. Rangsten, H. Bjorkman, K. Hjort: Microfluidic components in diamond, Proc. 10th Int. Conf. Solid State Sens. Actuators (IEEE, Sendai 1999) pp. 190–193

11.70 H. Bjorkman, P. Rangsten, K. Hjort: Diamond microstructures for optical microelectromechanical systems, Sens. Actuators **78**, 41–47 (1999)

11.71 M. Aslam, D. Schulz: Technology of diamond microelectromechanical systems, Proc. 8th Int. Conf. Solid State Sens. Actuators (IEEE, Stockholm 1995) pp. 222–224

11.72 R. Ramesham: Fabrication of diamond microstructures for microelectromechanical systems (MEMS) by a surface micromachining process, Thin Solid Films **340**, 1–6 (1999)

11.73 X. Yang, J.M. Yang, Y.C. Tai, C.M. Ho: Micromachined membrane particle filters, Sens. Actuators **73**, 184–191 (1999)

11.74 X.D. Wang, G.D. Hong, J. Zhang, B.L. Lin, H.Q. Gong, W.Y. Wang: Precise patterning of diamond films for MEMS application, J. Mater. Process. Technol. **127**, 230–233 (2002)

11.75 A.R. Krauss, O. Auciello, D.M. Gruen, A. Jayatissa, A. Sumant, J. Tucek, D.C. Mancini, N. Moldovan, A. Erdemire, D. Ersoy, M.N. Gardos, H.G. Busmann, E.M. Meyer, M.Q. Ding: Ultrananocrystalline diamond thin films for MEMS and moving mechanical assembly devices, Diam. Relat. Mater. **10**, 1952–1961 (2001)

11.76 X. Xiao, J. Birrell, J.E. Gerbi, O. Auciello, J.A. Carlisle: Low temperature growth of ultrananocrystalline diamond, J. Appl. Phys. **96**, 2232–2239 (2004)

11.77 T.A. Friedmann, J.P. Sullivan, J.A. Knapp, D.R. Tallant, D.M. Follstaedt, D.L. Medlin, P.B. Mirkarimi: Thick stress-free amorphous-tetrahedral carbon films with hardness near that of diamond, Appl. Phys. Lett. **71**, 3820–3822 (1997)

11.78 J.P. Sullivan, T.A. Friedmann, K. Hjort: Diamond and amorphous carbon MEMS, MRS Bulletin **26**, 309–311 (2001)

11.79 J.R. Webster, C.W. Dyck, J.P. Sullivan, T.A. Friedmann, A.J. Carton: Performance of amorphous diamond RF MEMS capacitive switch, Electron. Lett. **40**, 43–44 (2004)

11.80 K. Hjort, J. Soderkvist, J.-A. Schweitz: Galium arsenide as a mechanical material, J. Micromech. Microeng. **4**, 1–13 (1994)

11.81 K. Hjort: Sacrificial etching of III–V compounds for micromechanical devices, J. Micromech. Microeng. **6**, 365–370 (1996)

11.82 K. Fobelets, R. Vounckx, G. Borghs: A GaAs pressure sensor based on resonant tunnelling diodes, J. Micromech. Microeng. **4**, 123–128 (1994)

11.83 A. Dehe, K. Fricke, H.L. Hartnagel: Infrared thermopile sensor based on AlGaAs-GaAs micromachining, Sens. Actuators A **46/47**, 432–436 (1995)

11.84 A. Dehe, J. Peerlings, J. Pfeiffer, R. Riemenschneider, A. Vogt, K. Streubel, H. Kunzel, P. Meissner,

11.84 H.L. Hartnagel: III–V compound semiconductor micromachined actuators for long resonator tunable Fabry–Perot detectors, Sens. Actuators A **68**, 365–371 (1998)

11.85 T. Lalinsky, S. Hascik, Z. Mozolova, E. Burian, M. Drzik: The improved performance of GaAs micromachined power sensor microsystem, Sens. Actuators **76**, 241–246 (1999)

11.86 T. Lalinsky, E. Burian, M. Drzik, S. Hascik, Z. Mozolova, J. Kuzmik, Z. Hatzopoulos: Performance of GaAs micromachined microactuator, Sens. Actuators **85**, 365–370 (2000)

11.87 H.X. Tang, X.M.H. Huang, M.L. Roukes, M. Bichler, W. Wegscheider: Two-dimensional electron-gas actuation and transduction for GaAs nanoelectromechanical systems, Appl. Phys. Lett. **81**, 3879–3881 (2002)

11.88 T.S. Tighe, J.M. Worlock, M.L. Roukes: Direct thermal conductance measurements on suspended monocrystalline nanostructure, Appl. Phys. Lett. **70**, 2687–2689 (1997)

11.89 J. Miao, B.L. Weiss, H.L. Hartnagel: Micromachining of three-dimensional GaAs membrane structures using high-energy nitrogen implantation, J. Micromech. Microeng. **13**, 35–39 (2003)

11.90 C. Seassal, J.L. Leclercq, P. Viktorovitch: Fabrication of inp-based freestanding microstructures by selective surface micromachining, J. Micromech. Microeng. **6**, 261–265 (1996)

11.91 J. Leclerq, R.P. Ribas, J.M. Karam, P. Viktorovitch: III–V micromachined devices for microsystems, Microelectron. J. **29**, 613–619 (1998)

11.92 H. Yamaguchi, R. Dreyfus, S. Miyashita, Y. Hirayama: Fabrication and elastic properties of InAs freestanding structures based on InAs/GaAs(111) a heteroepitaxial systems, Physica E **13**, 1163–1167 (2002)

11.93 C. Lee, T. Itoh, T. Suga: Micromachined piezoelectric force sensors based on PZT thin films, IEEE Trans. Ultrason. Ferroelectr. Freq. Control **43**, 553–559 (1996)

11.94 B. Xu, L.E. Cross, J.J. Bernstein: Ferroelectric and antiferroelectric films for microelectromechanical systems applications, Thin Solid Films **377/378**, 712–718 (2000)

11.95 S.P. Beeby, A. Blackburn, N.M. White: Processing of PZT piezoelectric thick films on silicon for microelectromechanical systems, J. Micromech. Microeng. **9**, 218–229 (1999)

11.96 C. Shearwood, M.A. Harradine, T.S. Birch, J.C. Stevens: Applications of polyimide membranes to MEMS technology, Microelectron. Eng. **30**, 547–550 (1996)

11.97 F. Jiang, G.B. Lee, Y.C. Tai, C.M. Ho: A flexible micromachine-based shear-stress sensor array and its application to separation-point detection, Sens. Actuators **79**, 194–203 (2000)

11.98 D. Memmi, V. Foglietti, E. Cianci, G. Caliano, M. Pappalardo: Fabrication of capacitive micromechanical ultrasonic transducers by low-temperature process, Sens. Actuators A **99**, 85–91 (2002)

11.99 A. Bagolini, L. Pakula, T.L.M. Scholtes, H.T.M. Pham, P.J. French, P.M. Sarro: Polyimide sacrificial layer and novel materials for post-processing surface micromachining, J. Micromech. Microeng. **12**, 385–389 (2002)

11.100 T. Stieglitz: Flexible biomedical microdevices with double-sided electrode arrangements for neural applications, Sens. Actuators A **90**, 203–211 (2001)

11.101 T. Stieglitz, G. Matthias: Flexible BioMEMS with electrode arrangements on front and back side as key component in neural prostheses and biohybrid systems, Sens. Actuators B **83**, 8–14 (2002)

11.102 H. Lorenz, M. Despont, N. Fahrni, J. Brugger, P. Vettiger, P. Renaud: High-aspect-ratio, ultrathick, negative-tone-near-UV photoresist and its applications in MEMS, Sens. Actuators A **64**, 33–39 (1998)

11.103 H. Lorenz, M. Despont, N. Fahrni, N. LaBianca, P. Renaud, P. Vettiger: SU-8: A low-cost negative resist for MEMS, J. Micromech. Microeng. **7**, 121–124 (1997)

11.104 E.H. Conradie, D.F. Moore: SU-8 thick photoresist processing as a functional material for MEMS applications, J. Micromech. Microeng. **12**, 368–374 (2002)

11.105 C.T. Pan, H. Yang, S.C. Shen, M.C. Chou, H.P. Chou: A low-temperature wafer bonding technique using patternable materials, J. Micromech. Microeng. **12**, 611–615 (2002)

11.106 P.A. Stupar, A.P. Pisano: Silicon, parylene, and silicon/parylene micro-needles for strength and toughness, 11th Int. Conf. Solid State Sens. Actuators, Technical Digest (Springer, Berlin 2001) pp. 1368–1389

11.107 J.M. Zara, S.W. Smith: Optical scanner using a MEMS actuator, Sens. Actuators A **102**, 176–184 (2002)

11.108 H.S. Noh, P.J. Hesketh, G.C. Frye-Mason: Parylene gas chromatographic column for rapid thermal cycling, J. Microelectromech. Syst. **11**, 718–725 (2002)

11.109 T.J. Yao, X. Yang, Y.C. Tai: BrF_3 dry release technology for large freestanding parylene microstructures and electrostatic actuators, Sens. Actuators A **97/98**, 771–775 (2002)

11.110 X. Wang, J. Engel, C. Liu: Liquid crystal polymer (LCP) for MEMS: Processing and applications, J. Micromech. Microeng. **13**, 628–633 (2003)

11.111 C.J. Lee, S.J. Oh, J.K. Song, S.J. Kim: Neural signal recording using microelectrode arrays fabricated on liquid crystal polymer material, Mater. Sci. Eng. C **4**, 265–268 (2004)

11.112 F.F. Faheem, K.C. Gupta, Y.C. Lee: Flip-chip assembly and liquid crystal polymer encapsulation for variable MEMS capacitors, IEEE Trans. Microw. Theory Tech. **51**, 2562–2567 (2003)